The Complete
HVACR
Lab Manual

Second Edition

Eugene Silberstein & Jason Obrzut

 CENGAGE

Australia • Brazil • Canada • Mexico • Singapore • United Kingdom • United States

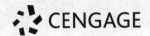

The Complete HVACR Lab Manual, Second Edition
Eugene Silberstein; Jason Obrzut

SVP, Higher Education Product Management: Erin Joyner

VP, Product Management, Learning Experiences: Thais Alencar

Product Director: Jason Fremder

Senior Product Manager: Vanessa Myers

Product Assistant: Bridget Duffy

Learning Designer: Elizabeth Berry

Senior Content Manager: Jenn Alverson

Digital Delivery Quality Partner: Elizabeth Cranston

VP, Product Marketing: Jason Sakos

Director, Product Marketing: Neena Bali

IP Analyst: Ashley Maynard

Production Service: Straive

Designer: Felicia Bennett

Cover Image Source: tashechka/Shutterstock.com

> For product information and technology assistance, contact us at
> **Cengage Customer & Sales Support, 1-800-354-9706**
> **or support.cengage.com.**
>
> For permission to use material from this text or product, submit all requests online at **www.copyright.com.**

Library of Congress Control Number: 2021921573

ISBN: 978-0-357-61873-8

Cengage
200 Pier 4 Boulevard
Boston, MA 02210
USA

Cengage is a leading provider of customized learning solutions with employees residing in nearly 40 different countries and sales in more than 125 countries around the world. Find your local representative at **www.cengage.com.**

To learn more about Cengage platforms and services, register or access your online learning solution, or purchase materials for your course, visit **www.cengage.com.**

Notice to the Reader

Publisher does not warrant or guarantee any of the products described herein or perform any independent analysis in connection with any of the product information contained herein. Publisher does not assume, and expressly disclaims, any obligation to obtain and include information other than that provided to it by the manufacturer. The reader is expressly warned to consider and adopt all safety precautions that might be indicated by the activities described herein and to avoid all potential hazards. By following the instructions contained herein, the reader willingly assumes all risks in connection with such instructions. The publisher makes no representations or warranties of any kind, including but not limited to, the warranties of fitness for particular purpose or merchantability, nor are any such representations implied with respect to the material set forth herein, and the publisher takes no responsibility with respect to such material. The publisher shall not be liable for any special, consequential, or exemplary damages resulting, in whole or part, from the readers' use of, or reliance upon, this material.

Printed in the United States of America
Print Number: 01   Print Year: 2022

Contents

PART I   SUBJECT AREA EXERCISES

Section 7   Electric Motors (MOT). **235**

Section 8   Heat Pump Systems (HPS). **325**

PART 2   EXERCISE CORRELATIONS

Preface

The idea for *The Complete HVACR Lab Manual* developed out of the need for a lab manual that comprehensively covered the knowledge and skills that are required of an HVACR technician entering the industry. It is designed to support the hands-on application and practice required within the HVACR program, and therefore will support any primary textbook or other learning resources that may be utilized within HVACR courses. Found within these pages are a great number of lab exercises divided among key subject areas:

- HVACR Core Concepts (HCC)
- Electrical Core Concepts (ECC)
- Refrigeration System Components (RCC)
- Tools and Safety (SFT)
- Refrigerants (REF)
- Air (AIR)
- Electric Motors (MOT)
- Heat Pump Systems (HPS)
- Heating (HTG)
- Mechanical System Troubleshooting (MST)
- Controls (CON)
- Domestic Appliances (DOM)
- Installation and Start-up (ISU)
- Building Science (BSC)
- Commercial and Industrial (COM)

FEATURES

As part of the development process for this lab manual, painstaking efforts were made to streamline exercises and focus on the most relevant and current material, while ensuring that there would not be any gaps or deficiencies in learning. Each section, or subject area and each lab exercise, therefore, follows the same basic outline.

Each section, or subject area, includes a framework for the corresponding exercises:

- **Subject Area Overview:** Provides a brief lesson covering the information within that subject area. It provides background knowledge for the lab exercises that follow, but it is intended to support, and not replace, the primary textbook or learning materials present in the course.
- **Key Terms:** Offers an opportunity for the students to review important terminology related to the subject area.
- **Review Test:** This test evaluates student knowledge of the entire section, or subject area. It can be administered as an open-book test to promote learning, or serve as an evaluation of student knowledge upon completion of the exercises within the section.

Likewise, each lab exercise guides students through each activity in a prescribed manner:

- **Objectives:** Outline the knowledge and skills students should achieve upon completion of the lesson.
- **Introduction:** Provides a brief overview of the activities covered in the lab exercise.
- **Tools and Materials:** Lists the specific requirements of the lab exercise.
- **Step-by-Step Procedures:** Takes the student through the lab activity in a clear, systematic manner.
- **Maintenance of Workstation and Tools:** Focus on promoting good work habits.
- **Summary Statement:** Requires students to highlight the key points of the lesson based on their response to a specific statement.
- **TakeAway:** Provides a brief statement that speaks to the importance of the exercise and how learning the material will benefit the student upon entering the workforce.
- **Questions:** Evaluate student knowledge of the concepts presented in that specific lab exercise.

INSTRUCTOR'S RESOURCES

Additional instructor resources for this product are available online. Instructor assets include a Solution and Answer Guide and a Transition Guide. Sign up or sign in at www.cengage.com to search for and access this product and its online resources.

A Note from the Authors

Quite often, lab manuals and workbooks are geared toward a specific book or a specific set of reference materials. This lab manual/workbook, which consists of over 200 exercises, is unique in the sense that, since it is not linked to a specific title, it can be effectively utilized as a standalone educational tool, as a supplement to manufacturer-specific training, or in conjunction with a formal HVACR training program. It includes valuable information for both instructors and students, and is sure to become a valuable asset to any HVACR program.

The exercises contained in this lab manual have been carefully scripted to maximize the student's exposure to the topic being covered and have been categorized into fifteen distinct sections. Each of the exercises contains follow-up questions that are intended to help reinforce concepts and provide a means to evaluate subject matter retention.

By dividing the material into content-specific sections, each concentrating on a different aspect of the industry, the user can easily identify exercises that relate to the specific area, regardless of which textbook or reference material is being used. Each section starts out with a "Test Your Knowledge" quiz that covers the basic concepts addressed in each section.

We hope that you benefit from the exercises contained in this manual and that you achieve the success you desire as you enter the exciting, ever-changing HVACR industry. We wish you the best of luck!

With warmest regards,

Eugene Silberstein and Jason Obrzut

About the Authors

EUGENE SILBERSTEIN, M.S., CMHE, BEAP

Since entering the HVACR industry in 1980, Eugene has taken on many roles ranging from field technician and system designer to company owner, teacher, administrator, consultant, and author. Eugene is presently the Director of Technical Education and Standards at the ESCO Institute.

Eugene has over 25 years of teaching experience and has taught air conditioning and refrigeration at many institutions including high school vocational programs, proprietary post-secondary institutions, and community colleges. In December of 2015, Eugene retired from his tenured teaching position at Suffolk County Community College, in Brentwood, New York, to join the ESCO Group and relocate to Southern California.

Eugene earned his dual Bachelor's Degree from The City College of New York (New York, NY) and his Masters of Science degree from Stony Brook University, (Stony Brook, NY) where he specialized in Energy and Environmental Systems, studying renewable and sustainable energy sources such as wind, solar, geothermal, biomass, and hydropower. He earned his Certified Master HVACR Educator (CMHE) credential from HVAC Excellence and the ESCO Group in 2010. Eugene also carries ASHRAE's BEAP credential, which classifies Eugene as a Building Energy Assessment Professional. He is also an active member of many industry societies.

Eugene served as the subject matter expert on over 15 HVACR-related educational projects. His most notable work is Cengage Learning's *Refrigeration and Air Conditioning Technology* title, which is presently in its 9th Edition (2021). This book is used in over 1000 schools both in this country and abroad to help individuals learn, and master, the skills required to service, install, design, service, and troubleshoot HVACR equipment. Other book credits include *Refrigeration and Air Conditioning Technology, 6th, 7th and 8th Editions (2008, 2012, 2017)*, *Residential Construction Academy: HVAC, 1st and 2nd Editions (2005, 2012)*, *Pressure Enthalpy Without Tears (2006)*, *Heat Pumps, 1st and 2nd Editions (2003, 2016)*, and *Psychrometrics Without Tears (2014)*. In addition to his book credits, Eugene has written the production scripts for over 30 HVACR educational videos and has written articles for the industry's newspapers, magazines, and other periodicals.

Eugene was selected as one of the top three HVACR instructors in the country for the 2005/2006, 2006/2007, and 2007/2008 academic school years by the Air Conditioning, Heating and Refrigeration Institute (AHRI) and the Air Conditioning, Heating and Refrigeration (ACHR) News.

JASON OBRZUT, CMHE

In 2002, Jason Obrzut graduated from the HVAC Technical Institute in Chicago, kick-starting his impressive career in the HVACR industry. Upon graduating from school, Jason began working as a service technician and, within a short time, became an installation supervisor, and ultimately a business owner. While working during the day, he returned to the HVAC Technical Institute as a part-time, evening instructor. Teaching quickly became his passion and working in the field took the proverbial backseat as his part-time night teaching assignment morphed into a full-time teaching position. While teaching, Jason developed, wrote, and implemented the curriculum that is currently being taught at the school. His dedication to his students and the institution became evident and, after a few short years, he assumed the role as the school's Director of Education in addition to being the center's lead HVAC/R instructor.

In 2016, The *ACHR News* recognized Jason as one of the "Top 40 Under 40" in the HVACR industry. He is affiliated with numerous HVACR industry organizations including Refrigeration Service Engineers Society (RSES), the ESCO Group, and HVAC Excellence. He is a member of the AHRI Safe Refrigerant Transition Task Force and is the chairman of the SRTTF Technician Outreach committee. At HVAC

Excellence's 2016 National HVAC Educators and Trainers Conference (NHETC), Jason participated in the first annual Teachers-N-Trainers (TnT) competition and took 1st place in a field of over 60 participants. Jason is one of only 125 instructors nationwide who have had the title of Certified Master HVACR Educator (CMHE) bestowed upon them. In addition, he possesses other industry certifications, such as Certified Subject Matter Educator (CSME) certificates from the ESCO Group. Jason has recently accepted a full-time position with the ESCO Institute, an industry partner of Cengage Learning, where he will be taking on a role as part of the content development team to help improve the quality of HVACR education and training. Jason is currently the Director of Industry Standards and Relations for ESCO.

Acknowledgments

The authors and the publisher would like to thank the following reviewers who graciously contributed their thoughts in the development of this lab manual:

Thomas Bush
South Florida State College
Avon Park, FL

Gene Smith
Athens Technical College
Athens, GA

John Christiansen
Bluegrass Community and Technical College
Lexington, KY

Brent Evans
William R. Britt Advance Technology Center at Johnston Community College
Smithfield, NC

Senobio Aguilera, HVACR SME
Riverside City College
Riverside, CA

Prof. Joseph Owens, CMS
Antelope Valley College
Lancaster, CA

Andy Smart
Wake Technical Community College
Raleigh, NC

PART

1

Subject Area Exercises

SECTION 1

HVACR Core Concepts (HCC)

SUBJECT AREA OVERVIEW

The core concepts of HVACR include matter, energy, heat, temperature, work, and pressure.

Matter is a term used to describe substances that occupy space, have mass, and are made up of atoms. Matter exists in three common forms: solids, liquids and gases. Solids exert pressure downward, liquids exert pressure downward and outward, and gases exert pressure outward in all directions.

An atom is the smallest amount of a substance that can exist while still retaining all of the properties of the substance. Atoms combine in different configurations to form molecules. Chemical reactions can cause atoms of different substances to combine with each other, forming compounds.

Mass and weight are not the same. Weight is defined as the combined effects of gravity and the mass of a substance. The density of a substance describes its mass- or weight-to-volume relationship. Specific gravity is a unitless quantity that compares the density of a substance to the density of water. A specific gravity of less than 1 indicates that the substance is less dense than water, while a specific gravity greater than 1 indicates that the substance is denser than water. A substance with a specific gravity of less than 1 will float on water, while a substance with a specific gravity greater than 1 will sink in water. Specific volume is the amount of space that a pound of vapor or gas will occupy.

Energy comes in many forms, including electrical, thermal or heat, chemical, and mechanical. Electricity is the driving force that allows pumps, compressors, fans, and other devices to operate. Heat is another type of energy that is important to the HVACR industry, especially because it is the transfer of thermal energy that allows us to create a wide range of conditions that benefit our very existence.

Fossil fuels are a popular source of heat energy. Examples of fossil fuels are natural gas, oil, and coal.

The term work can be simply explained as a force moving an object and is expressed in ft-lb. The formula for work is

$$\text{Work} = \text{Force} \times \text{Distance}$$

Power is the rate of doing work and is measured in horsepower, hp. One horsepower is the equivalent of doing 33,000 ft-lb of work in 1 minute. The unit of measurement for electrical power is the watt.

1 kilowatt (kW) = 1000 watts (W)

746 watts = 1 horsepower (hp)

3.413 British thermal units (Btu) = 1 watt-hour (Wh)

Gases follow three laws. These gas laws are Boyle's Law, Charles' Law, and Dalton's Law. Generally speaking, gases follow the General Law of Perfect Gases, which is

$$(P_1 \times V_1) \div T_1 = (P_2 \times V_2) \div T_2$$

Temperature can be thought of as the level of heat or the intensity of molecular activity. The Fahrenheit temperature scale is currently used by most people in the United States. The Celsius scale is used to some extent in the United States but is the scale of choice for the rest of the world. At standard atmospheric conditions, pure water boils at 212°F (100°C) and freezes at 32°F (0°C). Molecules in matter are constantly moving; the hotter the substance, the faster they move.

(continued)

The British thermal unit (Btu) describes the amount of heat it takes to raise the temperature of 1 lb of water 1°F. When there is a temperature difference between two substances, heat transfer will take place. Heat flows from a warmer substance to a cooler substance. Conduction is the transfer of heat from molecule to molecule within a substance. Convection is the movement of heat by a liquid or gas from one place to another. Radiation is the transfer of heat between surfaces or objects that are not otherwise connected without heating the medium between them.

Latent heat is the heat added to or rejected by a substance that results in a change of state of the substance without a change in temperature. Sensible heat is the heat added to or rejected by a substance that causes the temperature of a substance to change.

Specific heat is the amount of heat needed to raise the temperature of 1 lb of a substance by 1°F. The specific heat of water is 1, which means that the temperature of 1 lb of water will increase by 1°F if 1 Btu of heat energy is added to it. The specific heat of air is about 0.24 Btu/lb.

Pressure is the force applied to a specific unit of area. This is normally expressed in pounds per square inch (psi). The Earth's atmosphere has weight and applies a force of 14.696 psi at sea level at 70°F. Barometers measure atmospheric pressure, and gauges measure pressure in enclosed systems. Gauge pressures do not take the pressure of the atmosphere into account and are expressed in pounds per square inch gauge (psig). If the pressure of the atmosphere is taken into account, we express this pressure in the units of pounds per square inch absolute (psia).

KEY TERMS

Absolute temperature
Aneroid
Atmospheric pressure
Barometer
Bourdon tube
Boyle's Law
British thermal unit (Btu)
Charles' Law
Compound
Compound gauge
Conduction
Convection
Dalton's Law
Density
Forced convection

Gas
Global warming
Green
Greenhouse effect
Heat
Heat intensity
Heat island
Heat transfer
Hidden heat
High-pressure (high-side) gauge
High-side gauge
Horsepower (hp)
Kelvin scale
Latent heat

Latent heat of condensation
Latent heat of fusion
Latent heat of vaporization
Law of conservation of energy
Leadership in Energy and Environmental Design (LEED)
Liquid
Low-side gauge
Mass
Matter
Natural convection
Power

Pressure
Radiation
Rankine scale
Refrigeration
Sensible heat
Solid
Specific gravity
Specific heat
Specific volume
Standard condition
Sustainability
Sustainable design
Temperature
Watt (W)
Weight
Work

REVIEW TEST

Name	Date	Grade

Circle the letter that indicates the correct answer.

1. The standard atmospheric conditions for water to boil at 212°F are:
   A. 15.969 psi at a temperature of 70°F.
   B. 16.696 psi at a temperature of 68°F.
   C. 14.696 psi at a temperature of 70°F.
   D. 15.696 psi at a temperature of 68°F.

2. At standard conditions, water will boil at _____ on the Celsius scale.
   A. 100°C
   B. 212°C
   C. 32°C
   D. 0°C

3. If water weighs 8.33 pounds per gallon, it will take _____ Btu to raise the temperature of 4 gallons of water from 70°F to 72°F.
   A. 66.7
   B. 10
   C. 14.7
   D. 106

4. Heat energy flows from a warmer substance to a cooler substance.
   A. True
   B. False

5. When heat transfers by conduction, it:
   A. moves through space and air and heats only solid objects.
   B. is transported from one place to another by air or liquid.
   C. moves from molecule to molecule within the substance.
   D. All of the above.

6. When heat transfers by convection, it:
   A. moves through space and air and heats only solid objects.
   B. is transported from one place to another by air or liquid.
   C. moves from molecule to molecule within the substance.
   D. All of the above.

7. When heat transfers by radiation, it:
   A. moves through space and air and heats only solid objects.
   B. is transported from one place to another by air or liquid.
   C. moves from molecule to molecule within the substance.
   D. All of the above.

8. Sensible heat energy levels:
   A. can be measured with a thermometer.
   B. are known as hidden heat levels.
   C. are always in the comfort range.
   D. result in a change of state of the substance.

9. Latent heat energy levels:
   A. can be measured with a thermometer.
   B. are determined with a barometer.
   C. are always very warm.
   D. result in a change of state of the substance.

10. The atmosphere will support a column of mercury _____ in high at sea level.
    A. 32.00
    B. 28.69
    C. 14.696
    D. 29.92

11. A Bourdon tube can be found in a:
    A. mercury barometer.
    B. aneroid barometer.
    C. pressure gauge.
    D. mercury thermometer.

12. Psig is equal to:
    A. psia + 14.696.
    B. psia − 14.696.
    C. psia + 16.696.
    D. psia − 16.696.

13. 98°C is equal to:
   A. 214.6°F.
   B. 196.0°F.
   C. 119.8°F.
   D. 208.4°F.

14. 70°F is equal to:
   A. 21.1°C.
   B. 22.6°C.
   C. 98.6°C.
   D. 20.3°C.

15. The density of a substance:
   A. is used to measure its energy.
   B. describes its mass-to-volume relationship.
   C. is used to determine the transferability of heat.
   D. determines its ability to attract unlike magnetic charges.

16. How many ft-lb of work is done when a 500 lb condensing unit is lifted to the top of a 60 ft building?
   A. 560 ft-lb
   B. 440 ft-lb
   C. 8.33 ft-lb
   D. 30,000 ft-lb

17. Moving 66-lb object a distance of 1,000 feet in one minute is the equivalent of
   A. 0.5 hp
   B. 1.0 hp
   C. 1.5 hp
   D. 2.0 hp

18. A compound gauge can be used to measure pressures:
   A. above and below atmospheric pressure.
   B. that are only expressed in psia.
   C. that are only below 0 psig.
   D. that are only above 0 psig.

19. When a substance is heated, the molecules:
   A. move slower.
   B. move faster.
   C. move at the same speed.
   D. stop moving.

20. The latent heat of fusion for a water-to-ice transformation is:
   A. 1 Btu/lb.
   B. 144 Btu/lb.
   C. 288 Btu/lb.
   D. 970 Btu/lb.

21. If the specific gravity of copper is 8.91, its density is:
   A. 55.6 lb/ft^3.
   B. 238.8 lb/ft^3.
   C. 23.88 lb/ft^3.
   D. 556 lb/ft^3.

22. Specific volume is used to:
   A. compare the densities of various liquids.
   B. compare the densities of various solids.
   C. indicate the volume that a specific weight of gas will occupy.
   D. describe the energy characteristics of a substance.

23. Gas molecules push outward in:
   A. a downward direction.
   B. an upward direction.
   C. a sideways direction.
   D. All of the above.

24. An example of a fossil fuel is:
   A. natural gas.
   B. hydrogen.
   C. wood.
   D. liquid oxygen.

25. Work is equal to:
   A. force × power.
   B. force × energy.
   C. force × distance.
   D. distance × energy.

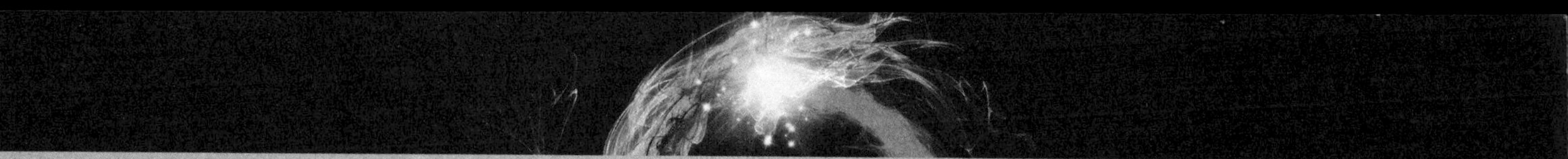

Temperature Conversions

Name	Date	Grade

OBJECTIVES: Upon completion of this exercise, you should be able to convert temperatures from one scale to another using mathematical equations.

INTRODUCTION: Using a calculator and the appropriate mathematical relationships, you will complete the temperature charts provided.

TOOLS, MATERIALS, AND EQUIPMENT: Calculator or temperature conversion app on a smart device.

⚠ **SAFETY PRECAUTIONS:** There are no safety precautions for this exercise.

STEP-BY-STEP PROCEDURES

Using your calculator or conversion app, complete the following chart by determining the correct values for all of the missing temperature readings.

Fahrenheit	Celsius	Rankine	Kelvin
		0	
32			
	100		
		600	
			0
200			
		100	
	50		
			300
	0		
150			
		50	
	−40		
−40			
212			
			373
	−273		
		672	
5			
			500

MAINTENANCE OF WORKSTATION AND TOOLS: With the exception of the calculator, there are no tools required for this exercise.

SUMMARY STATEMENT: Explain the differences between an absolute temperature scale, such as the Rankine scale, and a non-absolute temperature scale, such as the Fahrenheit scale. In your response, be sure to include the significance of the low-end values of the scales.

TAKEAWAY: There are many different scales used to measure temperature and their uses vary from one application to the next. A technician must be proficient at using these scales, understanding absolutes, and converting from one scale to another.

QUESTIONS

1. The mathematical relationship between a Fahrenheit temperature and the equivalent Celsius temperature is:

    A. 1:1.

    B. 2:1.

    C. 3:1.

    D. It cannot be determined, as the values are not proportional.

2. Which of the following temperature scales represent absolute values?

    A. Celsius and Kelvin

    B. Fahrenheit and Celsius

    C. Kelvin and Rankine

    D. Fahrenheit and Kelvin

3. Which of the following temperatures represents absolute zero?

    A. −460°F

    B. 0°C

    C. 672°R

    D. 273°K

4. Which of the following temperatures is the coldest?

    A. 32°F

    B. 672°R

    C. 100°C

    D. 373°K

5. Which of the following temperatures is the hottest?

    A. 212°F

    B. 492°R

    C. 273°K

    D. 0°C

6. Which of the following temperatures represent the same level of heat intensity?

    A. 0°C and 0°K

    B. 0°F and 0°R

    C. −40°C and −40°F

    D. 50°C and 50°R

The British Thermal Unit

Name	Date	Grade

OBJECTIVES: Upon completion of this exercise, you should be able to:

- estimate the heat output in Btu/h of a Bunsen burner.
- compare the rates of temperature rise of two different quantities of water.
- compare the temperature rise of water to the temperature rise of a saltwater solution.

INTRODUCTION: Using various pieces of laboratory equipment, you will be able to estimate the output of a heat source as well as measure and compare the rates of temperature rise for water and saltwater samples.

TOOLS, MATERIALS, AND EQUIPMENT:

Propane Bunsen burner
Propane cylinder
Flint striker
8 oz measuring cup
600 ml (about 20 oz) heat-resistant glass beaker
Digital scale
Stopwatch
Support stand
Support ring
Thermocouple thermometer
1 oz table salt
Safety glasses
Fire extinguisher
Tongs
Oven mitts
Stirring rod

Note: Alternate equipment and materials can be used to accomplish this exercise on the British thermal unit. Follow your instructor's directions and observe all safety precautions.

⚠ **SAFETY PRECAUTIONS:** This laboratory exercise utilizes an open flame so the potential for burn-related injuries is present. Always wear safety glasses or goggles, contain long hair, and do not wear loose clothing. Do not leave the lit burner unsupervised. Be very careful when handling objects that have been heated. Use tongs or oven mitts if it is necessary to move or handle a hot object. The best thing to do is wait until the object has cooled completely before handling it.

STEP-BY-STEP PROCEDURES

1. Connect the Bunsen burner to the propane cylinder.

2. Rest the burner assembly on the support stand.

3. Mount the support ring to the rod on the support stand so that the ring is about 2 in. above the top of the burner, Figure HCC-2.1.

FIGURE **HCC-2.1**

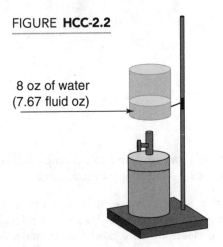

Support ring

2 in.

Propane Bunsen burner

Support stand

4. Place the empty glass beaker on the scale.

5. Zero the scale.

6. Add water to the beaker until there is 0.5 lb (8 oz) of water in the beaker. Note: 0.5 lb of water is the equivalent of 7.67 fluid oz.

7. Using the thermocouple thermometer, take a temperature reading of the water, making certain that the thermometer's sensing element is suspended in the water and not touching the sides of the beaker.

8. Record this temperature value in column 2 of the chart at the 0:00 time period.

9. Place the beaker on the support ring, Figure HCC-2.2.

FIGURE **HCC-2.2**

8 oz of water
(7.67 fluid oz)

10. Light the Bunsen burner with the flint striker, and start the stopwatch.

11. Obtain temperature readings of the water at 15-second time intervals, and record these temperature readings in column 2 of the chart. Make certain that the temperature probe is suspended in the water and does not touch the walls of the beaker. Once completed, there should be 21 entries in column 2.

12. Turn the Bunsen burner off.

13. Using the following formula, calculate the average temperature increase per minute: [Temperature at (5:00) − Temperature at (0:00)] ÷ 5 minutes

14. Record the average temperature rise per minute in the appropriate cell of column number 2.

15. Using the following formula, calculate the amount of heat energy, in Btu/h, that is being absorbed by the water: Btu/h = 1 Btu/(lb-°F) × °F/min × 60 min/h × 0.5 lb.

16. Record the amount of heat energy, in Btu/h, that is being absorbed in the appropriate cell of column number 2.

17. Does the amount of heat energy absorbed, as calculated in Step 15, accurately estimate the amount of heat produced by the burner? Explain your answer in detail, providing as many possible reasons for discrepancies as you can.

18. Empty the beaker and allow it to cool.

19. Place the empty glass beaker on the scale.

20. Zero the scale.

21. Add water to the beaker until there is 1.0 lb (16 oz) of water in the beaker. Note: 1.0 lb of water is the equivalent of 15.34 fluid oz, Figure HCC-2.3.

FIGURE **HCC-2.3**

16 oz of water
(15.34 fluid oz)

22. Using the thermocouple thermometer, take a temperature reading of the water, making certain that the thermometer's sensing element is suspended in the water and not touching the sides of the beaker.

23. Record this temperature value in column 3 of the chart at the 0:00 time period.

24. Place the beaker on the support ring.

25. Light the Bunsen burner with the flint striker, and start the stopwatch.

26. Obtain temperature readings of the water at 15-second time intervals, and record these temperature readings in column 3 of the chart. Make certain that the temperature probe is suspended in the water and does not touch the walls of the beaker. Once completed, there should be 21 entries in column 3.

27. Turn off the Bunsen burner.

28. Empty the beaker, and allow it to cool.

29. Using the following formula, calculate the average temperature increase per minute: [Temperature at (5:00) − Temperature at (0:00)] ÷ 5 minutes.

30. Record the average temperature rise per minute in the appropriate cell of column number 3.

31. Using the following formula, calculate the amount of heat energy, in Btu/h, that is being absorbed by the water: Btu/h = 1 Btu/(lb-°F) × °F/min × 60 min/h × 1 lb.

32. Record the amount of heat energy, in Btu/h, that is being absorbed in the appropriate cell of column number 3.

33. How do the "average temperature rise per minute" values in columns 2 and 3 relate to each other? Explain your answer in detail.

34. How do the "Btu/h absorbed" values in columns 2 and 3 relate to each other? Explain your answer in detail.

35. Place the empty glass beaker on the scale.

36. Zero the scale.

37. Add water to the beaker until there is 0.5 lb (8 oz) of water in the beaker. Note: 0.5 lb of water is the equivalent of 7.67 fluid oz.

38. Add one ounce of salt to the beaker.

39. Stir the saltwater solution until the salt has completely dissolved.

40. Using the thermocouple thermometer, take a temperature reading of the saltwater solution, making certain that the thermometer's sensing element is suspended in the water and not touching the sides of the beaker.

41. Record this temperature value in column 4 of the chart at the 0:00 time period.

42. Place the beaker on the support ring.

43. Light the Bunsen burner with the flint striker, and start the stopwatch.

44. Obtain temperature readings of the water at 15-second time intervals, and record these temperature readings in column 4 of the chart. Make certain that the temperature probe is suspended in water and does not touch the walls of the beaker. Once completed, there should be 21 entries in column 4.

45. Turn off the Bunsen burner.

46. Empty the beaker, and allow it to cool.

47. Using the following formula, calculate the average temperature increase per minute: [Temperature at (5:00) − Temperature at (0:00)] ÷ 5 minutes.

48. Record the average temperature rise per minute in the appropriate cell of column number 4.

49. Using the following formula, calculate the amount of heat energy, in Btu/h, that is being absorbed by the saltwater: Btu/h = 1 Btu/(lb-°F) × °F/min × 60 min/h × 0.5 lb.

50. Record the amount of heat energy, in Btu/h, that is being absorbed in the appropriate cell of column number 4.

51. How do the "average temperature rise per minute" values in columns 2 and 4 relate to each other? Explain your answer in detail.

52. Based on the amount of heat energy absorbed by the liquids in columns 2 and 3, perform calculations that will estimate the specific heat of the saltwater solution.

53. Is the specific heat calculated in Step #52 higher than, lower than, or equal to the specific heat of water (1.0)? Explain why this is so.

Column 1	Column 2	Column 3	Column 4
Time Min:Sec	0.5 lb of Water (7.67 Fluid oz)	1.0 lb of Water (15.34 Fluid oz)	0.5 lb of Water (7.67 Fluid oz) Mixed with 1 oz of Table Salt
0:00	°F	°F	°F
0:15	°F	°F	°F
0:30	°F	°F	°F
0:45	°F	°F	°F
1:00	°F	°F	°F
1:15	°F	°F	°F
1:30	°F	°F	°F
1:45	°F	°F	°F
2:00	°F	°F	°F
2:15	°F	°F	°F
2:30	°F	°F	°F
2:45	°F	°F	°F
3:00	°F	°F	°F
3:15	°F	°F	°F
3:30	°F	°F	°F
3:45	°F	°F	°F
4:00	°F	°F	°F
4:15	°F	°F	°F
4:30	°F	°F	°F
4:45	°F	°F	°F
5:00	°F	°F	°F
Average Temperature Rise Per Minute	°F/min	°F/min	°F/min
Btu/h Absorbed	Btu/h	Btu/h	Btu/h

MAINTENANCE OF WORKSTATION AND TOOLS:
- Make certain that all propane cylinders are closed tightly.
- Make certain that all beakers are emptied, cleaned, and dried before storing.
- Wipe up any water spills prior to leaving the work area.
- Turn off the thermocouple thermometer and digital scale.
- Return all tools and equipment to their proper storage location.

SUMMARY STATEMENT: Explain how the rate of temperature rise changes as the amount or composition of the substance changes. Provide detailed information from the exercise you just completed.

TAKEAWAY: Heat transfer and thermal measurements are critical to the understanding of thermodynamics. Key factors that affect both the rate and quantity of heat transfer are the fuel providing the heat, the medium being heated, and the quantity of the medium being heated.

QUESTIONS

1. How does the rate of temperature rise of the 8 oz water sample compare to the rate of temperature rise of the 16 oz water sample?
   A. The rate of temperature rise of the 8 oz water sample was greater than the rate of temperature rise of the 16 oz water sample.
   B. The rate of temperature rise of the 16 oz water sample was greater than the rate of temperature rise of the 8 oz water sample.
   C. The rate of temperature rise of the 8 oz water sample was the same as the rate of temperature rise of the 16 oz water sample.

2. How does the amount of heat energy absorbed by the water or saltwater solution relate to the amount of heat energy produced by the burner?
   A. The amount of heat produced by the burner is greater.
   B. The amount of heat absorbed by the water is greater.
   C. The amount of heat absorbed by the saltwater is greater.
   D. The amount of heat produced by the burner is equal to the amount of heat absorbed by the water.

3. With regards to this laboratory exercise, which of the following is an example of heat loss?
   A. The support ring getting hot.
   B. The glass beaker getting hot.
   C. The air around the beaker getting hot.
   D. All of the above are examples of heat losses.

4. Based on the results of this laboratory exercise, it can be concluded that:
   A. impurities added to water will increase the rate of temperature rise of the solution.
   B. impurities added to water will decrease the rate of temperature rise of the solution.
   C. impurities added to water will neither increase nor decrease the rate of temperature rise of the solution.

Exercise HCC-3

Heat Transfer by Conduction

Name	Date	Grade

OBJECTIVES: Upon completion of this exercise, you should be able to:

- determine the rate of temperature rise through a glass rod.
- determine the rate of temperature rise through a copper tube.
- compare the rates of temperature rise through various materials.

INTRODUCTION: Using various pieces of laboratory equipment, you will be able to measure and compare the rates of temperature rise for various materials.

TOOLS, MATERIALS, AND EQUIPMENT:

Propane Bunsen burner

Propane cylinder

Flint striker

Stopwatch

Support stand

Support clamp

12" long, 10 mm diameter, glass stirring rod

12" long piece of $\frac{3}{8}$ OD copper tubing, straightened

Thermocouple thermometer

Safety glasses

Fire extinguisher

Tongs

Oven mitts

Foam insulation tape

Note: Alternate equipment and materials can be used to accomplish this exercise on heat transfer through conduction. Follow your instructor's directions and observe all safety precautions.

⚠ **SAFETY PRECAUTIONS:** This laboratory exercise utilizes an open flame so the potential for burn-related injuries is present. Always wear safety glasses or goggles, contain long hair, and do not wear loose clothing. Do not leave the burner on unsupervised. Be very careful when handling objects that have been heated. Use tongs or oven mitts if it is necessary to move or handle a hot object. The best thing to do is wait until the object has cooled completely before handling it.

STEP-BY-STEP PROCEDURES

1. Connect the Bunsen burner to the propane cylinder.

2. Rest the burner assembly on the support stand.

3. Mount the support clamp to the rod on the support stand so that the clamp is about 6 inches above the top of the burner, Figure HCC-3.1.

4. Secure the thermocouple of the thermometer to one end of the glass stirring rod using foam insulation tape, approximately 1 in. from the end, in a manner similar to that shown in Figure HCC-3.2. Make certain that the thermocouple is insulated from the surrounding air. Do not simply hold the thermocouple to the surface of the material to measure its temperature, Figure HCC-3.3.

FIGURE **HCC-3.1**

FIGURE **HCC-3.2** Wrapping the temperature-sensing element will insulate it from the surrounding air, giving a more accurate temperature reading.

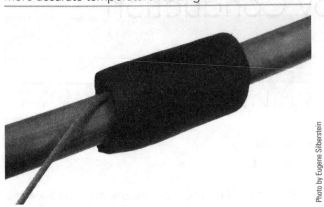

Photo by Eugene Silberstein

FIGURE **HCC-3.3** Simply holding the temperature-sensing element to a pipe will not yield accurate temperature measurements.

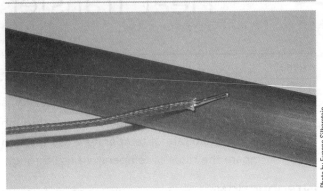

Photo by Eugene Silberstein

5. Secure the glass rod in the support clamp so that the end of the rod without the thermocouple connected to it is positioned over the burner, Figure HCC-3.4.

6. Using the thermocouple thermometer, take a temperature reading of the glass rod.

7. Record this temperature value in column 2 of the chart at the 0:00 time period.

8. Light the Bunsen burner with the flint striker, and start the stopwatch.

FIGURE **HCC-3.4**

Support clamp
Glass rod
6 in.
Propane Bunsen burner
Support stand
Thermocouple thermometer

9. Obtain temperature readings of the glass rod at 15-second time intervals, and record these temperature readings in column 2 of the chart. Once completed, there should be 21 entries in column 2.

10. Turn the Bunsen burner off.

11. Using the following formula, calculate the average temperature increase per minute: [Temperature at (5:00) − Temperature at (0:00)] ÷ 5 minutes.

12. Record the average temperature rise per minute in the appropriate cell of column number 2.

13. Using tongs and/or an oven mitt, remove the glass rod from the support clamp, and set it aside to cool completely.

14. Once the rod has cooled, remove the thermocouple from the glass rod.

15. Secure the thermocouple of the thermometer to one end of the copper tubing using foam insulation tape, approximately 1 in. from the end, in a manner similar to that used earlier with the glass rod.

16. Using the thermocouple thermometer, take a temperature reading of the copper tubing.

17. Record this temperature value in column 3 of the chart at the 0:00 time period.

18. Light the Bunsen burner with the flint striker, and start the stopwatch.

19. Obtain temperature readings of the copper tubing at 15-second time intervals, and record these temperature readings in column 2 of the chart. Once completed, there should be 21 entries in column 3.

20. Turn off the Bunsen burner.

21. Using the following formula, calculate the average temperature increase per minute: [Temperature at (5:00) − Temperature at (0:00)] ÷ 5 minutes.

22. Record the average temperature rise per minute in the appropriate cell of column number 3.

23. Using tongs and/or an oven mitt, remove the copper tubing from the support clamp, and set it aside to cool completely.

24. Once the copper tubing has cooled, remove the thermocouple.

25. How do the "average temperature rise per minute" values in columns 2 and 3 relate to each other? Explain your answer in detail.

26. Explain how your result from Step 25 might be used to discuss the differences between substances that are good conductors of heat and those that are poor conductors of heat.

Column 1	Column 2	Column 3
Time Min:Sec	Glass Rod	Copper Tubing
0:00	°F	°F
0:15	°F	°F
0:30	°F	°F
0:45	°F	°F
1:00	°F	°F
1:15	°F	°F
1:30	°F	°F
1:45	°F	°F
2:00	°F	°F
2:15	°F	°F
2:30	°F	°F
2:45	°F	°F
3:00	°F	°F
3:15	°F	°F
3:30	°F	°F
3:45	°F	°F
4:00	°F	°F
4:15	°F	°F
4:30	°F	°F
4:45	°F	°F
5:00	°F	°F
Average Temperature Rise Per Minute	°F/min	°F/min

MAINTENANCE OF WORKSTATION AND TOOLS:
- Make certain that all propane cylinders are closed tightly.
- Turn off the thermocouple thermometer.
- Return all tools and equipment to their proper storage location.

SUMMARY STATEMENT: Explain how the rate of temperature rise changes for different materials. Provide detailed information from the exercise you just completed.

TAKEAWAY: Different materials conduct heat at different rates, some better than others. Copper's ability to conduct and transfer heat has made it a staple design material in the HVACR industry.

QUESTIONS

1. How does the rate of temperature rise of the glass rod compare to the rate of temperature rise of the copper tubing?
   A. The rate of temperature rise of the glass rod was greater than the rate of temperature rise of the copper tubing.
   B. The rate of temperature rise of the copper tubing was greater than the rate of temperature rise of the glass rod.
   C. The rate of temperature rise of the glass rod was the same as the rate of temperature rise of the copper tubing.

2. Based on this exercise and your knowledge of electron theory, what can likely be concluded about the glass rod and the copper tubing?
   A. Glass is a good conductor, while copper is a good insulator.
   B. Glass is a good insulator, while copper is a good conductor.
   C. Glass and copper are both good insulators.
   D. Glass and copper are both good conductors.

3. Explain the relationship, if any, that exists between a substance being either a good conductor of heat and electricity or a good insulator.

Exercise HCC-4

Specific Gravity

Name	Date	Grade

OBJECTIVES: Upon completion of this exercise, you should be able to calculate the specific gravity of various substances and determine if these items will float on water or will sink.

INTRODUCTION: Using a calculator and the appropriate mathematical relationships, you will complete the charts provided.

TOOLS, MATERIALS, AND EQUIPMENT: Calculator or smart device.

⚠ **SAFETY PRECAUTIONS:** There are no safety precautions for this exercise.

STEP-BY-STEP PROCEDURES

Using your calculator, or a smart device, and information from Figure HCC-4.1, complete the following chart by determining the correct values for all of the missing readings.

FIGURE **HCC-4.1**

SUBSTANCE	DENSITY lb/ft^3	SPECIFIC GRAVITY
ALUMINUM	171	2.74
BRASS (RED)	548	8.78
COPPER	556	8.91
GOLD	1208	19.36
ICE @ 32°F	57.5	0.92
TUNGSTEN	1210	19.39
WATER	62.4	1
MARBLE	162	2.596

Substance	Density (lb/in.^{3})	Density (oz/in.^{3})	Density (lb/ft^3)	Specific Gravity	Substance	Sink or Float on Water
A			1208			
B				1		
C					Red brass	
D	0.3218					
E		0.5324				
F			171			
G				19.39		
H					Marble	

MAINTENANCE OF WORKSTATION AND TOOLS: With the exception of the calculator, there are no tools required for this exercise.

SUMMARY STATEMENT: Explain how the specific gravity of a substance changes as the density of the substance changes. Provide detailed information from the exercise you just completed.

TAKEAWAY: Specific gravity and density are directly related. The higher the specific gravity of a substance, the higher its density is and vice versa. Water is the generally accepted medium for comparison.

QUESTIONS

1. From the completed chart, which of the substances will float if placed in a container of water?
   A. Substances A, D, and E
   B. Substances A and B
   C. Substances B and H
   D. Only Substance E

2. From the completed chart, which of the substances is about three times as dense as water?
   A. Red brass
   B. Gold
   C. Aluminum
   D. Copper

3. Which of the following mathematical expressions best describes the conversion between $lb/in.^3$ and lb/ft^3?
   A. (12 in./ft) $\times$ (12 in./ft) $\times$ (12 in./ft)
   B. (12 in./ft) + (12 in./ft)
   C. (12 in./ft) $\times$ (12 in./ft)
   D. (12 in./ft) + (12 in./ft) + (12 in./ft)

4. From the completed chart, which two substances have the most similar densities?
   A. Gold and tungsten
   B. Aluminum and marble
   C. Red brass and copper
   D. Water and ice

Exercise HCC-5

Gas Laws

Name	Date	Grade

OBJECTIVES: Upon completion of this exercise, you should be able to calculate the missing values in a chart based on Boyle's Law and Charles' Law.

INTRODUCTION: Using a calculator and the appropriate mathematical relationships, you will complete the charts provided.

TOOLS, MATERIALS, AND EQUIPMENT: Calculator or smart device.

⚠ **SAFETY PRECAUTIONS:** There are no safety precautions for this exercise.

STEP-BY-STEP PROCEDURES

Using your calculator, or smart device, and the appropriate gas laws, complete the following chart by determining the correct values for all of the missing readings.

Line	P1	V1	T1	P2	V2	T2
1	50 psig	20 ft^3	100°F		50 ft^3	80°F
2	50 psia	2,000 ft^3	75°F	50 psia		130°F
3	40 psia	30 in.^{3}	80°F	50 psia		80°F
4		20 ft^3	500°R	200 psia	12 ft^3	600°R
5	100 psia	10 ft^3	600°R	100 psia	25 ft^3	
6	80 psia		400°R	150 psia	100 ft^3	500°R
7	150 psia	100 ft^3		300 psia	50 ft^3	700°R
8	80 psia	500 ft^3		160 psia	100 ft^3	240°R

MAINTENANCE OF WORKSTATION AND TOOLS: With the exception of the calculator, there are no tools required for this exercise.

SUMMARY STATEMENT: Based on the results of the exercise just completed, explain how the pressure, temperature, and volume of a gas are related to each other.

TAKEAWAY: Gas laws have many practical applications and are an important part of understanding how a system is designed and operates. Volume, temperature, and pressure are the variables used in the various calculations.

QUESTIONS

1. Which two properties of gases are directly proportional to each other?

    A. Pressure and temperature

    B. Pressure and volume

    C. Volume and temperature

    D. Both A and C are correct

2. Which two properties of gases are inversely proportional to each other?

    A. Pressure and temperature

    B. Pressure and volume

    C. Volume and temperature

    D. Both B and C are correct

3. State Boyle's Gas Law, Charles' Gas Law, and the General Law of Perfect Gases.

4. Explain why it is important to use absolute temperatures and pressures when using the various gas formulas.

Work, Power, and Heat Energy

Name	Date	Grade

OBJECTIVES: Upon completion of this exercise, you should be able to calculate the missing values in a chart based on the heat, power, and work formulas.

INTRODUCTION: Using a calculator and the appropriate mathematical relationships, you will complete the charts provided.

TOOLS, MATERIALS, AND EQUIPMENT: Calculator or smart device.

⚠ **SAFETY PRECAUTIONS:** There are no safety precautions for this exercise.

STEP-BY-STEP PROCEDURES

Using your calculator, or smart device, and the appropriate formulas for work, power, and/or conversions, complete the following chart by determining the correct values for all of the missing readings.

Line	Distance (ft)	Weight (lb)	Work (ft-lb)	Time (min)	Power (W)	Power (kW)	Heat (Btu/h)	hp
1	500	100		5				0.303
2		500	10,000	2	113.03			
3		250		1			85,000	
4	1000			2		25		
5		150	150,000					75
6	10,000			10		5		
7		500		1			20,000	
8	5000			5			200,000	

MAINTENANCE OF WORKSTATION AND TOOLS: With the exception of the calculator, there are no tools required for this exercise.

SUMMARY STATEMENT: Based on the results of the exercise just completed, explain how work, power, and heat are related to each other.

TAKEAWAY: The formulas for work and power can be used to calculate information such as hp, Btu/h, and kW. They are related and go hand in hand when servicing or installing equipment.

QUESTIONS

1. Which formula correctly relates heat energy to watts?
   A. 3.413 Btu = 1 W
   B. 3.413 Btu/h = 1 W
   C. 3.413 W = 1 Btu
   D. 3.413 W = 1 Btu/h

2. If the time it takes to move an object a certain distance increases, then:
   A. the amount of work done increases.
   B. the amount of work done decreases.
   C. the horsepower decreases.
   D. the horsepower increases.

3. Which of the following represents the greatest horsepower?
   A. Moving a 10 lb object 3300 ft in 2 minutes
   B. Moving a 10 lb object 33,000 ft in 2 minutes
   C. Moving a 10 lb object 3300 ft in 1 minute
   D. Moving a 1 lb object 3300 ft in 10 minutes

4. What would the Btu/h output of a 25 kW electric heater be?
   A. 0.853 Btu/h
   B. 85,325 Btu/h
   C. 85.32 Btu/h
   D. 8.53 Btu/h

Exercise HCC-7

Gauge Installation and Removal on Systems Equipped with Service Valves

Name	Date	Grade

OBJECTIVES: Upon completion of this exercise, you should be able to properly install and remove gauges from an operating air-conditioning or refrigeration system equipped with service valves without contaminating the system and with only minimal loss of system refrigerant.

INTRODUCTION: While servicing air-conditioning and refrigeration equipment, it may be necessary to install a set of gauges on the system in order to obtain the system's operating pressures. During the process of installing or removing gauges, it is important that system refrigerant not be released from the system and that contaminants are not permitted to enter the system. By following a detailed procedure for installing and removing gauges, the chances of releasing refrigerant or contaminating the system are greatly reduced.

TOOLS, MATERIALS, AND EQUIPMENT: Gauge manifold, refrigeration service wrench, adjustable wrench, safety glasses, gloves, operating air-conditioning or refrigeration system that is equipped with service valves.

⚠ **SAFETY PRECAUTIONS:** Be sure to protect yourself from refrigerant releases. Contact with released liquid refrigerant can cause frostbite, so be sure to wear appropriate pieces of personal protection equipment. The system's discharge line can be very hot, so be sure to avoid touching it. Check to make certain that the refrigerant hoses are rated appropriately for the system that you will be working on.

STEP-BY-STEP PROCEDURES

1. Position yourself at the unit as assigned by your instructor.

2. Energize the system and allow it to operate for five minutes.

3. Remove the stem caps from the suction and liquid line service valves, and make certain that the valves are fully backseated (stem is turned completely counterclockwise when looking at the end of the stem). If the stem caps are too tight, use an adjustable wrench to loosen the caps.

4. Remove the high- and low-side hoses from the blank ports on the gauge manifold.

5. Inspect the gauges for proper calibration. Calibrate as needed.

6. Make certain that the center hose on the gauge manifold is in place and that the hose connections on the manifold are thumb tight.

7. Remove the service port caps from the suction and liquid line service valves.

8. Place the service port caps on the blank ports on the gauge manifold.

9. Connect the high-side hose from the gauge manifold to the liquid line service port and connect the low-side hose from the gauge manifold to the suction service port. Make the hose connections thumb tight. Never use a wrench or pliers to tighten hoses as doing so can damage the O-rings in the hoses.

10. Using a refrigeration service wrench, crack the liquid service valve off the backseat (half to one turn clockwise from the fully backseated position).

11. Open the high-side and low-side valves on the gauge manifold.

12. Slightly loosen the low-side hose connection on the suction service valve to purge any air from the hoses and gauge manifold. This should take no more than two or three seconds.

13. Tighten the low-side hose connection on the suction service valve. Make the hose connection thumb tight.

14. Briefly loosen the center hose connection at the manifold's blank port to purge any air from the center hose. Retighten (thumb tight) the center hose's blank port connection.

15. Close the high-side and low-side valves on the gauge manifold.

16. Crack the suction line service valve off the backseat.

17. Replace the stem caps (thumb tight) on the liquid line and suction line service valves.

18. The gauges have now been properly installed on the system.

19. Record the operating pressures of the system here:

   High-side pressure: _____ psig   Low-side pressure: _____ psig

20. Have your instructor check your work and verify that the service valves and manifold valves are in the proper position.

21. Remove the stem caps from the liquid line and suction line service valves.

22. Backseat the liquid line service valve.

23. Replace the stem cap (thumb tight) on the liquid line service valve.

24. Open the high-side and low-side valves on the gauge manifold. The refrigerant in the high-side hose will flow through the manifold and enter the system through the low-side hose connection on the system. Wait until the high-side pressure and the low-side pressure have equalized before moving on to the next step.

   Note: Do not disconnect the hoses from a system if the suction pressure is at or below atmospheric (0 psig). De-energize the system and allow the lines to equalize.

25. Close the high-side and low-side valves on the gauge manifold.

26. Backseat the suction service valve.

27. Replace the stem cap (thumb tight) on the suction line service valve.

28. If the gauge manifold's hoses are equipped with manually operated low-loss fittings, close the valves on both the high- and low-side hoses. If the low-loss fittings on the hoses are self-sealing, continue to the next step.

29. Remove the high- and low-side hoses from the liquid and suction line service valves.

30. Remove the service port caps from the blank ports on the gauge manifold.

31. Place the service port caps on the suction line and liquid line service ports.

32. Connect the high-side and low-side hoses to the blank ports on the gauge manifold.

33. The gauges have now been properly removed from the system.

34. Have your instructor check your work, and verify that the service valves and manifold valves are in the proper position.

MAINTENANCE OF WORKSTATION AND TOOLS: Make certain that all service valves are fully backseated and that all stem caps are placed back (thumb tight) on the service valves. Make certain that both ends of all refrigerant hoses are connected to the manifold ports to prevent manifold and hose contamination.

SUMMARY STATEMENT: Explain how improperly installing gauges on an air-conditioning or refrigeration system can lead to system contamination. Provide as many examples of system contamination as you can think of as well as their effects on system operation and performance.

TAKEAWAY: The process of installing gauges on and removing gauges from an air-conditioning or refrigeration system should not have a negative effect on the system. Keeping the refrigerant contained in the system, as well as keeping the system contaminant free, should be top priority for the HVACR service technician.

QUESTIONS

1. Explain why the backseated position is the only service valve position that will result in no refrigerant loss when removing a refrigerant hose from the service port.

2. Explain why it is strongly recommended that the service port caps be placed on the manifolds blank ports when refrigerant hoses are connected to the system's service valves.

3. Explain what can happen if both manifold valves are in the open position and both (liquid line and suction line) service valves are cracked off the backseat on an operating air-conditioning or refrigeration system.

4. Explain the importance of replacing the stem caps on the service valves.

5. Explain why it is important to use only a refrigeration service wrench on the stems of service valves.

6. Explain why gauge calibration should only be done when the hoses are disconnected from the blank ports on the gauge manifold.

SECTION 2

Electrical Core Concepts (ECC)

SUBJECT AREA OVERVIEW

Without electricity, most of the things we rely on would simply not exist. Electricity plays an extremely important role in all aspects of modern society, including HVACR. Electricity allows for the operation of any heating, cooling, or refrigeration system. Because nearly 85% of the problems with HVACR systems are electrical in nature, it should be no surprise that a solid understanding of electrical concepts and acceptable practices is important for all technicians who install and service heating, cooling, and refrigeration equipment and control systems. Even the simplest HVACR system contains various types of electric circuits designed to perform specific tasks to effectively and efficiently control the system.

Matter is made up of atoms, which are made up of protons, neutrons, and electrons. Protons have a positive charge, neutrons have no charge, and electrons have a negative charge. An atom with more electrons than protons will have a negative charge, while an atom with fewer electrons than protons will have a positive charge. The law of charges states that opposite charges attract each other, and like charges repel each other. So an atom with a surplus of electrons (negative charge) will be attracted to an atom with a shortage of electrons (positive charge). An atom that is electrically neutral will have an equal number of protons and electrons.

Electricity is most commonly described as the flow of electrons. Electrons travel easily through good conductors, such as copper, aluminum, and silver. They do not travel easily through insulators, such as glass, air, rubber, and plastic. Substances that are good conductors of electricity are typically also good conductors of heat. Electricity can be produced in different ways, including chemically and magnetically.

The flow of electrons through a conductor is referred to as electric current. An electrical current that flows in only one direction is called direct current (DC). Current that reverses its direction is called alternating current (AC). Alternating current is used in most heating, cooling, and refrigeration equipment because its properties give it greater flexibility than direct current. When direct current is needed, for example, in control circuits, it can easily be produced by a rectifier circuit located within the equipment.

Voltage is a term that refers to the potential, or electrical pressure, that pushes electrons through an electric circuit. The higher the voltage, the greater the amount of current flow. Without voltage, there can be no current flow. Ampere is the unit used to measure the quantity of electrons moving past a given point in a specific period of time, and ohm is the unit used to measure the resistance, or opposition to electron flow. The relationship that exists among voltage, current, and resistance is referred to as Ohm's Law:

$$\text{Ohm's Law: Voltage } (E) = \text{Amperage } (I) \times \text{Resistance } (R)$$

Electric power is the rate at which electrons do work or the rate at which electrons are being used. The power consumed by an electric circuit is measured in watts. Electric power can be calculated by using the formula $P = IE$.

When installing HVACR systems, it is important that the correct power and ampacity be supplied to the unit. Equipment manufacturers supply a

(continued)

wire-sizing chart to determine the correct conductor that should be used to supply electrical power to the equipment. The correct wire size can be easily obtained from the table of recommended wire sizes from the installation instructions. The figures that are used in this table will correspond to the National Electrical Code®. Factors to be considered when sizing circuit conductors are voltage drop, insulation type, enclosure used, and safety. Wire-sizing charts are available in the National Electrical Code® and should be used when questions arise. The technician should make certain that the voltage drop in any circuit will maintain minimum standards.

An electric circuit is an arrangement that typically consists of power-passing devices (switches) and power-consuming devices (loads). Components in a circuit may be wired in series, in parallel, or in a combination of both. In a series circuit, there is only one possible path for the electric current to take. In a parallel circuit, there are multiple paths for the current to take. Most safety and operating controls are connected in series to control loads in electrical circuits. Parallel circuits are used to ensure that the correct voltage is supplied to the electrical loads in the system. Series and parallel circuits have different voltage, amperage, and resistance relationships that must be considered when creating, servicing, or troubleshooting electric circuits and components.

Whenever current flows through a conductor, a magnetic field is generated. When a conductor is shaped into a coil, the strength of the magnetic field is concentrated. This magnetic field is used in many control devices such as relays and contactors.

The digital multimeter (DMM) is a piece of instrumentation that is used to measure electrical characteristics such as voltage, amperage, and resistance. The HVACR technician will often be required to use electrical meters to perform the tasks that are required when installing, testing, and servicing equipment. As important as electricity is to the HVACR industry, it is not hard to see that it is of utmost importance for service and installation technicians to be able to determine the condition of electrical components and circuits in equipment and control systems.

In order to effectively troubleshoot the electrical devices and control systems of modern HVACR equipment, the technician must know how to read and interpret electrical wiring diagrams. These diagrams contain a wealth of information about the electrical installation and operation of the equipment. The schematic diagram is broken down into a circuit-by-circuit arrangement that allows the technician to easily identify the electrical circuit that is causing the problems. Once the inoperative circuit is located, the technician can determine which electrical component is at fault. The efficient use of wiring diagrams decreases the amount of time required to troubleshoot the system.

Symbols are used in wiring diagrams to represent the electrical components in the circuits. Technicians must be able to identify the components based on their symbols. Symbols are not standard across the industry, so the technician must have some means of knowing what a symbol represents. The legend of the wiring diagram is a listing of the components along with their symbols.

There are many types of electrical diagrams used in the refrigeration, heating, and air-conditioning industry. Types of wiring diagrams include the schematic, the pictorial, and the ladder. The schematic diagram shows each and every wire in the system along with other useful information such as wire color, terminal connection points, line and low-voltage wiring identification, and factory/field-installed wiring identification. The pictorial diagram can be extremely useful, as it often provides, in addition to the information on the schematic diagram, the location of each component in the system's control panel. In a ladder or line diagram, the power source is represented by two vertical lines, with the circuits of the equipment or control system connected between them. The ladder diagram is very useful for system troubleshooting. Since different manufacturers create their wiring diagrams differently, the features contained in a particular diagram will likely vary from system to system.

Electrical devices generally can be divided into two basic types: loads and switches. Loads are electrical devices that consume electrical energy to do work, while switches are used to control the current flow to the loads. In an electrical diagram, loads are controlled by switches to maintain a certain condition.

The HVACR industry has now incorporated electronic control circuits into most control systems. Solid-state control systems are configured as a group of circuits used to control a system. Improvements and innovations have revolutionized control systems and components in the industry, yielding smaller, more accurate, and more diversified electronic components. Electronic control systems and components have almost eliminated erratic control of individual zones in residential, commercial, and industrial applications. The HVACR technician must become proficient at troubleshooting both electromechanical circuitry and digital, electronic devices.

KEY TERMS

Alternating current
Ampacity
Ampere
Atom
Capacitance
Capacitor
Circuit breaker
Conductor
Current
Diode
Direct current
Electric energy

Electric power
Electricity
Electrodes
Electrolytes
Electron
Element
Field of force
Free electrons
Fuse
Insulator
Kilowatt-hour
Law of charges

Matter
Molecule
Neutron
Nucleus
Ohm
Ohm's Law
Power
Power factor
Proton
Rectifier
Semiconductor
Single-phase power

Solenoid
Solid state
Static electricity
Thermistor
Three-phase power
Transformer
Transistor
Volt
Voltage
Watt

ELECTRICAL CORE CONCEPTS: TEST YOUR KNOWLEDGE

Name	Date	Grade

Circle the letter that indicates the correct answer.

1. What type of charge does an electron have?
A. Negative
B. Positive
C. Neutral
D. Negative-positive

2. What part of an atom freely moves from one atom to another in a good conductor?
A. Electron
B. Proton
C. Neutron
D. Molecule

3. Examples of good insulators are:
A. copper, silver, gold.
B. steel, bronze, brass.
C. glass, rubber, plastic.
D. aluminum, nickel, magnesium.

4. Electricity can be produced from:
A. magnetism.
B. chemicals.
C. heat.
D. All of the above.

5. The unit of measurement for resistance is the:
A. volt.
B. ampere.
C. ohm.
D. milliampere.

6. An ampere is:
A. the difference in potential between two charges.
B. a measure of the quantity of electrons flowing past a point in a given period of time.
C. the unit of measure for resistance.
D. the unit of measure for capacitance.

7. Switches in electrical circuits are usually wired:
A. directly across line voltage.
B. near the power source.
C. in parallel with the load.
D. in series with the load.

8. Loads or power-consuming devices are usually wired:
A. with power being connected directly to the ground connection.
B. near the power source.
C. in parallel with each other.
D. in series with each other.

9. The mathematical relationship among the current, electromotive force, and resistance of an electric circuit is known as:
A. the Law of Charges
B. Ohm's Law
C. both A and B are correct.
D. neither A nor B is correct.

10. If there were a current flowing of 6 A in a 120 V circuit, what would the resistance be?
A. 10 Ω
B. 720 Ω
C. 17 Ω
D. 20 Ω

11. If the resistance in a 120 V electrical circuit were 30 Ω, what would the current be in amperes?
A. 100 A
B. 3600 A
C. 4 A
D. 6 A

12. What would the voltage be in a DC circuit if there were a current flow of 6 A and a load resistance of 8 Ω?
A. 120 V
B. 14 V
C. 115 V
D. 48 V

13. If a conductor carrying an alternating current were formed into a coil, the magnetic field that is generated:
  A. would become weaker and could not be used to open or close electrical switches or valves or perform other useful tasks.
  B. can be used to open or close switches or valves or perform other useful tasks.
  C. will cancel out and become zero.

14. A magnetic field can be used to:
  A. generate electricity.
  B. operate relays, contactors, and solenoids.
  C. cause electric motors to operate.
  D. All of the above.

15. The primary and secondary windings in a transformer are:
  A. wired in parallel.
  B. wired in series.
  C. not connected electrically.
  D. connected with a solenoid switching device.

16. The unit of measurement for the charge a capacitor can store is the:
  A. inductive reactance.
  B. microfarad.
  C. ohm.
  D. joule.

17. If an electrical circuit has a complete path, electrons will flow when:
  A. current is applied.
  B. voltage is applied.
  C. power is applied.
  D. resistance is applied.

18. Power is measured in:
  A. watts.
  B. volts.
  C. ohms.
  D. amperes.

19. Electrical utility companies bill their customers based on the number of:
  A. volts used per month.
  B. amperes used per month.
  C. watts used per month.
  D. kilowatt hours used per month.

20. A 10 kW electric strip heater will provide approximately how many btu/h of heat?
  A. 3413 Btu/h
  B. 10,000 Btu/h
  C. 34,130 Btu/h
  D. 100,000 Btu/h

21. Explain the term "semiconductor."

22. Explain P-material and N-material as they relate to semiconductors.

23. What is the intended function of a diode?

24. What is the intended function of a transistor?

25. Explain why solid state control devices are quickly replacing electromechanical devices.

26. What is a thermistor?

27. Explain the difference between a circuit breaker and a fuse.

28. Explain the function and application of a GFCI.

Taking Voltage and Amperage Readings

Name	Date	Grade

OBJECTIVES: Upon completion of this exercise, you should be able to demonstrate the skills needed to properly obtain voltage and amperage readings on actual operating equipment using the appropriate pieces of test instrumentation.

INTRODUCTION: You will be using a digital multimeter (DMM) to take voltage readings and a clamp-on ammeter to take AC amperage readings.

TOOLS, MATERIALS, AND EQUIPMENT: An operating air-conditioning system with a low-voltage control circuit, a digital multimeter (DMM) with insulated alligator clip test leads, a clamp-on ammeter, ¼" and ⁵⁄₁₆" nut drivers, and a straight-slot screwdriver.

⚠ **SAFETY PRECAUTIONS:** Working around live electricity can be very hazardous. Be sure to inspect the test leads on the meter to ensure that the insulation is not damaged before using the meter. For this exercise, it is beneficial to use test leads that have insulated alligator clamps on them. Whenever possible, make all meter connections with the POWER OFF. Have your instructor inspect and approve all connections before turning the power on. Your instructor should have given you thorough instruction regarding safe meter handling and use. Since meters vary greatly by make and model, make certain that you have been properly instructed as to how to properly handle and use the meter that you have been assigned to work with. Before working on electric circuits, make certain that you are not wearing metallic jewelry and that you are wearing proper eye protection. Always be aware of your surroundings to minimize the chance of getting shocked or receiving any other injury.

STEP-BY-STEP PROCEDURES

Taking Voltage Readings

1. With the main power to the unit off, locate the control transformer in the air-conditioning system.

2. Identify the wires or terminals on the transformer's secondary (24 V output) winding.

3. Insert the test leads into the appropriate jacks on the meter to read AC voltage. The black test lead should be inserted into the COMMON jack, and the red test lead should be inserted into the VOLTAGE jack.

4. Set the meter to read AC voltage. If the range on the AC voltage scale needs to be set, select the lowest range that is above 50 V AC.

5. Connect the insulated alligator clips to the transformer's secondary winding wires or terminals. Either lead can go on either terminal, as there is no polarity for AC voltage.

6. Have your instructor check the connections. Turn the power on. Record the AC reading: _____ V.

7. Turn the power off.

8. Remove the meter's test leads from the transformer.

9. You will now measure the voltage at the input (primary) side of the transformer. The meter should still be set up to measure AC voltage. Set the AC voltage range to the lowest setting that is above 250 V.

10. With the power off, fasten the meter's test leads to the two terminals on the primary side of the control transformer.

11. Ask your instructor to check the connections. With approval, turn the power on. Record the AC voltage: _____ V.

12. Turn the power off, and remove the test leads.

13. With the power still off, locate the L1 and L2 terminals on the contactor of an air-conditioning unit.

14. Set the meter to read AC voltage at a range that is higher than 500 V. (Always select a setting higher than the anticipated voltage.)

15. Connect the meter to the L1 and L2 terminals.

16. Have your instructor inspect the connections and, with approval, turn the power on. Record the voltage reading: _____ V.

17. Turn the power off, and remove the meter.

18. Connect the meter's test leads to the T1 and T2 terminals on the contactor.

19. At the thermostat, set the system to the COOL mode, and make certain that the thermostat is set to a temperature that is lower than the temperature in the lab.

20. Once the compressor cycles on, observe the voltage between the T1 and T2 terminals, and record the AC voltage here: _____ V.

21. With the meter still connected to the T1 and T2 terminals, turn the system off at the thermostat.

22. Once the compressor cycles off, observe the voltage between the T1 and T2 terminals, and record the AC voltage here: _____ V.

23. Turn the main power to the unit off, and remove the meter.

Taking Amperage Readings

1. With the power to the unit off, identify and remove the service panel that covers the blower motor on the air-conditioning system.

2. Without touching or removing any wires from their terminals, list the colors of all of the wires coming from the motor here:

   Wire color: _____   Wire color: _____
   Wire color: _____   Wire color: _____
   Wire color: _____   Wire color: _____

3. Set the meter to read AC amps, and set the range to the lowest setting above 10 amperes.

4. Clamp the jaws of the clamp-on ammeter around one of the wires on the blower motor.

5. Have your instructor check your setup, including any necessary settings on your meter.

6. With approval, after making certain you are at a safe distance from any rotating fans, blowers, and belts, turn the power on so the motor starts.

7. Observe the amperage reading, and record this reading, along with the wire color, here: Wire Color: _____, Amperage: _____ A.

8. Turn the power off, and remove the meter.

9. Repeat steps 4 through 8 for each of the wires connected to the blower motor, and record the amperage readings here:

    Wire color: _____, amperage: _____ A

    Wire color: _____, amperage: _____ A

    Wire color: _____, amperage: _____ A

    Wire color: _____, amperage: _____ A

    Wire color: _____, amperage: _____ A

    Wire color: _____, amperage: _____ A

10. Turn the main power to the air-conditioning system off.

11. Without disturbing the electrical connections, identify the wires leading to the system's compressor. Record the colors here:

    Wire color: _____

    Wire color: _____

    Wire color: _____

12. Set the meter to read AC amps, and set the range to the lowest setting above 40 amperes, or above the expected amperage draw of the compressor. Refer to the unit's nameplate for compressor amperage information.

13. Clamp the jaws of the clamp-on ammeter around one of the wires on the compressor.

14. Have your instructor inspect your meter, and with approval, turn the power so the compressor operates.

15. Observe the amperage reading and record this reading, along with the wire color here: Wire color: _____, amperage: _____ A

16. Very carefully, observe the amperage readings on the remaining compressor wires, and record that information here:

    Wire color: _____, amperage: _____ A

    Wire color: _____, amperage: _____ A

17. Turn the system power off, and remove the meter.

MAINTENANCE OF WORKSTATION AND TOOLS: Replace all panels on equipment with correct fasteners. Make certain that all meters have been turned off and that all test leads are accounted for. Return all meters and tools to their proper places.

SUMMARY STATEMENT: Explain how taking accurate voltage and amperage readings can help a technician in a number of ways, including system troubleshooting and avoiding personal injury.

TAKEAWAY: HVACR technicians rely on instrumentation to provide them with important information regarding system operation and performance. Since we draw conclusions based on the information obtained, it is important that this information be accurate. Obtaining useful system data relies on the proper and safe use of the test equipment.

QUESTIONS

1. Compare the voltage reading that was obtained from the line side of the contactor with the voltage reading taken on the primary side of the transformer. Explain your observation.

2. Compare the voltage reading that was obtained from the primary side of the transformer with the voltage reading taken on the secondary side of the transformer. Explain your observation.

3. Is the transformer in this system a step-up or step-down transformer? Explain your answer.

4. Explain the difference between a "function" setting and a "range" setting on a multimeter.

5. Explain how a clamp-on ammeter is able to read circuit amperage even though it is not connected electrically to the circuit.

6. Explain why some of the amperage readings obtained on the blower motor leads were different from each other (if they were).

7. Explain any difference in voltage that was observed across the T1 and T2 terminals on the contactor when the compressor was operating as compared to when it was not.

8. When measuring the voltage being supplied to a load, is the meter connected in series or parallel with the load? Explain your answer.

Measuring Electrical Characteristics and Performing Electrical Calculations

Name	Date	Grade

OBJECTIVES: Upon completion of this exercise, you should be able to demonstrate the skills needed to take current, voltage, and resistance readings. In addition, you should be able to determine the current, voltage, and resistance of a circuit using Ohm's Law.

INTRODUCTION: Ohm's Law is the relationship that exists among voltage, current, and resistance. In this exercise, you will be using a digital multimeter (DMM) to make resistance readings and verify your calculations using Ohm's Law.

TOOLS, MATERIALS, AND EQUIPMENT: An electric heating element, a digital multimeter (DMM) with insulated alligator clip test leads, a clamp-on ammeter, ¼" and ⁵⁄₁₆" nut drivers, and a multi-tip screwdriver.

⚠ **SAFETY PRECAUTIONS:** Use test leads that have insulated alligator clips on the ends to reduce the risk of electric shock. Make all meter connections with the power off. Have your instructor inspect all connections before turning the power on. Make certain that you completely understand how to use all of the features of the test instrumentation you have been assigned to work with. Electric heating elements can easily exceed temperatures of 500°F. Be sure to prevent coming in contact with the surfaces of the element. These elements remain hot for quite some time, even after they are de-energized. If the heating element being used has been removed from an appliance and is resting on a bench or other work surface, be sure to protect the area from the heat as well as from other metallic surfaces.

STEP-BY-STEP PROCEDURES

1. Position yourself at a heating appliance that contains an electric heating element.

2. With the power off, remove the service panels to access the electric heating element and its electrical connections on the unit's terminal board.

3. Disconnect the wires that feed power to the heaters. Be sure to label, cap, and/or tape the ends of the wires.

4. Set the digital multimeter (DMM) to read resistance.

5. Clip one lead of the meter to each terminal of the element. Do not turn the power on. Observe and record the resistance of the heating element: _____ ohms.

6. Disconnect the meter from the heating element.

7. Reconnect the heating element to its original circuit.

8. Turn the clamp-on ammeter on and set it to measure AC amps.

9. Place the clamp-on ammeter jaws around one wire that is connected to the heating element.

10. Have your instructor inspect and approve your setup.

11. Turn the power on and quickly observe the amperage draw of the heater.

12. Turn the power off.

13. Record the amperage draw of the heating element here: _____ A. You must take the reading quickly because, as the element heats up, its resistance increases.

14. Turn on the DMM and set it to read AC voltage. Set the range to the lowest setting over 500 V.

15. With the power to the heating element off, connect a meter lead to each terminal on the heating element.

16. Have your instructor inspect and approve your setup.

17. Turn the power to the heating element on. Observe and record the voltage: _____ V.

18. Turn the power off.

19. Enter the three obtained values here:

    Current (I): _____ amps
    Voltage (E): _____ volts
    Resistance (R): _____ ohms
    Check your readings by using Ohm's Law.

20. To check the amperage reading, you will use the following formula: $I = E/R$. Divide the voltage by the resistance and enter the result here: E/R = _____ amps.

21. Compare the current value entered in line 13 with the value entered in line 20.

22. Calculate how much higher or lower the calculated current value is than the measured value using the following formula:

$$\text{Percentage Change} = 100 \times (\text{Calculated Value} - \text{Measured Value})/\text{Measured Value}$$

MAINTENANCE OF WORKSTATION AND TOOLS: Turn all meters off, and make certain that all test leads are accounted for. Replace any service or access panels that were removed during the course of this exercise. Return all tools and equipment to their proper places. If any unmounted heating elements were used, allow them to completely cool before handling and storing them.

SUMMARY STATEMENT: Measured and calculated values often differ. Comment on the possible causes for any discrepancies that might be present between the calculated and measured current valves that were obtained during this exercise.

TAKEAWAY: Voltage, current, and resistance are interrelated by the laws of physics. Knowing the rules that govern the behavior of electricity allows us to better predict, understand, and work with this powerful force.

QUESTIONS

1. With the resistance of a circuit remaining the same, explain what will happen to the current flowing through the circuit if the voltage supplied to the circuit is increased?

2. With the voltage supplied to a circuit remaining the same, explain what will happen to the current flowing through the circuit if the resistance of the circuit is increased?

3. What type of load is the electric heating element that was used in this exercise?

4. Express Ohm's Law in three different ways, solving for voltage, current, and resistance.

5. Explain the concept of power factor and how it affected (or didn't affect) the results of the calculations in this exercise.

Exercise ECC-3

Electrical Load Identification

Name	Date	Grade

OBJECTIVES: Upon completion of this exercise, you should be able to classify various HVACR components as either loads or switches. If the device is a load, it must be determined if the load is either resistive or inductive. In addition, you should be able to determine what effect, if any, each of the devices will have on the circuit's power factor.

INTRODUCTION: In order to effectively evaluate air-conditioning system components, the technician must first be able to establish if the component is an electrical load or a switch. If this is not done properly, the technician will very likely reach inaccurate conclusions about the component, leading to an incorrect system diagnosis.

TOOLS, MATERIALS, AND EQUIPMENT: HVACR reference material.

SAFETY PRECAUTIONS: There are no safety precautions for this exercise.

STEP-BY-STEP PROCEDURES

Using your HVACR reference material, you are to complete the following chart by following these steps:

1. For each device listed, determine if the component is a switch or a load.

2. If the device is a switch, place a check mark (√) in the "switch" box that corresponds to that device.

3. If the device is a load, you must then determine if the load is resistive or inductive.

4. If the device is a resistive load, place a check mark (√) in the "RESISTIVE LOAD" box that corresponds to that device.

5. If the device is an inductive load, place a check mark (√) in the "INDUCTIVE LOAD" box that corresponds to that device.

6. The three right-hand columns will be completed by determining the device's effect on the power factor of the circuit.

    a. If the device will do nothing (or very little) to shift the voltage and current out of phase with each other, the power factor will remain high and will be very close to, or equal to, one. If such is the case, place a check mark (√) in the "HIGH" box that corresponds to that device.

    b. If the device has a significant effect on shifting the voltage and current out of phase with each other, the power factor will be reduced. If such is the case, place a check mark (√) in the "LOW" box that corresponds to that device.

    c. If the device is not a contributing factor that determines the power factor of the circuit, place a check mark (√) in the "N/A" box that corresponds to that device.

Device	Switch	Resistive Load	Inductive Load	Power Factor		
				N/A	Low	High (Close to 1)
High limit control						
Solenoid coil						
High-pressure control						
Motor winding						
Bimetal overload						
Electric heating element						
Line-voltage thermostat						
Relay coil						
N.C. set of contacts						
Circuit breaker						
Low-voltage thermostat						
Electric toaster heating elements						
Low-pressure control						
Contactor coil						
Incandescent light bulb						
Electromagnet						
N.O. set of contacts						
Transistor						
Aquastat						
Transformer winding						

MAINTENANCE OF WORKSTATION AND TOOLS: There are no workstation and tool maintenance issues for this exercise.

SUMMARY STATEMENT: Identify and comment on any relationships or correlations that exist between the boxes checked in the first three columns and the boxes checked in the last three columns.

TAKEAWAY: Electrical components play different roles in a circuit. It is important to understand how these components function as well as how they affect the circuit's power consumption in order to properly evaluate them.

QUESTIONS

1. Explain the characteristics of a resistive load.

2. Explain the characteristics of an inductive load.

3. Explain the concept of power factor and how it is calculated.

4. Explain how forming a wire into a coil will concentrate the magnetic field so it can do useful work.

5. Explain why the power factor for strictly resistive circuits is 1.

6. Explain what can be done in an electric circuit to offset the effects of the current lag caused by an inductive load.

Exercise ECC-4

Series Circuits

Name	Date	Grade

OBJECTIVES: Upon completion of this exercise, you should be able to construct various series circuits and take electrical measurements (voltage, current, and resistance) on the circuit.

INTRODUCTION: In this exercise, you will wire various series circuits based on sample diagrams and then take voltage, current, and resistance readings at various points in the circuit.

TOOLS, MATERIALS, AND EQUIPMENT: Wire cutters, wire strippers, crimping tools, screwdrivers, digital multimeters (DMM), clamp-on ammeters, safety goggles, and the following materials:

    14-gauge stranded wire
    Solderless connectors (forks)
    Wire nuts
    Four 75 W incandescent light bulbs
    Four 100 W incandescent light bulbs
    Four light sockets
    3-conductor (black, white, green) 14-gauge, SJ cable
    Male plug
    One single-pole, single-throw switch

⚠ **SAFETY PRECAUTIONS:** When working on electric circuits, the following safety issues should be adhered to:

- Whenever possible, work on circuits that are de-energized.
- Avoid coming in contact with bare conductors.
- Avoid becoming part of the active electric circuit.
- Use ohmmeters only on circuits that are de-energized.
- Be sure that the meters are set to the proper scale BEFORE taking readings.

STEP-BY-STEP PROCEDURES

1. Create a power cord by connecting the male plug to one end of a 3 ft length of SJ cable. Make certain that all three conductors are securely connected to the line, neutral, and ground prongs on the plug.

2. Remove about 6" of the casing from the loose end of the SJ cable.

3. Strip about ¼" of the insulation from the ends of the three conductors.

4. Connect solderless connectors (forks) to the ends of the three conductors. Set the power cord aside.

5. Using the digital multimeter (DMM) set to read resistance, take a resistance reading of the four 75 W light bulbs. Record the resistance value of each bulb in the left column of this table and calculate the average resistance. This is accomplished by adding the four resistance values together and then dividing this total by 4.

	75 W Bulbs	100 W Bulbs
Reading		
Reading		
Reading		
Reading		
Average		

6. Repeat step 5 with the four 100 W bulbs, and enter the information in the right column of the table. The information obtained in steps 7 through 16 will be entered in this table:

Series Circuit Connections	Calculated Resistance	Actual Resistance
Two 75 W bulbs		
Two 100 W bulbs		
Three 75 W bulbs		
Three 100 W bulbs		
Four 75 W bulbs		
Four 100 W bulbs		
One 75 W bulb and one 100 W bulb		
Two 75 W bulbs and two 100 W bulbs		

7. Using the readings and calculations from steps 5 and 6, calculate the value of the resistance of two 75 W light bulbs connected in series.

8. Repeat step 7 with two 100 W bulbs.

9. Using the readings and calculations from steps 5 and 6, calculate the value of the resistance of three 75 W light bulbs connected in series.

10. Repeat step 9 with three 100 W bulbs.

11. Using the readings and calculations from steps 5 and 6, calculate the value of the resistance of four 75 W light bulbs connected in series.

12. Repeat step 11 with four 100 W bulbs.

13. Using the readings and calculations from steps 5 and 6, calculate the value of the resistance of one 75 W bulb connected in series with one 100 W bulb.

14. Repeat step 13 with two 75 W bulbs and two 100 W bulbs.

15. Construct each of the preceding circuits, and measure the actual resistance of each circuit, using Figures ECC-4.1, ECC-4.2, and ECC-4.3 as reference.

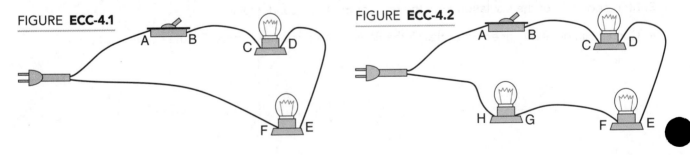

FIGURE **ECC-4.1**

FIGURE **ECC-4.2**

FIGURE **ECC-4.3**

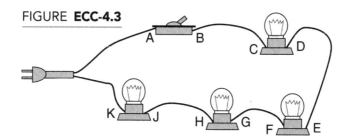

Note: Take the resistance readings between the line and neutral prongs on the plug, and make certain that the switch is in the ON position. Under no circumstances should the circuits be energized at this time.

16. Record the corresponding resistances in the appropriate cells in the "actual" column of the preceding table.

17. After each circuit has been constructed and has had the resistance readings taken, the circuit can be energized by first turning the switch to the OFF position and then plugging the power cord into an accessible outlet.

18. Measure the amperage of each circuit by taking a reading around one of the conductors connected to the switch.

19. Record the amperage of each circuit on the following chart:

Series Circuit Connections	Circuit Amperage
Two 75 W bulbs	
Two 100 W bulbs	
Three 75 W bulbs	
Three 100 W bulbs	
Four 75 W bulbs	
Four 100 W bulbs	
One 75 W bulb and one 100 W bulb	
Two 75 W bulbs and two 100 W bulbs	

20. After each circuit has been constructed and energized, take voltage readings at the power supply and across each load. Be sure to take notes about the brightness of the bulbs in each circuit. Record the data in the following chart:

Series Circuit Connections	Power Supply Voltage	Load Voltage
Two 75 W bulbs		75W: _____V, 75W: _____V
Two 100 W bulbs		100W: _____V, 100W: _____V
Three 75 W bulbs		75W: _____V, 75W: _____V, 75W: ____V
Three 100 W bulbs		100W: ____V, 100W: _____V, 100W: ___V
Four 75 W bulbs		75W: _____V, 75W: _____V, 75W: ____V, 75W: _____V
Four 100 W bulbs		100W: ____V, 100W: _____V, 100W: ___V, 100W: ____V
One 75 W bulb and one 100 W bulb		75W: _____V, 100W: ____V
Two 75 W bulbs and two 100 W bulbs		75W: _____V, 75W: _____V, 100W: ____V, 100W: ____V

21. Make comments about the brightness of the bulbs in this table:

Series Circuit Connections	Comments on Bulb Brightness
Two 75 W bulbs	
Two 100 W bulbs	
Three 75 W bulbs	
Three 100 W bulbs	
Four 75 W bulbs	
Four 100 W bulbs	
One 75 W bulb and one 100 W bulb	
Two 75 W bulbs and two 100 W bulbs	

MAINTENANCE OF WORKSTATION AND TOOLS:

Make certain that all light bulbs are properly stored to prevent breakage.

Make certain that all tools are properly stored.

Make certain that all materials (wire nuts, solderless connectors) are properly sorted and stored.

Make certain that the work area is cleaned.

Keep the power cord assembly as it will be used in other exercises.

SUMMARY STATEMENT: Explain why series circuits are not commonly used to control the operation of multiple loads. Provide examples and support for your statements by referencing specific points in the exercise just completed.

TAKEAWAY: In this exercise, it was seen that resistance and current are inversely related and that the voltage supplied to the circuit must be distributed among all the series-connected loads.

QUESTIONS

1. Compare the resistance of a 75 W bulb to that of a 100 W bulb. Which bulb has the higher resistance? Explain why this is so.

2. What happens to the power consumption of a circuit as its resistance increases? Explain your answer in detail, providing details from the exercise you just completed.

3. Compare and explain the difference in bulb brightness when one 75 W bulb was connected in series with one 100 W bulb. Be sure to include information from your responses to questions 1 and 2 to support your answer.

Parallel Circuits

Name	Date	Grade

OBJECTIVES: Upon completion of this exercise, you should be able to construct various parallel circuits and take electrical measurements (voltage, current, and resistance) on the circuit.

INTRODUCTION: In this exercise, you will wire various parallel circuits based on sample diagrams and then take voltage, current, and resistance readings at various points in the circuit.

TOOLS, MATERIALS, AND EQUIPMENT: Wire cutters, wire strippers, crimping tools, screwdrivers, digital multimeters (DMM), clamp-on ammeters, safety goggles, and the following materials:

14-gauge stranded wire

Solderless connectors (forks)

Wire nuts

Four 75 W incandescent light bulbs

Four 100 W incandescent light bulbs

Four light sockets

3-conductor (black, white, green) 14-gauge, SJ cable

Male plug

Four single-pole, single throw switches

 SAFETY PRECAUTIONS: When working on electric circuits, the following safety issues should be adhered to:

- Whenever possible, work on circuits that are de-energized.
- Avoid coming in contact with bare conductors.
- Avoid becoming part of the active electric circuit.
- Use ohmmeters only on circuits that are de-energized.
- Be sure that the meters are set to proper scale BEFORE taking readings.

PROCEDURES

1. Create a power cord by connecting the male plug to one end of a 3 ft length of SJ cable. Make certain that all three conductors are securely connected to the line, neutral, and ground prongs on the plug.

2. Remove about 6" of the casing from the loose end of the SJ cable.

3. Strip about ¼" of the insulation from the ends of the three conductors.

4. Connect solderless connectors (forks) to the ends of the three conductors. Set the power cord aside.

5. Using the digital multimeter (DMM) set to read resistance, take a resistance reading of the four 75 W light bulbs. Record the resistance value of each bulb in the left column of this table, and calculate the average resistance.

	75 W Bulbs	100 W Bulbs
Reading		
Reading		
Reading		
Reading		
Average		

6. Repeat step 5 with the four 100 W bulbs, and enter the information in the right column of the table. The information obtained in steps 7 through 15 will be entered in this table:

Parallel Circuit Connections	Calculated Resistance	Actual Resistance
Two 75 W bulbs		
Two 100 W bulbs		
Three 75 W bulbs		
Three 100 W bulbs		
Four 75 W bulbs		
Four 100 W bulbs		
One 75 W bulb and one 100 W bulb		
Two 75 W bulbs and two 100 W bulbs		

7. Using the readings and calculations from steps 5 and 6, calculate the value of the resistance of two 75 W light bulbs connected in parallel.

8. Repeat step 7 with two 100 W bulbs.

9. Using the readings and calculations from steps 5 and 6, calculate the value of the resistance of three 75 W light bulbs connected in parallel.

10. Repeat step 9 with three 100 W bulbs.

11. Using the readings and calculations from steps 5 and 6, calculate the value of the resistance of four 75 W light bulbs connected in parallel.

12. Repeat step 11 with four 100 W bulbs.

13. Using the readings and calculations from steps 5 and 6, calculate the value of the resistance of one 75 W bulb connected in parallel with one 100 W bulb.

14. Repeat step 13 with two 75 W bulbs and two 100 W bulbs.

15. Construct each of the above circuits and measure the actual resistance of each circuit, using Figures ECC-5.1, ECC-5.2, and ECC-5.3 as reference.

FIGURE **ECC-5.1**

FIGURE **ECC-5.2**

FIGURE **ECC-5.3**

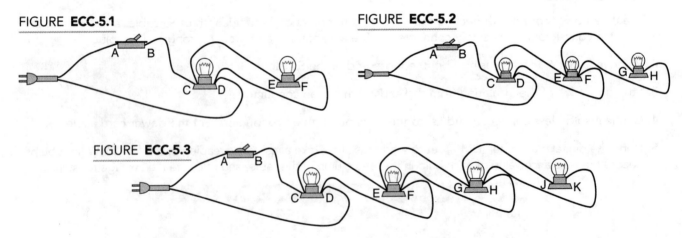

Note: Take the resistance readings between the line and neutral prongs on the plug, and make certain that the switch is in the ON position. Under no circumstances should the circuits be energized at this time.

16. Record the corresponding resistances in the appropriate cells in the "actual" column of the preceding table.

17. After each circuit has been constructed and has had the resistance readings taken, the circuit can be energized by first turning the switch to the OFF position and then plugging the power cord into an accessible outlet.

18. Measure the amperage of each circuit by taking a reading around one of the conductors connected to the switch.

19. Record the amperage of each circuit on the following chart:

Parallel Circuit Connections	Total Circuit Amperage
Two 75 W bulbs	
Two 100 W bulbs	
Three 75 W bulbs	
Three 100 W bulbs	
Four 75 W bulbs	
Four 100 W bulbs	
One 75 W bulb and one 100 W bulb	
Two 75 W bulbs and two 100 W bulbs	

20. Record the amperage of each branch of each circuit on the following chart:

Parallel Connections	Total Circuit Amperage (from previous table)	Branch 1	Branch 2	Branch 3	Branch 4
Two 75 W bulbs				N/A	N/A
Two 100 W bulbs				N/A	N/A
Three 75 W bulbs					N/A
Three 100 W bulbs					N/A
Four 75 W bulbs					
Four 100 W bulbs					
One 75 W bulb and one 100 W bulb				N/A	N/A
Two 75 W bulbs and two 100 W bulbs					

21. After each circuit has been constructed and energized, take voltage readings at the power supply and across each load. Be sure to take notes about the brightness of the bulbs in each circuit. Record the data in the following chart:

Series Circuit Connections	Power Supply Voltage	Load Voltage
Two 75 W bulbs		75 W: _____V, 75 W: _____V
Two 100 W bulbs		100 W: _____V, 100 W: _____V
Three 75 W bulbs		75 W: _____V, 75 W: _____V, 75 W: _____V
Three 100 W bulbs		100 W: _____V, 100 W: _____V, 100 W: _____V
Four 75 W bulbs		75 W: _____V, 75 W: _____V, 75 W: _____V, 75 W: _____V
Four 100 W bulbs		100 W: _____V, 100 W: _____V, 100 W: _____V, 100 W: _____V
One 75 W bulb and one 100 W bulb		75 W: _____V, 100 W: _____V,
Two 75 W bulbs and two 100 W bulbs		75 W: _____V, 75 W: _____V, 100 W: _____V, 100 W: _____V

22. Make comments about the brightness of the bulbs in this table:

Series Circuit Connections	Comments on Bulb Brightness
Two 75 W bulbs	
Two 100 W bulbs	
Three 75 W bulbs	
Three 100 W bulbs	
Four 75 W bulbs	
Four 100 W bulbs	
One 75 W bulb and one 100 W bulb	
Two 75 W bulbs and two 100 W bulbs	

MAINTENANCE OF WORKSTATION AND TOOLS:

Make certain that all light bulbs are properly stored to prevent breakage.

Make certain that all tools are properly stored.

Make certain that all materials (wire nuts, solderless connectors) are properly sorted and stored.

Make certain that the work area is cleaned.

Keep the power cord assembly, as it will be used in other exercises.

SUMMARY STATEMENT: Explain why parallel circuits are commonly used to control the operation of multiple loads. Provide examples and support for your statements by referencing specific points in the exercise just completed.

TAKEAWAY: In this exercise, it was seen that resistance and current are inversely related and that the voltage supplied to the circuit is the same voltage that is supplied to each branch of a parallel circuit. Although the individual branches of a parallel circuit share a common power supply, each branch has the ability to be controlled separately.

QUESTIONS

1. As more and more bulbs were connected in parallel, what happened to the total resistance of the circuit? Explain why this is the case.

2. As more and more bulbs were connected in parallel, what happened to the total amperage draw of the circuit? Explain why this is the case.

3. Compare and explain the differences in bulb brightness that were observed during this exercise.

Exercise ECC-6

Series and Parallel Circuits

Name	Date	Grade

OBJECTIVES: Upon completion of this lab, you should be able to recognize, describe, and construct series, parallel, and series-parallel circuits used in residential HVACR equipment. You should also be able to obtain and interpret voltage, current, and resistance readings from basic series, parallel, and series-parallel circuits and formulate well-thought-out responses to posed questions.

INTRODUCTION: In order to become a successful HVACR service technician, it is important to know how series, parallel, and series-parallel circuits are utilized in control systems and equipment in the industry and how to identify these circuits in wiring diagrams. This exercise provides a hands-on experience that will allow you to demonstrate the basic skills needed to safely and accurately manipulate these circuits.

TOOLS, MATERIALS, AND EQUIPMENT:

Screwdriver	Male plug
Wire-cutting pliers	14-gauge stranded wire
Crimpers	Solderless connectors (forks)
Light bulb sockets	Miscellaneous wood screws
Two 75 W incandescent light bulbs	VOM meter
Two 100 W incandescent light bulbs	Clamp-on ammeter
One single-pole, single-throw switch	One 12" × 12" piece of plywood
14-gauge, 3-conductor SJ cable	Terminal board with screw terminals for line, neutral, and ground connections

⚠ **SAFETY PRECAUTIONS:** In this lab exercise, you will be working with 115 V circuits. It is very important that you make wiring connections in a neat and orderly fashion to prevent the wires and electrical connections from accidentally touching each other or you. At no time should you try to connect any component into a circuit with the power on. Make certain that the electrical power is disconnected. Use caution when opening and closing switches in circuits that have power supplied to them. Do not touch any electrical connection when electrical power is supplied to the circuit. Light bulbs can be hot, even after they have been turned off. Allow sufficient time for the bulbs to cool down before handling them.

STEP-BY-STEP PROCEDURES

A. Series Circuits

1. Create a power cord by connecting the male plug to one end of a 3 ft length of SJ cable. Make certain that all three conductors are securely connected to the line, neutral, and ground prongs on the plug.

2. Remove about 6" of the casing from the loose end of the SJ cable.

3. Strip about ¼" of the insulation from the ends of the three conductors.

4. Connect solderless connectors (forks) to the ends of the three conductors. Set the power cord aside.

5. In the center of the plywood, mount two light bulb sockets side by side. They should be approximately 6 in. apart. Mount the terminal board 4 in. above the sockets. The plywood board should resemble Figure ECC-6.1.

6. Wire the board as shown in Figure ECC-6.2. The two light sockets form a series circuit, shown schematically in Figure ECC-6.3.

7. Connect the test cord with the 115 V plug to terminals L_1 and N of the terminal board.

8. Have the instructor check your series circuit.

9. Install two 75 W light bulbs in the light sockets.

10. Using a VOM multimeter set to read resistance, measure the resistance of the entire circuit between the L1 and neutral (L2) prongs on the plug. Record the resistance here: _____ ohms.

11. Plug the circuit into a 115 V receptacle. The light bulbs should illuminate. If they don't, ask your instructor for guidance.

FIGURE **ECC-6.1**

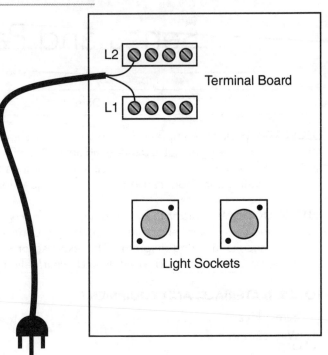

FIGURE **ECC-6.2**

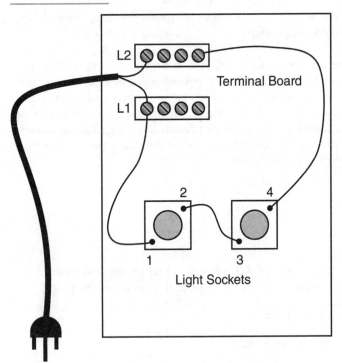

FIGURE **ECC-6.3**

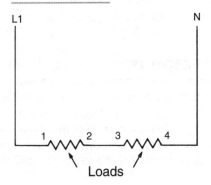

12. Comment on the brightness of the light bulbs. Are they brighter than you expected? Are they dimmer than you expected? Provide a possible explanation for the condition you observed.

_____

_____

_____

13. Using a VOM multimeter set to read AC voltage, measure the voltage (potential difference) between the following points in the circuit as indicated in Figure ECC-6.2:

Voltage between point 1 and point 4: _____ V
Voltage between point 1 and point 2: _____ V
Voltage between point 2 and point 3: _____ V
Voltage between point 3 and point 4: _____ V

14. Comment on the voltage readings obtained in step 13. _____

_____

_____

15. Using a clamp-on ammeter, measure the amperage draw of the circuit and record the amperage here: _____ A

16. Unplug the circuit.

17. Replace the right-hand bulb with a 100 W light bulb.

18. Plug the circuit into a 115 V receptacle. The light bulbs should illuminate. If they don't, ask your instructor for guidance.

19. Comment on the brightness of the light bulbs. Are they brighter than you expected? Are they dimmer than you expected? Is one bulb brighter than the other? Provide a possible explanation for the condition you observed.

_____

_____

_____

20. Using a VOM multimeter set to read AC voltage, measure the voltage (potential difference) between the following points in the circuit as indicated in Figure ECC-6.2:

Voltage between point 1 and point 4: _____ V
Voltage between point 1 and point 2: _____ V
Voltage between point 2 and point 3: _____ V
Voltage between point 3 and point 4: _____ V

21. Comment on the voltage readings obtained in step 13. _____

_____

_____

22. Using a clamp-on ammeter, measure the amperage draw of the circuit and record the amperage here: _____ A

23. Compare the amperage reading obtained in step 22 with the amperage reading obtained in step 15. Comment on these readings. Are they the same? Are they different? Which one was higher? Why?

_____

_____

24. Replace the left-hand bulb with a 100 W light bulb.

25. Plug the circuit into a 115 V receptacle. The light bulbs should illuminate. If they don't, ask your instructor for guidance.

26. Comment on the brightness of the light bulbs. Are they brighter than you expected? Are they dimmer than you expected? Is one bulb brighter than the other? Provide a possible explanation for the condition you observed.

_____

_____

_____

27. Using a VOM multimeter set to read AC voltage, measure the voltage (potential difference) between the following points in the circuit as indicated in Figure ECC-6.2:

Voltage between point 1 and point 4: _____ V

Voltage between point 1 and point 2: _____ V

Voltage between point 2 and point 3: _____ V

Voltage between point 3 and point 4: _____ V

28. Comment on the voltage readings obtained in step 13. _____

_____

_____

29. Using a clamp-on ammeter, measure the amperage draw of the circuit and record the amperage here: _____ A

30. Compare the amperage reading obtained in step 29 with the amperage reading obtained in step 15. Comment on these readings. Are they the same? Are they different? Which one was higher? Why?

_____

_____

31. Based on your observations, comment on the resistance of a 75 W bulb as compared to the resistance of a 100 W bulb.

_____

_____

B. Parallel Circuits

1. Wire the board as shown in ECC-6.4. The two light sockets are wired in parallel, as shown schematically in Figure ECC-6.5.

2. Have your instructor check your parallel circuit.

3. Install two 75 W light bulbs in the light sockets.

FIGURE **ECC-6.4**

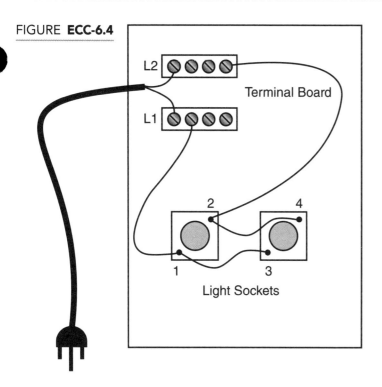

FIGURE **ECC-6.5**

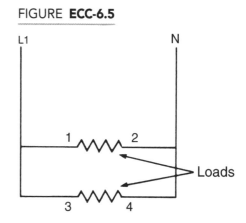

4. Using a VOM multimeter set to read resistance, measure the resistance of the entire circuit between the L1 and neutral prongs on the plug. Record the resistance here: _____ ohms.

5. Plug the circuit into a 115 V receptacle. The light bulbs should illuminate. If they don't, ask your instructor for guidance.

6. Comment on the brightness of the light bulbs. Are they brighter than you expected? Are they dimmer than you expected? Provide a possible explanation for the condition you observed.

_____

_____

_____

7. Using a VOM multimeter set to read AC voltage, measure the voltage (potential difference) between the following points in the circuit as indicated in Figure ECC-6.5:

Voltage between point 1 and point 2: _____ V

Voltage between point 3 and point 4: _____ V

8. Comment on the voltage readings obtained in step 6. _____

_____

_____

9. Using a clamp-on ammeter, measure the total amperage draw of the circuit and record the amperage here: _____ A

10. Using a clamp-on ammeter, measure the amperage flowing through one of the bulbs and record the amperage here: _____ A

11. Using a clamp-on ammeter, measure the amperage flowing through the other bulb and record the amperage here: _____ A

12. Compare and comment on the readings obtained in steps 9, 10, and 11. _____

_____

_____

13. Unplug the circuit.

14. Replace one of the 75 W bulbs with a 100 W bulb.

15. Using a VOM multimeter set to read resistance, measure the resistance of the entire circuit between the L1 and neutral prongs on the plug. Record the resistance here: _____ ohms

16. Plug the circuit into a 115 V receptacle. The light bulbs should illuminate. If they don't, ask your instructor for guidance.

17. Comment on the brightness of the light bulbs. Are they brighter than you expected? Are they dimmer than you expected? Provide a possible explanation for the condition you observed.

_____

_____

_____

18. Using a VOM multimeter set to read AC voltage, measure the voltage (potential difference) between the following points in the circuit as indicated in Figure ECC-6.5:

Voltage between point 1 and point 2: _____ V

Voltage between point 3 and point 4: _____ V

19. Comment on the voltage readings obtained in step 18. _____

_____

_____

20. Using a clamp-on ammeter, measure the total amperage draw of the circuit and record the amperage here: _____ A

21. Using a clamp-on ammeter, measure the amperage flowing through one of the bulbs and record the amperage here: _____ A

22. Using a clamp-on ammeter, measure the amperage flowing through the other bulb and record the amperage here: _____ A

23. Compare and comment on the readings obtained in steps 20, 21, and 22. _____

_____

_____

24. Compare and comment on the resistance readings obtained in steps 4 and 15. _____

_____

_____

25. Unplug the circuit.

C. Series-Parallel Circuits

1. The series-parallel circuit is a combination of the series and parallel circuits. The electrical circuitry of most air-conditioning equipment is composed of series-parallel circuits. An example of a series-parallel circuit is shown in Figure ECC-6.6.

FIGURE **ECC-6.6**

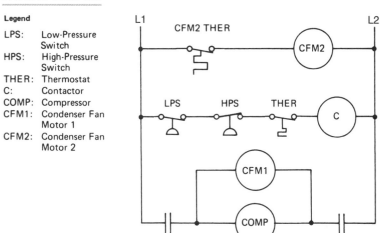

Legend

LPS: Low-Pressure Switch
HPS: High-Pressure Switch
THER: Thermostat
C: Contactor
COMP: Compressor
CFM1: Condenser Fan Motor 1
CFM2: Condenser Fan Motor 2

2. Wire the schematic diagrams of the following series-parallel circuits.

  a. Circuit A: Connect two light bulbs in parallel, and control each bulb by a switch in series with each light bulb.

  b. Wire the circuit.

  c. Have your instructor check your wiring.

  d. Operate the circuit.

    What happens when the switches are closed?

    What happens when the switches are opened?

  e. Circuit B: Connect two light bulbs in parallel and control both light bulbs with one switch.

  f. Wire the circuit.

  g. Have your instructor check your wiring.

  h. Operate the circuit.

    What happens when the switch is closed?

    What voltage is being supplied to the light bulbs?

i. Circuit C: Connect two light bulbs in series with one switch, and connect another switch in parallel with one light bulb.

j. Wire the circuit.

k. Have your instructor check your wiring.

l. Operate the circuit.

Explain the operation of the switch wired in parallel to the light bulb.

MAINTENANCE OF WORKSTATION AND TOOLS: Disconnect all wiring from components, remove the components from the plywood, and return all components and tools to their proper location(s). Turn all meters off, and make certain that all test leads are accounted for.

SUMMARY STATEMENT: Describe the application of series, parallel, and series-parallel circuits in HVACR control systems and equipment. Include examples of when a series circuit would be preferred over a parallel circuit and examples where a parallel circuit would be preferred over a series circuit.

TAKEAWAY: This exercise served as a hands-on opportunity to see first-hand the concepts outlined in Ohm's Law. In this exercise, it was seen that resistance and current are inversely related and that the voltage supplied to circuit loads varies depending on the circuit configuration. In addition, the current flow through circuits or circuit branches also varies depending on the circuit configuration.

QUESTIONS

1. Two 115 V light bulbs connected in series with 115 V would:
   A. burn out immediately
   B. burn correctly
   C. burn dimly
   D. do none of the above

2. Two 115 V light bulbs connected in parallel with 115 V would:
   A. burn out immediately
   B. burn correctly
   C. burn dimly
   D. do none of the above

3. A thermostat, high-pressure switch, and low-pressure switch are connected in series with a small compressor. If the low-pressure switch opened, the compressor would:
   A. continue to run
   B. stop immediately
   C. stop if all three safety switches opened
   D. do none of the above

4. Four light bulbs of the same wattage are connected in parallel with 115 V. The voltage supplied to each bulb is:
   A. 115
   B. 55
   C. 27.5
   D. 0

5. Four light bulbs are connected in series with 440 V. The voltage measured across each bulb is:

A. 440

B. 230

C. 110

D. 55

Answer the following questions using Figure ECC-6.7.

6. The compressor motor and the outdoor fan motor are connected in:

A. parallel

B. series

7. The high-pressure switch and low-pressure switch are connected in:

A. parallel

B. series

8. The indoor fan relay contacts are connected in _____ with the indoor fan motor.

A. series

B. parallel

9. The indoor fan relay coil and the contactor coil are connected in:

A. series

B. parallel

10. The compressor and compressor run capacitor are connected in:

A. series

B. parallel

FIGURE **ECC-6.7**

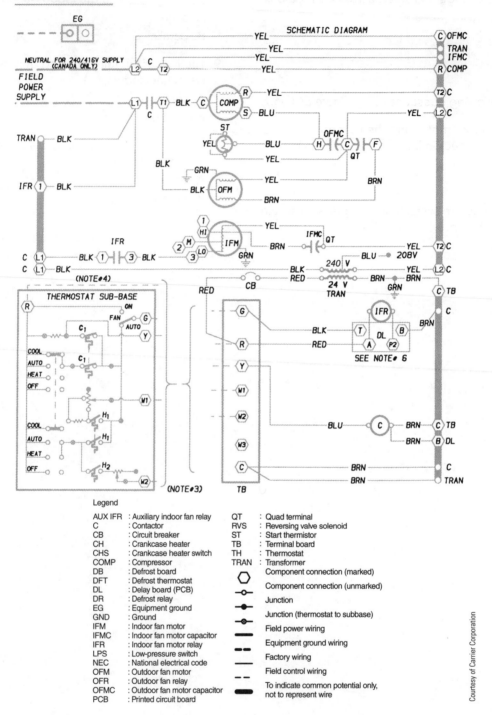

Legend

AUX IFR	: Auxiliary indoor fan relay	QT	: Quad terminal
C	: Contactor	RVS	: Reversing valve solenoid
CB	: Circuit breaker	ST	: Start thermistor
CH	: Crankcase heater	TB	: Terminal board
CHS	: Crankcase heater switch	TH	: Thermostat
COMP	: Compressor	TRAN	: Transformer
DB	: Defrost board		Component connection (marked)
DFT	: Defrost thermostat		Component connection (unmarked)
DL	: Delay board (PCB)		
DR	: Defrost relay		Junction
EG	: Equipment ground		Junction (thermostat to subbase)
GND	: Ground		
IFM	: Indoor fan motor		Field power wiring
IFMC	: Indoor fan motor capacitor		
IFR	: Indoor fan motor relay		Equipment ground wiring
LPS	: Low-pressure switch		Factory wiring
NEC	: National electrical code		
OFM	: Outdoor fan motor		Field control wiring
OFR	: Outdoor fan relay		
OFMC	: Outdoor fan motor capacitor		To indicate common potential only,
PCB	: Printed circuit board		not to represent wire

Electrical Symbols Used in the HVACR Industry

Name	Date	Grade

OBJECTIVES: Upon completion of this exercise, you should be able to identify commonly used electrical symbols used in schematic diagrams and be able to locate electrical components based on their symbols in wiring diagrams.

INTRODUCTION: A successful service technician must be able to recognize common electrical symbols in order to troubleshoot electrical systems. This exercise will help you recognize and sketch some commonly used symbols.

TOOLS, MATERIALS, AND EQUIPMENT:

Selection of common electrical components
Various air-conditioning appliances, such as window units, refrigerators, split system, and packaged unit.
Nut drivers
Screwdrivers
Flashlights

 SAFETY PRECAUTIONS: Make certain that the electrical source is disconnected when identifying electrical components on equipment. Make sure body parts do not come in contact with live electrical conductors. Keep hands and materials away from moving parts.

STEP-BY-STEP PROCEDURES

1. For each figure shown below, write the name of the device next to its associated symbol.

1. _____
2. _____
3. _____
4. _____
5. _____

6. _____
7. _____
8. _____
9. _____
10. _____

11. _____
12. _____
13. _____
14. _____
15. _____

16. _____
17. _____
18. _____
19. _____
20. _____

2. Obtain 10 different electrical components.

3. Write the name of each component and then determine if it is a switch or a load.

A.   F.

B.   G.

C.   H.

D.   I.

E.   J.

4. Using the piece of equipment you have been assigned to work on, make a list of all of the switches and loads that are contained in that unit:

Switches: **Loads:**

_____   _____

_____   _____

_____   _____

_____   _____

_____   _____

_____   _____

5. Move on to a second piece of equipment, and make a list of all of the switches and loads that are contained in that unit:

Switches: Loads:

_____   _____

_____   _____

_____   _____

_____   _____

_____   _____

_____   _____

MAINTENANCE OF WORKSTATION AND TOOLS: Clean and return all tools to their proper location(s). Return all electrical components to their proper location(s). Replace all service panel covers on equipment used in this exercise.

SUMMARY STATEMENT: Explain the importance of being able to identify and describe various electrical components based on their appearance, function, and location in a system.

TAKEAWAY: The majority of HVACR system failures are electrical in nature. It is, therefore extremely important for technicians to become as knowledgeable as possible in all aspects of electrical theory, including how to read and interpret electrical wiring diagrams. Understanding the symbols used in these diagrams is an important step in mastering the process.

QUESTIONS

1. Explain the difference between a relay with NO contacts and a relay with NC contacts.

2. Explain the action of a single-pole, double-throw switch.

3. What is the difference between a magnetic starter and a contactor?

4. How do the symbols for a magnetic starter and a contactor differ?

5. List at least five electrical loads that are used in the HVACR industry.

6. What is the purpose of the disconnect switch?

7. Explain the difference between heating and cooling thermostats.

8. How can you differentiate between a thermostat and a pressure switch without a legend?

9. Describe the controlling elements on a thermal and magnetic overload.

10. Explain the function of a multi-tap control transformer.

Exercise ECC-8

Reading Basic Schematic Diagrams

Name	Date	Grade

OBJECTIVES: Upon completion of this exercise, you should be able to interpret the wiring diagram of a piece of air-conditioning equipment and demonstrate knowledge of the system's sequence of operation.

INTRODUCTION: A schematic diagram shows a systematic layout of a control system. The diagram provides information regarding how, when, and why a system works as it does. Reading the diagram can be a simple process if you examine one specific branch of the circuit at a time, instead of trying to follow the entire diagram at once. All schematic diagrams are broken down into basic circuits, and each circuit usually contains one load. This exercise provides you with some specific diagrams that can be used to help master the process of reading wiring diagrams.

TOOLS, MATERIALS, AND EQUIPMENT: Selection of wiring diagrams.

STEP-BY-STEP PROCEDURES

1. Study Figure ECC-8.1, and then answer the following questions.

FIGURE **ECC-8.1**

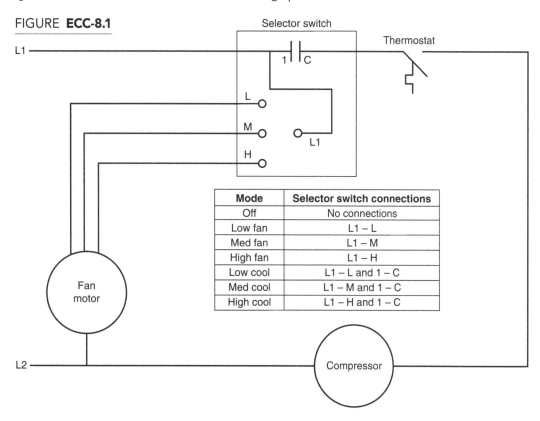

Mode	Selector switch connections
Off	No connections
Low fan	L1 – L
Med fan	L1 – M
High fan	L1 – H
Low cool	L1 – L and 1 – C
Med cool	L1 – M and 1 – C
High cool	L1 – H and 1 – C

2. The thermostat and contacts 1-C on the selector switch are wired in _____ with each other.

a. series

b. parallel

3. Explain how you arrived at the answer to the question in step 2.

4. The compressor and the fan motor are wired in _____ with each other.

    a. series

    b. parallel

5. Explain how you arrived at the answer to the question in step 4.

6. The L1-M contacts on the selector switch and the fan motor are wired in _____ with each other.

    a. series

    b. parallel

7. Explain how you arrived at the answer to the question in step 6.

8. The L1-M contacts on the selector switch and the thermostat are wired in _____ with each other.

    a. series

    b. parallel

9. Explain how you arrived at the answer to the question in step 8.

10. In order for the compressor to operate:

    a. the 1-C contacts must be closed.

    b. the thermostat contacts must be closed.

    c. both A and B

    d. neither A nor B

11. Based on the table of switch connections, which of the following is not possible?

    a. The compressor can operate without the fan motor.

    b. The fan motor operates whenever the compressor operates.

    c. The fan motor can operate when the compressor is not operating.

    d. The fan motor can operate even if the thermostat is open.

12. Which of the following statements is correct?

    a. The thermostat controls the operation of the fan motor.

    b. The 1-C contacts control the operation of the fan motor.

    c. The L, M, and H terminals control the speed of the compressor.

    d. Terminals 1 and L1 on the selector switch are electrically the same.

13. Which of the following ladder diagrams best represents the air conditioner wiring diagram provided in Figure ECC-8.1?

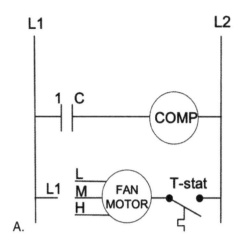

A.

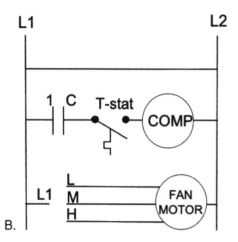

B.

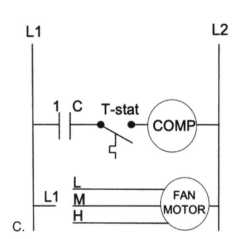

C.

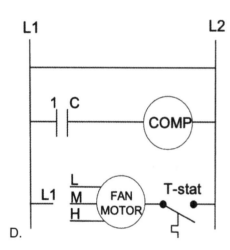

D.

14. Using the correct ladder diagram as a guide, determine the unit's sequence of operations in both fan only and cooling modes of operation.

15. Study Figure ECC-8.2 and then answer the following questions.

FIGURE **ECC-8.2**

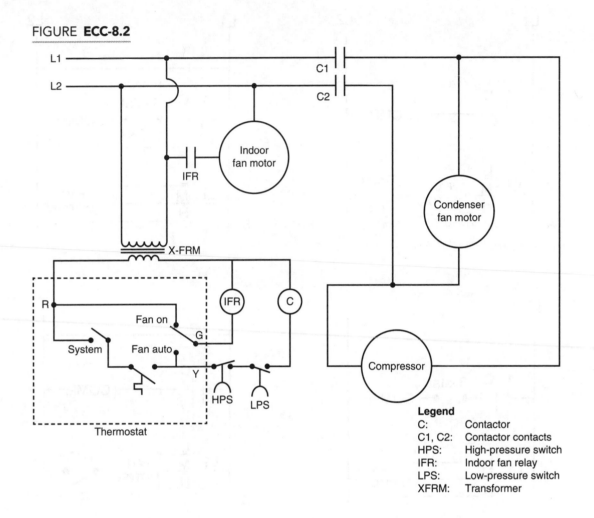

16. The compressor and the condenser fan motor are wired in _____ with each other.

    a. series

    b. parallel

17. Explain how you arrived at the answer to the question in step 16.

18. The transformer and the indoor fan motor are wired in _____ with each other.

    a. series

    b. parallel

19. Explain how you arrived at the answer to the question in step 18.

20. The indoor fan motor is controlled only by the:

    a. C1 and C2 contacts.

    b. IFR contacts.

    c. HPS and LPS.

    d. SYSTEM switch.

21. Which of the following statements is true regarding the condenser fan motor and the compressor if the system is operating correctly?

    a. The compressor will turn on first; then the condenser fan motor will turn on.

    b. The condenser fan motor will turn on first; then the compressor will turn on.

    c. The compressor and condenser fan motor will turn on and off at the same time.

    d. The compressor turns off first, and 2 seconds later, the condenser fan motor turns off.

22. Which of the following are all line voltage components?

    a. Thermostat, compressor, indoor fan motor

    b. Indoor fan motor, condenser fan motor, compressor

    c. Thermostat, high-pressure switch, low-pressure switch

    d. Indoor fan motor, high-pressure switch, low-pressure switch

23. Which components are wired in both the high-voltage and low-voltage circuits?

    a. Thermostat and indoor fan relay

    b. Thermostat and compressor

    c. Contactor and indoor fan relay

    d. High- and low-pressure switches

24. Compressor operation is controlled by the:

    a. C1 and C2 contacts.   **c.** fan switch.

    b. IFR relay.   **d.** condenser fan motor.

25. Which of the following ladder diagrams best represents the air conditioner wiring diagram provided in Figure ECC-8.2?

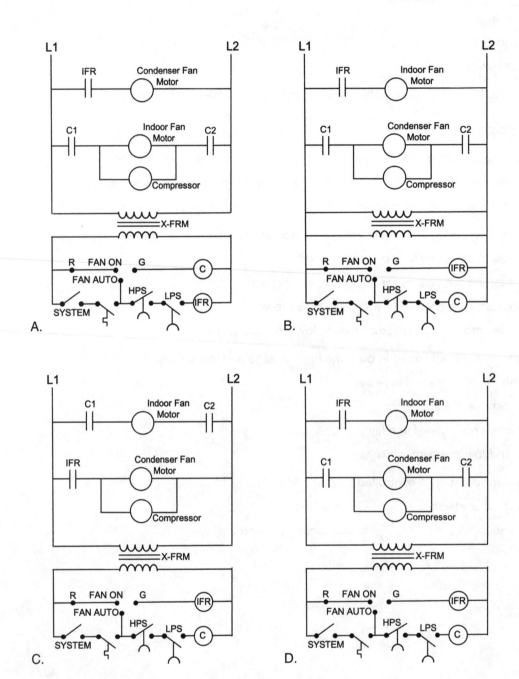

26. Using the correct ladder diagram as a guide, determine the unit's sequence of operations in both fan- only and cooling modes of operation.

MAINTENANCE OF WORKSTATION AND TOOLS: There is no maintenance required for his exercise.

SUMMARY STATEMENT: Explain how knowing the sequence of operations of a system helps a technician narrow down a system problem to effectively troubleshoot an air-conditioning system.

TAKEAWAY: Successful service technicians are able to look over a wiring diagram and, within a reasonably short period of time, be able to determine how the system is intended to operate. This is a valuable skill that takes time to master, but the investment is well worth it.

QUESTIONS

1. A schematic diagram shows the components in:
   A. an energized position
   B. a de-energized position

2. A schematic diagram is:
   A. an exact picture of a control panel with connecting wire
   B. an installation diagram
   C. a circuit-by-circuit layout of the control panel
   D. none of the above

3. A schematic diagram is most important for:
   A. servicing air-conditioning control systems
   B. locating components in the control panel
   C. installing air-conditioning equipment
   D. the homeowner

4. Most loads shown in a schematic diagrams are connected in _____ with each other.
   A. series
   B. parallel

5. Why is it important for the service technician to be able to read schematic diagrams?

6. Explain why the ladder diagram is often more useful for troubleshooting air-conditioning equipment than the schematic.

7. Explain why it is important to have at least one power-consuming device located on each completed rung of the ladder diagram.

Reading Advanced Schematic Diagrams

Name	Date	Grade

OBJECTIVES: Upon completion of this exercise, you should be able to interpret the operation of the equipment assigned by the instructor and write an operational sequence.

INTRODUCTION: A schematic diagram shows a systematic layout of a control system. It can tell you how, when, and why a system works as it does. Reading the diagram is easy if you take one circuit at a time instead of trying to follow the entire diagram at once. All schematic diagrams are broken down into basic circuits, and each circuit usually contains one load.

TOOLS, MATERIALS, AND EQUIPMENT: There are no tools, materials, or equipment required for this exercise. Selection of wiring diagrams.

STEP-BY-STEP PROCEDURES

PART I: Light Commercial Packaged Air Conditioner with Control Relay

Study Figure ECC-9.1, and then answer the following questions.

FIGURE **ECC-9.1**

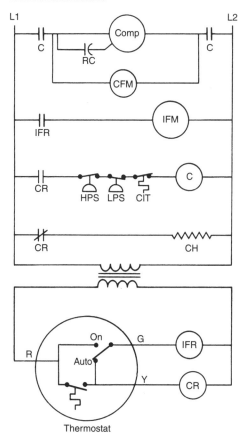

Legend

COMP:	Compressor
C:	Contactor
IFR:	Indoor fan relay
IFM:	Indoor fan motor
CR:	Control relay
HPS:	High-pressure switch
LPS:	Low-pressure switch
CR:	Control relay
CH:	Crankcase heater
TRANS:	Transformer
CIT:	Compressor internal thermostat
CT:	Cool thermostat

1. The crankcase heater is _____ when the compressor is operating.

    a. energized

    b. de-energized

2. What are the results if the HPS opens while the system is operating?

    a. The condenser fan motor is de-energized.

    b. The compressor is de-energized.

    c. The crankcase heater is energized.

    d. All of the above.

3. The IFR controls the _____.

    a. Comp

    b. CFM

    c. IFM

    d. CH

4. If the transformer was faulty, which of the following components would operate properly?

    a. IFM

    b. Comp

    c. CFM

    d. CH

5. Which of the following devices are connected in parallel?

    a. CFM and C contacts

    b. HPS and CIT

    c. CR normally closed contacts and CH

    d. IFM and C

6. The IFM would operate if the CR contacts failed to close.

    a. True

    b. False

7. The CIT would open on a (an) _____ in compressor motor temperature.

    a. increase

    b. decrease

8. Write an operational sequence for the light commercial air conditioner in Figure ECC-9.1.

PART II: Air-Cooled Packaged Unit with Remote Condenser

Study Figure ECC-9.2, and then answer the following questions.

FIGURE **ECC-9.2**

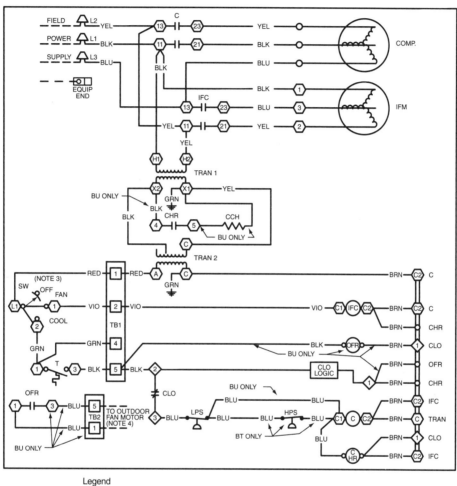

Legend

C	– Compressor contactor	SC	– Start capacitor
CH or	– Crankcase heater	SR	– Start relay
CCH		SW	– Switch
CHR	– Crankcase heater relay	T	– Thermostat
CLO	– Compressor lockout	TB	– Terminal block (board)
COMP	– Compressor	TC	– Thermostat cooling
CR	– Control relay	Tran	– Transformer
DU	– Dummy terminal		
Equip Gnd	– Equipment ground	⚡	Field splice
HPS	– High-pressure switch	▢	Terminal block (board)
IFC	– Indoor-fan contractor	◇	Terminal compressor lockout (CLO)
IFM	– Indoor-fan motor		
IP	– Internal protector	o	Terminal (unmarked)
LLS	– Liquid line solenoid	⬡	Terminal (marked)
LPS	– Low-pressure switch	──	Factory wiring
OFR	– Outdoor-fan relay	- - -	Field wiring
OL	– Overload	═══	Indicates common potential only;
QT	– Quadruple terminal		does not represent wire.
RC	– Run capacitor		
S	– Compressor solenoid		

(Reproduced with courtesy of Carrier Corporation, Syracuse, NY)

1. True or false.   The compressor and indoor fan motors are single-phase motors. Explain.

2. True or false.   The crankcase heater is de-energized when the compressor is operating. Explain.

3. True or false.   The HPS closes on a rise in pressure. Explain.

4. True or false.   The contacts of the compressor lockout are normally closed. Explain.

5. True or false.   The outdoor fan relay is connected in parallel with the contactor coil. Explain.

6. True or false.   The "off/fan/cool" switch will allow compressor operation without indoor fan operation. Explain.

7. True or false.   The "off/fan/cool" switch controls the operation of the indoor fan motor and compressor. Explain.

8. True or false.   The outdoor fan motor is supplied power from L1 and L2 in this schematic. Explain.

9. Write an operational sequence for the air-cooled packaged unit in Figure ECC-9.2.

PART III: Commercial Freezer with a Pump-Down Control

Study Figure ECC-9.3, and then answer the following questions.

FIGURE **ECC-9.3**

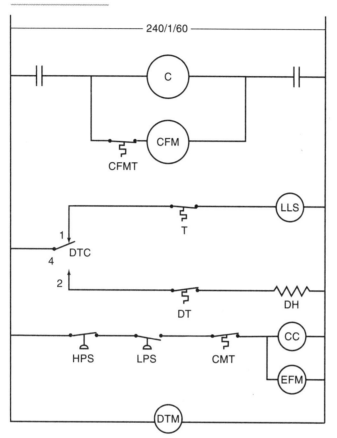

Legend

C:	Compressor
CC:	Compressor contactor
CFMT:	Condenser fan motor thermostat
CFM:	Condenser fan motor
DT:	Defrost timer motor
DTC:	Defrost timer contacts
T:	Thermostat
LLS:	Liquid line solenoid
DH:	Defrost heater
HPS:	High-pressure switch
LPS:	Low-pressure switch
CMT:	Compressor motor thermostat
EFM:	Evaporator fan motor

1. What controls the LLS?

   a. CFMT

   b. T

   c. DT

   d. LPS

2. Which of the following components operate continuously?

   a. CFM

   b. DTM

   c. CC

   d. LLS

3. Which of the following components operate at the same time?

   a. C

   b. LLS

   c. EFM

   d. All of the above

4. What component is energized if contacts 2 and 4 on the DTC are closed and the DT is closed?

    a. DH

    b. LLS

    c. CC

    d. C

5. If the compressor is operating and the condenser fan motor is not operating, which of the following conditions is most likely?

    a. The CFMT is open.

    b. The ambient temperature is low.

    c. All of the above.

    d. None of the above.

6. Which of the following switches would open when the LLS is de-energized?

    a. HPS

    b. LPS

    c. CMT

    d. T

7. What is the purpose of the DT?

    a. To prevent the temperature of the freezer from rising to high on the defrost cycle

    b. To prevent the compressor from operating on the defrost cycle

    c. To operate the EFM on the defrost cycle

    d. To start the DTM

8. Which of the following components are connected in parallel?

    a. C and CFM

    b. CC and EFM

    c. LLS and DH

    d. All of the above

9. Write an operational sequence for both the defrost and freezing mode of the commercial freezer in Figure ECC-9.3.

PART IV: Electric Air-Conditioning and Gas Heating Unit

Study Figure ECC-9.4, and then answer the following questions.

FIGURE **ECC-9.4**

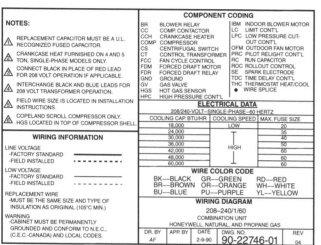

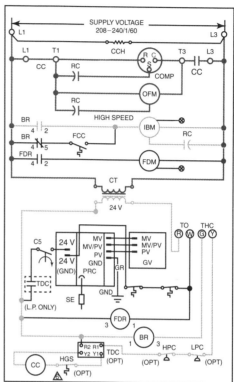

(Courtesy of Rheem Air-Conditioning Division, Fort Smith, AK)

1. The centrifugal switch is located in the forced draft motor. What is its purpose in the control circuit?

2. The thermostat closes, making an electrical connection between R and W. What two safety controls could interrupt the low-voltage supply to the PRC?

3. The thermostat controls the IBM during the cooling cycle by making an electrical connection between R and G, but no provision is made for the thermostat to control the IBM. How is the IBM energized during the heating cycle?

4. What control device is responsible for energizing and de-energizing the coil on the compressor contactor?

5. What is the supply voltage and control voltage of the unit?

6. Explain the operation of the crankcase heater.

7. True or false. The FDR and PRC are connected in series. Explain.

8. True or false. The HPC and LPC are connected in series. Explain.

9. True or false. The COMP and OFM are connected in parallel. Explain.

10. Write an operational sequence for the unit in Figure ECC-9.4.

PART V: Heat Pump System

Study Figure ECC-9.5, and then answer the following questions.

1. The transformer is located in _____.

   a. the outdoor unit   **c.** disconnect 1

   b. the indoor unit   **d.** none of the above

2. What does the first-stage cooling thermostat energize?

   a. MS   **c.** SC

   b. F   **d.** DFT

3. What does the second-stage cooling thermostat energize?

   a. MS   **d.** DFT

   b. F   **e.** a and b

   c. SC

4. What does first-stage heating thermostat energize?

   a. SC   **c.** F

   b. MS   **d.** b and c

5. What does the second-stage heating thermostat energize?

   a. SC   **d.** AH

   b. MS   **e.** c and d

   c. F

6. When the RHS-1 and RHS-2 switches are moved to the emergency heat position, which components are energized?

   a. AH   **c.** SC

   b. MS   **d.** Compressor

7. CR-A is energized by the _____.

   a. MS   **c.** F

   b. SC   **d.** closing of disconnect 1

8. In the defrost cycle, the outdoor fan motor is _____.

   a. energized   **b.** de-energized

9. The defrost cycle is terminated by _____.

   a. DFT contacts (3 and 5)

   b. DT

   c. DFT contacts (3 and 5) or DT

   d. ODS

FIGURE **ECC-9.5**

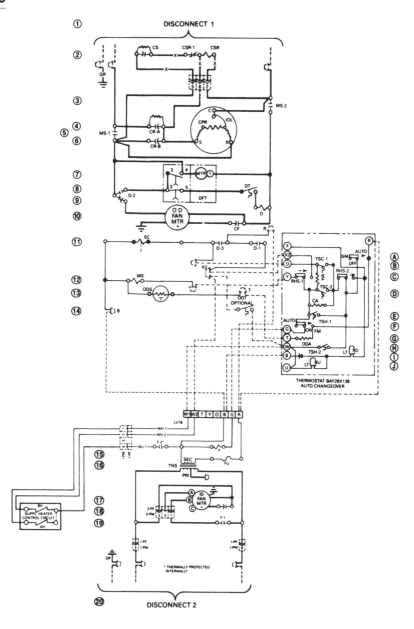

Legend

AH:	Supplementary Heat Contactor	(19)		LVTB:	Low-voltage Terminal Board	
BH:	Supplementary Heat Contactor	(18)		MS:	Compressor Motor Contactor	(5, 3, & 12)
CA:	Cooling Anticipator			MTR:	Motor	
CR:	Run Capacitor	(4 & 6)		ODA:	Outdoor Temperature Anticipator	(G)
CPR:	Compressor			ODS:	Outdoor Temperature Sensor	(13)
D:	Defrost Relay	(9)		ODT:	Outdoor Thermostat	
DFT:	Defrost Timer	(7)		RHS:	Resistance Heat Switch	(C)
DT:	Defrost Termination Thermostat	(8)		SC:	Switchover Valve Solenoid	(11)
F:	Indoor Fan Relay	(15)		SM:	System Switch	(A)
FM:	Manual Fan Switch	(F)		TNS:	Transformer	(16)
HA:	Heating Anticipator			TSC:	Cooling Thermostat	(B & D)
HTR:	Heater			TSH:	Heating Thermostat	(E & H)
IOL:	Internal Overload Protection					
LT:	Light					

MAINTENANCE OF WORKSTATION AND TOOLS: There is no maintenance required for this exercise.

SUMMARY STATEMENT: Explain how schematic diagrams can be used by the technician on a service call.

TAKEAWAY: Successful service technicians are able to look over a wiring diagram and, within a reasonably short period of time, be able to determine how the system is intended to operate. This is a valuable skill that takes time to master, but the investment is well worth it.

QUESTIONS

1. A schematic diagram shows the components in _____.
   A. an energized position
   B. a de-energized position

2. A schematic diagram is _____.
   A. an exact picture of a control panel with connecting wire
   B. an installation diagram
   C. a circuit-by-circuit layout of the control panel
   D. none of the above

3. A schematic diagram is most important for _____.
   A. servicing air-conditioning control systems
   B. locating components in the control panel
   C. installing air-conditioning equipment
   D. the homeowner

4. Most loads shown in schematic diagrams are connected in _____ with each other.
   A. series
   B. parallel

5. The thermostat used on the heat pump shown in Figure ECC-9.5 is a _____ thermostat.
   A. single-stage heating
   B. single-stage cooling
   C. two-stage heating, single-stage cooling
   D. two-stage heating, two-stage cooling

6. If 24 V are delivered to the MS coil in Figure ECC-9.5, what action can be expected?
   A. MS contacts close, starting the compressor
   B. MS contacts close, starting the outdoor fan motor
   C. MS contacts close, starting the compressor and outdoor fan motor
   D. F-1 contacts close, starting the indoor fan motor

7. The defrost termination thermostat (DT) _____.
   A. opens on a temperature rise
   B. closes on a temperature rise

8. What controls the IFM in Figure ECC-9.1?
   A. M
   B. IFR

9. The defrost timer (DFT) motor in Figure ECC-9.5 runs when _____.
   A. DFT contacts 3 and 5 are closed
   B. D coil is energized
   C. TS-1 is closed
   D. MS is energized

Exercise ECC-10

Constructing Basic Electric Circuits

Name	Date	Grade

OBJECTIVES: Upon completion of this exercise, you should be able to construct basic electric circuits following wiring diagrams.

INTRODUCTION: There will be times when you will have to construct basic electric circuits for control systems. You must keep in mind that a schematic diagram is laid out in a circuit-by-circuit arrangement. If this principle is followed, constructing circuits can be easy. This exercise provides several start-to-finish projects that will help strengthen your electrical ability.

TOOLS, MATERIALS, AND EQUIPMENT:

Small screwdriver	Thermostat
Wire-cutting pliers	Light bulb sockets
Wire strippers	Light bulbs to simulate loads
Terminal crimpers	Wire to make electrical connections
Supply power cords	Wire terminals (push-on and spade)
Transformer	12" × 18" piece of ½" plywood
Relays and contactors	Miscellaneous screws to attach components to plywood

 SAFETY PRECAUTIONS: Make certain all electrical components are securely attached to the plywood to prevent movement and possible electrical shock or short circuits. Allow your instructor to check your work before energizing a circuit or circuits. When a circuit or circuits are energized, make certain that electrical wiring connections are not touched; this could result in personnel injury or damage to equipment.

STEP-BY-STEP PROCEDURES

PART I: Diagram 1

1. Study the following wiring diagram for a circuit that uses a single-pole, single-throw switch to control the 24 V coil of a general purpose relay. When the coil on the relay is energized, the normally open contacts on the relay will close and energize a 115 V light bulb.

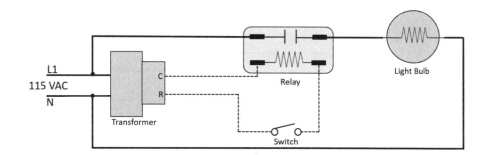

2. Make a list of the materials that are needed to complete your electrical circuit.

3. Obtain the necessary electrical components from your instructor.

4. Attach components securely to the plywood using screws.

5. Using the wiring diagram provided as a guide, make all of the necessary wiring connections.

6. Double-check your work.

7. Have your instructor check your circuit.

8. With the single-pole, single-throw switch in the OFF position, plug the circuit into a 115 V receptacle.

9. At this point, the light bulb should be off. If the light bulb is off, continue to step 10. If the light bulb is on, unplug the circuit and recheck all of your connections. Then, return to step 8.

10. Turn the single-pole, single-throw switch to the ON position. There should be an audible click, and the bulb should light up. If the bulb lights, turn the switch off and proceed to part II. If the bulb does not light, continue to step 11.

11. Is an audible click heard when the switch is turned on?   YES   NO

12. If YES, check the line voltage circuit, including the relay contacts and the bulb. If NO, check the low-voltage control circuit, including the switch and the coil of the relay.

13. Make any necessary repairs and/or adjustments to the circuit, and return to step 8.

PART II: Diagram 2

1. Study the following wiring diagram for a circuit that uses a single-pole, single-throw switch to control the 24 V coil of a general purpose relay. A 115 V light bulb will be connected through the normally open contacts on the relay. A second 115 V light bulb will be connected through the normally closed contacts on the relay. When the circuit is initially powered, with the single-pole, single-throw switch in the OFF position, one bulb will be illuminated and the other will be off. When the switch is turned on, the coil on the relay will be energized and the normally open contacts on the relay will close to energize a 115 V light bulb. At the same time, the normally closed contacts on the relay will open, de-energizing the first 115 V light bulb.

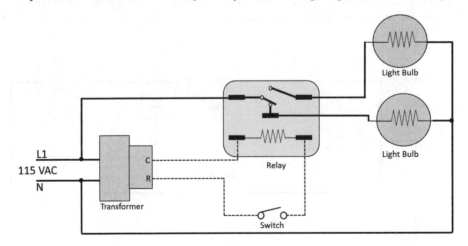

2. Make a list of the materials that are needed to complete your electrical circuit.

3. Obtain the necessary electrical components from your instructor.

4. Attach components securely to the plywood using screws.

5. Using the wiring diagram provided as a guide, make all of the necessary wiring connections.

6. Double-check your work.

7. Have your instructor check your circuit.

8. With the single-pole, single-throw switch in the OFF position, plug the circuit into a 115 V receptacle.

9. Observe the operation of the circuit. At this point, one light bulb should be on, and the other should be off. If one light bulb is on and the other is off, continue to step 10. If both bulbs are on or both bulbs are off, unplug the circuit and recheck all of your connections. Then return to step 8.

10. Turn the single-pole, single-throw switch to the ON position. There should be an audible click, and one bulb should light up while the other one turns off. If this changeover occurs, turn the switch off, and proceed to part III. If the changeover does not happen, continue to step 11.

11. Is an audible click heard when the switch is turned on?   YES   NO

12. If YES, check the line voltage circuit, including the relay contacts and the bulb. If NO, check the low-voltage control circuit, including the switch and the coil of the relay.

13. Make any necessary repairs and/or adjustments to the circuit, and return to step 8.

PART III: Diagram 3

1. Study the following wiring diagram that uses a thermostat to control four 115 V light bulbs representing the various functions of an air-conditioning system. The thermostat will be low voltage, with the following letter designations: R, representing the power source from the transformer; G, representing the fan function; Y, representing the cooling function; and W, representing the heating function. Each system function will require a 24 V relay to energize the light bulb representing that function.

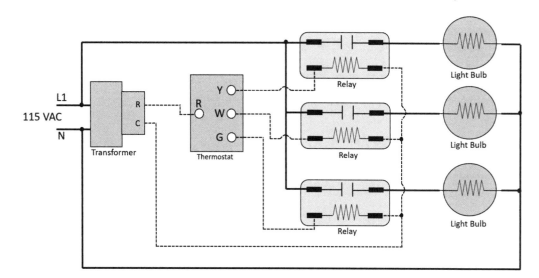

2. Make a list of materials that are needed to complete your electrical circuit.

3. Obtain the necessary components from your instructor.

4. Attach components securely to the plywood using screws.

5. Using the wiring diagram provided as a guide, make all of the necessary wiring connections.

6. Double-check your work.

7. Have your instructor check your wiring.

8. With the system switch set to OFF and the fan switch set to the AUTO position, plug the circuit into a 115 V receptacle.

9. Observe and comment on which bulbs are on and which are off: _____

_____

10. Move the fan switch to the ON position. Observe and comment on which bulbs are on and which are off:

_____

11. Move the fan switch to the AUTO position, and adjust the thermostat set point to a temperature that is above the temperature in the laboratory. Turn the system switch to the COOL position. Observe and comment on which bulbs are on and which are off: _____

_____

12. Adjust the thermostat set point to a temperature that is below the temperature in the laboratory. Observe and comment on which bulbs are on and which are off: _____

_____

13. Turn the system switch to the HEAT position. Observe and comment on which bulbs are on and which are off:

_____

14. Adjust the thermostat set point to a temperature that is above the temperature in the laboratory. Observe and comment on which bulbs are on and which are off: _____

_____

15. Remove all components from the board, and properly store them.

MAINTENANCE OF WORKSTATION AND TOOLS: Make certain that all components have been removed from the boards and that all components have been properly stored. Make certain that all tools have been accounted for and that they are stored away.

SUMMARY STATEMENT: Discuss any challenges that you faced in the creation of the actual circuit and how you met these challenges.

TAKEAWAY: Different skill sets are required to read wiring diagrams and to actually connect them. Continued practice in these areas will help ensure success in the HVACR industry.

QUESTIONS

1. What is the purpose of the transformer?

2. Is the transformer wired like a load or a switch? Explain your answer.

3. What is the purpose of a contactor or relay?

4. What are the two main electrical parts of a relay or contactor?

5. Is a contactor coil wired like a load or a switch? Explain your answer.

6. Are contactor contacts wired like a load or a switch? Explain your answer.

7. Why are safety controls wired in series?

8. Why are loads wired in parallel?

9. What is the basic difference between an operating control and a safety control?

Exercise ECC-11

Transformers

Name	Date	Grade

OBJECTIVES: Upon completion of this exercise, you should be able to install a control transformer in a circuit to energize a relay by closing a switch and install a transformer on a piece of equipment in the shop.

INTRODUCTION: Transformers are used primarily in the industry to supply a specific control voltage: 24 V in residential and 115 V in commercial and industrial equipment. Transformers are also used for many special applications, but this exercise will cover only the control transformer.

TOOLS, MATERIALS, AND EQUIPMENT:

24 V secondary, 115 V primary transformer
24 V secondary, 208 V primary transformer
24 V secondary, 230 V primary transformer
24 V secondary, 115/208/230 V primary transformer
115 V secondary, 208/230 V primary transformer

Relay with 24 V coil
Relay with 115 V coil
VOM meter
Basic electrical hand tools

⚠️ **SAFETY PRECAUTIONS:** Make certain that the electrical source is disconnected when making electrical connections. In addition:

- Make sure all electrical connections are tight.
- Make sure no bare current-carrying conductors are touching metal surfaces except ungrounded conductors.
- Make sure the correct voltage is being supplied to the circuit or equipment.
- Make sure body parts do not come in contact with live electrical conductors.
- Keep hands and materials away from moving parts.

STEP-BY-STEP PROCEDURES

PART I: Ohmic Values of Transformers

1. Obtain the following transformers from your instructor:
   24 V secondary, 115 V primary transformer
   24 V secondary, 208 V or 230 V primary transformer
   24 V secondary, 115/208/230 V primary transformer
   115 V secondary, 208/230 V primary

2. Using a VOM meter set to read resistance, take the resistance readings of the primaries and secondaries of the above transformers, and record the readings in the following chart.

Transformer	Primary Resistance	Secondary Resistance
24 V sec., 115 V prim.	_____	_____
24 V sec., 208/230 V prim.	_____	_____
24 V sec., 115/208/230 V prim.	_____	_____
115 V prim., 208/230 V sec.	_____	_____

3. Review the resistance readings of the transformers. Technicians must understand completely what the resistance should be on a good transformer so they can troubleshoot transformers effectively.

4. Keep the transformers used in this part of the lab for part II.

PART II: Wiring a Transformer

1. Using the transformers from part I of this exercise, connect the primary of each transformer to the correct voltage, the 24 V secondary to a 24 V relay coil, and the 115 V secondary to a 115 V relay coil. The relay solenoids should energize, closing the contacts of the relays. Have your instructor check the first relay before applying electrical energy. Figure ECC-11.1 shows an example of a 24 V transformer connection to a 24 V relay.

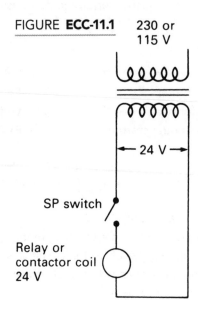

FIGURE **ECC-11.1**

2. Return all transformers to the proper location.

PART III: Installing a Transformer in an Air-Conditioning Unit

1. Obtain an air-conditioning unit from your instructor to replace the control transformer.

2. Determine what type of transformer is needed for replacement.

3. Disconnect the unit from the power source.

4. Remove the old transformer from the unit, paying attention to the wiring connections.

5. Install the new transformer.

6. Have your instructor check your installation.

7. Operate the unit, checking the current draw of all loads in the unit.

MAINTENANCE OF WORKSTATION AND TOOLS: Clean and return all tools and supplies to their proper location(s). Replace all equipment covers. Clean up the work area.

SUMMARY STATEMENT: What is the purpose of a 24 V control transformer in a residential air-conditioning system?

TAKEAWAY: Many air-conditioning systems operate on more than one voltage. Typically, one voltage is used in the power circuit, and another is utilized in the control circuit. Transformers are responsible for converting one voltage to another. Since transformers are found on the majority of air-conditioning systems, knowing how to properly work with them is a very important skill.

QUESTIONS

1. What are some advantages of using a low-voltage control system?

2. What is the resistance of the secondary of a 24 V transformer?

3. What is the resistance of a 115, 208, and 230 V primary, 24 V secondary transformer?

4. What is the purpose of using a fuse on the secondary side of the transformer?

5. What guidelines should be used when replacing a transformer in the field?

6. How are transformers rated?

7. What is the maximum load of a transformer used in an air-conditioning system?

8. What will happen if an undersized transformer is installed in a control system?

9. What size transformer is generally used on a heating-only gas furnace?

SECTION 3

Refrigeration System Components (RSC)

SUBJECT AREA OVERVIEW

Refrigeration systems incorporate a number of different components into their refrigerant circuit. The major components are the compressor, condenser, metering device, and evaporator. Depending on the application, the system may also incorporate components such as a receiver, liquid line filter drier, liquid line solenoid, evaporator pressure regulator (EPR), or crankcase pressure regulator (CPR). High-, medium-, and low-temperature refrigeration applications have different operating characteristics and may require additional refrigerant circuit components to operate as intended. Comfort cooling or air conditioning is similar to commercial refrigeration because the same basic system components are used. There are two basic types of air-conditioning units: the packaged-type system and the split system. The packaged-type system has all system components built into one cabinet. In the split system, the condenser and compressor are located outside, separate from the metering device and the evaporator, with field-installed refrigerant lines connecting the components. The evaporator may be located in an attic, basement, crawl space, or closet. The blower that moves air across the evaporator may also be part of the heating equipment. Proper system operation can be verified by monitoring the system's pressures and temperatures as well as calculating superheat, subcooling, condenser split (CS), and evaporator split (ES).

Refrigerant enters the evaporator as a mixture of liquid and vapor, though it is mostly liquid. As the mixture moves through the evaporator, more of the liquid boils to a vapor as a result of the latent heat absorbed into the refrigerant. Near the end of the evaporator, all of the liquid should be boiled off, leaving only refrigerant vapor. Older evaporators were designed as natural draft models, relying on natural convection. Modern designs use blowers to move air over the evaporator. This improves efficiency and allows evaporators to be much smaller. Some evaporators are designed with multiple circuits for the refrigerant to flow through. By using multiple circuits, the amount of pressure drop across the evaporator can be reduced. When the coil temperature is below freezing, a defrost cycle is needed to melt any ice that may accumulate during operation. Defrost should never be needed on an air-conditioning (cooling-only) system.

Condensers can be air-cooled, natural or mechanical draft, or water-cooled. In the condenser, heat absorbed by the refrigerant in the evaporator, suction line, and compressor is removed from the system, and the vapor refrigerant is condensed to a liquid, which is then cooled below its boiling point, or subcooled. The majority of the heat is given up during the change in state, from liquid to vapor. The condenser operates at higher temperatures and pressures than the evaporator. Air-cooled condensers are often located outside, where the heat can be rejected without making a difference to the outside air temperature. In a water-cooled condenser, water is used to absorb heat from the refrigerant. The water then can be wasted down the drain, or it can be pumped to a

(continued)

remote cooling tower where the heat is removed and the water reused. Normally, the condensing temperature of the refrigerant is approximately 10°F higher than the water leaving the system. In a recirculating water system, approximately 3 gallons of water per minute per ton of refrigeration is recirculated. Wastewater systems use about 1.5 gallons per minute.

The compressor compresses the refrigerant vapor from the suction line and raises its pressure so it can flow throughout the system. Compressors can be air-cooled, water-cooled, or cooled by suction gas. Refrigerant vapor enters the compressor through the suction valve and fills the cylinder(s). This refrigerant is cool but has absorbed heat in the evaporator and suction line. Most of this heat is latent heat absorbed while the refrigerant was changing state from a liquid to a vapor. A smaller amount may be sensible heat, called superheat. The compressor compresses this vapor, causing it to become very hot, at times as high as 200°F, and pumps this hot gas to the condenser. There are five major types of compressors used in the refrigeration and air-conditioning industry.

These are the reciprocating, screw, rotary, scroll, and centrifugal compressors. Newer compressors are often controlled by variable-frequency drives (VFDs), which offer better capacity control and improved efficiency.

The expansion device is often called the metering device, and certain types may be referred to as expansion valves. It can be an adjustable valve or a fixed-bore device, and it is one of the separation points between the high- and low-pressure sides of the system. The compressor is the other. The expansion device is located in the liquid line between the condenser and the evaporator, usually very close to the inlet of the evaporator. Its function is to meter the liquid refrigerant into the evaporator. This metering action causes a large temperature and pressure drop on the evaporator side of the device. The three types of devices used on modern equipment are the thermostatic expansion valve (TXV or TEV), the automatic expansion valve (AXV or AEV), and the fixed-bore type, such as a capillary tube, piston, or orifice. Newer equipment may also have an electronically controlled expansion valve called an EEV.

KEY TERMS

Absolute pressure
Air-cooled condenser
Aluminum fins
Automatic expansion valve (AXV or AEV)
Balanced-port TXV
Boiling temperature
Capillary tube
Centrifugal compressor
Clearance volume
Compressor
Condenser split (CS)
Condensing temperature
Cooling tower

Crankcase pressure regulator (CPR)
Cross liquid charge bulb
Cross vapor charge bulb
Distributor
Evaporator pressure regulator (EPR)
Evaporator split (ES)
Expansion (metering) device
External equalizer
Flooded evaporator
Forced-draft
Hot pulldown
Hermetic compressor

Hot gas defrost
King valve
Latent heat
Liquid charge bulb
Liquid slugging
Natural-draft
Off-cycle defrost
Oil separator
Open compressor
Piston
Pressure/temperature chart
Pumpdown
Reed valve
Refrigerant migration

Saturated vapor
Screw compressor
Scroll compressor
Service valve
Sight glass
Starved coil
Subcooled liquid
Suction-line accumulator
Thermostatic expansion valve (TXV or TEV)
Tube-within-a-tube coil
Valve plate
Vapor charge bulb
Wastewater system
Water-cooled condenser

REVIEW TEST

Name	Date	Grade

Circle the letter that indicates the correct answer.

1. The evaporator in a refrigeration system:
   A. rejects heat from the conditioned space.
   B. absorbs heat from the conditioned space.
   C. feeds a partial vapor to the compressor.
   D. feeds the liquid refrigerant to the compressor.

2. Refrigerant entering the evaporator is approximately _____ vapor.
   A. 50%
   B. 25%
   C. 10%
   D. 75%

3. An evaporator made up of two pieces of sheet metal that have an impression of refrigerant piping is called a _____ evaporator.
   A. finned-tube
   B. dry-type
   C. stamped
   D. liquid-to-liquid

4. Heat that is added to the refrigerant after it has all vaporized is called:
   A. superheat.
   B. latent heat.
   C. artificial heat.
   D. extended heat.

5. An evaporator with the coil on the suction side of the fan is called:
   A. forced draft.
   B. natural draft.
   C. induced draft.
   D. gravity flow.

6. The refrigerant leaving the compressor is often referred to as:
   A. hot liquid refrigerant.
   B. hot gas.
   C. a partial liquid.
   D. condensed vapor.

7. The design evaporator temperature in an air-conditioning system is:
   A. 40°F.
   B. 55°F.
   C. 75°F.
   D. 200°F.

8. Changing state from a liquid to a gas is often called:
   A. condensing.
   B. flooding.
   C. boiling.
   D. starving.

9. If the evaporator temperature is below 32°F, there must be some provision for:
   A. removing frost from the evaporator.
   B. adding refrigerant.
   C. increasing the rate of vaporization.
   D. flooding the evaporator.

10. The expansion device is located between the:
    A. compressor and condenser.
    B. evaporator and compressor.
    C. condenser and evaporator.

11. Off-cycle defrost occurs:
    A. with supplemental electric heat when the system is shut down.
    B. with hot gas when the system is operating.
    C. from the space temperature when the system is shut down.
    D. with hot liquid refrigerant from the compressor.

12. A saturated vapor condition exists when:
    A. there is still liquid refrigerant in the evaporator.
    B. all of the liquid has boiled to a vapor.
    C. there is excessive superheat.
    D. there is evaporator flooding.

13. Another name for an expansion device is:
    A. liquid-line receiver.
    B. suction-line filter drier.
    C. high-side service valve.
    D. metering device.

14. Two names for the interconnecting piping in a split air-conditioning system are the:
    A. discharge and distributor lines.
    B. fixed-bore and low ambient lines.
    C. thermostatic and pressure equalizer lines.
    D. suction (cool gas) and liquid lines.

15. Most of the heat absorbed by the suction gas is:
    A. heat of compression.
    B. sensible heat.
    C. latent heat.
    D. defrost heat.

16. Which of the following is not a part of a condensing unit?
    A. Condenser coil
    B. Condenser fan motor
    C. Compressor
    D. Evaporator fan motor

17. The reciprocating compressor is used most frequently in:
    A. large industrial refrigeration systems.
    B. air-conditioning systems in large buildings.
    C. small- and medium-size refrigeration systems and residential and light commercial air-conditioning.

18. The cooling for modern, reciprocating, fully hermetic compressors used in air-conditioning is controlled by or directly from:
    A. the suction gas.
    B. the discharge gas.
    C. a squirrel cage fan.
    D. a water-cooling system.

19. The screw-type compressor is used to a great extent in:
    A. domestic refrigerators.
    B. residential air-conditioning.
    C. small commercial refrigeration units.
    D. large commercial and industrial systems.

20. In a hermetic compressor:
    A. the motor and compressor operate in a refrigerant liquid atmosphere.
    B. the motor and compressor operate in a refrigerant vapor atmosphere.
    C. only the compressor operates in a refrigerant atmosphere.
    D. the belt drive must be adjusted precisely.

21. The crankshaft in a reciprocating compressor:
    A. serves as the crankcase oil heater.
    B. changes the rotary motion to a back-and-forth or reciprocating motion.
    C. operates the centrifugal fan.
    D. operates the ring valves.

22. The flapper or reed valve is used in the:
    A. scroll compressor.
    B. condenser.
    C. reciprocating compressor.
    D. oil pump.

23. A bimetal hermetic motor internal overload is:
    A. located in the motor rotor.
    B. embedded in the motor windings.
    C. located in the terminal box.
    D. fastened to the outside of the motor housing.

24. The shaft seal on an open-drive compressor:
    A. is normally constructed of brass.
    B. is the bearing surface for the shaft.
    C. provides the lubrication for the shaft.
    D. keeps the refrigerant vapor in and the atmosphere out of the compressor.

25. The clearance volume is the:
    A. amount of vapor refrigerant the compressor cylinder will hold during each stroke.
    B. volume by which the compressor is rated.
    C. space between the piston at top–dead center and the bottom of the valve plate of the compressor.
    D. space between the seat and the ring valve.

26. The discus valve design is more efficient than a flapper valve design because:
    A. it allows a cylinder to be designed with less clearance volume.
    B. it will operate with less spring tension.
    C. the reexpanded refrigerant in the cylinder is much greater.
    D. it will not allow liquid into the cylinder.

27. Liquid slugging will likely damage the reciprocating compressor because the:
- A. liquid does not mix well with the crankcase oil.
- B. compressor cannot compress liquid.
- C. electrical components are not designed to operate with a liquid refrigerant.
- D. arcing from the bimetal will deteriorate the refrigerant.

28. The TXV sensing bulb is mounted:
- A. on the liquid line between the condenser and the TXV.
- B. at the inlet to the condenser.
- C. on the suction line between the evaporator and compressor.
- D. on the compressor discharge line.

29. An external equalizer is used when:
- A. the pressure drop across the evaporator is excessive.
- B. the liquid-line pressure becomes too great.
- C. the receiver is too full of refrigerant.
- D. the pressure drop at the automatic expansion valve becomes too great.

30. When the load on the evaporator increases, the TXV:
- A. increases the flow of refrigerant.
- B. decreases the flow of refrigerant.
- C. maintains the flow of refrigerant.
- D. decreases the pressure across the evaporator.

31. The liquid charge bulb is:
- A. a sensing bulb on an automatic expansion valve.
- B. a bulb attached to one end of a capillary tube.
- C. a sensing bulb at the inlet of the expansion valve.
- D. a TXV-sensing bulb charged with a fluid similar to that of the system refrigerant.

32. Superheat is:
- A. the sensible heat absorbed by the refrigerant after it has boiled to a vapor.
- B. the heat of compression at the compressor.
- C. heat used to boil the liquid in the evaporator.
- D. latent heat given off by the condenser.

33. The thermostatic expansion valve:
- A. does not allow the pressure to equalize during the off cycle unless specifically designed to do so.
- B. normally allows the pressure to equalize during the off cycle.
- C. does not normally require the compressor to have a start capacitor.
- D. in an air-conditioning system is designed for the same temperature range as in commercial refrigeration.

34. A small amount of superheat in the suction line is desirable with a TXV to:
- A. keep the proper amount of refrigerant in the receiver.
- B. ensure the proper head pressure at the condenser.
- C. keep from starving the evaporator.
- D. ensure that no liquid refrigerant enters the compressor.

35. The thin metal disc connected to the needle in the TXV is called the:
- A. spring.
- B. diaphragm.
- C. seat.
- D. sensing bulb.

36. Liquid refrigerant should be completely boiled to a vapor:
- A. close to the inlet of the evaporator.
- B. approximately halfway through the evaporator.
- C. near the outlet of the evaporator.
- D. halfway through the suction line.

37. The condensing temperature of the refrigerant in a standard air-cooled condenser is approximately _____ higher than the ambient air.
- A. 10°F
- B. 30°F
- C. 55°F
- D. 100°F

38. In a water-cooled condenser, the condensing temperature is normally _____ higher than the leaving water.
- A. 40°F
- B. 35°F
- C. 25°F
- D. 10°F

39. The _____ of the air passing over the evaporator has the greatest effect on the film factor.
    A. quantity
    B. temperature
    C. humidity
    D. velocity

40. Refrigeration designed for the purpose of storing flowers or candy is considered _____ refrigeration.
    A. high-temperature
    B. medium-temperature
    C. low-temperature
    D. dry-type

41. For low-temperature refrigeration, _____ from the compressor discharge line may be used to defrost the evaporator.
    A. warm liquid
    B. hot gas
    C. moisture
    D. hot crankcase oil

42. Fins on a low-temperature evaporators are
    A. close together to allow for ice formation.
    B. close together to allow for dirt formation.
    C. far apart to allow for ice formation.
    D. far apart to allow for dirt accumulation.

43. The capillary tube is a type of:
    A. safety device.
    B. pressure gauge.
    C. filter.
    D. metering device.

44. High-efficiency condensers:
    A. can withstand higher pressures than standard condensers.
    B. have more surface area than standard condensers.
    C. have stainless steel coils.
    D. are located in a separate cabinet.

45. The refrigerant in the condenser of a refrigeration system:
    A. absorbs heat.
    B. gives up heat.
    C. changes to a vapor.
    D. is under lower pressure than in an evaporator.

Exercise RSC-1

Refrigeration System Components and Their Functions

Name	Date	Grade

OBJECTIVES: Upon completion of this exercise, you should be able to:
- list and explain the functions of the evaporator.
- list and explain the functions of the compressor.
- list and explain the functions of the condenser.
- list and explain the functions of the metering device.
- identify all of the electrical components used in a residential air-conditioning system.

INTRODUCTION: If a technician is to successfully install and troubleshoot refrigeration or air-conditioning systems, he or she must be familiar with and be able to identify the components used in both the refrigerant and electrical circuits.

TOOLS, MATERIALS, AND EQUIPMENT: Packaged air-conditioning unit, split-system air-conditioning unit, a straight-slot screwdriver, ¼" and 5⁄16" nut drivers, and HVACR reference materials.

 SAFETY PRECAUTIONS: Make certain that the electrical source is disconnected when examining electrical control panels and components.

STEP-BY-STEP PROCEDURES

Locating Components in an Air-Conditioning Unit

1. Your instructor will assign you an air-cooled unit (package or split system) on which you will locate and identify the following electrical components:

   - Condenser fan motor
   - Evaporator fan motor
   - Compressor motor
   - Transformer

   - Thermostat
   - Indoor fan relay
   - Contactor
   - Capacitor

2. Your instructor will assign you an air-cooled unit (package or split system) on which you will locate and identify the following refrigerant circuit components:

   - Condenser
   - Compressor
   - Evaporator
   - Metering device

MAINTENANCE OF WORKSTATION AND TOOLS: Replace any panels that were removed, and return all tools to their appropriate storage locations. Remember to clean your work area.

SUMMARY STATEMENT: State the operation of both the electrical and refrigerant circuits. Use an additional sheet of paper if more space is necessary for your answer.

TAKEAWAY: When learning any new system, electrical or refrigeration, it is important for a technician to be able to break the system down into its individual components. When looking at the function or purpose of each component, understanding the operation of the whole system becomes easier.

QUESTIONS

1. Explain the functions of each of the following:

- Evaporator

- Condenser

- Compressor

- Metering device

2. Explain the functions of each the following:

- Condenser fan motor

- Evaporator fan motor

- Compressor motor

- Transformer

- Thermostat

- Indoor fan relay

- Contactor

- Capacitor

Air-Conditioner Component Familiarization

Name	Date	Grade

OBJECTIVES: Upon completion of this exercise, you should be able to identify and describe various components in a typical air-conditioning system.

INTRODUCTION: You will remove panels and, when necessary, components from an air-conditioning system to identify and record specifications of various components.

TOOLS, MATERIALS, AND EQUIPMENT: Straight-slot and Phillips-head screwdrivers, ¼" and ⁵⁄₁₆" nut drivers, a set of open-end wrenches, a set of Allen wrenches, a flashlight, and an operating split-type air-conditioning system.

⚠ **SAFETY PRECAUTIONS:** Turn the power to the air-conditioning unit off before removing panels. Lock and tag the electrical panel where you have turned the power off, and keep the only key in your possession. When loosening or tightening bolts or nuts, be sure to use a wrench of the correct size.

STEP-BY-STEP PROCEDURES

1. Turn the power off and lock the panel. Remove enough panels and/or components to obtain the following information:

Indoor Fan Motor

Horsepower _____

Voltage _____

Locked-rotor amperage _____

Fan wheel width _____

Type of motor _____

Type of motor mount _____

Full-load amperage _____

Fan wheel diameter _____

Shaft diameter _____

Motor speed(s) _____

Indoor Coil

Material coil made of _____

No. of rows in coil _____

Suction-line size _____

Liquid-line size _____

Forced or induced draft coil _____

No. of circuits _____

Type(s) of metering device(s) _____

Type of field piping connection _____

Model number of liquid line filter drier _____

Provisions for condensate removal _____

Is there a trap in the condensate line? _____

Air filter location _____

Air filter size _____

Air filter type _____

Outdoor Unit

Top or side discharge _____

Type of refrigerant used in this unit _____ Type of field piping connection _____

Compressor suction-line size _____ Discharge-line size _____ Unit capacity _____

Type of compressor starting components _____

Type of compressor motor _____ Run capacitor (if there is one) rating _____

Model number of suction-line filter drier (if there is one) _____

Outdoor Fan Motor

Type of motor _____ Size of shaft _____ Motor speed _____ Number of fan blades _____

Horsepower _____ Voltage _____ Full-load amperage _____ Motor rotation _____

2. Replace all components you may have removed with the correct fasteners. Replace all panels.

MAINTENANCE OF WORKSTATION AND TOOLS: Ensure that the air-conditioning system is left ready for its next lab exercise. Replace all tools, equipment, and materials. Leave your workstation neat and orderly.

SUMMARY STATEMENT: Describe the function of the line- and low-voltage electrical circuits of a typical split air-conditioning system. Labeling the components and piping. Use an additional sheet of paper if more space is necessary for your answer.

TAKEAWAY: Each of the components in an air-conditioning system performs a specific function; together, these components make the system as a whole perform as designed. It is important to be aware of the specifications of a component in the event of a repair. The replacement must have the same specifications as the original for the system to operate as intended.

QUESTIONS

1. State two advantages of a top air-discharge condenser over a side air-discharge condenser.

2. Which of the interconnecting refrigerant lines is the largest, the suction or the liquid?

3. What is the purpose of the fins on the indoor and outdoor coils?

4. Why was R-22 the most popular refrigerant for many years for residential air-conditioning?

5. Where does the condensate on the evaporator coil come from?

6. Explain why refrigerants such as R-407C and R-410A are replacing R-22.

Exercise RSC-3

Evaporator Functions and Superheat

Name	Date	Grade

OBJECTIVES: Upon completion of this exercise, you should be able to take pressure and temperature readings and determine superheat of the refrigerant in the evaporator operating under normal and loaded conditions.

INTRODUCTION: You will be working with an evaporator in an operating system. Any typical evaporator in a refrigeration system operating in its normal environment can be used as long as there are gauge ports to take both high-side and low-side pressure readings. You will be taking pressure and temperature readings, which will be used to calculate superheat. For this exercise, an evaporator equipped with a TXV is recommended.

TOOLS, MATERIALS, AND EQUIPMENT: A gauge manifold, a straight-slot screwdriver, ¼" and ⁵⁄₁₆" nut drivers, light gloves, goggles, a thermometer that can be attached to the suction line, a small amount of pipe insulation for the thermometer, and an operating system.

⚠ **SAFETY PRECAUTIONS:** You will be working with refrigerant under pressure. Take care to not allow the refrigerant to get in your eyes or on your skin. Light gloves and goggles should be worn at all times when working with refrigerants. If a leak develops and liquid refrigerant escapes, DO NOT TRY TO STOP IT WITH YOUR HANDS. Your instructor must give you instruction in using the gauge manifold before you begin this exercise. Stay clear of all moving parts, especially the rotating evaporator fan blade or blower wheel.

STEP-BY-STEP PROCEDURES

1. Put on your goggles and gloves. Fasten the suction-gauge line to the suction-gauge port. The high-side line may be attached if desired; however, it is not necessary.

   FIGURE **RSC-3.1**

2. Purge air from gauge lines at the manifold.

3. Securely fasten the thermometer's sensing element to the suction line, close to the evaporator (electrical tape works well), and insulate the lead for about 4" in each direction, Figure RSC-3.1.

   Photo by Jason Obrzut

4. Start the unit, allow it to run for at least 15 minutes, and then record the following:

   Suction pressure: _____ psig

   Determine evaporator saturation temperature from the PT chart: _____ °F

   Suction-line temperature: _____ °F

   Determine evaporator superheat: Suction-line temperature − evaporator saturation temperature = _____ °F evaporator superheat.

5. Now, increase the load on the evaporator. If it is a reach-in box, open the door and place some room-temperature objects in the box. If it is an air conditioner, partially open the fan compartment door to increase the airflow across the evaporator. See Figure RSC-3.2. After 15 minutes, record the following:

   Suction pressure: _____ psig

   Evaporator saturation temperature (EST) from the PT chart: _____ °F

Suction-line temperature: _____ °F

Determine evaporator superheat: Suction-line temperature − evaporator saturation temperature = _____ °F evaporator superheat.

6. Remove the gauges and thermometer, and return the unit to normal running condition with all service and access panels in place.

MAINTENANCE OF WORKSTATION AND TOOLS: Wipe any oil off the gauges and replace all plugs and caps. Return all equipment and tools to their respective places.

SUMMARY STATEMENT: Describe what happened when a load was placed on the evaporator. Use an additional sheet of paper if more space is necessary for your answer.

FIGURE RSC-3.2

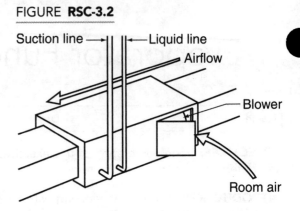

TAKEAWAY: Heat is absorbed in the evaporator of a refrigeration system. The amount of heat absorbed above the refrigerant's boiling point and after it has changed to a vapor is called superheat. The higher the heat load, the higher the temperatures, pressures, and superheat.

QUESTIONS

1. Is superheat latent heat or sensible? Explain your answer.

2. Why is it important to have superheat at the end of the evaporator?

3. What functions does the evaporator perform in a refrigeration system?

4. How does the evaporator remove humidity from the passing air?

5. What should be done with the condensate that forms on the evaporator?

6. Where does the refrigerant go when it leaves the evaporator?

7. How would the readings have differed if the evaporator was equipped with a fixed orifice metering device?

Exercise RSC-4

Evaluating Evaporator Performance

Name	Date	Grade

OBJECTIVES: Upon completion of this exercise, you should be able to evaluate the performance of a direct-expansion evaporator.

INTRODUCTION: You will use superheat and the coil-to-air temperature difference (evaporator split) to evaluate the operation of a direct-expansion evaporator.

TOOLS, MATERIALS, AND EQUIPMENT: Two digital thermometers, screwdriver, electrician's tape, some insulation, a service valve wrench, a gauge manifold, goggles, and a refrigeration system with a thermostatic expansion valve and service valves.

SAFETY PRECAUTIONS: You will be fastening gauges to the refrigeration system. Wear goggles and gloves. Your instructor must give you instruction in using the gauge manifold before you start this exercise.

STEP-BY-STEP PROCEDURES

1. Put on your goggles and gloves. Fasten the high- and low-side gauge lines to the service valves. Open the service valves by turning them off the backseat. Turn them clockwise.

2. Purge the air from the gauge lines by loosening the fittings slightly at the manifold.

3. Fasten a thermometer lead to the suction line of the evaporator where the line leaves the coil. This will be in the same area as the TXV's remote sensing bulb. Be sure the thermometer lead is insulated.

4. Insert another thermometer lead in the evaporator's return-air stream.

5. Start the system and allow it to pull down to within 10°F of its intended use—low, medium, or high temperature—and record the following before the thermostat shuts the compressor off:

   - Suction pressure: _____ psig

   - Suction-line temperature at the evaporator outlet: _____ °F

   - Evaporator saturation temperature (EST) (converted from suction pressure): _____ °F

   - Evaporator superheat (suction-line temperature − evaporator saturation temperature): _____ °F

   - Return-air temperature: _____ °F

   - Evaporator split (difference in temperature between return-air and evaporator saturation temperature): _____ °F

   - Discharge pressure: _____ psig

   - Condenser saturation temperature (converted from discharge pressure): _____ °F

6. Backseat the service valves, and remove the gauge lines.

7. Remove the thermometer leads and replace them in their case.

MAINTENANCE OF WORKSTATION AND TOOLS: Replace all panels and return all tools to their respective places. Make sure the refrigerated box is returned to its correct condition.

SUMMARY STATEMENT: Describe how well the evaporator is performing in this test.

TAKEAWAY: By monitoring a system's pressures, temperatures, evaporator split (ES), and superheat, a technician can evaluate the performance of an evaporator in a refrigeration unit. The actual measurements can be compared to a set of measurements that would be typical of the unit being evaluated.

QUESTIONS

1. What is a typical superheat for a typical direct-expansion evaporator performing in the correct temperature range?

2. What is meant by the term "direct-expansion" evaporator?

3. What is the function of a distributor in a multi-circuit evaporator?

4. What would the symptoms of a flooded evaporator be on a direct-expansion system?

5. What are the symptoms of a dirty evaporator coil?

6. Why is it necessary to insulate the bulb on the thermometer when measuring the superheat of an evaporator?

7. What are some typical problems when the evaporator superheat reading is too low?

8. What are some typical problems when the evaporator superheat reading is too high?

9. What does the term "starved evaporator" mean?

10. What effect(s) would a defective evaporator fan have on the system?

Exercise RSC-5

Checking an Evaporator During a Hot Pulldown

Name	Date	Grade

OBJECTIVES: Upon completion of this exercise, you should be able to check an evaporator while it is under excessive load and observe the changing conditions while the load is reduced.

INTRODUCTION: You will be assigned a packaged commercial refrigeration system with a thermostatic expansion valve that has been turned off long enough to be warm inside. You will then turn the unit on and use an ammeter, gauges, and thermometer to record the changes in operating conditions while the temperature is being lowered in the box.

TOOLS, MATERIALS, AND EQUIPMENT: A gauge manifold, a service valve wrench, four-lead thermometer, ammeter, goggles, gloves, straight-slot and Phillips-head screwdrivers, ¼" and ⁵⁄₁₆" nut drivers, and a commercial refrigeration system with thermostatic expansion valve.

⚠ **SAFETY PRECAUTIONS:** Wear goggles and gloves while working with pressurized refrigerant. Always take care when working with electricity or live circuits.

STEP-BY-STEP PROCEDURES

1. Put on your goggles and gloves. Connect the high- and low-side gauge lines.

2. If the system has service valves, turn the valve stems off the backseat far enough for the gauges to read. If the system has Schrader valves, the gauges will read when connected. Bleed air from the gauge lines.

3. Fasten the four thermometer leads in the following locations:

   - Suction line leaving the evaporator (before a heat exchanger, if any). Use Figure RSC-5.1 as an example of evaporator pressure and temperatures.

   - Inlet air to the evaporator.

   - Air entering the condenser.

   - Air leaving the evaporator (supply air).

4. Fasten the ammeter around the common wire on the compressor.

5. Record the following information two times—TEST 1 at 5 minutes after start-up and TEST 2 when the box temperature has been reduced to within 5° of its design temperature.

   - Type of refrigerant: R-_____

	TEST 1	**TEST 2**	
• Suction pressure	_____	_____	psig
• Suction-line temperature at the evaporator outlet	_____	_____	°F
• Evaporator saturation temperature (EST)	_____	_____	°F

FIGURE **RSC-5.1**

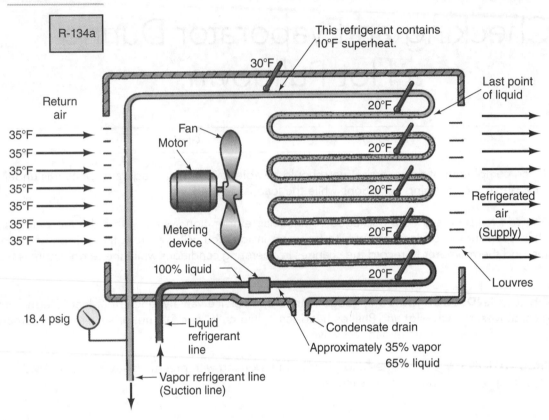

- Evaporator superheat _____ _____ °F
- Air temperature entering evaporator _____ _____ °F
- Evaporator split (air entering the evaporator − _____ _____ °F
  evaporator saturation temperature)
- Air temperature leaving the evaporator _____ _____ °F
- Delta T (air entering the evaporator − air leaving the evaporator) _____ _____ °F
- Discharge pressure _____ _____ psig
- Air temperature entering condenser _____ _____ °F
- Condenser saturation temperature (CST) _____ _____ °F
- Condenser split (condenser saturation temperature −
  entering air temperature) _____ _____ °F
- Compressor amperage _____ _____ A

6. Remove all instruments and gauges.

7. Replace all panels with the correct fasteners.

8. Return the system to normal operation.

MAINTENANCE OF WORKSTATION AND TOOLS: Return all tools to their respective workstations.

SUMMARY STATEMENT: Describe how the amperage, suction, and discharge pressure responded as the temperature in the refrigerated box was lowered. Use an additional sheet of paper if more space is necessary for your answer.

TAKEAWAY: During normal operation of a refrigeration system, the temperatures and pressures will drop as the temperature of the refrigerated space drops. By using calculations such as superheat and evaporator split as well as monitoring the pressures, a technician can evaluate how well a unit is running.

QUESTIONS

1. What is the function of the evaporator in a refrigeration system?

2. What is meant by the term "hot pulldown"?

3. What should the typical evaporator superheat be for the system that you were assigned in this exercise? _____ °F

4. What is the purpose of the fins on an evaporator?

5. Where does the condensate come from that forms on the evaporator?

6. State the mathematical formulas used to calculate:

- Evaporator split
- Superheat
- TD across the evaporator
- Condenser split

7. What is the difference between an induced draft and a forced draft evaporator?

8. Is the heat that is transferred to the evaporator sensible heat, latent heat, or both? Explain your answer.

Exercise RSC-6

Condenser Functions and Desuperheat

Name	Date	Grade

OBJECTIVES: Upon completion of this exercise, you should be able to take refrigerant and air temperature readings at the condenser. You will be able to determine condensing temperatures from the temperature/pressure chart using the compressor discharge pressure and determine the condensing superheat under normal and loaded conditions.

INTRODUCTION: You will connect a gauge manifold and secure temperature sensors to the air-cooled condenser of a typical refrigeration system to take pressure and temperature readings. You will also determine the superheat at the condenser under normal and loaded conditions.

TOOLS, MATERIALS, AND EQUIPMENT: A high-quality digital thermometer with a sensing bulb that can be strapped to the compressor discharge line, a small piece of insulation, a thermometer for measuring the condenser air inlet temperature, goggles, light gloves, a gauge manifold, a piece of cardboard, a straight-slot screwdriver, ¼" and ⁵⁄₁₆" nut drivers, a service valve wrench, a temperature/pressure conversion chart, and an operating split-type refrigeration unit.

⚠ **SAFETY PRECAUTIONS:** The discharge line may be hot, so give it time to cool with the compressor off before fastening the thermometer to it. Be very careful when working around the turning fan. Use care in fastening the refrigerant gauges. Wear goggles and light gloves. Do not proceed with this exercise until your instructor has given you instruction on the use of the gauge manifold.

STEP-BY-STEP PROCEDURES

1. With the compressor off and after the discharge line has cooled, fasten a temperature sensor to the discharge line and insulate it. Also, place a temperature sensor in the airstream entering the condenser, Figure RSC-6.1.

FIGURE **RSC-6.1** Thermometer leads secured to the compressor discharge line and where airflow enters the condenser.

Photo by Jason Obrzut

2. Put on your goggles and gloves. Fasten the high-pressure gauge to the high-pressure side of the system. You may also fasten a low-side gauge if you want, although you will not need it. Purge any air from your hoses.

3. Start the unit and record the following after 10 minutes:

Refrigerant type: R-_____

Discharge pressure: _____ psig

Discharge-line temperature: _____°F

Condenser saturation temperature (CST) (converted from pressure on the PT chart): _____°F

Discharge-line superheat (discharge-line temperature − condenser saturation temperature): _____°F

Air temperature entering the condenser: _____°F

Condenser split (condenser saturation temperature − condenser entering air temperature): _____°F

4. By blocking the condenser airflow, Figure RSC-6.2, increase the head (discharge) pressure by about 25 psig, and record the following:

Discharge pressure: _____ psig

Discharge-line temperature: _____°F

Condenser saturation temperature (CST) (converted from pressure on the PT chart): _____°F

Discharge-line superheat (discharge-line temperature − condenser saturation temperature): _____°F

Air temperature entering the condenser: _____°F

Condenser split (condenser saturation temperature − condenser entering air temperature): _____°F

FIGURE **RSC-6.2** Increasing the head pressure by partially blocking the condenser.

Photo by Jason Obrzut

MAINTENANCE OF WORKSTATION AND TOOLS: Stop the unit and remove the thermometer leads from the discharge line after it has cooled, remove the gauges, and wipe off any oil that may have accumulated on the unit. Return all tools to their appropriate locations, and replace all panels and caps on the unit. Be sure to leave the system as you found it.

SUMMARY STATEMENT: Describe why the compressor discharge line is so much hotter than the condensing temperature and why it does not follow the temperature/pressure relationship.

TAKEAWAY: The condenser performs specific functions in a refrigeration circuit. These functions can be observed and evaluated by measuring temperatures, pressures, and condenser split (CS).

QUESTIONS

1. What must be done to the hot gas leaving the compressor before it can start to condense?

2. On a properly operating system, the refrigerant leaving the condenser should be a (liquid, vapor, or solid) _____.

3. For a standard-efficiency system, what is the expected temperature difference between the condensing refrigerant and the air entering the condenser?

4. What happens to the temperature of the discharge line when the head pressure is increased?

5. Where does the condensing take place in a typical condenser?

6. What would happen to the head pressure if the fan motor on an air-cooled condenser stopped working?

7. What are the three functions of the condenser in a refrigeration system?

8. What is the function of the fins on an air-cooled condenser?

9. The head pressure of the refrigerant in a running refrigeration system is determined by the condensing temperature. True or false?

10. What effects would air or non-condensables inside the refrigerant piping circuit have on the operation of the condenser?

Exercise RSC-7

Air-Cooled Condenser Performance

Name	Date	Grade

OBJECTIVES: Upon completion of this exercise, you should be able to perform basic evaluation procedures on an air-cooled condenser.

INTRODUCTION: You will use a thermometer and a gauge manifold to calculate subcooling and the condenser split (CS) of the system. A typical split is about 30° but can vary depending on the type of unit and its efficiency.

TOOLS, MATERIALS, AND EQUIPMENT: A gauge manifold, a digital thermometer with two temperature sensing leads, goggles, light gloves, an adjustable wrench, a service valve wrench, and a refrigeration system with an air-cooled condenser and service valves.

⚠ **SAFETY PRECAUTIONS:** Wear goggles and gloves while connecting and disconnecting refrigerant hoses. Make sure you have had instruction on using a gauge manifold.

STEP-BY-STEP PROCEDURES

1. Put on your goggles and gloves. Fasten the high- and low-side gauge manifold hoses to the system's service valves.

2. Crack the service valves off the backseat (clockwise when looking at the end of the stem) enough to obtain a gauge reading.

3. Purge the air from the hoses by loosening the fittings slightly at the manifold.

4. Place one temperature sensor into the airstream entering the condenser, and fasten the other to the liquid line leaving the condenser.

5. Start the system and wait until it is within 10°F of the correct temperature range inside the box.

6. Record the following:

   - Type of box (low, medium, or high temperature): _____

   - Type of condenser (natural or mechanical draft): _____

   - Suction pressure: _____ psig

   - Head pressure: _____ psig

   - Condenser saturation temperature (CST) (converted from head pressure): _____°F

   - Condenser entering air temperature: _____°F

   - Condenser split (CS) (condenser saturation temperature − condenser entering air temperature): _____°F

   - Liquid line temperature _____°F

   - Condenser subcooling (condenser saturation temperature − liquid line temperature) _____°F

7. If the condenser split is more than 30°F, check for possible causes and note them here.

8. Remove the gauges.

MAINTENANCE OF WORKSTATION AND TOOLS: Return the refrigeration system to its proper condition. Clean the workstation and return all tools to their respective places.

SUMMARY STATEMENT: Describe how condenser split and subcooling can be used to evaluate the operation of a condenser. Use an additional sheet of paper if more space is necessary for your answer.

TAKEAWAY: When evaluating a system component, such as the condenser, it is important for technicians to gather as much information about its operating conditions as possible. Subcooling and condenser split are two important calculations done when executing an evaluation of an air-cooled condenser.

QUESTIONS

1. Name the three functions that a condenser performs.

2. What is considered the highest temperature that the discharge gas should reach upon leaving the compressor?

3. What may be used on an air-cooled condenser with a fan to prevent the head pressure from becoming too low in mild weather?

4. What effects would low head pressure have on a system with an air-cooled condenser?

5. Why must an air-cooled condenser be located where the air will not recirculate through it?

6. What is the difference between a natural draft and a mechanical draft condenser?

7. What are the symptoms of a dirty condenser?

8. What is the purpose of the fins on the condenser?

9. What materials are typical refrigeration condensers made of?

Exercise RSC-8

Evaluation of a Water-Cooled Condenser

Name	Date	Grade

OBJECTIVES: Upon completion of this exercise, you should be able to evaluate a simple water-cooled refrigeration system.

INTRODUCTION: You will use gauges and a thermometer to evaluate a simple water-cooled system. You will do this by evaluating the heat exchange that occurs in the condenser.

TOOLS, MATERIALS, AND EQUIPMENT: A gauge manifold, goggles, light gloves, an adjustable wrench, a service valve wrench, a digital thermocouple thermometer, and a water-cooled refrigeration system.

⚠ **SAFETY PRECAUTIONS:** Wear goggles and gloves while working with refrigerant under pressure.

STEP-BY-STEP PROCEDURES

1. Put on your goggles and gloves. Connect the high- and low-side gauges.

2. Crack the service valves off their backseat far enough for the gauges to register the pressure. Purge any air from your hoses.

3. Fasten one thermometer lead to the discharge line of the system.

4. Secure a thermometer or thermistor to the water line entering the condenser and another one to the water line leaving the condenser, Figure RSC-8.1. Make sure they are located in a manner that will give accurate water temperatures.

5. Start the system and allow it to run until it is operating at a steady state—that is, until the readings are not fluctuating. Record the following:

FIGURE **RSC-8.1**

Photo by Jason Obrzut

- Type of refrigerant: R-_____

- Suction pressure: _____ psig

- Discharge pressure: _____ psig

- Discharge-line temperature: _____°F

- Condenser saturation temperature (CST) (converted from discharge pressure): _____°F

- Condenser leaving-water temperature: _____°F

- Condenser entering-water temperature: _____°F

- Condenser split (temperature difference between the leaving water and the condensing refrigerant): _____°F

6. Disconnect the gauges and thermometer leads. Replace all panels that may have been removed with the correct fasteners.

MAINTENANCE OF WORKSTATION AND TOOLS: Return all tools to their respective places.

SUMMARY STATEMENTS: Describe the system's ability to transfer heat into the water. Was the system transferring the heat at the correct rate? Use an additional sheet of paper if more space is necessary for your answer.

TAKEAWAY: Water has a greater capacity than air to absorb and hold heat. For this reason, water-cooled systems are considered to be more efficient than air-cooled systems. Increasing or decreasing the flow of water through the condenser will affect system operation and efficiency.

QUESTIONS

1. Is a water-cooled condenser more efficient than an air-cooled condenser? Explain your answer.

2. Why are water-cooled condensers not as popular as air-cooled condensers in small equipment?

3. What is the typical condenser split for a water-cooled condenser that is operating efficiently?

4. Name two places to which the heat-laden water from a water-cooled condenser may be piped.

5. What extra maintenance procedures are required to be done on a water-cooled condenser that are not done on an air-cooled condenser?

6. Name the three functions that take place in the condenser.

7. What controls the water flow through a water regulating valve?

8. What happens to the condenser split of a water-cooled system if the water flow is reduced?

9. What happens to the condenser split of a water-cooled system if the condenser tubes are dirty?

10. What is the low form of plant life that grows in condenser water?

Determining the Correct Head Pressure

Name	Date	Grade

OBJECTIVES: Upon completion of this exercise, you should be able to determine the correct head pressure for air-cooled condensers under the existing operating conditions.

INTRODUCTION: You will use pressure and temperature measurements to determine the correct head pressure for an air-cooled condenser under typical operating conditions. If a system has an abnormal load on the evaporator, the head pressure will not conform exactly to the ambient temperature relationship.

TOOLS, MATERIALS, AND EQUIPMENT: A gauge manifold, goggles, light gloves, an adjustable wrench, a service valve wrench, straight-slot and Phillips-head screwdrivers, ¼" and ⁵⁄₁₆" nut drivers, a digital thermocouple thermometer, and an air-cooled refrigeration system with service valves or Schrader valve ports.

⚠ **SAFETY PRECAUTIONS:** Wear goggles and gloves while working with refrigerant under pressure. Always take care when working near electricity or live circuits.

STEP-BY-STEP PROCEDURES

1. Your refrigeration system should have been running for a period of time so that it is operating under typical conditions. Using the thermometer, measure the temperature of the air entering the condenser. Record here: _____°F.

2. Determine from the air entering the condenser what you think the maximum condenser saturation temperature should be. Air entering +30°F maximum = _____°F.

3. Convert the condensing temperature to head pressure using a temperature/pressure chart: _____ psig.

4. Put on your goggles and gloves. Now that you have determined what the head pressure should be, connect the high- and low-side gauge lines and record the following:

   • Actual head pressure: _____ psig

   • Actual condenser saturation temperature (CST): _____°F

   • Air temperature entering condenser: _____°F

   • Condenser split (condenser saturation temperature − temperature of the air entering the condenser): _____°F

   If the last item is over 30°F, list some possible causes based on your observations of the unit.

5. Remove the gauges and replace all panels with the correct fasteners.

MAINTENANCE OF WORKSTATION AND TOOLS: Return all tools to their respective places.

SUMMARY STATEMENTS: Describe the refrigerant cycle as it occurs in the condenser. Use an additional sheet of paper if more space is necessary for your answer.

TAKEAWAY: The relationship that exists between the condenser entering air and the condenser saturation temperature, condenser split (CS), can be used to effectively evaluate the operating efficiency of a condenser. It can also give a technician insight into what may be wrong in a system that is not operating properly.

QUESTIONS

1. State the three functions of a condenser.

2. Why does the head pressure rise when the evaporator has a high heat load?

3. What would be the condenser split (CS) for a high-efficiency condenser?

4. List some conditions that could cause a higher than normal condenser split.

5. List some conditions that could cause higher-than-normal head pressure.

6. Why must a minimum head pressure be maintained, particularly if equipment is located outside?

7. Name three methods used to control head pressure on air-cooled equipment.

8. What are some differences between a water-cooled and an air-cooled condenser?

9. What would be the symptoms for an air-cooled condenser where the air leaving the condenser is hitting a barrier and recirculating?

Exercise RSC-10

Compressor Operating Temperatures

Name	Date	Grade

OBJECTIVES: Upon completion of this exercise, you should be able to evaluate a suction gas-cooled hermetic compressor for the correct operating temperatures while it is running.

INTRODUCTION: You will use gauges and a thermometer to become familiar with how a suction gas-cooled hermetic compressor should feel while operating and where it should be hot, warm, and cool. You will also see the effects that different operating conditions have on the system.

TOOLS, MATERIALS, AND EQUIPMENT: A gauge manifold, straight-slot and Phillips-head screwdrivers, a service valve wrench, ¼" and ⁵⁄₁₆" nut drivers, goggles, gloves, a high-quality digital thermocouple thermometer, and a refrigeration system with a suction gas-cooled compressor.

⚠ **SAFETY PRECAUTIONS:** Wear goggles and gloves while working with refrigerant under pressure. Take care when working around the compressor as some areas of the compressor, including the discharge line, can be very hot.

STEP-BY-STEP PROCEDURES

1. The refrigeration system should have been off long enough to allow the compressor to cool down. It should have a welded hermetic compressor and a suction gas-cooled motor.

2. Put on your goggles and gloves. Connect the high- and low-side gauges to obtain system pressures. Remember to purge the air from the hoses.

3. Fasten one temperature sensor to the suction line entering the compressor, Figure RSC-10.1.

FIGURE **RSC-10.1** The correct location for the temperature sensors. The sensors may be fastened with electrical tape and should have some insulation.

Photo by Jason Obrzut

4. Fasten a second temperature lead to the compressor shell about 4" from where the suction line enters the shell.

5. Fasten a third temperature sensor to the bottom of the compressor shell where the oil would be.

6. Fasten the fourth temperature sensor to the discharge line leaving the compressor.

7. Record the following data before starting the system:

   - Type of refrigerant in the system: R-_____
   - Suction pressure: _____ psig
   - Evaporator saturation temperature (EST): _____ °F
   - Discharge pressure: _____ psig
   - Condenser saturation temperature (CST): _____ °F
   - Suction-line temperature: _____ °F
   - Upper-shell temperature: _____ °F
   - Lower-shell temperature: _____ °F
   - Discharge-line temperature: _____ °F

8. Start the system, let it run for about 15 minutes, and record the following information:

   - Suction pressure: _____ psig
   - Evaporator saturation temperature (EST): _____ °F
   - Discharge pressure: _____ psig
   - Condenser saturation temperature (CST): _____ °F
   - Suction-line temperature: _____ °F
   - Upper-shell temperature: _____ °F
   - Lower-shell temperature: _____ °F
   - Discharge-line temperature: _____ °F

9. Block the condenser airflow until the refrigerant is condensing at about 125°F. Provided here are four examples of the saturation pressure at 125°F for R-22, R-134a, R-404A and R-410A. If the system you are working on uses one of these refrigerants, circle the refrigerant and the corresponding pressure. If the refrigerant being used is not one of these, provide the information for that refrigerant in the space to the right.

   - R-22   278 psig
   - R-134a   185 psig
   - R-404A   332 psig
   - R-410A   446 psig

10. Let the system run for about 15 minutes at this condition.

    Note: You may have to adjust the thermostat to prevent the system from shutting off during this running time.

    Record the following information:

    - Suction pressure: _____ psig
    - Evaporator saturation temperature (EST): _____ °F
    - Discharge pressure: _____ psig

- Condenser saturation temperature (CST): _____ °F
- Suction-line temperature: _____ °F
- Upper-shell temperature: _____ °F
- Lower-shell temperature: _____ °F
- Discharge-line temperature: _____ °F

11. Shut the system off for 15 minutes, and record the following information again:

- Suction pressure: _____ psig
- Evaporator saturation temperature (EST): _____ °F
- Discharge pressure: _____ psig
- Condenser saturation temperature (CST): _____ °F
- Suction-line temperature: _____ °F
- Upper-shell temperature: _____ °F
- Lower-shell temperature: _____ °F
- Discharge-line temperature: _____ °F

12. If it was adjusted during this exercise, return the thermostat to the correct setting.

13. Remove the gauges and temperature sensors.

MAINTENANCE OF WORKSTATION AND TOOLS: Return all tools to their respective places. Replace all service and access panels. Leave the system in the manner in which you found it unless instructed otherwise.

SUMMARY STATEMENTS: Describe how the different temperatures of the compressor responded when the head pressure was raised and why.

Suction line:

Upper shell:

Lower shell:

Discharge line:

TAKEAWAY: A hermetic compressor does not allow a technician to view what is happening inside of it. Monitoring specific temperatures and pressures will allow a technician to determine what is occurring both in compressor and in the system itself.

QUESTIONS

1. What would happen to the discharge-line temperature with an increase in suction-line superheat?

2. What would happen to the discharge-line temperature with a decrease in suction-line temperature?

3. What would the result of an internal discharge-line leak be on a compressor's temperature?

4. What is the purpose of a crankcase heater on a compressor?

5. How is a typical hermetic compressor mounted inside the shell?

6. Is the compressor shell thought of as being on the high- or low-pressure side of the refrigeration system?

7. Which is typically larger on the compressor, the suction or the discharge line? Why is this so?

8. How do high discharge gas temperatures affect the oil in a compressor?

9. How is compression ratio calculated?

10. Which of the following terminal combinations would show an open circuit if a hermetic compressor with an internal overload device were to be open due to an overheated compressor?
   A. Common to run
   B. Common to start
   C. Run to start

Compressor Efficiency

Name	Date	Grade

OBJECTIVES: Upon completion of this exercise, you should be able to evaluate the performance of a hermetic compressor.

INTRODUCTION: You will use a gauge manifold to measure the suction and head pressure at, or close to, design conditions. You will compare the compressor's operating amperage to the compressor's nameplate rated-load amperage to help determine the compressor's operating characteristics. The system must be operated until the temperature in the conditioned space is lowered to a temperature that is close to design conditions.

TOOLS, MATERIALS, AND EQUIPMENT: A gauge manifold, goggles, gloves, an adjustable wrench, a digital multimeter (DMM), a clamp-on ammeter, a service valve wrench, and a refrigeration system with a known rated-load amperage (RLA) for the compressor motor.

⚠ **SAFETY PRECAUTIONS:** Wear goggles and gloves while working with refrigerant under pressure. Take care when working around the compressor, as some areas of the compressor and the compressor discharge line can be very hot.

STEP-BY-STEP PROCEDURES

1. Put on your goggles and gloves. With the power off, connect the gauges to the valve ports or service valves, and obtain system pressure readings. Remember to purge the air from your hoses.

2. Position the jaws of the clamp-on ammeter around the common or run wire leading to the compressor.

3. Start the system.

4. Allow the system to run until the conditioned space is near the design temperature, just before the thermostat would shut it off.

5. Lower the thermostat setting to ensure that the system continues to run.

6. Reduce the airflow across the condenser until the refrigerant is condensing at about 125°F. Provided here are four examples of the saturation pressure at 125°F for R-22, R-134a, R-404A, and R-410A. If the system you are working on uses one of these refrigerants, circle the refrigerant and the corresponding pressure. If the refrigerant being used is not one of these, provide the information for that refrigerant in the space to the right.

   - R-22   278 psig
   - R-134a   185 psig
   - R-404A   332 psig
   - R-410A   446 psig

7. Record the following information:

   - Suction pressure: _____ psig
   - Evaporator saturation temperature (EST): _____ °F

- Discharge pressure: _____ psig
- Condenser saturation temperature (CST): _____ °F
- Compressor rated-load amperage: _____ A
- Compressor actual amperage: _____ A
- Compressor rated voltage: _____ V
- Actual voltage at the unit: _____ V

8. If the amperage of the compressor is not within 10% of the rated-load amperage, list some possible causes based on your observation of the unit.

9. Remove gauges, making certain to replace all service port caps and properly prepare the gauge manifold for storage.

MAINTENANCE OF WORKSTATION AND TOOLS: Return all tools to their respective places, and replace all panels. Return the unit to its normal running condition.

SUMMARY STATEMENT: Based on the results of this exercise, summarize the relationship between a compressor's operating characteristics and the system's overall performance.

TAKEAWAY: On a system that is operating properly, the compressor should be operating within its design specifications. Monitoring the pressures, temperatures, voltage, and amperage at the compressor can give a technician valuable information about the operation of the system.

QUESTIONS

1. Name three types of compressors.

2. What are some factors that affect compression ratio?

3. Name two ways that small compressor motors are cooled.

4. Why should you never frontseat the discharge service valve while a compressor is running?

5. What effect does a rise in head pressure have on compressor amperage?

6. What would the symptoms be when a two-cylinder compressor has only one pumping?

7. Where is the rated-load amperage information for a compressor found?

8. What lubricates the compressor's moving parts?

Exercise RSC-12

Hermetic Compressor Changeout

Name	Date	Grade

OBJECTIVES: Upon completion of this exercise, you should be able to demonstrate how to change a hermetic compressor in a refrigeration system.

INTRODUCTION: You will recover the refrigerant charge, completely remove a compressor from a refrigeration system, and then replace it in the system. You will then leak-check, evacuate, and charge the system as though the compressor were a new one. It is recommended that you energize the system prior to this exercise to record its operating characteristics for comparison upon completion of the lab.

TOOLS, MATERIALS, AND EQUIPMENT: A torch arrangement, brazing rods or solder, leak detector (electronic and bubbles if possible), service valve wrench, tubing cutters, tube reamer, gauge manifold, goggles, gloves, line tap valves (if needed, such as for a small system), straight-slot and Phillips-head screwdrivers, small socket set, refrigerant recovery system, refrigerant scale, vacuum pump, cylinder of refrigerant, and an operating refrigeration system.

⚠ **SAFETY PRECAUTIONS:** Wear goggles and gloves while transferring liquid refrigerant. DO NOT VENT REFRIGERANT TO THE ATMOSPHERE. The use of a torch raises the risk of burns or starting a fire. Always practice safe brazing techniques.

STEP-BY-STEP PROCEDURES

1. Put on your goggles and gloves. Connect gauges to the system, and purge the air from the hoses.

2. Using a recovery machine, remove the refrigerant from the system. Allow the machine to run until the system has been brought down to the recommended vacuum level.

3. Turn off and lock out the power, discharge all capacitors, and disconnect the wiring from the compressor terminals, labeling each wire.

4. Remove the compressor hold-down nuts or bolts.

5. Use tubing cutters to cut the compressor out of the system. Cut the compressor suction, and discharge lines at least 2" from the shell so you have a stub of piping to work with when replacing the compressor.

6. Remove the compressor from the appliance, and set it aside.

7. Ream the compressor stubs and the piping in the system, and prepare them for brazing.

   Note: Do not allow any material to fall into the system or the compressor.

8. Reposition the compressor onto its mountings as though it were a new one.

   Note: To avoid contamination of the system, replace the compressor as quickly as possible. It is also recommended that the filter drier be replaced any time a system is opened.

9. Use sweat tubing connectors to fasten the suction and the discharge lines back together. A good grade of low-temperature solder may be what your instructor would recommend so that this process may be performed again on the same unit.

   Note: Brazing is the generally accepted method for joining copper using best practices in the field.

10. Pressurize the system with nitrogen to about 150 psi, and leak-check the connections you worked with. A refrigerant trace that matches the system charge can be used if necessary. Never exceed the low side's nameplate maximum pressure.

11. Evacuate the system down to the appropriate micron level. A triple evacuation is recommended. The motor terminal wiring and compressor mounting bolts may be refastened at this time.

12. Charge the system with the correct amount of refrigerant. A refrigerant scale is recommended for critically charged systems.

13. Start the system and check for normal operation.

    Note: Compare the operation of the unit to the system information obtained prior to this exercise if it is available.

14. Return the system to the condition specified by your instructor.

MAINTENANCE OF WORKSTATION AND TOOLS: Return all tools to their correct locations. Clean up any oil or other debris, and be sure to leave the system as you found it.

SUMMARY STATEMENT: Describe why a new filter-drier must be installed whenever the refrigerant piping circuit is opened for servicing. Use an additional sheet of paper if more space is necessary for your answer.

TAKEAWAY: Changing a compressor is a typical job performed in the HVACR field. It is important to understand all of the steps involved and to use best field practices when performing service or an installation.

QUESTIONS

1. How long did it take you to evacuate the system?

2. Why is nitrogen used to pressurize the system to check for leaks?

3. How long did it take to recover the refrigerant from the system?

4. Why is it illegal to vent refrigerant to the atmosphere?

5. If the lines were to be left open for an extended period of time, what effects would it have on the system?

6. Was this compressor mounted on external or internal springs?

7. How much refrigerant charge was used to recharge the unit?

8. What steps can be taken to speed up the refrigerant recovery or evacuation processes?

Exercise RSC-13

Evaluating a Thermostatic Expansion Valve

Name	Date	Grade

OBJECTIVE: Upon completion of this exercise, you will be able to describe the characteristics of a thermostatic expansion valve (TXV) and adjust it for more or less superheat.

INTRODUCTION: You will use a system with an adjustable thermostatic expansion valve and check the superheat when the valve is set correctly. You will then make adjustments to produce more and less superheat and watch the valve respond to adjustment.

TOOLS, MATERIALS, AND EQUIPMENT: A gauge manifold, two digital thermometers, goggles, gloves, an adjustable wrench, a service valve wrench, pipe insulation, and a refrigeration system with an adjustable thermostatic expansion valve.

NOTE: If an adjustable TXV is not available, this exercise can be completed by manipulating the TXV sensing bulb to make it open and close.

⚠️ **SAFETY PRECAUTIONS:** Wear goggles and gloves while connecting gauges to the system.

STEP-BY-STEP PROCEDURES

1. Put on your goggles and gloves. Connect the gauge manifold to the high- and low-pressure sides of the system. Purge the air from your hoses.

2. Fasten the temperature sensor to the suction line just as it leaves the evaporator. If there is a heat exchanger, make sure that the sensor is fastened between the evaporator and the heat exchanger. Be sure to insulate the temperature lead on the line. Fasten another temperature sensor to the discharge line.

3. Inspect the thermostatic expansion valve sensing bulb, and be sure that it is fastened correctly to the suction line at the outlet of the evaporator coil. It should be insulated.

4. Set the thermostat low enough to ensure that the unit will run for the duration of this exercise.

5. Energize the unit and allow the temperature to drop and then stabilize close to the system's design temperature (high, medium, or low temperature).

6. Record the following data when the system is operating in a stable manner with a cold refrigerated box temperature close to design temperatures:

   - Discharge pressure: _____ psig

   - Condenser saturation temperature (CST): _____ °F

   - Discharge line temperature: _____°F

   - Subcooling: _____°F

   - Suction-line temperature: _____ °F

- Suction pressure: _____ psig
- Evaporator saturation temperature (EST): _____ °F
- Superheat: _____ °F

7. Turn the thermostatic expansion valve adjustment two full turns clockwise for an increase in superheat, and wait for the system to stabilize. Record the following data:

   Note: If an non-adjustable TXV is used, submerge the remote sensing bulb in a cup of ice water, and record the data:

   - Discharge pressure: _____ psig
   - Condenser saturation temperature (CST): _____ °F
   - Discharge line temperature: _____ °F
   - Subcooling: _____ °F
   - Suction-line temperature: _____ °F
   - Suction pressure: _____ psig
   - Evaporator saturation temperature (EST): _____ °F
   - Superheat: _____ °F

8. Turn the valve adjustment back to where it was and then two full turns counter-clockwise toward a decrease in superheat, and wait for the system to stabilize. Record the following:

   Note: If an non-adjustable TXV is used, hold the remote sensing bulb firmly in your hand to heat it up. Record the data:

   - Discharge pressure: _____ psig
   - Condenser saturation temperature (CST): _____ °F
   - Discharge line temperature: _____°F
   - Subcooling: _____°F
   - Suction-line temperature: _____ °F
   - Suction pressure: _____ psig
   - Evaporator saturation temperature (EST): _____ °F
   - Superheat: _____ °F

9. Turn the valve adjustment back to the mid-position where it was in the beginning, and verify that the system is maintaining evaporator superheat at about 10°F–12°F for high-temperature evaporators (30°F), 5°F–10°F for medium-temperature evaporators (0°F–30°F), and 2°F–5°F for low-temperature evaporators (below 0°F).

   Note: If the TXV sensing bulb was removed from suction line, secure it back in place and insulate as required. Verify that the system is maintaining the proper superheat:

10. Remove the gauges and temperature sensors.

MAINTENANCE OF WORKSTATION AND TOOLS: Return all tools and equipment to their respective places. Replace all panels and fasteners in the proper manner. Leave the system in the manner you found it. Be sure to turn the thermostat back to normal.

SUMMARY STATEMENTS: Describe the change in superheat when the TXV was adjusted. Indicate how much change per turn was experienced.

TAKEAWAY: Thermostatic expansion valves adjust the flow of refrigerant to the evaporator depending on the heat load to maintain a constant evaporator superheat. Some TXVs can be adjusted to maintain more or less superheat by adjusting their stem. The adjustments will vary from one manufacturer to the next.

QUESTIONS

1. When a coil is starved of refrigerant, the superheat is _____.

2. When a coil is flooded with refrigerant, the superheat is _____.

3. The thermostatic expansion valve maintains a constant _____.

4. What is the function of an external equalizer on a TXV?

5. Does a TXV feed more or less refrigerant with an increase in evaporator heat load?

6. What are the three pressures that act to open or close a TXV?

7. What were the effects on the subcooling, if any, when the TXV was adjusted?

Capillary Tube System Evaluation

Name	Date	Grade

OBJECTIVES: Upon completion of this exercise, you should be able to evaluate a system using a capillary tube metering device.

INTRODUCTION: You will use a digital thermometer and gauge manifold to check the evaporator and capillary tube performance under light-load and full-load conditions for the correct charge under these changing conditions.

TOOLS, MATERIALS, AND EQUIPMENT: A gauge manifold, digital thermometer, goggles, gloves, straight-blade and Phillips-head screwdrivers, electrical tape, a short piece of insulation, a valve wrench, ¼" and ⁵⁄₁₆" nut drivers, and a refrigeration system with service valves and a capillary tube metering device.

⚠ **SAFETY PRECAUTIONS:** Wear goggles and gloves while working with refrigerant under pressure. Always take care when working around electricity or live circuits.

STEP-BY-STEP PROCEDURES

1. Put on your goggles and gloves. Connect the high- and low-side hoses on the gauge manifold to the unit. The connections will probably be Schrader-type connectors for a capillary tube system.

2. Purge a small amount of refrigerant from each hose to remove contaminants, especially air.

3. Fasten a temperature sensor to the suction line at the end of the evaporator but before any heat exchanger in the line. Connect another temperature sensor to the discharge line.

4. Insulate the temperature sensor on the suction line.

5. Start the system, let it run for 15 minutes, and record the following information:

   - Suction pressure: _____ psig

   - Suction-line temperature: _____ °F

   - Evaporator saturation temperature (EST) (converted from suction pressure): _____ °F

   - Superheat (subtract EST from the suction-line temperature): _____ °F

   - Discharge pressure: _____ psig

   - Condenser saturation temperature (CST) (converted from discharge pressure): _____ °F

   - Liquid-line temperature: _____ °F

   - Subcooling (CST – Liquid line temperature) _____ °F

6. Block the condenser airflow until the condensing temperature is about 125°F. Provided here are four examples of the saturation pressure at 125°F for R-22, R-134a, R-404A, and R-410A. If the system you are working on uses one of these refrigerants, circle the refrigerant and the corresponding pressure. If the refrigerant being used is not one of these, provide the information for that refrigerant in the space to the right.

   • R-22   278 psig

   • R-134a   185 psig

   • R-404A   332 psig

   • R-410A   446 psig

7. Allow the system to run for at least 15 minutes before taking the next set of readings. It may be necessary to adjust the thermostat to keep the unit running for the duration of this exercise. Take the following readings:

   • Suction pressure: _____ psig

   • Evaporator saturation temperature (EST): _____ °F

   • Suction-line temperature: _____ °F

   • Superheat (subtract the EST from the suction line temperature): _____ °F

   • Discharge pressure: _____ psig

   • Condenser saturation temperature (CST): _____ °F

   • Liquid line temperature: _____ °F

   • Subcooling (Subtract the liquid line temperature from the CST): _____ °F

8. Shut the unit off and remove the gauges. Replace all panels with the correct fasteners.

MAINTENANCE OF WORKSTATION AND TOOLS: Return all tools to their respective places. Ensure that your work area is left clean.

SUMMARY STATEMENTS: Describe what the superheat in the evaporator did when the head pressure was raised to the condensing temperature of 125°F. Explain why this happened. Use an additional sheet of paper if more space is necessary for your answer.

TAKEAWAY: Capillary tube metering devices do not open or close like a thermostatic expansion valve. They are rated for a specific capacity and will respond to changes in the heat load differently than a TXV. Technicians should be aware of how systems with different types of expansion devices operate.

QUESTIONS

1. Why is a system using a capillary tube thought of as a dry-type or direct-expansion system?

2. How would a system equipped with a thermostatic expansion valve differ from a system with a capillary tube when the heat load is increased?

3. How do the length and diameter of a capillary tube affect the operation of the refrigeration system?

4. What are the symptoms of a capillary tube metering device system with an overcharge of refrigerant?

5. What are the symptoms of a capillary tube metering device system with an undercharge of refrigerant?

6. What are the symptoms of a system operating with a restricted capillary tube?

7. What effect, if any, did the increased head pressure have on the subcooling?

Exercise RSC-15

Troubleshooting Exercise: Refrigeration System Operation

Name	Date	Grade

OBJECTIVES: Upon completion of this exercise, you should be able to evaluate a system to determine the cause for system malfunction.

INTRODUCTION: During this exercise, you will be working on a system that has had a specific fault placed in it by your instructor. You will locate the fault using a logical approach and test instruments. THE INTENT IS NOT FOR YOU TO MAKE THE REPAIR, BUT TO LOCATE THE PROBLEM AND ASK YOUR INSTRUCTOR FOR DIRECTION BEFORE COMPLETION.

TOOLS, MATERIALS, AND EQUIPMENT: A digital multimeter (DMM), ammeter, gauge manifold, thermometer, goggles, gloves, straight-slot and Phillips-head screwdrivers, electrical tape, short piece of insulation, service valve wrench, and ¼" and ⁵⁄₁₆" nut drivers.

⚠ **SAFETY PRECAUTIONS:** Wear goggles and gloves while transferring refrigerant. Always take care when working around electricity or live circuits. Be careful when working around a running compressor. Parts of the compressor, especially the discharge line, may be very hot.

STEP-BY-STEP PROCEDURES

1. Start the unit and allow it to run for about 15 minutes to establish stable conditions.

2. Touch-test the suction line leaving the evaporator. Describe how it feels compared to your hand temperature: _____

3. Touch-test the compressor housing. How does it feel compared to your hand temperature? _____

4. Secure a temperature probe on the compressor suction line 6" before the compressor, one on the compressor discharge line 6" from the compressor, and one on the system's liquid line 6" from the outlet of the condenser. Suction-line temperature: _____ °F. Compressor discharge-line temperature: _____ °F. Liquid line temperature: _____ °F. Also, record the temperature of the ambient air entering the condenser. _____ °F.

5. Describe the problem with the evidence you have up to now: _____

6. Put on your goggles and gloves. Connect gauges to the system and record the pressures. Suction pressure: _____ psig. Discharge pressure: _____ psig.

7. Calculate the following:

   • Superheat: _____ °F

   • Condenser split: _____ °F

   • Subcooling: _____ °F

8. Record what the pressures should be under normal conditions. Suction pressure: _____ psig. Discharge pressure: _____ psig.

9. What would you recommend as a repair for this system?

MAINTENANCE OF WORKSTATION AND TOOLS: Your instructor will direct you how to leave the unit. Return all tools to their proper places.

SUMMARY STATEMENT: Describe how the problem with this unit affected system performance. Use an additional sheet of paper if more space is necessary for your answer.

TAKEAWAY: When troubleshooting a live refrigeration system, technicians need to observe the unit and make accurate temperature and pressure measurements. From their observations and the information gathered from their test instruments, they can take an informed, logical approach to finding the fault.

QUESTIONS

1. Were the system temperatures observed for this exercise within range for the unit you were assigned? If not, were they high or low?

2. What would be the normal discharge-line temperature for this unit while operating normally? _____ °F

3. If the discharge line is overheating, what else is overheating?

4. What are the effects on the system when the compressor discharge line is excessively hot?

5. What may cause a discharge line to overheat?

6. What is the maximum discharge-line temperature recommended by most manufacturers? _____ °F

7. What cools the compressor in the unit used for this exercise?

8. What should the suction pressure be for this system when operating correctly?

9. What is the typical condenser split (CS) for the unit used in this exercise?

10. What problems may cause an evaporator to freeze?

Exercise RSC-16

Pumping Refrigerant into the Receiver

Name	Date	Grade

OBJECTIVES: Upon completion of this exercise, you should be able to pump the refrigerant into the receiver so that the low-pressure side of the system may be serviced.

INTRODUCTION: You will work with a refrigeration or air-conditioning system equipped with a receiver. You will pump the refrigerant into the receiver, storing all of the refrigerant on the high-pressure side of the system. This will allow the low-pressure side of the system to be opened for service.

TOOLS, MATERIALS, AND EQUIPMENT: A gauge manifold, goggles, gloves, an adjustable wrench, a service valve wrench, and a refrigeration system with a receiver and service valves.

⚠ **SAFETY PRECAUTIONS:** Wear goggles and gloves while connecting gauges and turning valve stems. DO NOT ALLOW THE COMPRESSOR TO OPERATE IN A VACUUM.

STEP-BY-STEP PROCEDURES

1. Put on your goggles and gloves. Connect the gauge manifold to the high- and low-pressure sides of the system at the liquid-line and suction-line service valves.

2. Slowly crack the service valves off the backseat until system pressure readings are observed on the gauges. Purge any air from the refrigerant hoses.

3. Start the system and observe the gauge readings.

4. When the system has stabilized, slowly frontseat the liquid-line (receiver) service valve, Figure RSC-16.1. Watch the low-pressure gauge, and do not let the low-side pressure go into a vacuum.

5. If the system has a low-pressure control, the control will shut the compressor off before the pressure reaches 0 psig. You may use a short piece of wire to jump out the low-pressure control to keep the compressor on. Shut the system down when the low-side pressure reaches a level close to 0 psig on your gauge manifold.

6. When the system has been pumped down to 0 psig and the compressor is shut off, the pressure on the low-pressure side of the system may increase above 0. The compressor will have to be started again to pump it back to 0 psig. You may have to pump the low side down as many as three times to completely remove all refrigerant from the low-pressure side of the system.

7. The system should now hold the 0 psig. This isolates the low side of the system for service and stores the refrigerant in the receiver and condenser. If the pressure keeps increasing and will not stay at 0 psig, refrigerant is leaking into the low side. Inspect the system and list some possible causes for this condition.

FIGURE **RSC-16.1.**

Photo by Jason Obrzut

8. Slowly open the liquid line service valve and allow refrigerant to flow back into the low side.

9. Once the system pressures have equalized, start the system, and verify that it is operating correctly.

10. Remove the gauge manifold from the system, and replace all valve stem covers and service port caps.

MAINTENANCE OF WORKSTATION AND TOOLS: Return all tools to their respective places, and clean up the workstation. Leave the system as you found it or as your instructor directs you.

SUMMARY STATEMENT: Describe what components may be serviced when the low side of the system is at 0 psig. (*Hint:* Look at the EPA rules.) Use an additional sheet of paper if more space is necessary for your answer.

TAKEAWAY: Refrigeration units that incorporate a receiver, and the appropriate valves, can have the system charge pumped into the high-pressure side of the refrigerant circuit for storage, making complete system recovery unnecessary. This is beneficial when the low-pressure side of the system requires a repair. Make sure that the compressor does pull down below 0 psig during the process, and that any air in the system is evacuated prior to reintroducing the charge.

QUESTIONS

1. What other substance circulates in the refrigeration system with the refrigerant?

2. What would happen if there were more refrigerant in the system than the condenser and receiver could hold?

3. If air enters the low side of the refrigeration system during servicing, how could it be removed?

4. What causes the pressure to rise after the system is pumped down, causing you to have to pump it down again?

5. Could you pump the refrigerant into the receiver if the compressor valves leaked?

6. If an expansion valve is changed using the pumpdown method, is it always necessary to evacuate the system before start-up?

7. What is the name of the valve on the receiver that is used to stop refrigerant flow for pump-down purposes?

8. Can the gauges be removed while the refrigerant is pumped into the condenser and receiver?

9. Is the crankcase of the compressor considered to be on the high- or low-pressure side of the system?

Exercise RSC-17

Changing a Thermostatic Expansion Valve

Name	Date	Grade

OBJECTIVES: Upon completion of this exercise, you should be able to change a thermostatic expansion valve with a minimum loss of refrigerant.

INTRODUCTION: You will change the TXV on a working unit. Because there may not be a unit with a defective valve, you will go through the exact changeout procedure by removing the valve and then replacing it. The system should have service valves and a TXV that is installed with flare fittings. You will be working on a system equipped with a receiver and performing a pump down instead of a recovery.

TOOLS, MATERIALS, AND EQUIPMENT: A gauge manifold, goggles, gloves, leak detector (electronic and bubble solution will do), a service valve wrench, two 8" adjustable wrenches, an assortment of flare nut wrenches, a vacuum pump, flat-slot and Phillips-head screwdrivers, and an operating refrigeration system equipped with a TXV and a receiver or condenser large enough to hold the system charge.

⚠ **SAFETY PRECAUTIONS:** Watch the high-pressure gauge while pumping the system down to be sure the condenser will hold the complete charge. You may need to jump out or lower the settings on the low-pressure control to pump the unit down. DO NOT JUMP OUT THE HIGH-PRESSURE CONTROL. As always, gloves and goggles should be worn when working with refrigerants.

STEP-BY-STEP PROCEDURES

1. Put on your goggles and gloves. Connect gauges to the system service valves and purge air from the hoses.

2. Front seat the liquid line service valve at the receiver or close the liquid-line service valve at the condensing unit.

3. Jump or adjust the settings on the low-pressure control if there is one.

4. Start the unit and allow it to run until the low-side pressure drops to 0 psig. Turn the unit off and allow it to stand for a few minutes. If the pressure rises, pump it down again. You will normally need to pump the system down to 0 psig three times to remove all refrigerant from the low side. Remember, you are pumping the filter drier and the liquid line down also.

5. Remove enough panels from the evaporator to comfortably work on the TXV.

6. Remove the TXV sensing bulb from the evaporator outlet. Take care not to damage or kink the capillary line connecting the bulb to the valve.

7. Using the adjustable wrenches or the flare nut wrenches, loosen the inlet and outlet connections to the TXV. Then disconnect the external equalizer line if the valve has one. Remove the valve from the system.

8. When the valve is completely clear of the evaporator compartment, you may then replace it as though the valve were a new one. THIS SHOULD BE DONE QUICKLY SO THE SYSTEM IS NOT OPEN ANY LONGER THAN ABSOLUTELY NECESSARY, OR THE DRIER SHOULD BE CHANGED. In a real-world situation, the filter-drier should be changed along with the TXV.

9. When the valve is replaced, pressurize the system and leak-check. Pressure-testing can be done with pure nitrogen or a mixture of nitrogen and a trace gas of the refrigerant normally used in the system. The trace gas of refrigerant is used for electronic leak detection purposes. The mixture of nitrogen and refrigerant trace gas should never be recovered with a recovery unit; it can be blown to the atmosphere when the service technician finishes leak checking. Any HCFC or HFC refrigerant can be used as a trace gas for leak checking. CFCs cannot be used as trace gases. It is best to use the refrigerant that is normally used in the system when leak checking with a trace gas, as long as it is not a CFC refrigerant. This ensures that the oil in the compressor's crankcase and in the system will be compatible with the refrigerant used as a trace gas.

10. Connect the vacuum pump and evacuate down to the appropriate micron level. A triple evacuation procedure is recommended.

11. Replace all service and access panels.

12. After evacuation, move the liquid line service valve toward the backseated position, but leaving it cracked off the back seat.

13. With both service valves cracked off the back seat, start the system and verify that the charge is correct.

14. Return the unit to the condition your instructor advises.

MAINTENANCE OF WORKSTATION AND TOOLS: Return all tools to their correct places.

SUMMARY STATEMENT: Describe the differences between a system recovery and a system pump down. List the pros and cons for each procedure. Use an additional sheet of paper if more space is necessary for your answer.

TAKEAWAY: Depending on the components installed on the refrigeration system, the system may be pumped down, storing the refrigerant on the high-pressure side of the system when making a repair. The TXV has a number of connections that must be made to the system including: refrigerant inlet, refrigerant outlet, external equalizer, and sensing bulb.

QUESTIONS

1. How was the sensing element fastened to the suction line on the unit you worked on?

2. Was the valve sensing bulb insulated?

3. What is the purpose of the insulation on the sensing bulb?

4. What type of charge was used in the bulb of the TXV used in this exercise?

5. Did this unit have a low-pressure control?

6. What are the functions of the low-pressure control?

7. What are the functions of the high-pressure control?

8. What is the purpose of the external equalizer line on a TXV?

9. What would be the symptom of a TXV that had lost its charge in the sensing element?

10. What are the benefits of a triple evacuation?

Exercise RSC-18

Adjusting the Crankcase Pressure Regulator (CPR) Valve

Name	Date	Grade

OBJECTIVES: Upon completion of this exercise, you should be able to adjust a CPR valve in a typical refrigeration system.

INTRODUCTION: You will use gauges and an ammeter to adjust a CPR valve properly to prevent compressor overloading on the start-up of a hot refrigerated box (sometimes called a hot pulldown). This will more than likely be a low-temperature system.

TOOLS, MATERIALS, AND EQUIPMENT: Two gauge manifolds, goggles, gloves, clamp-on ammeter, valve wrench, large straight-slot screwdriver, Allen wrench set, Phillips-head and medium-size straight-slot screwdrivers, and refrigeration system with CPR (crankcase pressure regulator).

⚠ **SAFETY PRECAUTIONS:** Wear goggles and gloves while working with refrigerant under pressure.

STEP-BY-STEP PROCEDURES

1. Put on your goggles and gloves. Connect the gauge manifold to the high- and low-side service ports to obtain the system's operating pressures.

2. Connect the other low-side gauge line to the connection on the CPR valve.

   Note: This connection is on the evaporator side of the valve.

3. Purge each hose to remove any air or contaminates from the hoses.

4. Make sure the refrigerated box is under a load from being off, or shut it off long enough for the load to build up. The box temperature must be at least 10°F above its design temperature.

5. Clamp the ammeter on the common wire to the compressor. Make sure that the wire is isolated from other system loads to avoid inaccurate current readings.

6. Record the compressor rated-load amperage (RLA) (from the nameplate) here: _____ A.

7. Start the system and record the following information:
   - Suction pressure at the evaporator (from the gauge attached to the CPR valve): _____ psig
   - Suction pressure at the compressor: _____ psig
   - Difference in suction pressure: _____ psig
   - Actual amperage of the compressor: _____ A

8. Remove the protective cover from the top of the CPR valve.

9. Using the correct screwdriver or Allen wrench, adjust the valve until the actual compressor amperage matches the recommended running load amperage.

10. If the valve is set correctly, turn the adjusting stem until the amperage is too high, and then adjust it to the correct reading, as in step 9.

11. Remove the gauges and replace all panels with their correct fasteners.

MAINTENANCE OF WORKSTATION AND TOOLS: Return all tools to their respective places. Leave the system in the condition you found it.

SUMMARY STATEMENT: Describe the action of the CPR valve and why this type of valve is necessary for a low-temperature system when it is energized with a high heat load. Use an additional sheet of paper if more space is necessary for your answer.

TAKEAWAY: Refrigeration systems can be equipped with a number of different controls or valves. A CPR is one such valve, typically used on low-temperature systems. These valves and controls must be adjusted for the systems that they are installed on.

QUESTIONS

1. The CPR valve acts as a restriction in the suction line until the evaporator is operating within its correct temperature range. True or false?

2. Where is a CPR valve installed in the refrigerant circuit?

3. On what types of systems are CPR valves typically found?

4. Does the oil circulating through the system pass through the CPR valve?

5. Is a CPR valve an electromechanical control?

6. Which of the following is used to correctly adjust a CPR valve: amperage or pressure?

7. What would be the symptoms if the CPR valve were to close and not open?

8. In what direction did the valve adjustment need to be turned to increase the amperage draw of the compressor?

9. Explain the relationship between the spring pressure and the pressure at the inlet of the compressor.

Application and Evaluation of a Refrigeration System

Name	Date	Grade

OBJECTIVES: Upon completion of this exercise, you should be able to determine the application and the approximate temperature range of a refrigeration system by its features.

INTRODUCTION: You will be assigned a refrigeration system by your instructor. You will record information about the components of the system, from which you will be able to determine its application. You will look for evaporator location, fin spacing, type of defrost, and style of box.

TOOLS, MATERIALS, AND EQUIPMENT: Phillips-head and straight-slot screwdrivers, ¼" and ⁵⁄₁₆" nut drivers, a flashlight, an adjustable wrench, and a refrigeration system.

⚠ **SAFETY PRECAUTIONS:** Care must be taken to avoid injury to your hands when disassembling any piece of equipment. Safety goggles and gloves are recommended.

STEP-BY-STEP PROCEDURES

1. Make sure the power to your refrigeration box is off and the panel is locked and tagged. Fill in the following information:

   - Type of box (open, chest, or upright): _____

   - If closed, are doors single, double, triple pan, or metal? _____

   - Package or split system? _____

2. Fill in the following information from the unit nameplate:

   - Manufacturer: _____

   - Model number: _____

   - Serial number: _____

   - Type of refrigerant: _____

   - Quantity of refrigerant (if known): _____

3. From the compressor nameplate:

   - Manufacturer: _____

   - Model number: _____

   - Rated-load amps (RLA) (if available): _____

   - Locked-rotor amperes (LRA): _____

4. From the evaporator examination:

- Evaporator location: _____
- Fin spacing: _____
- Forced or induced draft: _____
- Number of fans: _____

5. From the condenser examination:

- Condenser location: _____
- Natural or mechanical draft: _____
- Number of fans: _____
- Type of head pressure control (if any): _____

6. Control data:

- Type of temperature control: _____
- Type of defrost (off-cycle, other): _____
- Does unit have a defrost timer? _____
- Defrost time interval(s): _____

MAINTENANCE OF WORKSTATION AND TOOLS: Replace all panels with the correct fasteners.

SUMMARY STATEMENT: From information gathered in this exercise, state the application of the refrigeration system you worked on. What are typical uses for this unit in the field? Why? Use an additional sheet of paper if more space is necessary for your answer.

TAKEAWAY: By making specific observations of the components of a unit, a technician should be able to determine the temperature application of a system as well as its function in the field.

QUESTIONS

1. How is ice melted with off-cycle defrost?

2. What types of systems would use an off-cycle defrost?

3. What additional components are necessary when an electric defrost is used?

4. What are the differences between electric and hot gas defrost cycles?

5. What is the function of the defrost termination/fan delay (DTFD) switch?

6. What would the symptoms be for a medium-temperature cooler that did not get enough off-cycle time for correct defrost?

7. What keeps frost and sweat from forming on the doors and panels of refrigerated boxes?

8. How is condensate dealt with in package-type refrigeration systems?

SECTION 4

Safety and Tools (SFT)

SUBJECT AREA OVERVIEW

The HVACR technician is exposed to many hazards on a daily basis that can cause severe personal injury or death. The single best thing that an individual can do to prevent becoming injured on the job is to be properly trained to handle and work with, on, and around electricity, chemicals, and rotating machinery in many different environments. However, even if trained, accidents and injuries do occur. HVACR technicians are exposed to extreme temperatures and pressures, as well as a myriad of chemicals and other harmful substances. Proper personal protection from these conditions, as well as others, must be worn. It is also important that technicians be prepared to react in the event an accident or injury occurs.

Unfortunately, the number one cause for accidents on the job is carelessness. It is important to always be alert and coherent at all times. Never work if under the influence of alcohol or drugs, whether legal or not; that can cause you to become impaired. It is important to be aware of your surroundings on the job and attempt to identify any potentially dangerous situations before they cause injury. Many injuries occur from using improper lifting techniques. Be sure to lift with your legs, not your back, and never lift more than you can do so safely and comfortably.

Other injuries in the HVACR industry result from improper use of tools and fastening devices. Technicians use many hand tools common to most trades, and it is important to know how to use them correctly. There are also many tools that are designed specifically for this field, and they must be used properly. It is important to never exceed the working limits of any tool. Some pieces of test instrumentation that are used must be calibrated periodically to ensure that they perform as intended.

HVACR technicians should also be aware of different types of fasteners and fastening systems so that the fastener most appropriate for the job is used to securely install all equipment and materials. In addition to knowing about the various types of fasteners, it is also important to know their limitations to ensure that they are strong enough for a particular application.

KEY TERMS

Air-acetylene torch
Brazing
Calibration
Compound gauge
Electrical test equipment
Frostbite
Gauge manifold
High-pressure gauge
Hollow wall anchor
Micron gauge
Oxy-acetylene torch
Pin rivet
Pressure test instruments
Recovery equipment
Refrigeration service wrench
Soldering
Solderless connectors
Temperature-sensing instruments
Toggle bolts
Vacuum pump
Wire nuts

REVIEW TEST

Name	Date	Grade

Circle the letter that indicates the correct answer.

1. An example of a pressure vessel is:
   A. a small ship containing refrigerant cylinders.
   B. the compressor in a refrigerant system.
   C. a cylinder containing refrigerant for a refrigeration system.
   D. a portable dolly.

2. When transferring refrigerant from a cylinder to a refrigeration system, you should always wear:
   A. gloves and goggles.
   B. heavy protective clothing.
   C. rubber gloves.
   D. industrial-type shoes and insulated gloves.

3. Before using a voltmeter, always:
   A. zero the ohms indicator.
   B. touch the two leads to ensure continuity.
   C. set the range selector to a setting higher than the voltage you will be reading.
   D. turn off the power.

4. Properly grounded power tools should be used:
   A. so that if the motor or other electrical part of the tool becomes shorted to the frame, the electric current will follow the grounding circuit.
   B. so that they will not create sparks at the plug.
   C. so that the technician will not be burned from the tool overheating.
   D. so that the technician will not have to use a nonconducting ladder.

5. Nitrogen must be used with a regulator:
   A. to reduce the very high pressure in the cylinder to a safer pressure.
   B. to help keep the nitrogen from exploding.
   C. because it would create a fire hazard otherwise.
   D. because there is not always a protective cap available.

6. What two types of injury are caused by electrical energy?
   A. Shock and burns
   B. Nausea and choking
   C. Possible loss of limb and hearing
   D. None of the above

7. A crimping tool is used to:
   A. cut and strip wire.
   B. crimp tubing.
   C. staple low-voltage wire.
   D. fasten insulation to duct.

8. The gauge manifold includes:
   A. a compound gauge, high-pressure gauge, manifold, valves, and hoses.
   B. two compound gauges, hoses, and valves.
   C. two high-pressure gauges, a manifold, valves, and hoses.
   D. two compound gauges, two low-pressure gauges, valves, and hoses.

9. The vacuum pump is used to:
   A. vacuum dust from the evaporator coil.
   B. evacuate nitrogen and carbon dioxide from the area around the refrigeration system.
   C. remove air and non-condensable gases.
   D. vacuum dust from the condenser coil.

10. Thermocouple thermometers are classified as a(an):
    A. dial-type instrument.
    B. glass-stem-type instrument.
    C. electronic-type instrument.
    D. electromechanical-type instrument.

11. The pin rivet is used to join:
    A. a piece of sheet metal to concrete.
    B. two pieces of sheet metal together.
    C. a wood member to concrete.
    D. two pieces of wood together.

12. Wood screws can have:
    A. flat heads.
    B. round heads.
    C. oval heads.
    D. round, flat, or oval heads.

13. Anchor shields, wall anchors, and toggle bolts may all be used in:
    A. wood.
    B. masonry walls.
    C. hollow walls.
    D. solid concrete walls.

14. Pin rivets are installed using a:
    A. drill screw.
    B. pin rivet gun.
    C. pin rivet drill.
    D. hammer.

15. Threaded rod can be connected to steel members by using:
    A. anchor shields.
    B. wall anchors.
    C. beam clamps.
    D. toggle bolts.

16. Which of the following can be used with threaded rod to support an air handler?
    A. Steel angle and/or channel
    B. Pressure-treated lumber.
    C. Plastic sheeting.
    D. Galvanized sheet metal.

17. When a stranded wire needs to be connected under a screw terminal, which of the following can be used?
    A. Female quick connector
    B. Spade tongue connector
    C. Butt connector
    D. Male quick connector

18. Three commonly used reference points for checking temperature-measuring instruments are:
    A. 0°F, 100°F, 110°F.
    B. −460°F, 0°F, 32°F.
    C. 32°F, 98.6°F, 212°F.
    D. 32°F, 100°F, 250°F.

19. To check the temperature leads of an electronic thermometer for low temperatures, they should be stirred in a container of:
    A. crushed ice and water with half a teaspoon of salt.
    B. crushed ice and water with a pinch of sugar.
    C. crushed ice and pure water.
    D. cubed ice and water.

20. To check the temperature leads of an electronic thermometer for high temperatures, they should be:
    A. placed on the bottom of a pan of boiling water.
    B. suspended in a pan of boiling water.
    C. suspended in the vapor just above a pan of boiling water.
    D. suspended in a pan of boiling water when at least 1000 ft above sea level.

21. Of the following, the instrument that is used for leak detection purposes is the:
    A. micron gauge.
    B. halide torch.
    C. U-tube manometer.
    D. nitrogen flow indicator.

22. The most accurate thermometer is the:
    A. glass stem thermometer.
    B. electronic thermometer.
    C. dial-type thermometer.
    D. pocket thermometer.

23. Which of the following gauges is considered to be the most accurate gauge for checking very low vacuum levels when evacuating a system?
    A. U-tube manometer
    B. Compound manifold gauge
    C. Micron gauge
    D. Nitrogen regulator

24. Which of the following will find the smallest leak?
    A. Soap bubbles
    B. Electronic leak detector
    C. Halide torch
    D. Compound manifold gauge

Exercise SFT-1

Safety in the Laboratory

Name	Date	Grade

OBJECTIVES: Upon completion of this exercise, you should be able to identify potential safety hazards in the laboratory.

INTRODUCTION: The primary goal of all HVACR service technicians is to remain safe on the job. One important way to help avoid personal injury is to be aware of any unsafe conditions that have the potential to cause injury. In this exercise, you will be walking into a laboratory setting where you will identify as many unsafe conditions as possible.

TOOLS, MATERIALS, AND EQUIPMENT: The tools, materials, and equipment used in the exercise are not listed here, as it is your job to identify unsafe conditions in the laboratory.

⚠ **SAFETY PRECAUTIONS:** While walking through the laboratory as part of this exercise, you are to simply evaluate your surroundings. It is important that you do not touch or otherwise interact with anything that you observe.

STEP-BY-STEP PROCEDURES

1. Make certain that you are properly dressed and prepared to enter the laboratory.

2. Once your instructor tells you to do so, enter the laboratory.

3. Without touching anything or talking to your classmates, evaluate your surroundings, and identify any unsafe conditions or safety violations that are present in the laboratory.

4. Create a list of the safety issues here, and explain why each is a concern and what can be done to remedy the situation.

SAFETY CONCERN #1 _____

_____

SAFETY CONCERN #2 _____

_____

SAFETY CONCERN #3 _____

_____

SAFETY CONCERN #4 _____

_____

SAFETY CONCERN #5 _____

_____

SAFETY CONCERN #6 _____

SAFETY CONCERN #7 _____

_____

SAFETY CONCERN #8 _____

_____

SAFETY CONCERN #9 _____

_____

SAFETY CONCERN #10 _____

_____

MAINTENANCE OF WORKSTATION AND TOOLS: At the conclusion of the exercise, your instructor will give you instructions on how to properly clean up and remedy the safety concerns that were staged as part of this learning experience.

SUMMARY STATEMENT: Explain the importance of being aware of your surroundings on the job. Based on your observations in the exercise, provide some examples of how being careless can result in injury.

TAKEAWAY: An important way to help avoid personal injury is to be aware of any unsafe conditions that have the potential to cause injury. By identifying these unsafe conditions early, actions can be taken to greatly reduce the possibility of injury.

QUESTIONS

1. Why is wearing personal protection equipment (PPE) an important first step in staying safe on the job?

2. Explain the importance of being alert at all times while in the laboratory setting or out in the field.

3. Explain how drug and/or alcohol use can increase the chance of getting injured on the job.

4. Explain the purpose of and process of performing cardio pulmonary resuscitation (CPR).

5. Explain how to properly lift a heavy object.

6. Explain how to properly handle, transport, and store pressurized vessels or tanks.

7. Describe the dangers associated with working on live electrical circuits.

8. Describe the precautions that should be taken when working on or around rotating machinery.

Exercise SFT-2

Hand Tool Identification

Name	Date	Grade

OBJECTIVES: Upon completion of this exercise, you should be able to identify basic hand tools based on their intended function and description.

INTRODUCTION: HVACR technicians work with many different hand tools on a daily basis. It is important that technicians use tools only for the function they are intended to perform. Exceeding the safe operating range of a tool or using a tool for a purpose that it was not intended to be used for can result in damage to the tool and severe personal injury. In this exercise, you will identify various tools based on the description of the tasks they are intended to perform.

TOOLS, MATERIALS, AND EQUIPMENT: Tin snips, Allen keys, tap, hacksaw, crimping tool, awl, file, level, die, keyhole saw, wire stripper, Torx driver, nut driver, socket set, utility knife.

⚠ **SAFETY PRECAUTIONS:** Always use a tool only for the task it was intended. Although you will not be using hand tools in this exercise, your instructor will at some point likely demonstrate the use of these and other tools. Be sure to pay close attention to these demonstrations, as they are intended to help keep you safe in the laboratory and out in the field.

STEP-BY-STEP PROCEDURES

1. Read through the list of tools in the left-hand column.

2. Study the lettered tool descriptions in the right-hand column.

3. In the spaces provided to the right of the tools, enter the letter that corresponds to the best description of what the specific tool is intended to do. Note: Each letter will be used only once.

Tin snips	_____	(A) Used to cut small holes in sheetrock
Allen keys	_____	(B) Used to remove insulation from conductors
Tap	_____	(C) Tightens/loosens screws with six-sided heads
Hacksaw	_____	(D) Tightens/loosens screws with star-shaped heads
Crimping tool	_____	(E) Ensures horizontal installation
Awl	_____	(F) Tightens/loosens screws with recessed six-sided heads
File	_____	(G) Used to cut thin sheets of metal
Level	_____	(H) Used to create female pipe threads
Die	_____	(I) Used to cut metal stock
Keyhole saw	_____	(J) Used to create male pipe threads
Wire stripper	_____	(K) Used to scratch/score lines in metal or wood
Torx driver	_____	(L) Ratcheting tool to tighten/loosen hex-headed bolts

Nut driver	_____	(M) Used to cut various non-metallic objects
Socket set	_____	(N) Used to remove sharp edges from metallic surfaces
Utility knife	_____	(O) Used to secure connectors onto sections of wire

MAINTENANCE OF WORKSTATION AND TOOLS: With the exception of the tools that are used for demonstration purposes, there are no maintenance issues associated with this exercise. At your instructor's request, make certain that all tools are properly stored and that the work area has been properly cleaned.

SUMMARY STATEMENT: Explain the importance of knowing what tools are to be used for various tasks and the potential results if tools are used improperly.

TAKEAWAY: An important way to help avoid personal injury is to use tools only for their intended purpose. In order to work safely, effectively, and efficiently, it is important to know what a tool is used for and, just as important, how to use it properly.

QUESTIONS

1. Explain the difference between a screwdriver and a nut driver.

2. Explain the importance of using sharp blades on saws as opposed to using older, dull blades.

3. Explain why screwdrivers should never be used as pry bars or chisels.

4. Explain the difference between a pair of slip-joint pliers and locking pliers.

5. Explain the difference between an open-end wrench and a box wrench.

6. Explain the importance of keeping tool handles free from grease.

7. Explain the importance of using tools with insulated handles, especially when working on or around electric circuits.

8. Explain the importance of inspecting tools prior to use and not using tools that are damaged in any way.

9. Explain why tools with loose handles should not be used.

HVACR Tool Identification

Name	Date	Grade

OBJECTIVES: Upon completion of this exercise, you should be able to identify some specialty tools used in the HVACR industry based on their intended function and description.

INTRODUCTION: HVACR technicians work with many different hand tools on a daily basis. It is important that technicians use tools only for the function they are intended to perform. Exceeding the safe operating range of a tool or using a tool for a purpose that it was not intended to be used can result in damage to the tool as well as cause severe personal injury. In this exercise, you will identify various specialty tools based on the description of the tasks they are intended to perform.

TOOLS, MATERIALS, AND EQUIPMENT: Electronic leak detector, tubing cutter, swaging tool, flaring tool, reamer, chain vise, gauge manifold, micron gauge, recovery unit, vacuum pump, service wrench, striker, digital multimeter (DMM), blower door test, duct pressurization test, thermal imager, low-loss fittings, clamp-on ammeter, flow hood, air-acetylene torch.

 SAFETY PRECAUTIONS: Always use a tool only for the task it was intended. Although you will not be using tools in this exercise, your instructor will at some point likely demonstrate the use of these and other tools. Be sure to pay close attention to these demonstrations, as they are intended to help keep you safe in the laboratory and out in the field.

STEP-BY-STEP PROCEDURES

1. Read through the list of tools in the left-hand column.

2. Study the lettered tool descriptions in the right-hand column.

3. In the spaces provided to the right of the tools, enter the letter that corresponds to the best description of what the specific tool is intended to do. Note: Each letter will be used only once.

Electronic leak detector _____	(A) Used to remove burrs from the end of tubing
Tubing cutter _____	(B) Used to obtain system operating pressures
Swaging tool _____	(C) Used to dehydrate and degas a system
Flaring tool _____	(D) Used to light torches
Reamer _____	(E) Used to measure circuit current
Chain vise _____	(F) Used to measure deep vacuum levels
Gauge manifold _____	(G) Used to measure house leakage
Micron gauge _____	(H) Used to turn valve stems
Recovery unit _____	(I) Used to expand the end of soft-drawn piping

Vacuum pump	_____	(J) Used to measure air distribution system leakage
Striker	_____	(K) Used to hold sections of piping material in place
Digital multimeter (DMM)	_____	(L) Used to take heat-signature photos
Blower door test	_____	(M) Used to measure voltage and resistance
Duct pressure test	_____	(N) Used to cut sections of soft-drawn piping
Thermal imager	_____	(O) Used to minimize refrigerant loss
Low-loss fittings	_____	(P) Used to locate the point of refrigerant loss
Clamp-on ammeter	_____	(Q) Used to measure supply/return air volumes
Flow hood	_____	(R) Used to melt filler materials
Air-acetylene torch	_____	(S) Used to create a bell shape on the end of tubing
Service wrench	_____	(T) Used to remove refrigerant from a system

MAINTENANCE OF WORKSTATION AND TOOLS: With the exception of the tools that are used for demonstration purposes, there are no maintenance issues associated with this exercise. At your instructor's request, make certain that all tools are properly stored and that the work area has been properly cleaned.

SUMMARY STATEMENT: Explain the importance of knowing what tasks tools are to be used for and the potential results if tools are used improperly.

TAKEAWAY: An important way to help avoid personal injury is to use tools only for their intended purpose. In order to work safely, effectively, and efficiently, it is important to know what a tool is used for and, just as important, how to use it properly.

QUESTIONS

1. Explain the difference between an air-acetylene torch and an oxy-acetylene torch.

2. Explain the importance of using a striker as opposed to other possible methods to light a torch.

3. Describe the color-coding used on the gauge manifold.

4. Describe a compound gauge.

5. Describe the various methods of detecting leaks in an air-conditioning system.

6. Explain the process and importance of refrigerant recovery.

7. Explain the process and importance of system evacuation.

8. Explain the importance of proper tool maintenance.

Exercise SFT-4

Fastener Identification

Name	Date	Grade

OBJECTIVES: Upon completion of this exercise, you should be able to identify various types of fasteners used in the HVACR industry.

INTRODUCTION: In this matching exercise, you will identify the various types of fasteners.

TOOLS, MATERIALS, AND EQUIPMENT: None.

SAFETY PRECAUTIONS: There are no safety precautions for this exercise.

STEP-BY-STEP PROCEDURES

Under each figure, you are to enter the letter that corresponds to the correct name for each fastening device.

A. Hollow wall anchor
B. Toggle bolt
C. Finishing nail
D. Hairpin
E. Pin rivet
F. Roofing nail

G. Self-drilling drywall anchor
H. Anchor shield
I. Masonry nail
J. Cotter pin
K. Lag shield anchor
L. Lag bolt

FIGURE **SFT-4.1**

FIGURE **SFT-4.2**

FIGURE **SFT-4.3**

FIGURE **SFT-4.4**

Courtesy Duro Dyne Corp.

_____

_____

FIGURE **SFT-4.5**

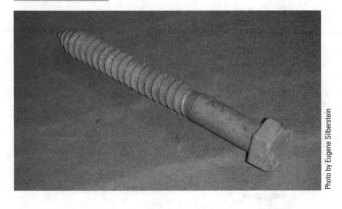

Photo by Eugene Silberstein

FIGURE **SFT-4.6**

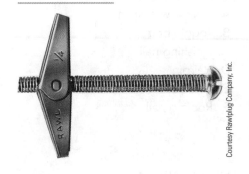

Courtesy Rawlplug Company, Inc.

_____

_____

FIGURE **SFT-4.7**

Courtesy Rawlplug Company, Inc.

FIGURE **SFT-4.8**

Courtesy Rawlplug Company, Inc.

_____

_____

FIGURE **SFT-4.9**

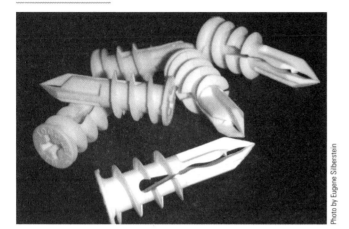

Photo by Eugene Silberstein

FIGURE **SFT-4.10**

FIGURE **SFT-4.11**

Photo by Eugene Silberstein

FIGURE **SFT-4.12**

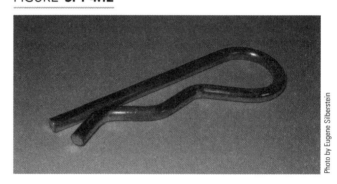

Photo by Eugene Silberstein

MAINTENANCE OF WORKSTATION AND TOOLS: There are no requirements for workstation maintenance associated with this exercise.

SUMMARY STATEMENT: Explain why it is important to use the proper fastener for a particular job. In your answer, be sure to explain what can possibly happen if the wrong fastener is used.

TAKEAWAY: HVACR technicians must be aware of the different types of fasteners and fastening systems that are available. In addition to knowing about the various types of fasteners, it is also important to know their limitations to ensure that they are strong enough for the particular application.

QUESTIONS

1. Of the following, which fastener is intended for use on hollow walls?
   A. Finishing nail
   B. Toggle bolt
   C. Pin rivet
   D. Anchor shield

2. Of the following, which fastener is intended to join two pieces of sheet metal?
   A. Pin rivet
   B. Lag shield anchor
   C. Cotter pin
   D. Hairpin

3. Which of the following fasteners is the only one that does not utilize a screw-type mechanism?
   A. Toggle bolt
   B. Lag shield anchor
   C. Hollow wall anchor
   D. Pin rivet

4. Which of the following devices is commonly used to secure one end of a threaded rod to a structure?
   A. Toggle bolt
   B. Beam clamp
   C. Cotter pin
   D. Pin rivet

5. Which of the following methods is commonly used to fabricate custom hanging arrangements for air handlers and similar pieces of equipment?
   A. Threaded rod and angle steel
   B. Threaded rod and toggle bolts
   C. Steel channel and pipe clamps
   D. Toggle bolts and pipe clamps

Machine and Set Screw Identification

Name	Date	Grade

OBJECTIVES: Upon completion of this exercise, you should be able to identify various types of heads found on machine screws and various types of set screws.

INTRODUCTION: Based on appearance, you will identify the various types of machine screw heads and set screws.

TOOLS, MATERIALS, AND EQUIPMENT: Various types of machine screws with a variety of heads that include flat, round, fillister, oval, pan, and hexagon. Various types of set screws with a variety of points that include flat, cone, oval, cup, dog, and half dog. If these items are not available, utilize the following figures.

⚠ **SAFETY PRECAUTIONS:** There are no safety precautions for this exercise.

STEP-BY-STEP PROCEDURES

1. Under each machine screw in Figure SFT-5.1, enter the letter that corresponds to the correct name for each head.
   A. Round head
   B. Hexagon head
   C. Fillister head
   D. Oval head
   E. Pan head
   F. Flat head

2. Under each set screw in Figure SFT-5.2, enter the letter that corresponds to the correct name for each point.
   A. Half dog point
   B. Cone point
   C. Cup point
   D. Dog point
   E. Flat point
   F. Oval point

FIGURE **SFT-5.1**

MACHINE SCREWS

FIGURE **SFT-5.2**

SET SCREWS

Set screws styles
(Head and headless)

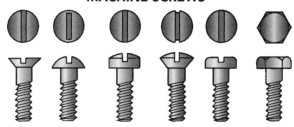

MAINTENANCE OF WORKSTATION AND TOOLS: If used, place all screws back in their proper storage place, and make certain that the area is left clean.

SUMMARY STATEMENT: Why do you think that there are different types of machine screw heads? Provide as many reasons as you can and, if appropriate, provide specific examples.

TAKEAWAY: HVACR technicians must be aware of the different types of screws that are available. Different screws have different applications, so it is important to use the right screw for the right job.

QUESTIONS

1. Which of the following is true regarding a ⁵⁄₁₆-18 UNC-2 machine screw?
   A. The machine screw is 5" long.
   B. The outside diameter of this machine screw is 2 inches.
   C. There are 18 threads per inch on this machine screw.
   D. The "class of fit" for this machine screw is 16.

2. Set screws are typically used to:
   A. join two sections of sheet metal together.
   B. hold pulleys securely on motor shafts.
   C. secure a pipe section to a wall.
   D. provide a self-tapping action through a piece of metal.

3. Cone and oval points are most likely found on set screws with which types of heads?
   A. No heads at all
   B. Only square heads
   C. Only hexagonal heads
   D. Both square and hexagonal heads

Calibrating Thermometers

Name	Date	Grade

OBJECTIVES: Upon completion of this exercise, you should be able to check the accuracy of some typical thermometers and calibrate them if they are designed to be calibrated.

INTRODUCTION: You will use a variety of thermometers and check their calibration using different temperature standards.

TOOLS, MATERIALS, AND EQUIPMENT: One each of the following if available to you in your lab: a dial-type thermometer with immersion stem, a glass stem thermometer, an electronic thermometer, a room thermostat cover with a thermometer in it, a heat source, ice, and water.

SAFETY PRECAUTIONS: You will be working with hot water (212°F) and a heat source. Do not spill the water on yourself or get burned by the heat source. Be sure to wear gloves and proper eye protection.

STEP-BY-STEP PROCEDURES

For Immersion-Type Thermometers

1. From several different thermometers, list the type and the range (high and low reading). Example: glass stem, mercury-type thermometer with a high of 250°F and a low of −20°F:
   A. Type: _____, High mark: _____ °F, Low mark: _____ °F
   B. Type: _____, High mark: _____ °F, Low mark: _____ °F
   C. Type: _____, High mark: _____ °F, Low mark: _____ °F
   D. Type: _____, High mark: _____ °F, Low mark: _____ °F

   Notice that some of these thermometers may indicate as low as 32°F and may be checked in ice and water. Some of them may indicate as high as 212°F and may be checked in boiling water. Later we will deal with thermometers that may not be immersed in water, such as the one on a thermostat cover. All thermometers should be checked at two reference points and somewhere close to the range in which they will be used.

2. In a mixture of ice and water, submerge all of the thermometers that are submersible and that indicate down to freezing, and record their actual readings below. Make sure that both the ice and water reach to the bottom of the container, stir the mixture, and allow time for the thermometers to indicate.
   A. Type: _____, Reading: _____ °F
   B. Type: _____, Reading: _____ °F
   C. Type: _____, Reading: _____ °F
   D. Type: _____, Reading: _____ °F

3. Use a container of rapidly boiling water for the thermometers that are submersible and that indicate as high as 212°F and submerge them in the boiling water. DO NOT LET THEM TOUCH THE BOTTOM OF THE PAN. Stir them in the water, and allow time for them to indicate. Record the readings below:

   A. Type: _____, Reading: _____ °F

   B. Type: _____, Reading: _____ °F

   C. Type: _____, Reading: _____ °F

   D. Type: _____, Reading: _____ °F

4. If the range of the thermometer does not reach a low of 32°F or a high of 212°F, record its high and low indications. For example, if it does not go to 212°F but only to 120°F, record 120°F.

   A. Low reading: _____ °F, High reading: _____ °F

   B. Low reading: _____ °F, High reading: _____ °F

   C. Low reading: _____ °F, High reading: _____ °F

   D. Low reading: _____ °F, High reading: _____ °F

5. Examine each of the thermometers checked in steps 2 and 3 for its method of calibration. Submerge each again, and change the calibration to the correct reading of 32°F or 212°F. AFTER A THERMOMETER HAS BEEN CALIBRATED, CHECK IT AGAIN. Body heat may have caused an error, and the calibration may have to be changed again. If a thermometer cannot be calibrated, a note may be placed with it to indicate its error, for example, "This thermometer reads 2°F high."

STEP-BY-STEP PROCEDURES

For Dial-Type Thermometers

1. Place the thermostat cover in a moving airstream at room temperature, such as in a return-air grille or a fan inlet.

2. Place one of the calibrated thermometers next to the room thermostat cover. The moving airstream will cause both temperature indicators to reach the same temperature quickly.

3. Change the calibration of the room thermostat cover to read the same as the calibrated thermometer.

4. Allow the thermometers to remain in the airstream for another short period of time because body heat may have caused an error.

5. Recalibrate if necessary.

MAINTENANCE OF WORKSTATION AND TOOLS: Return all tools and equipment to their respective places, and clean the work area.

SUMMARY STATEMENT: Describe the purpose of three different temperature-measuring instruments, and include how their sensing elements work.

TAKEAWAY: HVACR technicians rely on the readings they obtain from equipment to properly evaluate and troubleshoot the system. Properly calibrated instruments help ensure that the conclusions reached by the technician are accurate.

QUESTIONS

1. Why should the thermometer element not touch the bottom of the pan when checking it in boiling water?

2. Why should the thermometer be stirred when checking it in ice and water?

3. What can be done if a thermometer is not correct and cannot be calibrated?

4. What is the purpose of calibrating a thermometer?

5. Name three handy reference temperatures that may be used to check a thermometer.

6. Why were the thermostat cover and the temperature lead placed in a moving airstream in step 1 of the procedures for dial-type thermometers?

7. What tool was used to adjust the calibration in the thermostat cover?

8. Why is it important to keep the calibrated thermometer in the cover of the thermostat when calibrating it?

9. Why should a thermometer be rechecked after calibration?

Exercise SFT-7

Checking the Accuracy or Calibration of Electrical Instruments

Name	Date	Grade

OBJECTIVES: Upon completion of this exercise, you should be able to check the accuracy of and calibrate an ohmmeter, an ammeter, and a voltmeter.

INTRODUCTION: You will use field and bench techniques to verify the accuracy of and calibrate an ohmmeter, a voltmeter, and an ammeter so that you will have confidence when using them. Reference points that should be readily available to you will be used.

TOOLS, MATERIALS, AND EQUIPMENT: A calculator; a resistance heater (an old duct heater will do); several resistors of known values (gold band electronic resistors); an ammeter; a digital multimeter (DMM); 24 V, 115 V, and 230 V power sources; and various live components in a system available in your lab.

⚠ **SAFETY PRECAUTIONS:** You will be working with live electrical circuits. Caution must be used while taking readings. Do not touch the system while the power is on. Working around live electrical circuits will probably be the most hazardous part of your job as an HVACR technician. You should perform the following procedures only after you have had proper instruction and only under the close personal supervision of your instructor. You must follow the procedures as they are written here, as well as the procedures given by your instructor.

STEP-BY-STEP PROCEDURES

Ohmmeter Test

1. Set the DMM to read resistance. Touch the leads of the meter together to ensure that the leads are good and the reading is very close to 0 ohms.

2. Using several different resistors of known values, record the resistances:

   Meter reading: _____ ohms   Resistor value: _____ ohms
                                              Difference in reading: _____ ohms
                                              Percentage difference: _____ %

   Meter reading: _____ ohms   Resistor value: _____ ohms
                                              Difference in reading: _____ ohms
                                              Percentage difference: _____ %

   Meter reading: _____ ohms   Resistor value: _____ ohms
                                              Difference in reading: _____ ohms
                                              Percentage difference: _____ %

   Meter reading: _____ ohms   Resistor value: _____ ohms
                                              Difference in reading: _____ ohms
                                              Percentage difference: _____ %

Meter reading: _____ ohms   Resistor value: _____ ohms
                                      Difference in reading: _____ ohms
                                      Percentage difference: _____ %

Voltmeter Test

In this test, you will use the best voltmeter you have and compare it to others. Your instructor will make provisions for taking voltage readings across typical low-voltage and high-voltage components. You must perform these tests in the following manner and follow any additional safety precautions from your instructor.

Use only meter leads with insulated alligator clips.

You and your instructor should ensure that all power to the unit you are working on is turned off.

Fasten alligator clips to terminals indicated by your instructor.

Make sure that all range selections have been set properly on your meter.

When your instructor has approved all connections, turn the power on and record the measurement.

1. Take a voltage reading of both a high- and low-voltage source using your best meter (we will call it a STANDARD). Your instructor will show you the appropriate terminals.
   Record the voltages here:

   Low-voltage reading: _____ V   High-voltage reading: _____ V

2. Use your other meters such as the digital multimeter (DMM) and the ammeter's voltmeter feature, and compare their readings to the standard meter. Record the following:

   Standard: Low V reading: _____ V   High V reading: _____ V

   DMM: Low V reading: _____ V   High V reading: _____ V

   Ammeter: Low V reading: _____ V   High V reading: _____ V

   Record any differences in the readings here:

   DMM reads _____ V high, or _____ V low.

   Ammeter reads _____ V high, or _____ V low.

Ammeter Test

(Follow all procedures indicated in the voltmeter test.)

1. WITH THE POWER OFF, connect the electric resistance heater to the power supply.

2. Fasten the leads of the most accurate voltmeter to the power supply directly at the heater.

3. Take an ohm reading WITH THE POWER OFF: _____ ohms

4. Clamp the ammeter around one of the conductors leading to the heater.

5. Change the selector switch to read amps. HAVE YOUR INSTRUCTOR CHECK THE CONNECTIONS, and then turn the power on long enough to record the following. (The reason for not leaving the power on for a long time is that the resistance of the heater will change as it heats, and this will change the ampere reading.)

   Volt reading _____ V,   Ampere reading _____ A

   Volt reading _____ V ÷ Ohm reading _____ Ω = Ampere reading _____ A

6. Use a bench meter, if possible, and compare to the clamp-on ammeter.

Bench meter reading: _____ A   Clamp-on ammeter reading: _____ A

Difference between the two: _____ A

7. Create a double conductor loop through the jaws of the ammeter, as shown in Figure SFT-7.1, and comment on the relationship between the reading obtained on the ammeter with that on the bench ammeter.

_____

_____

8. Create a triple conductor loop through the jaws of the ammeter, and comment on the relationship between the reading obtained on the ammeter with that on the bench ammeter.

_____

_____

FIGURE **SFT-7.1**

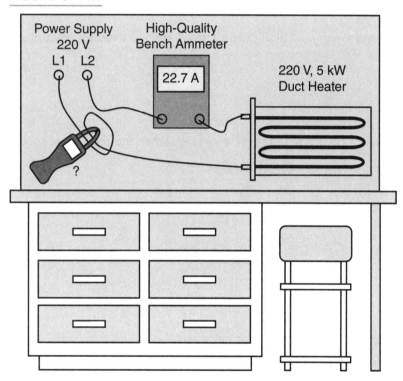

MAINTENANCE OF WORKSTATION AND TOOLS: Put each meter back in its case, along with any accessories that belong to it. Turn all electrical switches off. Return all tools and equipment to their respective places.

SUMMARY STATEMENT: Discuss the results of the accuracy calculations that were made. In your discussion, include possible causes for inaccuracies and other discrepancies that surfaced during the exercise.

TAKEAWAY: HVACR technicians rely on the readings they obtain from equipment to properly evaluate and troubleshoot the system. Properly calibrated instruments help ensure that the conclusions reached by the technician are accurate.

QUESTIONS

1. Describe a digital multimeter.

2. Why does the amperage change on an electrical heater as it heats up?

3. What does the gold band mean on a high-quality carbon resistor used for electronics?

4. What should the resistance be in a set of ohmmeter leads when the probes are touched together?

5. What should be done with a meter with a very slight error?

6. What should be done with a meter that has a large error and cannot be calibrated?

7. Why must a technician be sure that meters and instruments are reading correctly and are in calibration?

SECTION 5

Refrigerants (REF)

SUBJECT AREA OVERVIEW

Refrigeration, by definition, is the process of removing heat from a conditioned space and transferring to a space where it is less objectionable. Comfort cooling, food storage, and dehumidification are just a few refrigeration applications. Bacterial growth that causes food spoilage slows at lower temperatures. Refrigeration applications fall into three main temperature ranges. They are high temperature (box temperatures between 45°F and 72°F), medium temperature (box temperatures between 35°F and 45°F), and low temperature (box temperatures between −10°F and 0°F).

The refrigeration cycle can be described as a vapor-compression, repeating cycle. The boiling point of liquids can be manipulated by changing the pressure acting on the liquid. When the pressure is increased, the liquid's boiling point is raised. When the pressure is decreased, the boiling point is lowered. A compressor is a pump that compresses refrigerant vapor and pumps the refrigerant through the refrigeration circuit. Refrigerant flow is created by the difference in pressure between the inlet and outlet of the compressor. The compressed refrigerant is pumped to the condenser where heat is rejected and the vapor condenses into a liquid. From the condenser, the liquid refrigerant flows to the metering device, which regulates the amount of refrigerant entering the evaporator. The metering device causes a drop in pressure and temperature, which allows the refrigerant to absorb heat as it moves through the evaporator. The refrigerant completely vaporizes or boils within the evaporator. This vapor proceeds through the suction line back to the compressor, where it is compressed, and the cycle repeats itself.

During the boiling and condensing processes, which take place in the evaporator and condenser, respectively, a refrigerant is said to be saturated. When a refrigerant is saturated, it is comprised of a mixture of liquid and vapor. During this phase, the refrigerant will follow a specific temperature–pressure relationship. Manufacturers produce literature for each refrigerant detailing this relationship, called pressure–temperature charts. These charts can be used to verify proper operation of a refrigeration system or diagnose a fault. Applications on smart devices can also be used to locate pressure–temperature relationships for many different refrigerants.

The American Society of Heating, Refrigerating and Air-Conditioning Engineers (ASHRAE) designates refrigerants with an "R" and a number, determined by their characteristics and chemical composition. They can either be a single compound or a blend of two or more refrigerants. Refrigerants should not be allowed to escape into the atmosphere. Chlorofluorocarbon (CFC) and hydrochlorofluorocarbon (HCFC) refrigerants deplete the stratospheric ozone layer, allowing harmful ultraviolet wavelengths from the sun to reach the Earth. Hydrofluorocarbon (HFC) refrigerants do not deplete the ozone layer but do contribute to global warming. In fact, CFCs, HCFCs, and HFCs all contribute to global warming. Refrigerants should always be recovered and/or stored while a refrigeration system is being serviced or disposed of.

One ton of refrigeration is the amount of heat it would take to melt 1 ton of ice in 24 hours. Because it takes 144 Btu of heat energy to melt 1 pound of ice at 32°F, it will take 288,000 Btu to melt 1 ton, or 2000 lb, of ice in a 24-hour period. This equates to 12,000 Btu/h (288,000 ÷ 24), or 200 Btu/min (12,000 ÷ 60).

KEY TERMS

Azeotropic
Condenser
Critical pressure
Critical temperature
Dehumidifying
Desuperheating
Discharge line
Enthalpy
Entropy

Evaporator
Fixed-bore
Kinetic displacement
   compressor
Metering device
Near-azeotropic
Net refrigeration effect
   (NRE)
One ton of refrigeration

Orifice
Ozone layer
Pure compound
Reciprocating compressor
Refrigerant
Refrigerant blend
Refrigeration
Rotary compressor
Saturated

Saturation pressure
Scroll compressor
Subcooled
Temperature glide
Temperature–pressure
   relationship
Vapor pressure
Vapor pump

REVIEW TEST

Name	Date	Grade

Circle the letter that indicates the correct answer.

1. Ice melts in ice boxes because:
   A. heat flows through the insulated walls.
   B. heat flows through the doorway when it is opened.
   C. heat is moved into the box when warm food is placed in it.
   D. All of the above.

2. High-temperature refrigeration is that produced by:
   A. heat from a furnace.
   B. an air-conditioning system.
   C. the vegetable cooling system of a refrigerator.
   D. the freezer section of a refrigerator.

3. A ton of refrigeration is equal to:
   A. 1 ton of ice.
   B. 288,000 Btu/24 h.
   C. 144 Btu/h.
   D. a large refrigerator.

4. The evaporator in a refrigeration system:
   A. meters the refrigerant.
   B. condenses the refrigerant.
   C. is the component where refrigerant boils and absorbs heat.
   D. compresses the vaporized refrigerant.

5. The compressor in a refrigeration system:
   A. creates a pressure differential.
   B. pumps heat-laden liquid through the system.
   C. is located at the outlet of the condenser.
   D. is located in the system's liquid line.

6. The condenser:
   A. rejects the heat from the refrigerant.
   B. meters the refrigerant throughout the system.
   C. causes superheat in the refrigerant.
   D. is located between the evaporator and the compressor.

7. Only liquid refrigerant should enter the metering device.
   A. True
   B. False

8. Reciprocating compressors have:
   A. a piston and a cylinder.
   B. two scrolls that mesh together.
   C. a rotor to compress the refrigerant.
   D. a large fan-type compressor component.

9. R-22 is commonly found in older _____ systems.
   A. residential air-conditioning
   B. light commercial refrigeration
   C. industrial refrigeration
   D. household refrigeration

10. R-134a is commonly found in _____ systems.
   A. medium-temperature
   B. residential air-conditioning
   C. low-temperature
   D. humidification

11. The cylinder color code for R-410A is:
   A. rose.
   B. white.
   C. yellow.
   D. green.

12. The cylinder color code for R-134a is:
   A. red.
   B. yellow.
   C. light blue.
   D. green.

13. The cylinder color code for R-404A is:
   A. orchid.
   B. orange.
   C. purple.
   D. green.

14. The cylinder color code for R-22 is:
   A. orchid.
   B. white.
   C. purple.
   D. green.

15. Good ventilation is important when working around refrigeration equipment because modern refrigerants:
A. are toxic and can poison you.
B. have a very unpleasant odor.
C. may get on your skin and cause a rash.
D. are heavier than air and may displace the oxygen around you.

16. Saturated refrigerants will have a direct relationship between their:
A. oil pressure and oil viscosity.
B. refrigerant boiling/condensing temperature and pressure.
C. wet bulb and condenser.
D. compressor and suction valve.

17. It is believed that certain refrigerants such as CFCs and HCFCs, when allowed to escape into the atmosphere, will:
A. destroy the hydrogen layer and allow more of the sun's rays to overheat the earth.
B. deplete the stratospheric ozone layer and allow harmful ultraviolet rays from the sun to reach the Earth.
C. reduce the nitrogen content of the air surrounding the earth.
D. reduce the carbon dioxide necessary for plant life.

18. Recovering refrigerant is the process of:
A. releasing refrigerant to the atmosphere.
B. remanufacturing refrigerant.
C. passing refrigerant through a filter drier.
D. removing and storing refrigerant.

19. Enthalpy describes:
A. the moisture content in air.
B. the amount of refrigerant needed in a system.
C. the amount of heat a substance contains from some starting point.
D. the amount of harmful ultraviolet rays in the atmosphere.

20. A near-azeotropic refrigerant blend replacing R-22 in residential and commercial air-conditioning applications is:
A. R-404A.
B. R-407C.
C. R-410A.
D. R-134a.

21. A refrigerant that has properties very similar to R-12 and is used primarily in medium- and high-temperature refrigeration applications, refrigerators and freezers, and automotive air-conditioning is:
A. R-22.
B. R-134a.
C. R-404A.
D. R-407C.

22. A near-azeotropic refrigerant blend, with a small temperature glide, that has replaced R-502 is:
A. R-404A.
B. R-407C.
C. R-134a.
D. R-500.

23. The cylinder color code for R-404A is:
A. bright green.
B. rose.
C. aqua.
D. orange.

24. Near-azeotropic refrigerant blends all exhibit some:
A. temperature glide and fractionation.
B. oil problems.
C. high boiling points.
D. low condensing pressures.

25. The quantity of heat expressed in Btu/lb that the refrigerant absorbs from the refrigerated space to produce useful cooling is:
A. a calorie.
B. enthalpy.
C. the net refrigeration effect.
D. a degree.

26. The amount of heat, in Btu/lb, that is given off by the system in the discharge line and the condenser is referred to as the:
A. net refrigeration effect.
B. heat of work.
C. heat of compression.
D. total heat of rejection.

27. When the portion of a pressure/enthalpy diagram is beneath the saturation curve, then:
   A. the refrigerant is at a constant temperature.
   B. the refrigerant is at a constant pressure.
   C. the refrigerant follows a temperature–pressure relationship.
   D. the refrigerant is 100% vapor.

28. According to the pressure/enthalpy chart, as refrigerant flows through the compressor:
   A. the pressure, temperature, and heat content of the refrigerant all increase.
   B. the pressure and temperature of the refrigerant increase, while the heat content of the refrigerant decreases.
   C. the pressure and heat content of the refrigerant increase, while the temperature of the refrigerant decreases.
   D. the pressure, temperature, and heat content of the refrigerant all decrease.

29. Lines of constant pressure on a pressure/enthalpy diagram are referred to as:
   A. isotherms.
   B. isometrics.
   C. isobars.
   D. glide lines.

30. Lines of constant temperature on a pressure/ enthalpy diagram are referred to as:
   A. isotherms.
   B. isometrics.
   C. isobars.
   D. fractionation lines.

Refrigerant Familiarization

Name	Date	Grade

OBJECTIVES: Upon completion of this exercise, you should be able to:

- determine, recognize, and explain the safety classification of refrigerants.
- determine the refrigerant's empirical formula (whether the refrigerant is a CFC, HCFC, HFC, or a HC).
- identify the refrigerant cylinder's color code.

INTRODUCTION: Today, in the HVACR industry, there are many different refrigerants available. These refrigerants are classified mainly by the applications for which they are best suited and their effects on the environment. This exercise serves to help you better understand some of the properties of some commonly used refrigerants.

TOOLS, MATERIALS, AND EQUIPMENT: An HVACR textbook or smart device to search for the required information.

⚠ **SAFETY PRECAUTIONS:** There are no safety precautions for this exercise.

STEP-BY-STEP PROCEDURES

For the following refrigerants, determine their empirical formula, cylinder color, and safety classification. If it is a single-compound refrigerant, record whether it is ethane-based or methane-based in the last column. If it is a blended refrigerant, list the refrigerants in the blend in the last column.

	Refrigerant Class	Cylinder Color	Safety Classification	Single Compound or Blend	Refrigerant Composition
R-134a	HFC	Light Sky Blue	A1		
R-22					
R-404A					
R-32					
R-407C					
R-124					
R-401C					
R-125					
R-407A					
R-502					
R-407B					
R-143a					
R-402A					
R-600a					
R-152a					
R-406A					
R-500					

(Continued)

	Refrigerant Class	Cylinder Color	Safety Classification	Single Compound or Blend	Refrigerant Composition
R-290	_____	_____	_____	_____	_____
R-123	_____	_____	_____	_____	_____
R-11	_____	_____	_____	_____	_____
R-12	_____	_____	_____	_____	_____

MAINTENANCE OF WORKSTATION AND TOOLS: None required.

SUMMARY STATEMENT: Explain the differences between CFCs, HCFCs, HFCs, and HCs. Provide information pertaining to their uses and safety.

TAKEAWAY: There are numerous refrigerants in use today. Their chemical compositions vary greatly as do their applications and uses. It is important for an HVACR technician to be aware of the properties and characteristics of various refrigerants to ensure that they are used properly and safely in the field.

QUESTIONS

1. What elements are found in each of the following: CFCs, HCFCs, HFCs, and HCs?

2. What types of oils are used with each of the refrigerants from question 1?

3. What is the difference between an A1 classification and a B1 classification?

4. What organization provides these classifications?

5. Of the refrigerant types listed in this exercise, which one(s) have been or are being phased out?

Exercise REF-2

Refrigerant Pressure/ Temperature Relationships

Name	Date	Grade

OBJECTIVES: Upon completion of this exercise, you should be able to:

- understand what is meant by a refrigerant's pressure–temperature relationship.
- find the refrigerant's corresponding temperature or pressure when given a refrigerant's pressure or temperature at saturation.

INTRODUCTION: Using a pressure/temperature chart, you will complete the pressure/temperature relationships for various given refrigerants.

TOOLS, MATERIALS, AND EQUIPMENT: A pressure/temperature chart or optional smart device app.

⚠ **SAFETY PRECAUTIONS:** There are no safety precautions in this exercise.

STEP-BY-STEP PROCEDURES

For the following refrigerants at saturation, determine their pressure/temperature relationships in psig and degrees Fahrenheit.

Given a temperature, determine the corresponding pressure.

R-22 at 40°F has a corresponding pressure of _____ psig.

R-22 at 100°F has a corresponding pressure of _____ psig.

R-134a at 40°F has a corresponding pressure of _____ psig.

R-134a at 40°F has a corresponding pressure of _____ psig.

R-502 at 10°F has a corresponding pressure of _____ psig.

R-502 at 110°F has a corresponding pressure of _____ psig.

R-404A at −10°F has a corresponding pressure of _____ psig.

R-404A at 120°F has a corresponding pressure of _____ psig.

R-410A at 45°F has a corresponding pressure of _____ psig.

R-410A at 115°F has a corresponding pressure of _____ psig.

Given a pressure, determine the corresponding temperature.

R-22 at 226.4 psig has a corresponding temperature of _____ °F.

R-22 at 77.6 psig has a corresponding temperature of _____ °F.

R-134a at 198.7 psig has a corresponding temperature of _____ °F.

R-134a at 31.3 psig has a corresponding temperature of _____ °F.

R-502 at 9.2 psig has a corresponding temperature of _____ °F.

R-502 at 408.4 psig has a corresponding temperature of _____ °F.

R-404A at 3.8″ Hg has a corresponding temperature of _____ °F.

R-404A at 148.5 psig has a corresponding temperature of _____ °F.
R-410A at 88.0 psig has a corresponding temperature of _____ °F.
R-410A at 317.1 psig has a corresponding temperature of _____ °F.

MAINTENANCE OF WORKSTATION AND TOOLS: None required.

SUMMARY STATEMENT: State the relationship between temperature and pressure as applied to refrigerants at saturation.

TAKEAWAY: Changes in the pressure of a refrigerant system will cause changes in its temperature and vice versa. P/T charts are provided by the manufacturers of refrigerants to aid in designing, servicing, and installing refrigeration equipment. Understanding a P/T chart is fundamental to understanding any refrigerant system's operation.

QUESTIONS

1. What is meant by the term "saturation"?

2. At atmospheric pressure (0 psig), which refrigerant has the lowest saturation temperature?

3. How can we use the temperature–pressure relationship to identify the refrigerant in an unmarked cylinder?

Exercise REF-3

Refrigerant Temperature Applications

Name	Date	Grade

OBJECTIVES: Upon completion of this exercise, you should be able to describe and categorize different refrigeration applications based on the refrigerants and oils used in each, as well as their average temperatures and temperature range classifications.

INTRODUCTION: Some refrigerants can be used in multiple units across different temperature applications, while others are used exclusively in one type of unit or application. In this exercise, you will become more familiar with refrigerants and their uses by filling in the chart provided.

TOOLS, MATERIALS, AND EQUIPMENT: You will need an HVACR textbook or smart device to search for the required information.

SAFETY PRECAUTIONS: There are no safety precautions for this exercise.

STEP-BY-STEP PROCEDURES:

Using your textbook, class notes, and any other materials available to you, fill out the following chart with the type of refrigerant, type of oil, temperature application, and the average temperature used for each unit.

Type of Unit	Refrigerant(s) Used	Type of Oil(s)	Low-, Medium-, or High-Temperature Application	Average °F
Walk-in cooler				
Walk-in freezer				
Reach-in cooler				
Reach-in freezer				
Domestic refrigerator				
Residential split-system air conditioner				
Commercial PTAC				
Ice maker				
Floral cabinet				
Mini-split ductless air conditioner				
Residential split-system heat pump				

MAINTENANCE OF WORKSTATION AND TOOLS: There is no maintenance required for this exercise.

SUMMARY STATEMENT: State the different refrigerants commonly used in each temperature application (high, medium, and low), and generalize the types of units/systems typically found in each application.

TAKEAWAY: Refrigeration systems can operate with any one of many different refrigerants. One refrigerant can be used in different types of systems, and one system type can often utilize more than one specific refrigerant. For these reasons, technicians must take care not to mix refrigerants or use the wrong refrigerant or oil in the system that they are servicing or installing. Being knowledgeable in the uses and applications for the most commonly employed refrigerants will help prevent refrigerant-related problems from occurring in the field.

QUESTIONS

1. What should be done to a system that has been found to contain mixed refrigerants?

2. Can a system that is designed to work with one refrigerant be converted to work with a different refrigerant? Explain your answer.

3. Can a refrigerant be used with more than one type of oil? Explain your answer.

4. Explain what is meant by the term "drop-in refrigerant replacement," and identify any drop-ins that are presently used in the industry.

5. Can different types of refrigeration oil be mixed together? Explain why or why not.

SECTION 6

Air (AIR)

SUBJECT AREA OVERVIEW

In order to effectively service and troubleshoot air-conditioning systems, it is important for technicians to first possess a general understanding of air as well as the factors that affect human comfort. Four factors that can affect our comfort are temperature, humidity, air movement, and air cleanliness. We are typically comfortable when heat from our body is transferred to our surroundings at the correct and acceptable rate. The study of air and its properties is called psychrometrics. Air is made up primarily of nitrogen and oxygen but also contains water. Water in the air can be in both the liquid and vapor states. The moisture content in air is expressed in terms including absolute and relative humidity. Other characteristics of air include dry-bulb temperature, wet-bulb temperature, dew point, specific volume, and enthalpy.

Maintaining indoor air at acceptable quality levels is a major concern for the HVACR service technician. Maintaining the quality of air involves a combination of filtering the air and controlling the contamination level in the air. Methods to control indoor air contamination include eliminating the source of the contamination, providing adequate ventilation, and providing a means for cleaning the air. Ventilation is the process of supplying and removing air by either natural or mechanical means to and from the occupied space. Air cleaning can be accomplished by utilizing mechanical filters, electronic air cleaners, and ion generators. Dirty ducts may also be a source of air contamination and may need to be cleaned periodically.

Ensuring the correct amount of airflow to each of the areas being conditioned is also an important task that the successful service technician must perform. When correct volumes of air are delivered to the occupied space, the treated air mixes with the room air and creates a comfortable atmosphere. It is important to note that different spaces have different air quantity requirements. The air distribution, or duct, system is responsible for delivering the correct amounts of air to the different spaces. The pressure in a duct system is measured with a manometer and is expressed in inches of water column (in. wc). The three pressures associated with duct systems are total pressure, velocity pressure, and static pressure. Air moves through a duct system with the aid of a blower. The most common type of blower used to move air through the duct system is the forward-curved centrifugal blower, also referred to as a squirrel cage blower. This type of blower is designed to effectively overcome the duct pressures.

The duct system is typically made up of a supply duct and a return duct. The supply duct distributes air to the registers or diffusers in the conditioned space. A well-designed air distribution system will have balancing dampers that are used to adjust the airflow to the various parts of the structure. The return duct is responsible for delivering air from the conditioned space to the air-conditioning equipment and may have individual room returns, or it may have a central return. In a central return system, larger grilles are located in a central location so that air from the individual rooms can flow easily to the return duct system. Air distribution systems should be balanced to ensure that the correct amount of

(continued)

air is delivered to each area in the conditioned space.

Air distribution systems can be zoned so that different areas in the structure can be maintained at different temperatures, based on the comfort requirements of the occupants. Zoning is accomplished by using dampers located in the air distribution system. Zone dampers open and close based on the sensed temperature in a particular zone.

KEY TERMS

Absolute humidity
Carbon monoxide (CO)
Comfort
Dew point temperature
Diffuser
Dust mites
Electrostatic precipitators
Enthalpy

Exhaust
Friction loss
Grille
Humidistat
Ion generators
Mold
Nitrogen dioxide
Plenum

Propeller fan
Psychrometric chart
Psychrometrics
Radon
Register
Relative humidity
Sick building syndrome (SBS)

Squirrel cage blower
Static pressure
Sulfur dioxide
Total pressure
Velocity pressure
Velometer
Ventilation

REVIEW TEST

Name	Date	Grade

Circle the letter that indicates the correct answer.

1. The human body is comfortable when:
   A. heat is transferring from the surroundings to the body at the correct rate.
   B. heat is transferring from the body to the surroundings at the correct rate.

2. To be comfortable, lower temperatures can often be offset by:
   A. lower humidity.   C. more air movement.
   B. higher humidity.   D. less activity.

3. The generalized comfort chart can be used as a basis to determine:
   A. temperature and humidity combinations that produce comfort.
   B. the percentages of the components that make up the air in a structure.
   C. the amount of fresh-air makeup necessary for comfort.
   D. the exhaust requirements for a structure.

4. Air is made up primarily of:
   A. oxygen and carbon dioxide.
   B. nitrogen and carbon dioxide.
   C. oxygen and ozone.
   D. nitrogen and oxygen.

5. The dry-bulb temperature is the:
   A. same as the wet-bulb depression.
   B. total of the sensible and latent heat.
   C. sensible heat temperature.
   D. sensible heat loss and humidity.

6. The wet-bulb temperature:
   A. is the same as the wet-bulb depression.
   B. is the total of the sensible and latent heat.
   C. takes into account the moisture content of the air.
   D. is always higher than the dry-bulb temperature.

7. The wet-bulb depression is:
   A. the difference between the dry-bulb and the wet-bulb temperature.
   B. the same as the wet-bulb temperature.

   C. the same as the dry-bulb temperature.
   D. the saturation of air with moisture.

8. The dew point temperature is the:
   A. temperature when the air is saturated with moisture.
   B. temperature when the moisture begins to condense out of the air.
   C. difference between the dry-bulb and the wet-bulb temperatures.
   D. temperature at which the relative humidity is 50%.

9. If the dry-bulb temperature is 76°F and the wet-bulb temperature is 68°F, the wet-bulb depression is:
   A. 6°F.   C. 10°F.
   B. 8°F.   D. 12°F.

10. Fresh-air intake is necessary to keep the indoor air from becoming:
   A. too humid.
   B. superheated.
   C. oxygen starved and stagnant.
   D. too dry.

11. Infiltration is the term used when air comes into a structure:
   A. through an air makeup vent.
   B. around windows and doors.
   C. through a ventilator.

12. If an air sample has a wet-bulb temperature of 65°F and a dry-bulb temperature of 80°F, the relative humidity will be closest to:
   A. 44%.   C. 52%.
   B. 28%.   D. 60%.

13. A furnace puts out 122,500 Btu/h of sensible heat and the air temperature rise across the furnace is 55°F. The airflow across the furnace is:
   A. 2062 cfm.   C. 2525 cfm.
   B. 1850 cfm.   D. 1750 cfm.

14. What is the absolute humidity of an air sample that has the ability to hold 100 grains of moisture and is presently holding only 50 grains?
   A. 50%
   B. 50 grains
   C. 100%
   D. 100 grains

15. As air is heated:
   A. the relative humidity and the absolute humidity both drop.
   B. the relative humidity rises and the absolute humidity decreases.
   C. the relative humidity falls and the absolute humidity increases.
   D. the relative humidity falls and the absolute humidity remains the same.

16. Dust mites are:
   A. usually found in damp areas in basements.
   B. often found in dust blown through vacuums with HEPA filters.
   C. microscopic spider-like insects.
   D. related to mold spores.

17. Mold:
   A. is usually found in dusty, dry areas.
   B. is usually found in areas where there is moisture and/or high humidity.
   C. is seldom found in areas where it can be seen.
   D. spores are not harmful to humans.

18. The first step with regard to mold cleanup should be to:
   A. remove the source of moisture.
   B. vacuum the area.
   C. check for radio activity.
   D. check the filter media in the area.

19. Carbon monoxide:
   A. has a bluish color.
   B. can easily be seen.
   C. is heavier than air.
   D. is produced by incomplete combustion in appliances that use the combustion process.

20. Disturbed materials containing asbestos:
   A. should be brushed down and swept up by the HVACR technician.
   B. should be removed by a certified or licensed contractor.
   C. may be removed when found as part of a mold remediation program.
   D. should be vacuumed up when found.

21. Ventilation is:
   A. a problem to deal with.
   B. generally not recommended.
   C. never used in a residence.
   D. the process of supplying or removing air by natural or mechanical means.

22. Duct cleaning is accomplished by:
   A. polishing the outside of the duct.
   B. washing out with a hose and water.
   C. using compressed air.
   D. using a vacuum system and brushing.

23. If the relative humidity is 60%, it means that:
   A. 40% of the air is moisture.
   B. 60% of the air is moisture.
   C. the air has 60% of the moisture it has the capacity to hold.
   D. the air has 40% of the moisture it has the capacity to hold.

24. Humidification generally relates to a process of causing water to _____ into the air.
   A. condense
   B. evaporate
   C. be ionized
   D. be filtered

25. Atomizing humidifiers discharge _____ into the air.
   A. water vapor
   B. electronic ions
   C. trivalent oxygen
   D. tiny water droplets

26. In the ionizing section of an electronic air cleaner, the particles in the air are:
   A. trapped by a filter.
   B. charged with an electrical charge.
   C. washed with a special cleaner.
   D. treated to remove many of the odors.

27. Extended surface filters often use _____, which provides air cleaning efficiencies up to three times greater than those of fiberglass.
   A. trivalent oxygen
   B. charcoal
   C. stainless steel
   D. nonwoven cotton

28. The pressure in a duct system is measured in:

A. psi.

B. in. wc.

C. cfm.

D. tons of refrigeration.

29. Velocity pressure is the:

A. same as total pressure.

B. total pressure − the static pressure.

C. total pressure + the static pressure.

D. same as the static pressure.

30. Squirrel cage blowers are used in air-conditioning systems primarily in _____ applications.

A. exhaust

B. ducted

C. condenser

D. compressor cooling

31. The fittings between the main trunk and branch ducts in a duct system are called:

A. takeoff fittings.

B. diffusers.

C. spiral metal fittings.

D. boots.

32. One reason for friction loss in duct systems is the:

A. humidity in the air.

B. lack of humidity in the air.

C. rubbing action of the air on the walls of the duct.

D. thickness of the sheet metal used to fabricate the duct.

33. To measure airflow by traversing the duct is to:

A. measure the duct to determine the cross-sectional area.

B. determine the cfm on each side of the duct.

C. determine the total pressure at the center of the duct.

D. measure the velocity of the airflow in a pattern across a cross section of the duct and averaging the measurements.

34. Static pressure in a duct system is:

A. equal to the weight of the air standing still.

B. the force per unit area in a duct.

C. equal to the velocity and weight of the moving air.

D. determined with a venturi system.

Exercise AIR-1

Comfort Conditions

Name	Date	Grade

OBJECTIVES: Upon completion of this exercise, you should be able to take wet-bulb and dry-bulb temperature readings and then, by using a psychrometric chart, determine the relative humidity. Then, using this information, you should be able to determine the level of comfort based on the American Society of Heating, Refrigerating and Air-Conditioning Engineers (ASHRAE) generalized comfort chart.

INTRODUCTION: With a sling, or digital, psychrometer, you will take several dry- and wet-bulb temperature readings in a conditioned space. By referring to a psychrometric and generalized comfort chart, you will determine whether the conditioned space is within accepted comfort limits.

TOOLS, MATERIALS, AND EQUIPMENT: A sling, or digital, psychrometer and comfort and psychrometric charts, Figures AIR-1.1 and AIR-1.2. Additionally, smart devices equipped with a psychrometric app can be used if permitted by the instructor.

FIGURE **AIR-1.1**

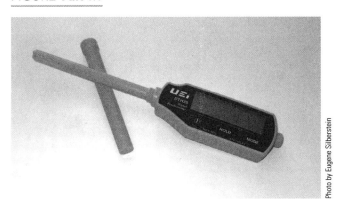

Photo by Eugene Silberstein

⚠ **SAFETY PRECAUTIONS:** The psychrometer is a fragile instrument. Be careful not to bang it, drop it, or break it. Keep it away from other objects when using it.

STEP-BY-STEP PROCEDURES

1. Review procedures for using the psychrometer in different areas, outside and in classrooms.

2. Record the following for 10 wet-bulb/dry-bulb readings.

FIGURE **AIR-1.2**

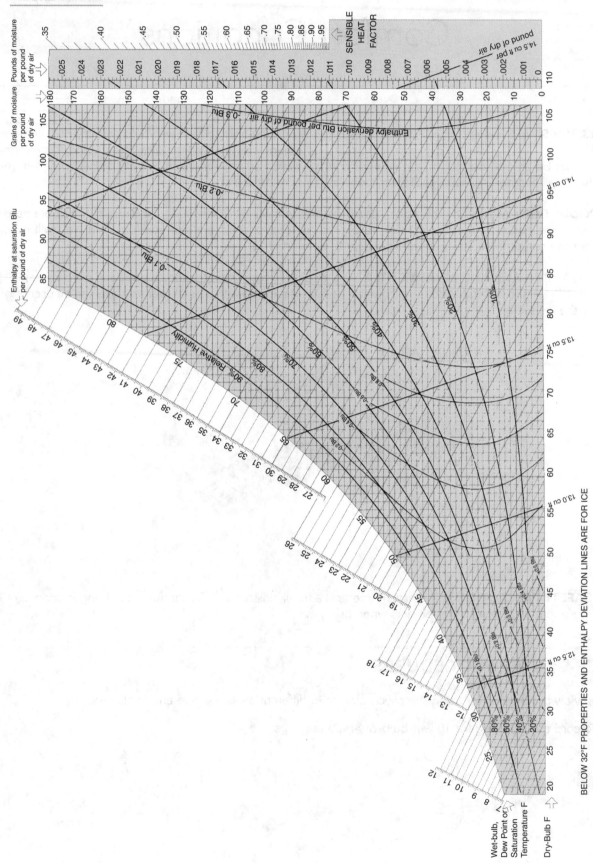

	Dry-Bulb Temperature	Wet-Bulb Temperature	Relative Humidity (from psychrometric chart)	Within Generalized Comfort Zone (yes or no)
1.	_____	_____	_____	_____
2.	_____	_____	_____	_____
3.	_____	_____	_____	_____
4.	_____	_____	_____	_____
5.	_____	_____	_____	_____
6.	_____	_____	_____	_____
7.	_____	_____	_____	_____
8.	_____	_____	_____	_____
9.	_____	_____	_____	_____
10.	_____	_____	_____	_____

MAINTENANCE OF WORKSTATION AND TOOLS: Properly store the psychrometer.

SUMMARY STATEMENT: Explain how different combinations of humidity and temperature can create a comfortable environment.

TAKEAWAY: There is no single set of conditions that will make all individuals comfortable. Since everyone has a unique definition of comfortable, comfort is often described as an acceptable range, within which most people will be comfortable. To ensure that the majority of people are comfortable in an occupied space, every effort should be made to ensure that the air conditions in the space fall within the comfort zones provided by ASHRAE. If conditions fall outside of these zones, the majority of the occupants will likely not be happy, and system adjustments will need to be made.

QUESTIONS

1. What causes the wet-bulb thermometer to read a lower temperature than the dry-bulb thermometer?

2. Why must the sling psychrometer be swung in order to obtain the wet-bulb reading?

3. List as many factors as you can that affect comfort.

4. How does air movement improve comfort?

5. Explain some of the differences between the summer and winter comfort chart readings?

6. Name three different kinds of air filters.

7. What does it mean when it is said that air has a relative humidity (RH) of 70%?

8. Name two ways that humidity can be added to air.

9. Name three ways that the human body gives up heat.

10. Explain the difference between absolute humidity and relative humidity.

11. Under what conditions will the wet-bulb temperature be higher than the dry-bulb temperature?

Exercise AIR-2

Heat Recovery Ventilators (HRVs) and Energy Recovery Ventilators (ERVs)

Name	Date	Grade

OBJECTIVES: Upon completion of this exercise, you should be able to observe the operation of a heat recovery ventilator and/or an energy recovery ventilator.

INTRODUCTION: Heat recovery ventilators are used to transfer sensible heat between a ventilation and an exhaust airstream. By measuring the conditions of the exhaust and ventilation air streams both before and after flowing through the heat recovery ventilator, the amount of heat energy transferred can be determined.

TOOLS, MATERIALS, AND EQUIPMENT: A heat recovery ventilator (HRV) and/or an energy recovery ventilator (ERV), hand tools, digital psychrometer, psychrometric chart, calculator, and safety goggles.

⚠ **SAFETY PRECAUTIONS:** Energy recovery ventilators and heat recovery ventilators have blowers installed in them. Be sure to keep your hands out of the blower compartments while the blowers are turning.

STEP-BY-STEP PROCEDURES

1. Obtain the following information from the ERV or HRV:
   A. Brand: _____
   B. Model: _____
   C. ERV or HRV?: _____
   D. Airflow through the device: _____ cfm

2. Obtain the following readings from the indoor air and the outside air:
   A. Indoor air dry-bulb temperature: _____ °F
   B. Indoor air wet-bulb temperature: _____ °F
   C. Indoor air relative humidity: _____ %
   D. Indoor air enthalpy: _____ Btu/lb
   E. Outdoor air dry-bulb temperature: _____ °F
   F. Outdoor air wet-bulb temperature: _____ °F
   G. Outdoor air relative humidity: _____ %
   H. Outdoor air enthalpy: _____ Btu/lb

3. Turn the HRV on and obtain the following readings. The exhaust airstream conditions are to be measured at the outlet of the HRV before the airstream has been able to mix with the outside air. The ventilation airstream conditions are to be measured at the outlet of the HRV before the airstream has been able to mix with the inside air:
   A. Exhaust airstream dry-bulb temperature: _____ °F
   B. Exhaust airstream wet-bulb temperature: _____ °F

C. Exhaust airstream relative humidity: _____ %

D. Exhaust airstream enthalpy: _____ Btu/lb

E. Ventilation airstream dry-bulb temperature: _____ °F

F. Ventilation airstream wet-bulb temperature: _____ °F

G. Ventilation airstream relative humidity: _____ %

H. Ventilation airstream enthalpy: _____ Btu/lb

4. Plot the outside air conditions and the ventilation airstream conditions on a psychrometric chart.

5. Using the plotted points from step 4, determine the following, where ΔT is the change in temperature, Δh is the change in enthalpy, and ΔW is the change in absolute humidity:

A. ΔT between the outside air and the ventilation airstream: _____ °F

B. Δh between the outside air and the ventilation airstream: _____ Btu/lb

C. ΔW between the outside air and the ventilation airstream: _____ gr/lb

Note: If the unit used is an HRV, the ΔW will be zero and does not need to be calculated.

6. Using the information from step 5 and from step 1D, perform the following calculations, where Q_T is the total heat, Q_S is the sensible heat, and Q_L is the latent heat:

A. $Q_T = 4.5 \times cfm \times \Delta h =$ _____ Btu/h

B. $Q_S = 1.08 \times cfm \times \Delta T =$ _____ Btu/h

C. $Q_L = 0.68 \times cfm \times \Delta W =$ _____ Btu/h

Note: If the unit used is an HRV, Q_L will be zero and will not need to be calculated.

7. Referring to the results from step 6, comment on the relationship among Q_T, Q_S, and Q_L.

8. Plot the inside air conditions and the exhaust airstream conditions on a second psychrometric chart.

9. Using the plotted points from step 8, determine the following, where ΔT is the change in temperature, Δh is the change in enthalpy, and ΔW is the change in absolute humidity:

A. ΔT between the inside air and the exhaust airstream: _____ °F

B. Δh between the inside air and the exhaust airstream: _____ Btu/lb

C. ΔW between the Inside air and the exhaust airstream: _____ gr/lb

Note: If the unit used is an HRV, the ΔW will be zero and does not need to be calculated.

10. Using the information from step 9 and from step 1D, perform the following calculations, where Q_T is the total heat, Q_S is the sensible heat, and Q_L is the latent heat:

A. $Q_T = 4.5 \times cfm \times \Delta h =$ _____ Btu/h

B. $Q_S = 1.08 \times cfm \times \Delta T =$ _____ Btu/h

C. $Q_L = 0.68 \times cfm \times \Delta W =$ _____ Btu/h

Note: If the unit used is an HRV, Q_L will be zero and will not need to be calculated.

11. Referring to the results from step 10, comment on the relationship among Q_T, Q_S, and Q_L.

12. Compare and comment on the obtained values for Q_T, Q_S, and Q_L in steps 6 and 10.

MAINTENANCE OF WORKSTATION AND TOOLS: Be sure to replace all panels on the HRV or ERV units. Replace all tools in their proper location, and clean the work area.

SUMMARY STATEMENT: Explain in detail the operation of the HRV or ERV and how they are used to save heating- and cooling-related energy costs.

TAKEAWAY: ERVs and HRVs are intended to limit the amount of energy lost when ventilating a structure. Properly operating and well-maintained equipment will help ensure that the amount of heat energy recovered is maximized.

QUESTIONS

1. Explain the difference between absolute humidity and relative humidity.

2. Explain the differences between an ERV and an HRV.

3. Explain why it is important for the two blowers in an ERV or HRV to move the same volume of air into and out of the structure.

Plotting an Air-Conditioning Process on a Psychrometric Chart (Part 1 of a Two-Part Exercise)

Name	Date	Grade

OBJECTIVES: Upon completion of this exercise, you should be able to take dry-bulb and wet-bulb temperature readings on an operating air-conditioning system and then plot these points on a psychrometric chart. From these plotted points you will be able to determine the change in enthalpy, the change in temperature, and the change in moisture content that took place during the conditioning process.

INTRODUCTION: Air-conditioning systems perform two functions. They not only cool an airstream but also dehumidify, or remove moisture from, an airstream. By obtaining temperature readings at various points in the system, you should be able to determine, with a reasonable degree of accuracy, the amount of heat, in Btu/lb, that the airstream gave up to the air-conditioning system. In addition, you will be able to determine the change in temperature between the return and supply airstreams, as well as the change in absolute humidity and relative humidity between the return and supply airstreams.

TOOLS, MATERIALS, AND EQUIPMENT: An operating air-conditioning system, hand tools, digital psychrometer, psychrometric chart, calculator, and safety goggles. Additionally, a smart device equipped with a psychrometric app can be used to aid with this exercise if permitted by the instructor.

⚠ **SAFETY PRECAUTIONS:** Air-conditioning systems have blowers installed in them. Be sure to keep your hands out of the blower compartments while the blowers are turning. Avoid coming in contact with the evaporator coil on an air-conditioning system, as the fins are very sharp.

STEP-BY-STEP PROCEDURES

1. Obtain the following information from the air-conditioning system:
   A. Indoor unit brand: _____
   B. Indoor unit model: _____
   C. Nameplate system capacity: _____ Btu/h, _____ tons
   D. Indoor unit serial number: _____
   E. Outdoor unit brand (if a split-type system): _____
   F. Outdoor unit model (if a split-type system): _____
   G. Outdoor unit serial number (if a split-type system): _____

2. Using a psychrometer, obtain the following readings from the return air entering the evaporator coil of the air-conditioning system:
   A. Return air dry-bulb temperature: _____ °F
   B. Return air wet-bulb temperature: _____ °F

3. Plot the point that corresponds to the return air on a psychrometric chart, and determine the following information:
   A. Enthalpy of the return air: _____ Btu/lb
   B. Relative humidity of the return air: _____ %
   C. Absolute humidity of the return air: _____ grains/lb
   D. Dew point temperature of the return air: _____ °F
   E. Specific volume of the return air: _____ ft³/lb

4. Obtain the following readings from the supply air leaving the evaporator coil of the air-conditioning system:
   A. Supply air dry-bulb temperature: _____ °F
   B. Supply air wet-bulb temperature: _____ °F

5. Plot the point that corresponds to the supply air on a psychrometric chart, and determine the following information:
   A. Enthalpy of the supply air: _____ Btu/lb
   B. Relative humidity of the supply air: _____ %
   C. Absolute humidity of the supply air: _____ grains/lb
   D. Dew point temperature of the supply air: _____ °F
   E. Specific volume of the supply air: _____ ft³/lb

6. Using the plotted points from steps 3 and 5, determine the following, where ΔT is the change in dry-bulb temperature, Δh is the change in enthalpy, and ΔW is the change in absolute humidity:
   A. ΔT between the return and supply airstreams: _____°F
   B. Δh between the return and supply airstreams: _____Btu/lb
   C. ΔW between the return and supply airstreams: _____ grains/lb

MAINTENANCE OF WORKSTATION AND TOOLS: Be sure to replace all panels on the air-conditioning system. Replace all tools in their proper location and clean the work area.

SUMMARY STATEMENT: Compare the location of the supply air point to the location of the return air point on the psychrometric chart. Explain, in detail, how the location of these two points confirms that an air-conditioning system both cools and dehumidifies.

TAKEAWAY: An air-conditioning system is responsible not only for cooling the air but also for dehumidifying the air. Properly taking, working with and interpreting supply and return air readings is a valuable exercise that will help the student of air conditioning more fully and completely understand the properties and behavior of air.

QUESTIONS

1. Explain the difference between absolute humidity and relative humidity.

2. Explain the difference between dry-bulb temperature and wet-bulb temperature.

3. Explain why knowing only the wet-bulb temperature of an airstream provides information regarding the enthalpy, or heat content, of the airstream.

Exercise AIR-4

Calculating Total Heat, Sensible Heat, and Latent Heat (Part 2 of a Two-Part Exercise)

Name	Date	Grade

OBJECTIVES: Upon completion of this exercise, you should be able to calculate the total, latent, and sensible heat capacities from an operating air-conditioning system.

INTRODUCTION: Air-conditioning systems perform two functions. They not only cool an airstream but also dehumidify, or remove moisture from, an airstream. Once temperature readings have been obtained from various points in the system, you can then determine, with a reasonable degree of accuracy, the amount of heat, in Btu/h, that the airstream gave up to the air-conditioning system. In addition to determining the total amount of heat transferred from the airstream to the refrigerant, you will also be able to determine how much of this heat was sensible heat and how much was latent.

TOOLS, MATERIALS, AND EQUIPMENT: An operating air-conditioning system, hand tools, digital psychrometer, psychrometric chart, calculator, and safety goggles. Additionally, a smart device equipped with a psychrometric app can be used to aid with this exercise if permitted by the instructor.

⚠️ **SAFETY PRECAUTIONS:** Air-conditioning systems have blowers installed in them. Be sure to keep your hands out of the blower compartments while the blowers are turning. Avoid coming in contact with the evaporator coil on an air-conditioning system, as the fins are very sharp.

STEP-BY-STEP PROCEDURES

1. Obtain the information from Exercise AIR-3.

2. Calculate the airflow rate that is moving through the air-conditioning system. If you have not yet learned how to determine the actual airflow through an air-conditioning system, ask your instructor to provide you with a value that can be used for calculation purposes.

3. Using the information from the previous step as well as from step 6 in Exercise AIR-3, perform the following calculations, where Q_T is the total heat, Q_S is the sensible heat, and Q_L is the latent heat:
   A. $Q_T = 4.5 \times cfm \times \Delta h =$ _____ Btu/h
   B. $Q_S = 1.08 \times cfm \times \Delta T =$ _____ Btu/h
   C. $Q_L = 0.68 \times cfm \times \Delta W =$ _____ Btu/h

4. Referring to the results from step 3, comment on the relationship that exists among Q_T, Q_S, and Q_L.

5. Calculate the sensible heat ratio for this system.

6. Using the psychrometric chart with the two plotted points from Exercise AIR-3, connect the two dots that represent the return and supply air conditions. This is called the process line.

7. On the chart, locate the reference point, defined as 80°F @50% R.H.

8. Draw a line that passes through the reference point and is parallel to the process line. Make certain that the line just drawn extends all the way to the right of the psychrometric chart.

9. Read the value for the sensible heat ratio (SHR) as shown on the chart.

10. Compare and comment on the value calculated in step 5 and the value obtained in step 9.

11. Compare and comment on the value for Q_T obtained in step 3A with the nameplate capacity of the unit (step 1C of Exercise AIR-3).

MAINTENANCE OF WORKSTATION AND TOOLS: Be sure to replace all panels on the air-conditioning system. Replace all tools in their proper location, and clean the work area.

SUMMARY STATEMENT: Explain how the psychrometric chart functions as a service tool that a technician can use to evaluate and troubleshoot an air-conditioning system.

TAKEAWAY: The psychrometric chart visually and graphically summarizes the properties of air. Understanding the layout and format of the chart will empower the HVACR technician to better understand the properties of air, ultimately making the technician more successful.

QUESTIONS

1. Explain why the relative humidity of an airstream increases when the airstream is cooled and dehumidified.

2. What happens to the density of an air sample when the air sample is cooled and dehumidified. Explain your answer.

Humidification

Name	Date	Grade

OBJECTIVES: Upon completion of this exercise, you should be able to describe the various components and the application of a humidifier.

INTRODUCTION: Your instructor will furnish you with a humidifier and you will perform a typical service inspection on the unit.

TOOLS, MATERIALS, AND EQUIPMENT: Soap, chlorine solution (if possible, to kill the algae), goggles, plastic or rubber gloves, straight-slot and Phillips-head screwdrivers, ¼" and $^5/_{16}$" nut drivers, two small (6" or 8") adjustable wrenches, and a pair of slip-joint pliers.

⚠ **SAFETY PRECAUTIONS:** Turn off the power and the water to the furnace and humidifier (if the humidifier is installed). Lock and tag the electrical panel. Do not put your hands around your face while cleaning a humidifier, as algae might be present. Algae can contain harmful germs.

STEP-BY-STEP PROCEDURES

1. Put on safety goggles and wear gloves. Place the humidifier on the workbench. If it is installed, remove it from the system using the preceding cautions.

2. Provide the following humidifier information:
   - Manufacturer: _____
   - Model number: _____
   - Serial number: _____
   - Principle of operation: _____

3. Remove any necessary panels for service.

4. Clean any surfaces that are dirty. If there is mineral buildup, it may not be removed easily. Do not force the cleaning; if it cannot be removed, let it stay.

5. State the condition of the pads or drum if the unit operates on that principle.

6. Is there algae present in the water basin (if the unit has a basin)? Algae will be a soft brown, green, or black substance.

7. If there is algae, clean the algae with strong soap. A chlorine solution is preferred to kill the algae. Wear gloves and goggles.

8. Check the float for free movement if the unit has a float.

9. Assemble the unit and place it back into the system, if appropriate.

MAINTENANCE OF WORKSTATION AND TOOLS: Return all tools to their respective places. Leave your work area clean.

SUMMARY STATEMENT: Describe the complete operation of the humidifier that you worked with, including the type of controls. Use an additional sheet of paper if more space is necessary for your answer.

TAKEAWAY: If not properly maintained, humidifers can become breeding grounds for harmful bacteria and other contaminants and irritants. Humidifiers should be inspected and serviced on a regular basis to minimize the possibility of sickening the occupants of a structure.

QUESTIONS

1. How does moisture added to the air in a residence affect the thermostat setting?

2. What happens to the relative humidity level in the air if it is heated?

3. How is moisture added to air using pads and rotating drums in a humidifier?

4. What is the algae that forms in a humidifier?

5. What may be used to remove algae and kill germs?

6. What is the purpose of the drain line on a humidifier with a flat chamber?

7. What is the effect of mineral deposits on the pads or rotating drum of a humidifier?

Exercise AIR-6

Heat Recovery Ventilators (HRVs)

Name	Date	Grade

OBJECTIVES: Upon completion of this exercise, you should be able to observe the operation of a heat recovery ventilator (HRV).

INTRODUCTION: Heat recovery ventilators are used to transfer sensible heat between a ventilation and an exhaust airstream. By measuring the temperature of the exhaust and ventilation air streams both before and after flowing through the heat recovery ventilator, the amount of heat energy transferred can be determined.

TOOLS, MATERIALS, AND EQUIPMENT: A heat recovery ventilator (HRV), hand tools, thermocouple thermometer, calculator, and safety goggles.

⚠ **SAFETY PRECAUTIONS:** Heat recovery ventilators have blowers installed in them. Be sure to keep your hands out of the blower compartments while the blowers are turning.

STEP-BY-STEP PROCEDURES

1. Obtain the following information from the HRV:
   A. Brand: _____
   B. Model: _____
   C. Airflow through the device: _____ cfm

2. Obtain the following readings from the indoor air and the outside air:
   A. Indoor air dry-bulb temperature: _____ °F
   B. Outdoor air dry-bulb temperature: _____ °F

3. Turn the HRV on, and obtain the following readings. The exhaust airstream temperature is to be measured at the outlet of the HRV (outside the structure) before the airstream has been able to mix with the outside air. The ventilation airstream temperature is to be measured at the outlet of the HRV (inside the structure) before the airstream has been able to mix with the inside air:
   A. Exhaust airstream dry-bulb temperature: _____ °F
   B. Ventilation airstream dry-bulb temperature: _____ °F

4. Determine the following values, where ΔT is the change in temperature:
   A. ΔT between the inside air and the exhaust airstream: _____ °F.

   Note: This is calculated as |2A − 3A|. The ΔT should always be positive, so you are using the absolute value here.

   B. ΔT between the outside air and the ventilation airstream: _____ °F.

   Note: This is calculated as |2B − 3B|. The ΔT should always be positive, so you are using the absolute value here.

5. Using the information from steps 1C and 4, perform the following calculations, where Q_S is the amount of sensible heat transferred:

   A. Q_S of the outgoing (Exhaust) airstream $= 1.08 \times cfm \times \Delta T$ (from 4A) = _____ Btu/h

   B. Q_S of the incoming (Ventilation) airstream $= 1.08 \times cfm \times \Delta T$ (from 4B) = _____ Btu/h

6. Referring to the results from step 5, comment on the relationship between the results from steps 5A and 5B.

MAINTENANCE OF WORKSTATION AND TOOLS: Be sure to replace all panels on the HRV unit. Replace all tools in their proper location, and clean the work area.

SUMMARY STATEMENT: Explain in detail the operation of the HRV when it is used to save heating- and cooling-related energy costs.

TAKEAWAY: Heat recovery ventilators allow structures to be ventilated while reducing the heat-related losses associated with exhausting conditioned air from the space. Properly sized, installed, and maintained HRV equipment helps reduce energy costs while reducing the wear and tear on the air-conditioning equipment.

QUESTIONS

1. Explain why the amount of incoming air must always be equal to the amount of outgoing air.

2. Explain why an HRV has air filters at both ends of the appliance.

3. Assume that the HRV used moves 100 cfm of air and is operating in the colder winter months. The ΔT between the inside air and the exhaust airstream is 25°F. If the cost of fuel oil is $3.00/gallon and the heating system is operating at 80% efficiency, how much money is being saved each hour the HRV is operating?

Exercise AIR-7

Electronic Air Cleaners

Name	Date	Grade

OBJECTIVES: Upon completion of this exercise, you should be able to disassemble, reassemble, and inspect the component parts of an electronic air cleaner.

INTRODUCTION: Electronic air cleaners have multiple filtration stages and are intended to remove small particulate matter. In order for these devices to operate effectively, they must be properly maintained and inspected on a regular basis. In this exercise, you will perform an inspection on an electronic air cleaner.

TOOLS, MATERIALS, AND EQUIPMENT: An electronic air cleaner, hand tools, and safety goggles.

⚠ **SAFETY PRECAUTIONS:** Electronic air cleaners are powered by line voltage. Be sure that the unit is not energized when it is opened for inspection.

STEP-BY-STEP PROCEDURES

1. Obtain the following information from the electronic air cleaner:
   A. Brand: _____
   B. Model: _____
   C. Filter size: _____

2. If the electronic air cleaner is installed in an air-conditioning or heating system, energize the equipment, and make certain that the power switch on the electronic air cleaner is in the ON position.

3. If the electronic air cleaner is equipped with a TEST button, push the button in. You should hear a snapping sound, which indicates that the electronic air cleaner is functioning. If there is no snapping sound, disconnect all power to the unit, and lock out the circuit. Ask your instructor for assistance to determine why the electronic air cleaner is not being powered.

4. Once it has been determined that the electronic air cleaner is functioning properly, disconnect all power to the unit, and lock out the circuit.

5. Carefully remove the front cover from the electronic air cleaner and set it aside.

6. On the upstream (inlet) side of the electronic air cleaner, you should see the prefilter. This is where the larger particles of dirt, such as lint, are captured by the cleaner. Slide out this prefilter. Describe the materials used to make this filter here: _____

7. Is the prefilter a permanent filter or a disposable filter? How do you know? _____
   _____

8. Clean the prefilter either by vacuuming it (if it is not very dirty) or by soaking the filter (if there is an indication that there is grease, or other tough dirt on the filter).

9. Set the filter aside.

10. Carefully slide out the electronic cell(s) from the housing of the electronic air cleaner, making note of the direction of the arrow that is located on the cells.

    Note: Be sure to handle these cells very carefully to prevent damage.

11. Visually inspect the cells for any damaged ionizing wires or bent collector plates.

12. Clean the electronic cells as needed. If they are not very dirty, a quick wipe with a slightly damp cloth will be sufficient. If the cells are dirty, they can be soaked in warm soapy water. After they have soaked and the water has cooled, the cells can be rinsed and set aside to dry completely.

13. Some electronic air cleaners have a postfilter. This filter is located on the downstream (outlet) side of the electronic air cleaner. If the electronic air cleaner being worked on has a postfilter, slide it out to the casing and inspect it.

14. With the housing of the electronic air cleaner completely empty now, visually inspect the interior surfaces of the cleaner. If there are traces of dirt, wipe the interior of the casing with a slightly damp rag and allow it to dry.

15. Once all of the pieces of the electronic air cleaner have dried, slide them back into the housing of the cleaner, making certain that each component has been replaced in the right place and is facing in the right direction.

16. Replace the cover on the electronic air cleaner.

17. Restore power to the system, and check the operation of the electronic air cleaner once again.

MAINTENANCE OF WORKSTATION AND TOOLS: Be sure to replace all panels on the electronic air cleaner. Replace all tools in their proper location, and clean the work area.

SUMMARY STATEMENT: Explain in detail the operation of an electronic air cleaner, making certain to include information about the prefilter, electronic cells, and postfilter (if the unit has one).

TAKEAWAY: Preventive maintenance helps ensure that air-conditioning equipment operates in an effective and efficient manner. Maintaining indoor air quality and filtration devices helps keep air contaminant levels to a minimum, which can help reduce the possibility of respiratory irritation and infection.

QUESTIONS

1. Explain the purpose of the ionizing section of an electronic air cleaner.

2. Explain how an electronic air cleaner is typically wired.

3. Explain the purpose of the prefilter on an electronic air cleaner.

Exercise AIR-8

Using Duct Charts to Evaluate Duct Systems

Name	Date	Grade

OBJECTIVES: Upon completion of this exercise, you should be able to use various charts to evaluate the duct systems.

INTRODUCTION: Using Figures AIR-8.1 and AIR-8.2, you will evaluate the characteristics of a duct system.

TOOLS, MATERIALS, AND EQUIPMENT: Other than a calculator, there are no tools and materials required for this exercise.

⚠ **SAFETY PRECAUTIONS:** There are no safety precautions for this exercise.

STEP-BY-STEP PROCEDURES

1. Using the friction chart (Figure AIR-8.1), complete this chart by filling in the missing information.

Duct Diameter (inches)	Friction Rate	Airflow (cfm)	Air Velocity (ft/min)
10	0.1		
12		1000	
8			700
	0.1	1000	
	0.16		800
		500	700
14	0.08		
9		200	
6			500
	0.09	800	
	0.1		600
		700	1300
8	0.1		
5		100	
7			800
	0.07	600	
	0.08		1000
		400	800

FIGURE **AIR-8.1**

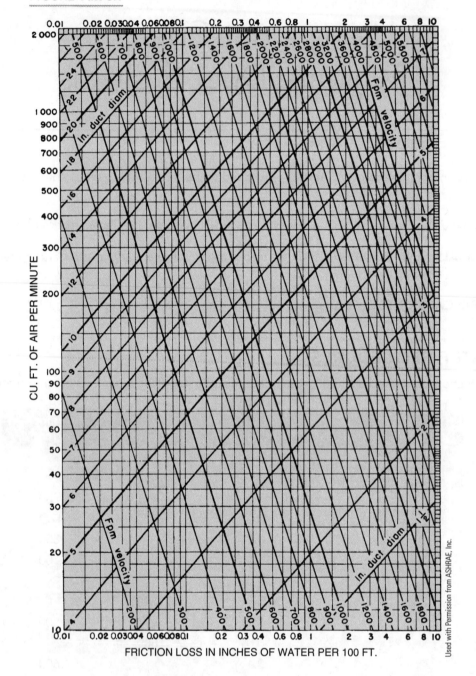

CU. FT. OF AIR PER MINUTE

FRICTION LOSS IN INCHES OF WATER PER 100 FT.

2. Using the duct conversion chart (Figure AIR-8.2), complete this chart by filling in the missing information:

FIGURE **AIR-8.2**

Lgth Adj.[b]	6	7	8	9	10	11	12	13	14	15	16	17	18	19	20	22	24	26	28	30	Lgth Adj.[b]
6	6.6																				6
7	7.1	7.7																			7
8	7.6	8.2	8.7																		8
9	8.0	8.7	9.3	9.8																	9
10	8.4	9.1	9.8	10.4	10.9																10
11	8.8	9.5	10.2	10.9	11.5	12.0															11
12	9.1	9.9	10.7	11.3	12.0	12.6	13.1														12
13	9.5	10.3	11.1	11.8	12.4	13.1	13.7	14.2													13
14	9.8	10.7	11.5	12.2	12.9	13.5	14.2	14.7	15.3												14
15	10.1	11.0	11.8	12.6	13.3	14.0	14.6	15.3	15.8	16.4											15
16	10.4	11.3	12.2	13.0	13.7	14.4	15.1	15.7	16.4	16.9	17.5										16
17	10.7	11.6	12.5	13.4	14.1	14.9	15.6	16.2	16.8	17.4	18.0	18.6									17
18	11.0	11.9	12.9	13.7	14.5	15.3	16.0	16.7	17.3	17.9	18.5	19.1	19.7								18
19	11.2	12.2	13.2	14.1	14.9	15.7	16.4	17.1	17.8	18.4	19.0	19.6	20.2	20.8							19
20	11.5	12.5	13.5	14.4	15.2	16.0	16.8	17.5	18.2	18.9	19.5	20.1	20.7	21.3	21.9						20
22	12.0	13.0	14.1	15.0	15.9	16.8	17.6	18.3	19.1	19.8	20.4	21.1	21.7	22.3	22.9	24.0					22
24	12.4	13.5	14.6	15.6	16.5	17.4	18.3	19.1	19.9	20.6	21.3	22.0	22.7	23.3	23.9	25.1	26.2				24
26	12.8	14.0	15.1	16.2	17.1	18.1	19.0	19.8	20.6	21.4	22.1	22.9	23.5	24.2	24.9	26.1	27.3	28.4			26
28	13.2	14.5	15.6	16.7	17.7	18.7	19.6	20.5	21.3	22.1	22.9	23.7	24.4	25.1	25.8	27.1	28.3	29.5	30.6		28
30	13.6	14.9	16.1	17.2	18.3	19.3	20.2	21.1	22.0	22.9	23.7	24.4	25.2	25.9	26.6	28.0	29.3	30.5	31.7	32.8	30
32	14.0	15.3	16.5	17.7	18.8	19.8	20.8	21.8	22.7	23.5	24.4	25.2	26.0	26.7	27.5	28.9	30.2	31.5	32.7	33.9	32
34	14.4	15.7	17.0	18.2	19.3	20.4	21.4	22.4	23.3	24.2	25.1	25.9	26.7	27.5	28.3	29.7	31.0	32.4	33.7	34.9	34
36	14.7	16.1	17.4	18.6	19.8	20.9	21.9	22.9	23.9	24.8	25.7	26.6	27.4	28.2	29.0	30.5	32.0	33.3	34.6	35.9	36
38	15.0	16.5	17.8	19.0	20.2	21.4	22.4	23.5	24.5	25.4	26.4	27.2	28.1	28.9	29.8	31.3	32.8	34.2	35.6	36.8	38
40	15.3	16.8	18.2	19.5	20.7	21.8	22.9	24.0	25.0	26.0	27.0	27.9	28.8	29.6	30.5	32.1	33.6	35.1	36.4	37.8	40
42	15.6	17.1	18.5	19.9	21.1	22.3	23.4	24.5	25.6	26.6	27.6	28.5	29.4	30.3	31.2	32.8	34.4	35.9	37.3	38.7	42
44	15.9	17.5	18.9	20.3	21.5	22.7	23.9	25.0	26.1	27.1	28.1	29.1	30.0	30.9	31.8	33.5	35.1	36.7	38.1	39.5	44
46	16.2	17.8	19.3	20.6	21.9	23.2	24.4	25.5	26.6	27.7	28.7	29.7	30.6	31.6	32.5	34.2	35.9	37.4	38.9	40.4	46
48	16.5	18.1	19.6	21.0	22.3	23.6	24.8	26.0	27.1	28.2	29.2	30.2	31.2	32.2	33.1	34.9	36.6	38.2	39.7	41.2	48
50	16.8	18.4	19.9	21.4	22.7	24.0	25.2	26.4	27.6	28.7	29.8	30.8	31.8	32.8	33.7	35.5	37.2	38.9	40.5	42.0	50
52	17.1	18.7	20.2	21.7	23.1	24.4	25.7	26.9	28.0	29.2	30.3	31.3	32.3	33.3	34.3	36.2	37.9	39.6	41.2	42.8	52
54	17.3	19.0	20.6	22.0	23.5	24.8	26.1	27.3	28.5	29.7	30.8	31.8	32.9	33.9	34.9	36.8	38.6	40.3	41.9	43.5	54
56	17.6	19.3	20.9	22.4	23.8	25.2	26.5	27.7	28.9	30.1	31.2	32.3	33.4	34.4	35.4	37.4	39.2	41.0	42.7	44.3	56
58	17.8	19.5	21.2	22.7	24.2	25.5	26.9	28.2	29.4	30.6	31.7	32.8	33.9	35.0	36.0	38.0	39.8	41.6	43.3	45.0	58
60	18.1	19.8	21.5	23.0	24.5	25.9	27.3	28.6	29.8	31.0	32.2	33.3	34.4	35.5	36.5	38.5	40.4	42.3	44.0	45.7	60
62		20.1	21.7	23.3	24.8	26.3	27.6	28.9	30.2	31.5	32.6	33.8	34.9	36.0	37.1	39.1	41.0	42.9	44.7	46.4	62
64		20.3	22.0	23.6	25.1	26.6	28.0	29.3	30.6	31.9	33.1	34.3	35.4	36.5	37.6	39.6	41.6	43.5	45.3	47.1	64
66		20.6	22.3	23.9	25.5	26.9	28.4	29.7	31.0	32.3	33.5	34.7	35.9	37.0	38.1	40.2	42.2	44.1	46.0	47.7	66
68		20.8	22.6	24.2	25.8	27.3	28.7	30.1	31.4	32.7	33.9	35.2	36.3	37.5	38.6	40.7	42.8	44.7	46.6	48.4	68
70		21.1	22.8	24.5	26.1	27.6	29.1	30.4	31.8	33.1	34.4	35.6	36.8	37.9	39.1	41.2	43.3	45.3	47.2	49.0	70
72			23.1	24.8	26.4	27.9	29.4	30.8	32.2	33.5	34.8	36.0	37.2	38.4	39.5	41.7	43.8	45.8	47.8	49.6	72
74			23.3	25.1	26.7	28.2	29.7	31.2	32.5	33.9	35.2	36.4	37.7	38.8	40.0	42.2	44.4	46.4	48.4	50.3	74
76			23.6	25.3	27.0	28.5	30.0	31.5	32.9	34.3	35.6	36.8	38.1	39.3	40.5	42.7	44.9	47.0	48.9	50.9	76
78			23.8	25.6	27.3	28.8	30.4	31.8	33.3	34.6	36.0	37.2	38.5	39.7	40.9	43.2	45.4	47.5	49.5	51.4	78
80			24.1	25.8	27.5	29.1	30.7	32.2	33.6	35.0	36.3	37.6	38.9	40.2	41.4	43.7	45.9	48.0	50.1	52.0	80
82				26.1	27.8	29.4	31.0	32.5	34.0	35.4	36.7	38.0	39.3	40.6	41.8	44.1	46.4	48.5	50.6	52.6	82
84				26.4	28.1	29.7	31.3	32.8	34.3	35.7	37.1	38.4	39.7	41.0	42.2	44.6	46.9	49.0	51.1	53.2	84
86				26.6	28.3	30.0	31.6	33.1	34.6	36.1	37.4	38.8	40.1	41.4	42.6	45.0	47.3	49.6	51.7	53.7	86
88				26.9	28.6	30.3	31.9	33.4	34.9	36.4	37.8	39.2	40.5	41.8	43.1	45.5	47.8	50.0	52.2	54.3	88
90				27.1	28.9	30.6	32.2	33.8	35.3	36.7	38.2	39.5	40.9	42.2	43.5	45.9	48.3	50.5	52.7	54.8	90
92					29.1	30.8	32.5	34.1	35.6	37.1	38.5	39.9	41.3	42.6	43.9	46.4	48.7	51.0	53.2	55.3	92
96					29.6	31.4	33.0	34.7	36.2	37.7	39.2	40.6	42.0	43.3	44.7	47.2	49.6	52.0	54.2	56.4	96

Length of One Side of Rectangular Duct (a), in.

Duct Diameter (inches)	Length of One Side of Rectangular Duct	Length of Second Side of Rectangular Duct	Square Dimensions (If duct is to be square)
10	12		
12	12		
21	30		
17.5	20		
24	17		
13	16		
22	24		
26	20		
30.5	20		

MAINTENANCE OF WORKSTATION AND TOOLS: There are no tools required for this exercise.

SUMMARY STATEMENT: Describe the relationships that exist, based on Figure AIR-8.1, among the diameter of a round duct, the friction rate, the velocity of the air, and the volume of air delivered in cfm.

TAKEAWAY: The air distribution portion of an air-conditioning system is one of the most important factors that affects overall system performance. Improperly designed, sized, or installed duct systems will negatively affect the overall operation of the system. This exercise is one of many that will help you improve your understanding of air distribution and, ultimately, its importance in the system.

QUESTIONS

1. Explain how the velocity of air through a duct system affects the noise levels of the air.

2. Describe some of the factors that might be considered when selecting among round, square, and rectangular duct runs.

3. Explain how the friction rate is calculated.

Airflow Measurements

Name	Date	Grade

OBJECTIVES: Upon completion of this exercise, you should be able to use basic airflow-measuring instruments to measure airflow from registers and grilles.

INTRODUCTION: You will use an anemometer such as that in Figure AIR-9.1 to record the velocity of the air leaving a supply register. You will then measure the size of the register or use the manufacturer's data to calculate, in cfm, the volume of air leaving the register.

TOOLS, MATERIALS, AND EQUIPMENT: An anemometer, or equivalent measuring device, and eye protection.

⚠ **SAFETY PRECAUTIONS:** Do not allow air to blow directly into your eyes because it may contain dust.

STEP-BY-STEP PROCEDURES

1. Put on your safety glasses. Start the system blower.

2. Take air velocity measurements from a rectangular air discharge register that has a flat face, such as a high-side wall, floor, or ceiling type.

3. Measure the velocity in several places on the register for an average reading. Record the readings here in fpm, feet per minute: _____, _____, _____, _____, _____.

   • Calculate the average value of the obtained readings, and record the result here: _____.

FIGURE **AIR-9.1**

Courtesy Dwyer Instruments, Inc.

- The volume of air in cfm may be found by multiplying the velocity of the air in feet per minute (fpm) by the free area of the register in square feet. The free area of the register may be found by the following method:

  Measure the open area of the front of the register where the air flows out. For example, the register may measure 8" × 10" = 80 sq. in. Divide by 144 sq. in. to get the number of sq. ft (0.56 sq. ft register area).

  *This value represents the face area, not the free area.

  *The register's blades take up some of the area that you measured. For a register with movable blades, multiply the area × 0.75 (75% free area). For a register with stamped blades, multiply the area × 0.60 (60% free area).

  _____ fpm × _____ sq. ft = _____ cfm

Note: If the manufacturer's information is available for the resisters, the free air values can be obtained from this documentation.

4. Repeat the previous measurements for each outlet in the system. Record this information here:

   Supply register 1: _____ fpm × _____ sq. ft = _____ cfm

   Supply register 2: _____ fpm × _____ sq. ft = _____ cfm

   Supply register 3: _____ fpm × _____ sq. ft = _____ cfm

   Supply register 4: _____ fpm × _____ sq. ft = _____ cfm

   Supply register 5: _____ fpm × _____ sq. ft = _____ cfm

   Supply register 6: _____ fpm × _____ sq. ft = _____ cfm

   Supply register 7: _____ fpm × _____ sq. ft = _____ cfm

   Supply register 8: _____ fpm × _____ sq. ft = _____ cfm

   Supply register 9: _____ fpm × _____ sq. ft = _____ cfm

   Supply register 10: _____ fpm × _____ sq. ft = _____ cfm

5. Add up all of the individual cfm values from the pervious step and record the total airflow here:
   _____ cfm

6. Repeat steps 2, 3, 4, and 5 with the return grille(s). Keep in mind that the airflow through the return grille is in the opposite direction as that through the supply registers. Record the results here:

   Return grille 1: _____ fpm × _____ sq. ft = _____ cfm

   Return grille 2: _____ fpm × _____ sq. ft = _____ cfm

   Return grille 3: _____ fpm × _____ sq. ft = _____ cfm

   Total return airflow: _____ cfm

   If the system has one or two return air inlets, you may arrive at the total airflow by using the preceding method with these grilles. Remember, the air is flowing toward the return air grille.

MAINTENANCE OF WORKSTATION AND TOOLS: Return all tools to their respective places.

SUMMARY STATEMENT: Compare and comment on the results obtained for the total cfm in steps 5 and 6. Be sure to provide reasons for any discrepancies that may exist between the two values.

TAKEAWAY: Using proper instrumentation to measure the properties of airflow through a duct system is an important part of ensuring proper system operation. Knowing how to take these measurements correctly and also knowing how to properly interpret the readings obtained are highly valued skills that the successful HVACR technician possesses.

QUESTIONS

1. How is a stamped grille different from a register?

2. When comparing a stamped grille to a register, which is more desirable for directing airflow?

3. When comparing a stamped grille to a register, which is typically the most expensive?

4. What is the difference between a register and a grille?

5. Describe what free area in a register or grille means.

6. What problem would excess velocity from the supply registers cause?

Exercise AIR-10

Measuring the Amperage of Furnace or Air Handler Blower Motors

Name	Date	Grade

OBJECTIVES: Upon completion of this exercise, you should be able to measure and evaluate the amperage draw of a furnace or air handler blower motor under various conditions.

INTRODUCTION: The amperage draw of a motor will be affected by the amount of air that a blower moves. During this exercise, you will simulate a number of conditions that commonly arise in an air-conditioning or heating system and evaluate the airflow based on the amperage draw of the blower motor.

TOOLS, MATERIALS, AND EQUIPMENT: An operating air handler, hand tools, clamp-on ammeter, duct tape, utility knife, tape measure, safety goggles, pencil, and corrugated cardboard.

⚠ **SAFETY PRECAUTIONS:** Be sure to exercise caution when working with the utility knife. Always cut away from yourself to avoid getting cut. Be sure to wear safety goggles to avoid having dust or dirt blown into your eyes. When working on the blower, make certain that the power has been disconnected to the unit.

STEP-BY-STEP PROCEDURES

1. Obtain the blower motor information from the unit's nameplate, and record it here:
   A. Make of the motor: _____
   B. R.L.A.: _____ A
   C. F.L.A: _____ A
   D. L.R.A.: _____ A
   E. Voltage rating of motor: _____ V

2. With the power disconnected to the air handler, inspect the air filter and the blower wheel to make certain that they are both clean. Clean them if necessary.

3. With the power still disconnected, open the service disconnect switch, and locate one of the conductors that supplies power to the blower motor.

   Note 1: This should be at the disconnect switch, not in the air handler itself. It is important that the service panels be in place for this exercise.

   Note 2: The conductor should be supplying power ONLY to the blower motor for best results.

4. Clamp an ammeter around this conductor, and turn the meter on.

5. Energize the blower.

6. Measure the amperage draw of the blower motor, and record this value here: _____ A

7. Compare this amperage reading to the nameplate amperage ratings you obtained in step 1, and comment here:

8. Measure the voltage being supplied to the blower, and record this value here: _____ V

9. Compare this voltage to the nameplate voltage ratings you obtained in step 1, and comment here:

10. Disconnect the power to the air handler, and lock out the disconnect switch.

11. Measure the return air opening on the air handler.

12. Mark out these measurements on a piece of corrugated cardboard, and carefully cut the cardboard to size.

13. Position the cardboard over the inlet of the air handler so that the cardboard covers 50% of the return opening. Tape the cardboard in place.

14. Restore power to the unit, and energize the blower.

15. Measure the amperage draw of the blower motor, and record this value here: _____ A

16. Compare this amperage to the amperage obtained in step 6, and explain the difference here:

17. Disconnect the power to the air handler, and lock out the disconnect switch.

18. Reposition the cardboard over the inlet of the air handler so that the cardboard covers 100% of the return opening. Tape the cardboard in place.

19. Restore power to the unit, and energize the blower.

20. Measure the amperage draw of the blower motor, and record this value here: _____ A

21. Compare this amperage to the amperage obtained in step 15, and explain the difference here:

22. Disconnect the power to the air handler and lock out the disconnect switch.

23. Remove the cardboard from the inlet of the air handler.

24. Repeat the previous steps 11 through 21, but this time block off the discharge (outlet side) of the air handler. Record your data here:

    Amperage draw of blower motor with discharge blocked 50%: _____ A

    Amperage draw of blower motor with discharge blocked 100%: _____ A

25. Disconnect the power to the air handler, and lock out the disconnect switch.

26. Using duct tape, wrap approximately one-third of the squirrel cage blower on the air handler, Figure AIR-10.1.

FIGURE **AIR-10.1**

Blower wheel

Duct tape

27. Replace any panels that might have been removed to tape the blower.

28. Restore power to the unit, and energize the blower.

29. Measure the amperage draw of the blower motor, and record this value here: _____ A

30. Disconnect the power to the air handler, and lock out the disconnect switch.

31. Remove the duct tape from the blower.

32. Remove the clamp-on ammeter from the service disconnect.

33. Close the service disconnect.

MAINTENANCE OF WORKSTATION AND TOOLS: Make certain that all equipment has been properly closed and secured with all service panels replaced and all screws accounted for and in place. Return all tools to their proper location, and clean the work area.

SUMMARY STATEMENT: Explain how airflow through an air handler affects the amperage draw of the motor.

TAKEAWAY: Improper amperage draw of a motor is not always an electrical concern. As was seen in this exercise, the amount of air that is moved by the blower will affect the amperage draw of the motor. Be sure to establish correct airflow before concluding that the motor is not operating properly.

QUESTIONS

1. Explain how reduced airflow through a duct system can cause a blower motor to fail.

2. Explain why it is important to check the air filters on an air handler on a regular basis.

3. Explain what would happen if a service panel on the inlet side of a blower comes loose.

Determining Airflow (cfm) by Using the Air Temperature Rise Across an Electric Furnace

Name	Date	Grade

OBJECTIVES: Upon completion of this exercise, you should be able to calculate the quantity of airflow in cubic feet per minute (cfm) in an electric furnace by measuring the air temperature rise, amperage, and voltage.

INTRODUCTION: You will measure the air temperature at the electric furnace supply and return ducts and the voltage and amperage at the main electrical supply. By using the formulas furnished, you will calculate the cubic feet per minute being delivered by the furnace.

TOOLS, MATERIALS, AND EQUIPMENT: A digital multimeter (DMM), a clamp-on ammeter, electronic thermometer, calculator, Phillips-head and straight-slot screwdrivers, ¼" and ⁵⁄₁₆" nut drivers, amemometer, safety glasses, and operating, single-phase electric furnace.

SAFETY PRECAUTIONS: Use proper procedures in making electrical measurements. Your instructor should check all measurement setups before you turn the power on. Use digital multimeter leads with alligator clips. Keep your hands away from all moving or rotating parts. Lock and tag the panel where you turn the power off. Keep the single key in your possession.

STEP-BY-STEP PROCEDURES

1. Turn the power to the furnace off, and lock out the panel. Under the supervision of your instructor and with the power off, connect the digital multimeter so that it will measure the voltage supplied to the electric furnace. Position the clamp-on ammeter so it will measure the amperage draw of the furnace. Figure AIR-11.1 provides a sample calculation.

FIGURE **AIR-11.1**

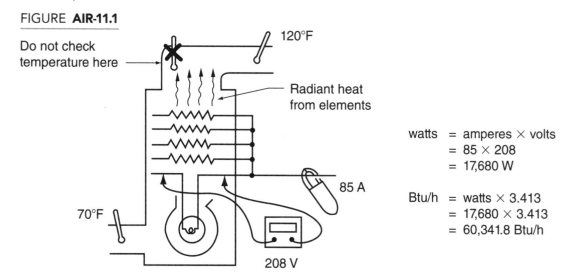

$$\text{watts} = \text{amperes} \times \text{volts}$$
$$= 85 \times 208$$
$$= 17{,}680 \text{ W}$$

$$\text{Btu/h} = \text{watts} \times 3.413$$
$$= 17{,}680 \times 3.413$$
$$= 60{,}341.8 \text{ Btu/h}$$

2. Turn the unit on. Wait for all elements to be energized. Record the current and voltage:

Current: _____ A

Voltage: _____ V

(Amperes × Volts = Watts)

(_____ × _____ = _____ Watts)

3. Convert watts to Btu per hour, Btu/h. This is the furnace input. Watts × 3.413 = Btu/h.

_____ × 3.413 = _____ Btu/h

4. Find the change in temperature (ΔT) from the return duct to the supply duct. Place a temperature sensor in the supply duct (around the first bend in the duct to prevent radiant heat from the elements from hitting the probe) and a temperature sensor in the return duct. Record the temperatures:

Supply _____°F (Supply − Return = Change in temperature or ΔT)

Return duct _____°F (_____ − _____ = _____°F ΔT)

5. To calculate the cubic feet per minute (cfm) of airflow, use the following formulas and procedures:

$$\text{cfm airflow} = \frac{\text{Total heat input (step 3)}}{1.08 \times \Delta T \text{ (step 4)}} = \frac{\_\_\_\_ \text{ Btu/h}}{1.08 \times \_\_\_ \text{ °F}} = _____$$

Note: 1.08 is a constant used for standard air conditions. Its derivation is: Specific Heat of air (0.24 btu/lb) × Density of standard air (0.075lb/ft³) × 60 minutes = 1.08

6. Allow the furnace to cool down and then turn it off. Replace all panels on the unit with the correct fasteners.

7. To verify the cfm value that was determined in step 5, you will use an anemometer to determine the velocity of the air moving through the furnace.

8. With the system operating, traverse the return duct to obtain several air velocity readings, in feet per minute, in the return duct, and record them here:

Measurement 1: _____ fpm   Measurement 2: _____ fpm

Measurement 3: _____ fpm   Measurement 4: _____ fpm

Measurement 5: _____ fpm   Measurement 6: _____ fpm

If you are unable to access the return duct, take the air velocity measurements at the return grill on the system.

9. Calculate the average air velocity, and record the result here: _____ fpm

10. Measure the cross-sectional dimensions of the return duct in inches, and record the measurements here: _____ in. x _____ in. If you are taking measurements at a return grille, skip to step 15.

11. Determine the cross-sectional area of the return duct in square inches and square feet. Record these values here: _____ sq. in., _____ sq. ft

12. Multiply the area of the duct (in square feet) from step 11 and the average velocity from step 9, to determine the airflow in cfm. Record this value here: _____ cfm

13. Compare this value to the cfm value that was calculated in step 5. Describe any difference between these two values, providing any possible reasons for the discrepancy.

_____

_____

_____

14. Proceed to the Maintenance of Workstation and Tools portion of this exercise.

15. Look up the technical data for the return grille, and determine the Ak value for the grille, which is the grille's, net area in square feet. Record this value here: Ak = _____ sq. ft.

If the technical data for the grille is not available, obtain the value that you are to use from your instructor.

16. Multiply the Ak value entered in step 15 and the average velocity from step 9 to determine the airflow in cfm. Record this value here: _____ cfm

17. Compare this value to the cfm value that was calculated in step 5. Describe any difference between these two values, providing any possible reasons for the discrepancy.

_____

_____

_____

MAINTENANCE OF WORKSTATION AND TOOLS: Return all tools and instruments to their proper places. Leave the furnace as you are instructed.

SUMMARY STATEMENT: Describe why it is necessary for a furnace to provide the correct cfm. Use an additional sheet of paper if more space is necessary for your answer.

TAKEAWAY: During service, it may be necessary to verify the airflow (cfm) being delivered by a furnace. Taking accurate measurements and applying the appropriate formulae are critical to obtaining an accurate air volume flow.

QUESTIONS

1. The kW rating of an electric heater is the _____ (temperature range, power rating, voltage rating, or name of the power company).

2. The Btu/h rating is a rating of the _____ (time, wire size, basis for power company charges, or heating capacity).

3. The factor for converting watts to Btu/h is _____.

4. One kW is (1000 W, 1200 W, 500 W, or 10,000 W) _____.

5. The typical air temperature rise across an electric furnace is normally considered (high or low) compared to a gas furnace.

6. The air temperature at the outlet grilles of an electric heating system is considered (warmer or cooler) compared to a gas furnace.

7. The circulating blower motor normally runs on (high or low) speed in the heating mode.

8. What is the current rating of a 7.5 kW heater operating at 230 V?

9. A 12 kW heater will put out _____ Btu per hour.

Electric Motors (MOT)

SUBJECT AREA OVERVIEW

The electric motor is the most important load in a refrigeration system. The electric motor changes electrical energy into mechanical energy, which is then used to drive compressors, fans, pumps, dampers, and other devices. There are many different types of electric motors, each with different running and starting characteristics, based on the many different functions they perform. Some motors, for example, operate under heavy loads, but others do not.

The basic types of motors used in the HVACR industry include the shaded pole, split-phase, permanent split capacitor (PSC), induction-start–induction-run (ISIR), capacitor-start–induction-run (CSIR), capacitor-start–capacitor-run (CSCR), electronically commutated (ECM), and three-phase. Most single-phase electric motors have a start and run winding, while three-phase motors have three or more sets of run windings. Three-phase motors have a very high starting torque and are desirable for use when three-phase power is available. The ECM is a high-efficiency, variable-speed motor that can be programmed to maintain constant airflow over a wide range of external static pressures.

Single-phase motors often utilize additional components that assist in starting and running the motor. These devices include start capacitors, run capacitors, current magnetic relays (CMRs), potential magnetic relays (PMRs), positive temperature coefficient (PTC) devices, relays, and contactors. Capacitors are devices that are used to increase the starting torque or running efficiency of a motor and are classified as either start or run devices, depending on the function they serve. PMRs and CMRs are control relays that will open and close their contacts in order to add or remove motor windings or starting components from the active electrical circuit to ensure proper motor operation. Relays and contactors are generally described as devices that open and close sets of electrical contacts in response to some control signal. These devices, unlike the CMR and PMR, typically do not monitor or respond to the motor's operational state, but simply start and stop the device. Relays and contactors perform the same basic function, but the contacts on contactors are designed to handle larger amounts of current. Three-phase motors utilize motor starters, which are contactors that have overload protection built into them.

Most motors need overload protection in addition to the circuit protection provided by fuses or circuit breakers. In their simplest form, overloads are intended to protect motors from high-heat/amperage conditions and can be mechanical, electrical, or electronic devices. These additional overload protection devices can be either internal or externally located. In addition, these devices can be configured as either manually or automatically resetting devices. When an overload device has tripped, it is important for the technician to determine the reason why the device tripped as opposed to simply resetting it.

All rotating electric devices have some type of bearing to allow for smooth and easy rotation. The two types of bearings used in the HVACR industry are ball bearings and sleeve bearings. Ball bearings are the most efficient because they produce less friction and are used on heavy loads. The sleeve bearing is the most popular because of its low price.

A motor drive is the connection between an electric motor and a component that requires rotation. There are basically two types of motor drives

(continued)

used in the HVACR industry. These are the direct-driven drive and the belt-driven drive. The direct-drive method couples the motor directly to the device requiring rotation, so the rotational speed of the driven device is the same as the speed of the motor. The belt-driven assembly requires the use of belts and pulleys to connect the motor and the driven device. The speed of a belt-driven device can be changed by changing the size of the drive or driven pulley.

One of the most important tasks performed by an HVACR technician is the diagnosis of motors and their associated starting components. Successful troubleshooting relies on an understanding of how electrical components operate, how to use electric meters to obtain system readings, and how to interpret those readings. Troubleshooting electric motors can generally be divided into two categories, electrical problems and mechanical problems. Electrical problems are those with the electrical continuity of the motor windings or their starting components. Mechanical problems are those that are related to the mechanical parts of the motor such as bearings.

Mechanical motor problems normally occur in the bearings. Bearings may wear due to lack of lubrication or due to grit that has found its way into them. Most motor bearings can be replaced, but it may be less expensive to replace the entire motor, especially on smaller systems, than to replace the bearings. Motor bearings can often be quickly diagnosed by turning the shaft manually; if the shaft is easily turned, the bearings are probably okay.

Electrical problems may consist of an open winding, a short circuit from the winding to ground, or a short circuit from winding to winding. There should be no circuit from any winding to ground. A megohmmeter, which has the capacity to detect very high resistances, may be used to determine high-resistance ground circuits. Motor windings should have a measurable resistance. Readings of either 0 Ω or infinite resistance are sure signs that there is a problem with the motor.

Motor-related electrical problems can also lie with the motor's starting components. Before condemning a motor, it is important to rule out the starting components as the cause for failure. The relays, contactors, starters, and capacitors should be checked, and if they are all good, the motor should be replaced. It is good practice to replace all starting components when replacing a motor.

KEY TERMS

Back electromotive force (BEMF)
Ball bearing
Bearings
Belly-band-mount motor
Capacitor
Capacitor-start motor
Capacitor-start–capacitor-run
Centrifugal switch
Coil
Contactor
Contacts
Continuous coil voltage
Cradle-mount motor
Current magnetic relay (CMR)
Current overload
Direct drive
Drop-out voltage
Electromagnet
Electromotive force (EMF)

Electronic relay
Electronically commutated
End bell
End mount
Frequency
Full-load amperage (FLA)
Fuses
Hermetic compressor
Housing
Ignition module
Induced magnetism
Induction motor
Inductive load
Internal compressor overload
Inverter
Leads
Line break overload
Locked-rotor amperage
Magnetic field
Magnetic overload

Magnetic starter
Magnetism
Mechanical linkage
Microfarad
Motor
Overload
Permanent split-capacitor
Pick-up voltage
Pilot duty overload
Potential magnetic relay (PMR)
Pressure switch
Primary control
Pulse-width modulation (PWM)
Push-button station
Relay
Resilient-mount motor
Resistive load
Rigid-mount motor
Rotor
Run windings

Run capacitor
Service factor
Service factor amperes (SFA)
Shaded-pole motor
Single-phase
Single-phase motor
Sleeve bearing
Slip
Solder pot
Solid-state relay
Split-phase motor
Squirrel cage rotor
Start capacitor
Starting relays
Stator
Switch frequency
Three-phase motor
Thrust surface
Torque
Transformer
V-belt

REVIEW TEST

Name	Date	Grade

Circle the letter that indicates the correct answer.

1. A two-pole, split-phase motor operates at a speed that is slightly slower than:
   A. 1800 rpm.
   B. 3600 rpm.
   C. 4800 rpm.
   D. 5200 rpm.

2. The start winding of a single-phase motor is wired:
   A. in series with the run winding.
   B. in parallel with the run winding.
   C. in parallel with the start capacitor.
   D. in parallel with the run capacitor.

3. The amount by which the current leads or lags the voltage in an AC circuit is referred to as the:
   A. phase angle.
   B. phase delay.
   C. inductive angle.
   D. capacitive delay.

4. Torque is best described as:
   A. a rotational force.
   B. a linear force.
   C. a radial force.
   D. a parallel force.

5. In an AC capacitive circuit:
   A. the current leads the voltage.
   B. the current lags the voltage.
   C. the voltage leads the current.
   D. the voltage and current are in phase with each other.

6. In an AC inductive circuit:
   A. the current leads the voltage.
   B. the voltage lags the current.
   C. the voltage leads the current.
   D. the voltage and current are in phase with each other.

7. The start winding has:
   A. fewer turns than the run winding.
   B. more turns than the run winding.
   C. the same number of turns as the run winding.
   D. no turns at all.

8. The term "slip," as it applies to motors, is used to indicate:
   A. the amount of horsepower lost due to a motor's internal resistance.
   B. the amount of efficiency lost due to a lack of starting components.
   C. the difference between a motor's calculated and actual speed.
   D. the difference between a motor's starting and running amperage.

9. The frequency of AC power sources used in the United States is:
   A. 50 Hz (cycles per second).
   B. 60 Hz (cycles per second).
   C. 120 Hz (cycles per second).
   D. 208 Hz (cycles per second).

10. The amperage that a split-phase motor will draw when it is initially energized, before the rotor starts to turn, is referred to as the:
   A. locked rotor amperage (LRA).
   B. full load amperage (FLA).
   C. running load amperage (RLA).
   D. rated load amperage (RLA).

11. How does a split-phase motor's locked rotor amperage (LRA) compare to the motor's full load amperage (FLA)?
   A. The locked rotor amperage is about two to four times greater than the motor's full load amperage.
   B. The locked rotor amperage is about five to seven times greater than the motor's full load amperage.
   C. The full load amperage is about two to four times greater than the motor's locked rotor amperage.
   D. The full load amperage is about five to seven times greater than the motor's locked rotor amperage.

12. A four-pole, split-phase motor:
    A. operates faster than a two-pole, split-phase motor.
    B. operates slower than a two-pole, split-phase motor.
    C. has two start windings and one run winding.
    D. has two run windings and one start winding.

13. Which of the following motors has the highest starting torque?
    A. Capacitor-start–capacitor-run (CSCR)
    B. Induction-start–induction-run (ISIR)
    C. Permanent split capacitor (PSC)
    D. Shaded pole

14. ECM motors are best described as:
    A. three-phase shaded-pole motors.
    B. single-phase DC motors.
    C. single-phase shaded pole motors.
    D. three-phase DC motors.

15. The direction of rotation of a shaded pole motor can be reversed by:
    A. reversing the electrical connections on the power supply.
    B. reversing the electrical connections at the motor.
    C. physically reversing the position of the motor's stator.
    D. physically reversing the position of the motor's rotor.

16. A three-phase power supply can be found in most, if not all:
    A. single-family homes.
    B. mobile homes.
    C. factories and hospitals.
    D. apartments.

17. The shaded-pole motor:
    A. operates with low starting torque.
    B. uses both start and run capacitors.
    C. has greater efficiency than most split-phase motors.
    D. is designed for use with three-phase power supplies.

18. Three-phase motors have:
    A. only start windings.
    B. a start capacitor.
    C. no start capacitor and no start winding.
    D. both a start capacitor and a start winding.

19. On a three-phase motor, each winding is how far out of phase with the others?
    A. 90 electrical degrees
    B. 120 electrical degrees
    C. 180 electrical degrees
    D. 360 electrical degrees

20. Three-phase motors typically have:
    A. low starting torque and high operating efficiency.
    B. low starting torque and low operating efficiency.
    C. high starting torque and low operating efficiency.
    D. high starting torque and high operating efficiency.

21. Which of the following devices can be used to remove the start winding from the active electric circuit once the motor has started and is operating close to its design speed?
    A. Start capacitor
    B. Run capacitor
    C. Centrifugal switch
    D. Dual-voltage switch

22. The current magnetic relay (CMR) is comprised of a:
    A. low-resistance coil and a normally closed set of contacts.
    B. low-resistance coil and a normally open set of contacts.
    C. high-resistance coil and a normally closed set of contacts.
    D. high-resistance coil and a normally open set of contacts.

23. The potential magnetic relay (PMR) is comprised of a:
    A. low-resistance coil and a normally closed set of contacts.
    B. low-resistance coil and a normally open set of contacts.
    C. high-resistance coil and a normally open set of contacts.
    D. high-resistance coil and a normally closed set of contacts.

24. The term "microfarad" is used to represent:
    A. one million farads.
    B. one thousand farads.
    C. one one-millionth of a farad.
    D. one one-thousandth of a farad.

25. The purpose of a capacitor is to:
A. change the charge of electrons.
B. temporarily store electrons.
C. permanently store electrons.
D. produce additional electrons.

26. A run capacitor:
A. is encased in a hard plastic casing and is classified as a dry-type capacitor.
B. is used to reduce the phase angle between the voltage and the current.
C. is only in the active electric circuit long enough to help the motor start.
D. is wired in parallel with both the start and run windings of the motor.

27. If a 5-microfarad capacitor and a 20-microfarad capacitor are wired in parallel with each other, the total effective capacitance will be:
A. 4 µF.
B. 5 µF.
C. 20 µF.
D. 25 µF.

28. If a 5-microfarad capacitor and a 20-microfarad capacitor are wired in series with each other, the total effective capacitance will be:
A. 4 µF.
B. 5 µF.
C. 20 µF.
D. 25 µF.

29. A PTC starting device is a type of:
A. capacitor.
B. inductor.
C. thermistor.
D. transistor.

30. The pilot-duty relay is used primarily to:
A. open and close contacts that control other control devices.
B. directly control single-phase, fractional horsepower motors.
C. directly control three-phase motors rated over 5 horsepower.
D. allow a larger load to control the operation of a smaller load.

31. If a general purpose relay fails, how is it generally repaired?
A. The coil is replaced.
B. The contacts are replaced.
C. Both the coil and the contacts are replaced.
D. The entire device is replaced.

32. A three-phase motor starter is best described as a:
A. general-purpose relay with built-in overload protection.
B. single-pole contactor with built-in overload protection.
C. two-pole, general-purpose relay with built-in overload protection.
D. three-pole contactor with built-in overload protection.

33. Inherent motor protection devices are located:
A. at the circuit breaker.
B. at the space thermostat.
C. within the motor winding.
D. within the motor's starter.

34. Sleeve bearings are preferred when:
A. the load on the bearing is low and low noise levels are desired.
B. the load on the bearing is high and low noise levels are desired.
C. the load on the bearing is high and the motor operates in a dirty environment.
D. the load on the bearing is low and the motor operates in a moist environment.

35. The bearing typically preferred for use on the bottom of a vertically mounted motor is the:
A. sleeve bearing.
B. ball bearing.
C. thrust bearing.
D. bronze bearing.

36. A direct-drive assembly connects the motor to the driven device using:
A. pulleys.
B. belts and pulleys.
C. couplings.
D. belts and couplings.

37. The device that corrects for a minor shaft misalignment between the drive and the driven device is the:
    A. flexible coupling.
    B. drive pulley.
    C. driven pulley.
    D. rigid coupling.

38. A motor that is turning at a speed of 1800 rpm with a 4" pulley connected to it will cause a driven device to turn at 1200 rpm if the pulley connected to the driven device has what diameter?
    A. 3 in.
    B. 4 in.
    C. 5 in.
    D. 6 in.

39. If a motor is rated at 230 V, what is the acceptable range of the supply voltage?
    A. 220 to 240 V
    B. 207 to 253 V
    C. 227 to 233 V
    D. 190 to 270 V

40. A loose-fitting belt on a belt-driven assembly will cause:
    A. excessive bearing wear.
    B. motor overheating.
    C. the pulleys to wobble.
    D. worn pulley grooves.

41. Dirty and/or pitted contactor contacts can cause:
    A. the motor to operate at an increased speed.
    B. the voltage supplied to the motor to drop.
    C. the resistance across the contacts to decrease.
    D. the coil of the contactor to burn out.

42. An open winding in an electric motor means that:
    A. a winding is making contact with the motor's frame.
    B. a winding is making contact with another winding.
    C. the contacts on the motor starting switch are open.
    D. a conductor in one of the motor windings is broken.

43. If there is a decrease in the resistance of the run winding:
    A. the motor will not start and will draw locked rotor amperage.
    B. the motor will run, but it will draw high running amperage.
    C. the centrifugal switch will prevent the motor from operating.
    D. the motor will run, but it will draw low running amperage.

44. The voltage rating of a replacement capacitor must be:
    A. equal to the voltage rating of the capacitor being replaced.
    B. greater than or equal to the voltage rating of the capacitor being replaced.
    C. greater than the voltage rating of the capacitor being replaced.
    D. within ±10% of the voltage rating of the capacitor being replaced.

45. A "megger" is best described as a(n):
    A. ammeter that can measure very low currents.
    B. voltmeter that can measure very high AC voltages.
    C. ohmmeter that can measure very high resistances.
    D. voltmeter that can measure very high DC voltages.

46. If a 230 V fan motor is grounded, it will usually:
    A. trip the circuit breaker.
    B. run normally.
    C. run at a faster speed.
    D. run at a slower speed.

47. If a service technician measures 50 Ω between the compressor's start terminal and the shell of the compressor, what can be correctly concluded?
    A. The winding is open.
    B. The winding is shorted.
    C. The winding is grounded.
    D. The winding is good.

Identification of Electric Motors

Name	Date	Grade

OBJECTIVES: Upon completion of this exercise, you should be able to identify different types of electric motors, state their typical applications, and identify the appropriate wiring diagram showing how they are wired into circuits.

INTRODUCTION: You will be provided an air-conditioning unit and a refrigeration unit. You will remove enough panels on each system to identify the motors. You will identify the diagram that correctly indicates how the motor is wired into the circuit.

TOOLS, MATERIALS, AND EQUIPMENT: An air-conditioning unit, refrigeration unit, DMM, straight-slot and Phillips-head screwdrivers, ¼" and ⁵⁄₁₆" nut drivers, and a flashlight.

SAFETY PRECAUTIONS: Turn the power off before removing any panels. Lock and tag the distribution box where you turned the power off, and keep the single key in your possession. Properly discharge any capacitors in the circuit. Wear safety glasses.

STEP-BY-STEP PROCEDURES
Air-Conditioning Unit

1. Turn the power to the unit off. Lock and tag the disconnect switch. Keep the key in your possession.

2. Remove panels as necessary to expose the indoor fan motor.

3. Using a flashlight, if needed, obtain the following information:

   - Number of wires on the motor: _____

     How many wires are connected electrically? _____
     - Wire color: _____ Connected? YES _____ NO _____
     - Wire color: _____ Connected? YES _____ NO _____
     - Wire color: _____ Connected? YES _____ NO _____
     - Wire color: _____ Connected? YES _____ NO _____
     - Wire color: _____ Connected? YES _____ NO _____
     - Wire color: _____ Connected? YES _____ NO _____
     - Wire color: _____ Connected? YES _____ NO _____
     - Wire color: _____ Connected? YES _____ NO _____

   - Is there a run capacitor? _____

   - Is there a start capacitor? _____

   - Motor voltage: _____

   - Type of motor (capacitor-start or other type): _____

4. Replace the panels with the correct fasteners.

5. Remove the panel to the condenser section, exposing the outdoor fan motor.

6. Using a good light, obtain the following information:

- Number of wires on the motor: _____

  How many wires are connected electrically? _____

  Wire color: _____ Connected? YES _____ NO _____
  Wire color: _____ Connected? YES _____ NO _____
  Wire color: _____ Connected? YES _____ NO _____
  Wire color: _____ Connected? YES _____ NO _____
  Wire color: _____ Connected? YES _____ NO _____

- Is there a run capacitor? _____

- Is there a start capacitor? _____

- Motor voltage: _____

- Type of motor (capacitor-start or other type): _____

7. Follow all wires going to the compressor, and describe the following:

- Number of wires: _____

- Is there a run capacitor? _____

- Is there a start capacitor? _____

- Is there a start assist (relay or a PTC device)? _____

- Type of compressor motor (capacitor-start or other type): _____

8. Replace the panels with the correct fasteners.

Refrigeration Unit

9. Turn the power off. Lock and tag the box. Keep the key in your possession.

10. Remove the panel to the evaporator section, exposing the indoor fan motor.

11. Using a good light, obtain the following information:

- Number of wires on the motor: _____

  How many wires are connected electrically? _____

  Wire color: _____ Connected? YES _____ NO _____
  Wire color: _____ Connected? YES _____ NO _____
  Wire color: _____ Connected? YES _____ NO _____
  Wire color: _____ Connected? YES _____ NO _____
  Wire color: _____ Connected? YES _____ NO _____
  Wire color: _____ Connected? YES _____ NO _____
  Wire color: _____ Connected? YES _____ NO _____
  Wire color: _____ Connected? YES _____ NO _____

- Is there a run capacitor? _____

- Is there a start capacitor? _____

- Motor voltage: _____

- Type of motor (capacitor-start or other type): _____

12. Replace the panels with the correct fasteners.

13. Remove the panel to the condenser section, exposing the outdoor fan motor.

14. Using a good light, obtain the following information:

- Number of wires on the motor: _____

  How many wires are connected electrically? _____

    Wire color: _____ Connected? YES _____ NO _____
    Wire color: _____ Connected? YES _____ NO _____
    Wire color: _____ Connected? YES _____ NO _____
    Wire color: _____ Connected? YES _____ NO _____
    Wire color: _____ Connected? YES _____ NO _____

- Is there a run capacitor? _____

- Is there a start capacitor? _____

- Motor voltage: _____

- Type of motor (capacitor-start or other type): _____

15. Replace the panels using the correct fasteners.

16. Follow all wires going to the compressor, and describe the following:

- Number of wires on the motor: _____

  How many wires are connected electrically? _____

    Wire color: _____ Connected? YES _____ NO _____
    Wire color: _____ Connected? YES _____ NO _____
    Wire color: _____ Connected? YES _____ NO _____
    Wire color: _____ Connected? YES _____ NO _____
    Wire color: _____ Connected? YES _____ NO _____

- Is there a run capacitor? _____

- Is there a start capacitor? _____

- Motor voltage: _____

- Type of motor (capacitor-start or other type): _____

17. Replace the panels with the correct fasteners.

18. Identify which of the following represents a wiring diagram of a typical shaded-pole motor.

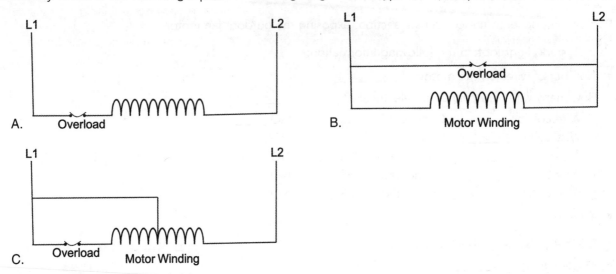

19. Identify which of the following represents a wiring diagram of a typical three-speed permanent split-capacitor motor.

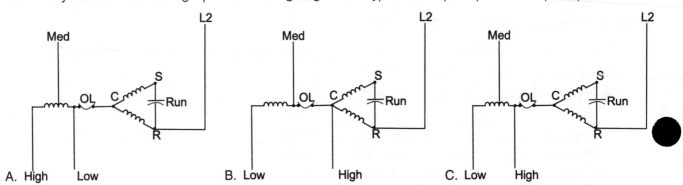

20. Identify which of the following represents a wiring diagram of a single-phase compressor motor with a start relay and a start and run capacitor.

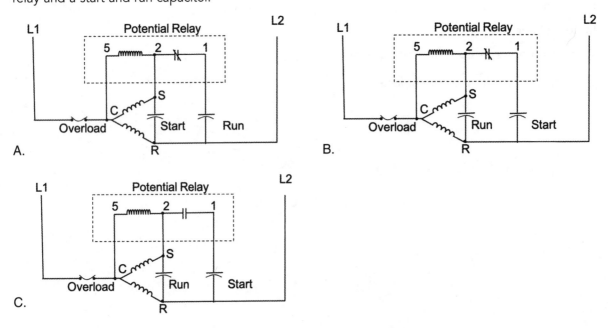

MAINTENANCE OF WORKSTATION AND TOOLS: Return all tools to their respective places. Make certain that all access and service panels are properly secured and that all screws and fasteners are in place.

SUMMARY STATEMENTS: Describe the difference between a shaded-pole and a permanent split-capacitor (PSC) fan motor. Explain which of the two is more efficient and why.

TAKEAWAY: In order to properly evaluate a motor, a technician must first be able to determine the type of motor being evaluated and how it is supposed to operate. This exercise provides introductory exposure to various types of motor and their starting components.

QUESTIONS

1. What is the reason for a start capacitor on a motor?

2. Must a motor with a start capacitor have a start relay? Why or why not?

3. Name two methods used in hermetic compressor motors to take the start winding out of the circuit when the motor is up to speed.

4. Explain the differences between start and run capacitors.

5. How are capacitors rated?

6. Describe a PTC device.

7. How is a typical open electric motor cooled?

8. How is a typical hermetic compressor motor cooled?

Compressor Winding Identification

Name	Date	Grade

OBJECTIVES: Upon completion of this exercise, you should be able to identify the terminals to the start and run windings in a single-phase hermetic compressor.

INTRODUCTION: In order to properly evaluate the windings of a motor, a digital multimeter or similar test instrument must be used. In this exercise, you will use a digital multimeter (DMM) to measure the resistance of the start and run windings of a single-phase hermetic motor.

TOOLS, MATERIALS, AND EQUIPMENT: A single-phase hermetic compressor, DMM, straight-slot and Phillips-head screwdrivers, and ¼" and ⁵⁄₁₆" nut drivers.

⚠ **SAFETY PRECAUTIONS:** Compressors are heavy. Lift them carefully, and make certain that the compressor does not fall over, as the oil contained in the compressor can spill out. If the compressor being used is still mounted in a system, be sure to de-energize any piece of equipment you are servicing. Lock the distribution box where you turned the power off, and keep the single key in your possession. Properly discharge all capacitors.

STEP-BY-STEP PROCEDURES

1. With the compressor completely de-energized, as indicated in the safety precautions, carefully remove the terminal cover from the compressor.

2. If there are wires connected to the compressor terminals, use wire markers or labels to identify the wiring connections before disconnecting them.

3. Double-check your work and have your instructor check it to ensure that you have properly identified the connections. Wiring the compressor improperly can cause permanent damage to the compressor, so be very careful.

4. Using a pair of needle-nose pliers, carefully remove the wires from the compressor terminals.

5. Connect the test leads to the DMM, and set the meter to measure resistance.

6. Check the test leads by touching them together. The meter should indicate continuity and very low resistance. If not, report this to your instructor before proceeding.

7. Select the terminal configuration from Figure MOT-2.1 that best matches the compressor used.

FIGURE **MOT-2.1**

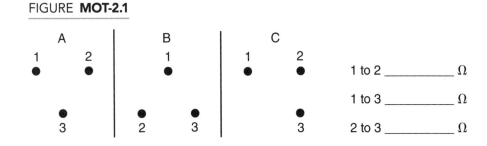

8. Take and record the following measurements:

    1 to 2: _____ Ω

    1 to 3: _____ Ω

    2 to 3: _____ Ω

9. Based on your resistance readings, identify the start winding: terminal _____ to _____.

10. Based on your resistance readings, identify the run winding: terminal _____ to _____.

MAINTENANCE OF WORKSTATION AND TOOLS: Return all equipment and tools to their respective places. Replace all panels with correct fasteners, if appropriate. Replace the cover on the compressor's terminal block using the correct clip.

SUMMARY STATEMENT: Describe the function of the start and run winding in a single-phase compressor. Use an additional sheet of paper if more space is necessary for your answer.

TAKEAWAY: In order for split-phase motors to start, there must be an imbalance in magnetic fields. This imbalance is created by the different resistances in the start and run windings of the motor. This exercise provided the opportunity to measure the resistances of these windings and verify that the two windings are different.

QUESTIONS

1. Name the three terminals on a single-phase compressor. Briefly explain why they are so named.

2. Name two types of relays used to start a single-phase compressor motor. Explain their operation.

3. Why does a hermetic compressor not use a centrifugal switch like an open motor?

4. Which winding has higher resistance, the run or the start? Why is this so?

Exercise MOT-3

Motor Application and Features

Name	Date	Grade

OBJECTIVES: Upon completion of this exercise, you should be able to fully describe a basic motor used for a fan in an air handler.

INTRODUCTION: You will remove a motor from an air handler, inspect it, describe the type of motor, and describe its features, such as the speeds, insulation class, shaft size, rotation, mount characteristics, voltage, amperage, phase, and bearing type.

TOOLS, MATERIALS, AND EQUIPMENT: An air-conditioning air handler, DMM, adjustable wrench, straight-slot and Phillips-head screwdrivers, a flashlight, Allen wrench set, and ¼" and ⁵⁄₁₆" nut drivers.

⚠ **SAFETY PRECAUTIONS:** Turn the power off to any unit you are servicing. Lock and tag the power distribution panel where you turned the power off, and keep the single key in your possession. Do not force any parts that you are disassembling or assembling. Properly discharge all capacitors. Be sure to wear safety glasses.

STEP-BY-STEP PROCEDURES

1. Make sure that the power is off to the air handler to which you have been assigned, the power panel is locked and tagged, and you have the key. Remove the panels. Properly discharge all capacitors.

2. Carefully remove the blower assembly. Label all wires you disconnect to indicate their proper positions.

3. Remove the motor from the mounting, and obtain the following information:

   - Number of motor speeds: _____
   - Insulation class of motor: _____
   - Diameter of the motor shaft: _____ in.
   - Rotation, looking at the shaft: _____

   Note: If the motor has two shafts, look at the end where the motor leads enter.

   - Length of the shaft(s): _____
   - Type of motor mount: _____
   - Voltage rating: _____ V
   - Full-load amperage: _____ A
   - Locked-rotor amperage: _____ A
   - Phase of motor: _____
   - Type of motor (PSC, shaded-pole, ECM, etc.): _____
   - Type of bearings (sleeve or ball): _____
   - Number of motor leads (wires): _____

4. Replace the motor in its mount.

5. Make certain that the blower is attached to the motor shaft and that the blower and shaft rotate freely without contacting the blower housing.

6. Reinstall the blower assembly in unit. If any wires were removed, replace them according to the labels and drawing you created.

7. Replace panels with the correct fasteners.

MAINTENANCE OF WORKSTATION AND TOOLS: Return all tools to their respective places. Make sure that your work area is clean.

SUMMARY STATEMENTS: Describe how the motor was used in the preceding exercise. Explain whether it is a belt- or direct-drive motor, and state which is better for the application. Use an additional sheet of paper if more space is necessary for your answer.

TAKEAWAY: This exercise provided the opportunity to not only to obtain valuable information from the motor but to actually remove and reinstall a motor on an operational piece of equipment.

QUESTIONS

1. Why are sleeve bearings preferable for some fan applications?

2. What are the advantages of a ball-bearing motor?

3. What does "resilient mount" mean when applied to a motor?

4. Why must resilient-mount motors have a ground strap?

Exercise MOT-4

Identifying Types of Electric Motors

Name	Date	Grade

OBJECTIVES: Upon completion of this exercise, you should be able to identify the most common types of motors used in the HVACR industry.

INTRODUCTION: The technician will often have to replace electric motors when repairing HVACR equipment. Not all electric motors are the same, as they are selected to meet the specific needs of the system they are installed in. It is important for the technician to know what type of electric motor is being replaced so that an acceptable replacement can be installed. There are seven general types of electric motors commonly used in the industry:

1. Shaded-pole
2. Split-phase
3. Permanent split capacitor (PSC)
4. Capacitor-start–induction-run (CSIR)
5. Capacitor-start–capacitor-run (CSCS)
6. Three-phase
7. Electronically commutated motor (ECM)

Often, the best source of information about a motor is the nameplate if it is still readable. The unit's wiring diagram and product literature can also be used as reference material if available. On other occasions, the technician will be required to determine the specifications and details of the motor being replaced by examining the faulty motor.

TOOLS, MATERIALS, AND EQUIPMENT:

Selection of electric motors, both installed in equipment and removed from equipment, motor manufacturers' wiring diagrams and specification sheets, wire-cutting pliers, nut drivers, and screwdrivers.

⚠ **SAFETY PRECAUTIONS:** When examining and working on motors that are installed in equipment, make certain that the electrical power source is disconnected and that all capacitors, if any are present, are properly discharged.

STEP-BY-STEP PROCEDURES

1. Make certain that all power to the motor is disconnected and that all capacitors are properly discharged.

2. Using the wiring diagram(s) and any other available information about the equipment and the motor, provide the following information:
   Motor voltage: _____
   Rated-load amperage (RLA): _____
   Locked-rotor amperage (LRA): _____
   Motor frame type: _____

Number of speeds: _____
Types of motor capacitors: _____
Capacitor ratings: _____
Number of motor shafts: _____
Motor rotation direction: _____
Is motor's rotation reversible? _____
What type of starting devices does this motor utilize? _____

3. Based on your inspection of the motor and its starting components, determine the type of motor being examined, and enter the motor type here: _____

4. Explain how you were able to reach the conclusion about the motor type: _____

_____

_____

5. Obtain a second motor.

6. Make certain that all power to the motor is disconnected and that all capacitors are properly discharged.

7. Using the wiring diagram(s) and any other available information about the equipment and the motor, provide the following information:
Motor voltage: _____
Rated-load amperage (RLA): _____
Locked-rotor amperage (LRA): _____
Motor frame type: _____
Number of speeds: _____
Types of motor capacitors: _____
Capacitor ratings: _____
Number of motor shafts: _____
Motor rotation direction: _____
Is motor's rotation reversible? _____
What type of starting devices does this motor utilize? _____

8. Based on your inspection of the motor and its starting components, determine the type of motor being examined, and enter the motor type here: _____

9. Explain how you were able to reach the conclusion about the motor type: _____

_____

_____

10. Obtain a third motor.

11. Make certain that all power to the motor is disconnected and that all capacitors are properly discharged.

12. Using the wiring diagram(s) and any other available information about the equipment and the motor, provide the following information:

Motor voltage: _____

Rated-load amperage (RLA): _____

Locked-rotor amperage (LRA): _____

Motor frame type: _____

Number of speeds: _____

Types of motor capacitors: _____

Capacitor ratings: _____

Number of motor shafts: _____

Motor rotation direction: _____

Is motor's rotation reversible? _____

What type of starting devices does this motor utilize? _____

13. Based on your inspection of the motor and its starting components, determine the type of motor being examined, and enter the motor type here: _____

14. Explain how you were able to reach the conclusion about the motor type: _____

_____

_____

MAINTENANCE OF WORK STATION AND TOOLS: Clean and return all tools to their proper location(s). Replace all equipment covers. Clean up the work area.

SUMMARY STATEMENT: Explain, in detail, why it is important for the HVACR technician to know the type of motor that is being replaced.

TAKEAWAY: This exercise provides another opportunity to practice obtaining information from a motor. Not only will you be obtaining technical data from the motor; you will begin to develop the skills needed to identify various motor types.

QUESTIONS

1. What type of motor would most likely be found in a hermetic compressor used on a refrigeration system that is equipped with a fixed-bore metering device? Explain your answer.

2. What challenges are likely to be faced when a technician needs to determine a motor type if no information is available?

3. Why are three-phase motors easier to identify than single-phase motors?

4. What type of motor would most likely be used to turn a small propeller fan?

5. How would the technician determine if a hermetic compressor motor was a PSC or CSR (CSCR) motor?

6. What are some key facts that a technician could use in identifying PSC motors?

Exercise MOT-5

Single-Phase Electric Motors

Name	Date	Grade

OBJECTIVES: Upon completion of this exercise, you should be able to correctly make the electrical connections to shaded-pole, split-phase, capacitor-start, and permanent split-capacitor motors; operate each motor; and complete the associated data sheet.

INTRODUCTION: The service technician will be required to install or replace all types of single-phase motors in HVACR equipment. The installation of single-phase motors will include the electrical connections, direction of rotation, and final check to determine if the motor is operating properly.

TOOLS, MATERIALS, AND EQUIPMENT:

Split-phase motor for disassembly; shaded-pole motor; split-phase motor; capacitor-start motor; permanent split-capacitor motor; power supply to operate motors; digital multimeter (DMM); clamp-on ammeter; and miscellaneous hand tools, electrical tools, wires, capacitors, connectors, and electrical tape.

⚠ **SAFETY PRECAUTIONS:** Make certain that the electrical source is disconnected when making electrical connections. In addition:

- Make sure all electrical connections are tight.
- Make sure no bare conductors are touching metal surfaces except the grounding conductor and the connection terminals.
- Make sure the correct voltage is being supplied to the motor.
- Keep hands and other materials away from the rotating shaft.
- Make sure body parts do not come in contact with live electrical circuits.
- Make certain that all capacitors are properly discharged before working on motors.

STEP-BY-STEP PROCEDURES

1. Obtain a split-phase motor from your instructor.

2. Mark the end bell location on the winding section of the motor to ensure correct placement when reassembled.

3. Remove bolts connecting the end bells and the winding section of the motor.

4. Carefully disassemble the motor fully.

5. Make a list of the motor components you see inside the motor, explaining the function that each of these components performs:

MOTOR COMPONENT	FUNCTION
_____	_____
_____	_____
_____	_____
_____	_____
_____	_____
_____	_____
_____	_____

6. Describe the motor's starting components:

_____

_____

_____

_____

7. Carefully reassemble the motor, making sure the marks on the stator and end bells line up properly. Do not force the motor back together. If you are having difficulty, ask your instructor for assistance.

8. Obtain a two-speed shaded-pole motor from your instructor.

9. Measure the resistance of the windings of the motor with an ohmmeter.

10. Record the resistance readings here:

   _____ to _____: _____ Ω
   _____ to _____: _____ Ω
   _____ to _____: _____ Ω

11. Make the proper electrical connections to the motor.

12. Have your instructor check your wiring.

13. Operate the motor on both speeds, and record its current draw in each speed.
   Current draw (speed 1) _____
   Current draw (speed 2) _____

14. Remove the electrical wiring from the motor, and return the motor to the proper location.

15. Obtain an open two-speed PSC motor from your instructor.

16. Measure the resistance of the windings of the motor with an ohmmeter.

17. Record the resistance readings here:

   _____ to _____: _____ Ω
   _____ to _____: _____ Ω
   _____ to _____: _____ Ω

18. Select and obtain the proper capacitor.

19. Record the data from the capacitor here: _____

20. Make the proper electrical connections to the motor.

21. Have your instructor check your wiring.

22. Operate the motor on both speeds, and record its current draw in each speed.
Current draw (speed 1) _____
Current draw (speed 2) _____

23. Remove the electrical wiring from the motor, and return the motor to the proper location.

24. Obtain an open split-phase or capacitor-start motor from your instructor.

25. Measure the resistance of the windings of the motor with an ohmmeter.

26. Record the resistance readings here:
_____ to _____: _____ Ω
_____ to _____: _____ Ω
_____ to _____: _____ Ω

27. Make the proper electrical connections.

28. Have your instructor check your wiring.

29. Operate the motor and record its current draw.
Current draw: _____

30. Remove the electrical wiring from the motor, and return the motor to the proper location.

MAINTENANCE OF WORK STATION AND TOOLS: Clean and return all tools to their proper location(s). Replace all equipment covers. Clean up the work area.

SUMMARY STATEMENT: Explain why it is important to use the correct type of single-phase motor for the application in which it is being used.

TAKEAWAY: Pumps, blowers, fans, and compressors all rely on motors to operate. It is extremely important for an HVACR service technician to become completely familiar with the wiring and operation of all types of motors. This exercise provided the opportunity to continue building a solid understanding of motors, their operation, and their wiring.

QUESTIONS

1. What would be the results of using a shaded-pole motor for a high starting torque application?

2. What is the difference between a split-phase motor and a capacitor-start motor?

3. What are some popular applications for the PSC motor?

4. How are PSC motors reversed? Can all PSC motors be reversed?

5. Explain how the speed of a PSC motor is changed.

6. Which winding in a split-phase motor has the greatest resistance?

7. What is the purpose of the run capacitor in a PSC motor?

8. What is the synchronous speed of a six-pole, single-phase motor?

Exercise MOT-6

Three-Phase Electric Motors

Name	Date	Grade

OBJECTIVES: Upon completion of this exercise, you should be able to correctly wire and operate a three-phase motor. You should also be able to alter the motor's wiring to reverse the direction of the motor's rotation.

INTRODUCTION: Three-phase motors are commonly used in the HVACR industry in light commercial, commercial, and industrial applications. Three-phase motors have more torque than single-phase motors. This is because three legs of power each 120 electrical degrees out of phase with each other are supplied to these motors. Three-phase motors do not require starting components because of this increased level of torque. Three-phase motors are rugged, reliable, and more dependable than other types of motors. Many three-phase motors utilized in the HVACR industry are dual voltage and are often classified as 220 V/440 V or 230 V/460 V. A 220 V/440 V motor, for example, can operate, depending on how it is wired, at either 220 V or 440 V.

TOOLS, MATERIALS, AND EQUIPMENT:

Dual-voltage, three-phase motor, motor wiring diagram, power supply to operate motor, digital multimeter (DMM), clamp-on ammeter, and miscellaneous electrical hand tools.

SAFETY PRECAUTIONS: Make certain that the electrical source is disconnected when making electrical connections. In addition:

- Make sure all connections are tight.
- Make sure no bare current-carrying conductors are touching metal surfaces.
- Make sure the correct voltage is being supplied to the motor.
- Make sure body parts do not come in contact with live electrical conductors.
- Keep hands and materials away from the rotating shaft.

STEP-BY-STEP PROCEDURES

1. Obtain a dual-voltage, three-phase motor from your instructor.

2. Using the wiring diagram of the motor, identify the motor windings and the wire numbers that correspond to them.

3. Measure the resistance of the windings of the three-phase motor using an ohmmeter.

4. Record the resistance readings here:

_____ to _____ : _____ Ω
_____ to _____ : _____ Ω
_____ to _____ : _____ Ω
_____ to _____ : _____ Ω
_____ to _____ : _____ Ω
_____ to _____ : _____ Ω

5. Study the schematic wiring diagram on the motor, if available.

6. Make the proper electrical connections for the motor to operate at the lower voltage following the wiring diagram on the motor. If no diagram is available, refer to Figures MOT-6.1 and MOT-6.2 for "wye" and "delta" wiring diagrams of three-phase motors. Do not connect the motor to the power source at this time.

FIGURE **MOT-6.1**

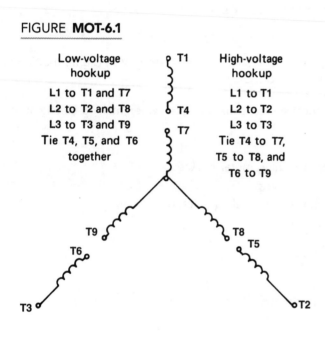

Low-voltage hookup	High-voltage hookup
L1 to T1 and T7	L1 to T1
L2 to T2 and T8	L2 to T2
L3 to T3 and T9	L3 to T3
Tie T4, T5, and T6 together	Tie T4 to T7, T5 to T8, and T6 to T9

FIGURE **MOT-6.2**

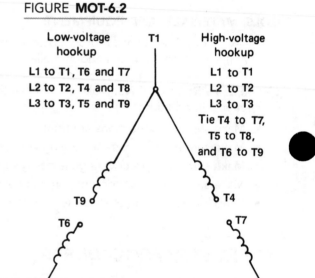

Low-voltage hookup	High-voltage hookup
L1 to T1, T6 and T7	L1 to T1
L2 to T2, T4 and T8	L2 to T2
L3 to T3, T5 and T9	L3 to T3
	Tie T4 to T7, T5 to T8, and T6 to T9

7. Have your instructor inspect your wiring.

8. Using a digital multimeter set to read resistance, measure the resistance of the motor windings between the L1, L2, and L3 power leads. Record these resistances here:

   L1 to L2: _____ Ω
   L1 to L3: _____ Ω
   L2 to L3: _____ Ω

9. Before energizing the motor, measure the voltage being supplied to the motor across all three phases. Record these voltages here:

   L1 to L2: _____ V
   L1 to L3: _____ V
   L2 to L3: _____ V

10. Using the information from step 9, calculate the voltage imbalance, and record it here:

    Voltage imbalance: _____

11. Explain why (or why not) the voltage imbalance calculation results indicate that the power supply is properly balanced to adequately energize the motor.

12. Connect the motor to the power supply.

13. Operate the motor, paying attention to the rotation. Record the direction of motor rotation here:
    _____

14. Measure the current flow through each of the three power lines. Record these measurements here:

    L1 amperage: _____A
    L2 amperage: _____A
    L3 amperage: _____A

15. De-energize the circuit.

16. Switch the connections of any two power leads (L1, L2, L3) feeding the motor. This will reverse the motor's direction of rotation.

17. Operate the motor, paying attention to the rotation. Record the direction of motor rotation here:
    _____

18. Measure the current flow through each of the three power lines. Record these measurements here:

    L1 amperage: _____A
    L2 amperage: _____A
    L3 amperage: _____A

19. Remove the motor from the voltage source.

20. Make the electrical connections necessary for the motor to operate on high voltage.

21. Have your instructor check the wiring of the three-phase motor on high voltage.

22. Using a digital multimeter set to read resistance, measure the resistance of the motor windings between the L1, L2, and L3 power leads. Record these resistances here:

    L1 to L2: _____ Ω
    L1 to L3: _____ Ω
    L2 to L3: _____ Ω

23. Disconnect the connections made for high-voltage wiring, and return the motor to the proper location.

MAINTENANCE OF WORK STATION: Clean and return all tools to their proper location(s). Replace all equipment covers. Clean up the work area.

SUMMARY STATEMENT: Explain how the motor can be used on low and high voltage by making adjustments to the electrical connections.

TAKEAWAY: Three-phase motors are very efficient and widely used in the HVACR industry. Although they often have more leads than single-phase motors, they do not require starting components, such as relays and capacitors. This exercise provides an overview of these popular motors and gives you the ability to manipulate the winding configurations in a controlled laboratory environment.

QUESTIONS

1. Why are no starting components needed on three-phase motors?

2. Explain what changes must be made to the winding wiring to allow a dual-voltage, three-phase motor to operate at different voltages.

3. How can the direction of rotation of a three-phase motor be reversed?

4. What are some advantages of using three-phase motors as opposed to using single-phase motors?

Exercise MOT-7

Electronically Commutated Motors (ECMs)

Name	Date	Grade

OBJECTIVES: Upon completion of this exercise, you should be able to disassemble an ECM and examine the three parts (controller, rotor, and stator), change the controller, use a digital multimeter (DMM) to determine if the motor section is good, and correctly install an ECM. You will operate an HVACR system with an ECM and observe the operation and speed of the motor.

INTRODUCTION: The service technician will be required to install, replace, and troubleshoot ECMs in various pieces of HVACR equipment. The ECM is a variable-speed motor that helps provide a more comfortable environment for the consumer at a reduced operating cost. The controller and motor can be separated and the faulty section can be replaced without replacing both components. The technician must follow the motor manufacturer's instructions when evaluating and troubleshooting these motors.

TOOLS, MATERIALS, AND EQUIPMENT:

ECMs, operating HVACR systems equipped with ECMs, digital multimeter (DMM), clamp-on ammeter, miscellaneous electrical supplies, miscellaneous electrical hand tools, velometer (anemometer), clean (new) air filters for the operating HVACR systems, and dirty (clogged) air filters for the operating HVACR systems.

⚠ **SAFETY PRECAUTIONS:** Make certain that the electrical source is disconnected when making electrical connections. In addition:

- Make sure power is off before inserting or removing power connector on the ECM.
- Make sure all electrical connections are tight.
- Make sure no bare conductors are touching metal surfaces except the grounding conductor and the connection terminals.
- Make sure the correct voltage is being supplied to the equipment.
- Keep hands and other materials away from the rotating shaft.
- Make sure body parts do not come in contact with live electrical circuits.
- Wear safety glasses.

STEP-BY-STEP PROCEDURES

1. As per your instructor's instructions, position yourself at an operating HVACR system equipped with an ECM.

2. Make certain that the air filter that is installed on this system is clean.

3. Energize the system and set the thermostat to call for blower operation.

4. Measure and record the airflow and amperage draw of the ECM on the assigned unit at four different times, according to the following guidelines:

Immediately after start-up: _____ cfm; _____ A
5 minutes after start-up: _____ cfm; _____ A
10 minutes after start-up: _____ cfm; _____ A
15 minutes after start-up: _____ cfm; _____ A

5. Turn the system off.

6. Provide an explanation for any discrepancies that are present among the four sets of readings obtained in step 4.

7. Remove the clean air filter, and replace it with a clogged/dirty filter.

8. Energize the system.

9. Measure and record the airflow and the amperage of the ECM on the assigned unit at four different times, according to the following guidelines:

5 minutes after start-up: _____ cfm; _____ A
10 minutes after start-up: _____ cfm; _____ A
15 minutes after start-up: _____ cfm; _____ A
Immediately after start-up: _____ cfm; _____ A

10. Turn the system off.

11. Provide an explanation for any discrepancies that are present among the four sets of readings obtained in step 9.

_____

_____

12. Compare the data obtained in step 9 to the data obtained in step 4, and comment on the similarities and/or differences that exist. Provide explanations for these similarities and/or differences.

_____

_____

13. Obtain an ECM from your instructor.

14. Record the information from the motor's nameplate here:

_____

_____

_____

_____

_____

_____

_____

15. Examine the motor, and locate motor, the motor control, and bolts that hold the motor assembly together.

16. Mark the mating portions of the motor assembly to ensure proper alignment upon motor reassembly.

17. Remove the bolts at the back (opposite end of shaft) of the control housing. Prevent the motor and control from falling when the bolts are removed. This should mechanically free the control module from the motor.

18. Carefully remove the plug that connects the motor to the control module. Do not pull on the wire, only the molded plastic plug. The control module and motor should now be completely separated. If you are having difficulty, ask your instructor for assistance. Do not force anything, as this can potentially damage the motor.

19. For future reference, label the wires connected to the control and power connectors to indicate where the wires go and if they are in the low-voltage or line-voltage circuit.

20. Measure the resistance of the motor windings. Record the obtained readings here:

Resistance of motor windings 1 to 2 _____ Ω
Resistance of motor windings 1 to 3 _____ Ω
Resistance of motor windings 2 to 3 _____ Ω

21. Comment on the relationship that exists among the resistances of the motor windings:

_____

_____

_____

22. Check to determine if the motor is grounded by measuring the resistance between the motor windings and motor housing. Record the resistance readings here:

Resistance of motor terminal 1 to housing _____ Ω
Resistance of motor terminal 2 to housing _____ Ω
Resistance of motor terminal 3 to housing _____ Ω

23. Explain how the readings taken in step 22 indicate whether or not the motor is grounded.

_____

_____

_____

_____

24. Mark the two end bells of the motor and the stator for correct placement when reassembly occurs.

25. Disassemble the motor by removing the remaining bolts. Examine the stator and rotor. Notice the three permanent magnets attached to the rotor by heavy-duty adhesive.

26. Assemble the motor, being careful to properly align all markings and not to overtighten the bolts.

MAINTENANCE OF WORK STATION: Clean and return all tools and motors to their proper location(s). Replace all equipment covers. Clean up the work area.

SUMMARY STATEMENT: Explain how the ECM differs from a traditional split-phase motor as far as construction, operation, speed control, and energy savings.

TAKEAWAY: Electronically commutated motors are fast becoming the motor of choice for both newly manufactured equipment and for motor replacement on existing systems. Becoming familiar with the operating, configuration, wiring, and troubleshooting of these motors is extremely important to help ensure the success of the service technician.

QUESTIONS

1. How can a technician distinguish the power connection from the control connection on an ECM?

2. How is the speed of an ECM controlled?

3. If you checked an ECM and determined that the motor and bearing portions of the motor are good, what is most likely the problem with the motor?

4. What type of bearings are typically utilized in an ECM?

5. What precautions should a technician take when disconnecting the control module from an ECM?

6. What is the advantage of using an ECM over a conventional PSC motor in an air-moving application?

Hermetic Compressor Motors

Name	Date	Grade

OBJECTIVES: Upon completion of this exercise, you should be able to determine the C, S, and R terminals of a single-phase compressor; correctly wire a PSC hermetic compressor motor; and determine the condition of a hermetic compressor motor.

INTRODUCTION: Most compressor motors used in the HVACR industry are enclosed in a sealed housing. This type of compressor is called a *hermetic* compressor. The electrical connections of a hermetic motor are made via insulated terminals that extend through the hermetic casing. Single-phase hermetic compressors typically have three terminals, whereas three-phase hermetic compressors could have more, depending on the motor's winding configuration.

TOOLS, MATERIALS, AND EQUIPMENT:

Hermetic compressors, power source to operate compressors, run capacitors for operating compressors, digital multimeter (DMM), clamp-on ammeter, miscellaneous electrical supplies, and miscellaneous electrical hand tools.

⚠ **SAFETY PRECAUTIONS:** Make certain that the electrical source is disconnected when making electrical connections. Do not allow any of the oil from the hermetic compressor to spill or come in contact with your skin. In addition:

- Make sure all connections are tight.
- Make sure no current-carrying conductors are touching metal surfaces except the grounding conductor.
- Make sure the correct voltage is being supplied to the unit.
- Make sure body parts do not come in contact with live electrical conductors.
- Keep hands and materials away from the rotating parts.

STEP-BY-STEP PROCEDURES

1. Obtain a single-phase hermetic compressor from your instructor.

2. If the compressor is installed in a unit, make certain the unit is de-energized and that all capacitors are properly discharged. Once it is determined that the power is off, mark and remove the wires from the compressor terminal, creating a wiring diagram, if necessary, to ensure proper wire placement at the conclusion of the exercise.

3. Measure the resistance of the motor windings in the hermetic compressor. Make certain that good contact is made between the terminals of the compressor. Record the resistance values here:

R to C: _____ Ω

R to S: _____ Ω

S to C: _____ Ω

The internal windings of a single-phase hermetic compressor motor are shown in Figure MOT-8.1. Figure MOT-8.2 shows the resistance values of the single-phase hermetic compressor motor windings used for the following example.

a. Find the largest reading between any two terminals. The remaining terminal is the common terminal. In Figure MOT-8.2 the reading between A and C is the largest (12 Ω), so B is the common terminal.

FIGURE **MOT-8.1**

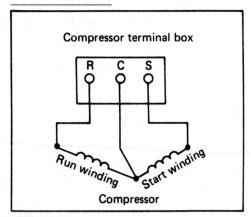

FIGURE **MOT-8.2**

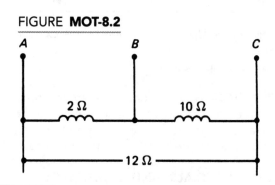

Legend

R: Run Terminal
C: Common Terminal
S: Start Terminal

b. Next, compare the resistance between the common terminal and terminal A with the resistance between the common terminal and terminal C. The largest reading (10 Ω between terminals B and C) represents the start winding. Since B is the common terminal, it can be concluded that C is the start terminal.

c. The remaining terminal is the run terminal. The resistance reading between the common terminal B and terminal A is the lowest (2 Ω), so terminal A is the run terminal.

4. Identify the terminals on the compressor you are working on as terminals 1, 2, and 3.

5. Using the data obtained in step 3 and the information in the example provided, label the resistances of the windings in the appropriate space.

Resistance between terminals 1 and 2: _____ Ω

Resistance between terminals 1 and 3: _____ Ω

Resistance between terminals 2 and 3: _____ Ω

6. Determine the common, start, and run terminals of the hermetic compressor.

7. Return the compressor to its proper location, if necessary.

8. Per your instructor's instructions, position yourself at an air-conditioning system that is equipped with a single-phase hermetic compressor.

9. Make certain the unit is de-energized and that all capacitors are properly discharged. Once it is determined that the power is off, label and remove the wires from the compressor terminals. Labeling the wires will help ensure proper wire placement at the conclusion of the exercise.

FIGURE **MOT-8.3**

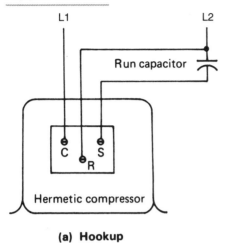

(a) Hookup

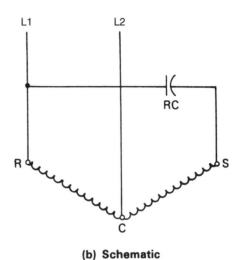

(b) Schematic

Legend
C: Common Terminal
R: Run Winding Terminal
S: Start Winding Terminal
RC: Run Capacitor

10. Identify the common, start, and run terminals.

11. Make the proper electric connections in order to operate the single-phase hermetic compressor using the appropriate run capacitor. Figure MOT-8.3 shows the diagram and wiring of a PSC hermetic compressor.

12. Connect the PSC hermetic compressor to the appropriate power supply.

13. Have your instructor check your wiring.

14. Energize the PSC hermetic compressor motor.

15. Measure the current draw of the PSC hermetic compressor, and record it here: _____ A

16. De-energize the compressor.

17. Discharge any capacitors, and disconnect the wiring from the compressor.

18. Replace any service panels from the unit that might have been removed.

19. Obtain a hermetic compressor from your instructor.

20. Identify the compressor terminals as terminals 1, 2, and 3.

21. Measure the resistance of the windings and record the values here:
    Resistance between terminals 1 and 2: _____ Ω
    Resistance between terminals 1 and 3: _____ Ω
    Resistance between terminals 2 and 3: _____ Ω

22. Verify/identify the common, start, and run terminals on the compressor.

23. Measure the resistance for each of the terminals to the shell of the compressor and record the values here:

    Common to ground: _____ Ω
    Start to ground: _____ Ω
    Run to ground: _____ Ω

24. Based on the readings obtained in steps 21 and 23, comment on the condition of the compressor, and recommend whether or not this compressor is suitable for installation in an operating system.

    _____

    _____

    _____

25. Return the hermetic compressor to its proper location, if necessary.

MAINTENANCE OF WORK STATION AND TOOLS: Clean and return all tools and equipment to their proper location(s). Replace all equipment covers if any removed. Clean up the work area.

SUMMARY STATEMENT: What are the physical characteristics of a hermetic compressor motor? Why is it important for service technicians to be able to determine the common, start, and run terminals?

TAKEAWAY: Since it is not possible to physically inspect the motor on a hermetically sealed compressor, it is important to be able to evaluate the motor using electrical test instrumentation. Proper use of the instruments and properly interpreting the information that is obtained will help ensure that the system diagnosis is correct.

QUESTIONS

1. Why are all starting components except the motor winding external to a hermetic compressor motor?

2. Find the common, start, and run terminals of the hermetic compressors shown in Figure MOT-8.4.

FIGURE **MOT-8.4**

a.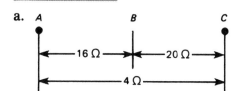

b.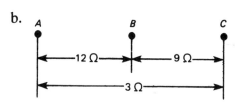

c.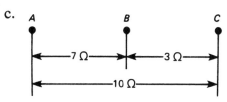

3. Explain, in detail, some common electrical failure categories for hermetic compressor motors?

4. Provide a complete set of resistance readings that might be obtained if a hermetic compressor had a faulty (stuck open) internal overload?

5. Explain how some mechanical failures in hermetic compressors might seem like electrical problems.

Exercise MOT-9

Identification of a Relay, Contactor, and Starter

Name	Date	Grade

OBJECTIVES: Upon completion of this exercise, you should be able to identify the differences among relays, contactors, and starters and describe the individual parts of each.

INTRODUCTION: You will disassemble, inspect the parts, and reassemble a relay, a contactor, and a starter. You will compare the features of each.

TOOLS, MATERIALS, AND EQUIPMENT: Straight-slot and Phillips-head screwdrivers, a digital multimeter (DMM), relay, contactor, and starter.

⚠ **SAFETY PRECAUTIONS:** These devices have small parts, often under spring pressure. Wear safety glasses to protect your eyes from small parts that might become airborne.

STEP-BY-STEP PROCEDURES

1. Disassemble the relay. Lay the parts out on the bench.

2. Explain how the mechanics of the relay cause the position of the contacts to change.

3. Reassemble the relay, being careful not to force the casing back together. If you are having difficulty, ask your instructor for assistance.

4. Disassemble the contactor. Lay the parts out in order on the workbench.

5. Explain how the mechanics of the contactor cause the position of the contacts to change.

6. Reassemble the contactor, being careful not to force the device back together. If you are having difficulty, ask your instructor for assistance.

7. Once reassembled, make certain that the contacts close when the armature is pushed down. Make certain that all screws, especially those that hold the stationary contacts in place, are tight.

8. Disassemble the starter. Lay the parts out on the bench.

9. Explain how the mechanics of the starter cause the position of the contacts to change.

10. Reassemble the starter, being careful not to force the casing back together. If you are having difficulty, ask your instructor for assistance.

11. Once reassembled, make certain that the contacts close when the armature is pushed down. Make certain that all screws, especially those that hold the stationary contacts in place, are tight.

MAINTENANCE OF WORKSTATION AND TOOLS: Return all tools and parts to their proper places. Make sure that your work area is left clean.

SUMMARY STATEMENT: Provide examples of applications that would be appropriate for relays, contactors, and starters, making certain to provide explanations regarding why the specific device is best suited for the applications you are describing.

TAKEAWAY: Relays, contactors, and starters all perform very similar functions, but they are each geared toward specific applications. Knowing what each of these devices is intended to be used for in addition to knowing what unique features each of them has will help ensure that the right part is used for the right job.

QUESTIONS

1. Explain the differences between a relay and a contactor.

2. Explain the differences between a contactor and a starter.

3. Explain the differences between a relay and a potential relay.

4. Describe the materials that are used to manufacture the contacts on a relay or contactor.

5. Explain how to best repair a set of contactor contacts that has become pitted.

6. Explain the function and operation of the heaters on a starter.

7. Explain how the heaters on a starter are selected.

Exercise MOT-10

Contactors and Relays

Name	Date	Grade

OBJECTIVES: Upon completion of this exercise, you should be able to install contactors, relays, and overloads in HVACR equipment and evaluate their schematic diagrams.

INTRODUCTION: Contactors and relays are used in the HVACR industry to stop and start loads. The major difference between a contactor and a relay is the ampacity that the component can safely carry. Large loads, such as compressors usually are controlled by contactors, while smaller loads, such as fractional horsepower blower motors, are controlled by relays. The technician must be able to install contactors and relays in HVACR equipment.

TOOLS, MATERIALS, AND EQUIPMENT:

Relays and contactors with 24, 115, and 230 V coils, 24 V control transformer, digital multimeter (DMM), terminal boards, switches, basic electrical hand tools, light bulbs, light sockets, wire, wire nuts, and electrical connectors.

⚠ **SAFETY PRECAUTIONS:** Make certain that the electrical source is disconnected when making electrical connections. In addition:

- Make sure all connections are tight.
- Make sure no bare current-carrying conductors are touching metal surfaces except the grounding conductor.
- Make sure the correct voltage is being supplied to the circuits.
- Make sure body parts do not come in contact with live electrical conductors.
- Keep hands and materials away from moving parts.

STEP-BY-STEP PROCEDURES

1. Using the component symbols provided in Figure MOT-10.1, determine which of the following wiring diagrams allows for the single-pole switch to control the relay that, in turn, energizes the light bulb.

FIGURE **MOT-10.1**

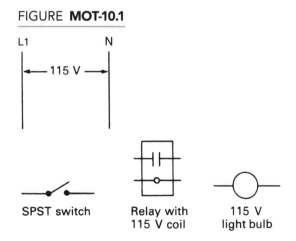

| SPST switch | Relay with 115 V coil | 115 V light bulb |

(continued)

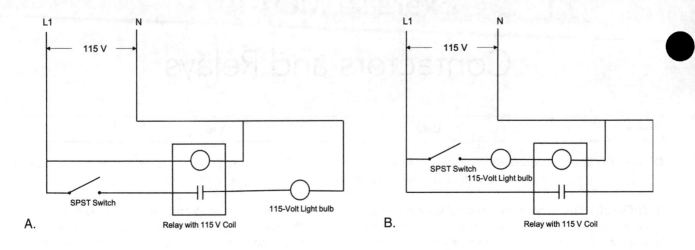

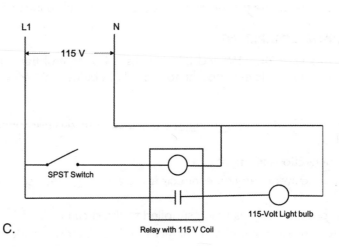

2. Have your instructor verify that you have selected the correct diagram.

3. Using the diagram, wire the circuit on plywood board.

4. Before energizing the circuit, have your instructor check your work.

5. Connect your circuit to a 115 V power source.

6. Operate the circuit by closing the switch.

7. Record what happens when the switch is closed. Be sure to discuss what you saw as well as what you heard.

8. Disconnect the power source from the circuit.

9. Using the component symbols provided in Figure MOT-10.2, determine which of the following wiring diagrams allows for the single-pole switch to control the contactor that turns on the 115 V light bulb.

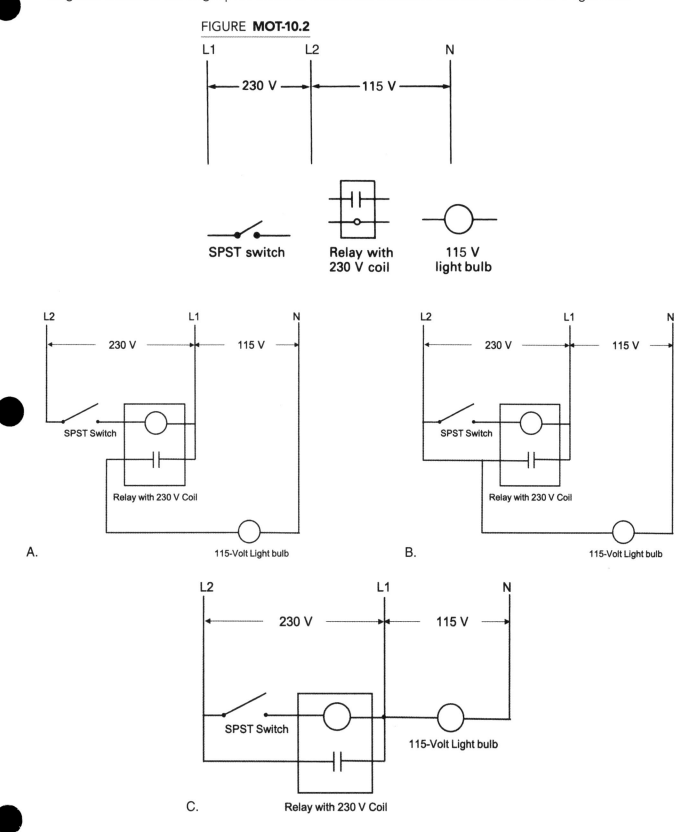

FIGURE **MOT-10.2**

A.

B.

C.

10. Have your instructor verify that you have selected the correct diagram.

11. Using the diagram, wire the circuit on the plywood board.

12. Before energizing the circuit, have your instructor check your work.

13. Connect your circuit to the appropriate power source.

14. Operate the circuit by closing the switch.

15. Record what happens when the switch is closed. Be sure to discuss what you saw as well as what you heard.

16. Disconnect the power source from the circuit.

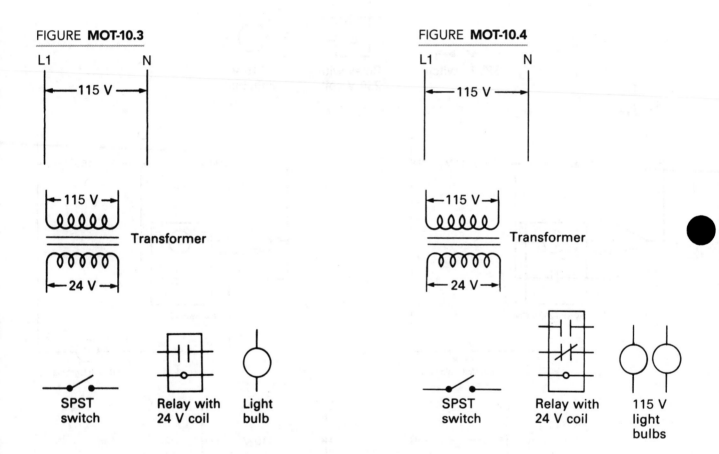

FIGURE **MOT-10.3**

FIGURE **MOT-10.4**

17. Using the component symbols provided in Figure MOT-10.3, determine which of the following wiring diagrams allows for the single-pole switch to control the relay that, in turn, powers the 115 V light bulb.

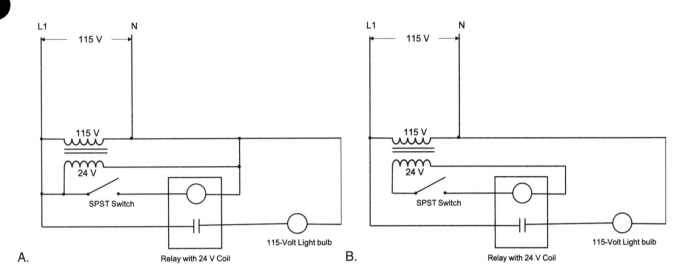

A.

B.

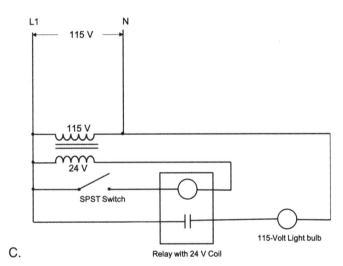

C.

18. Have your instructor verify that you have selected the correct diagram.

19. Using the diagram, wire the circuit on the plywood board.

20. Before energizing the circuit, have your instructor check your work.

21. Connect your circuit to the appropriate power source.

22. Operate the circuit by closing the switch.

23. Record what happens when the switch is closed. Note what you saw as well as what you heard.

24. Disconnect the power source from the circuit.

25. Using the component symbols provided in Figure MOT-10.4, determine which of the following wiring diagrams allows for the circuit to function as follows: When the SPST switch is in the off position, one light bulb is on. When the switch is flipped to the on position, the bulb that was on turns off N, and the bulb that was off turns on.

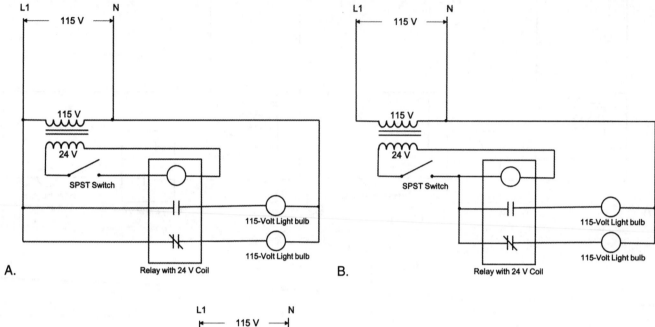

A.   B.

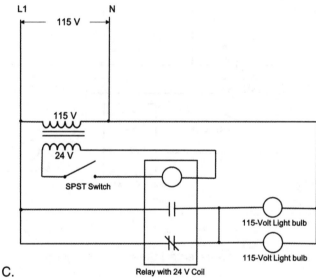

C.

26. Have your instructor verify that you have selected the correct diagram.

27. Using the correct diagram, wire the circuit on the plywood board.

28. Before energizing the circuit, have your instructor check your work.

29. Connect your circuit to the appropriate power source.

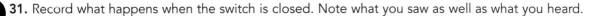

30. Operate the circuit by closing the switch.

31. Record what happens when the switch is closed. Note what you saw as well as what you heard.

32. Disconnect the power source from the circuit and remove the components from the board as instructed.

MAINTENANCE OF WORK STATION AND TOOLS: Clean and return all tools and parts to their proper location(s). Replace all equipment covers. Clean up the work area.

SUMMARY STATEMENT: Explain why some circuits utilize different power supplies of different voltage. What are the benefits of using multiple voltages in the same circuit?

TAKEAWAY: A solid understanding of control wiring is one of the most desirable traits that an HVACR service technician can possess. This exercise provides a number of opportunities for you to work toward developing and improving this skill set.

QUESTIONS

1. Why would a relay typically be the control device of choice for a small residential blower motor?

2. If information on a contactor stated 40 A resistive load and 30 A inductive load, could the contactor be used on a hermetic compressor that draws 40 A? Explain why or why not.

3. What is the difference between a sliding armature and a swinging armature?

4. What type of relay would be used to control a blower motor on a residential furnace if the blower needs to operate at high speed during cooling operation and low speed during heating operation? Explain your answer, including reference to the circuit constructed in this exercise that would best be used to facilitate this changeover.

Exercise MOT-11

Capacitors

Name	Date	Grade

OBJECTIVES: Upon completion of this exercise, you should be able to determine the condition of start and run capacitors. You will be able to use the capacitor rules to select replacement capacitors.

INTRODUCTION: Many electric motors used in the HVACR industry require capacitors for adequate starting torque and/or to increase running efficiency. The technician must be able to determine the condition of capacitors and, utilizing a set of capacitor rules, properly size and replace a capacitor. Very often, this will allow the technician to make the necessary repairs without having to make a trip to the local parts supply house.

TOOLS, MATERIALS, AND EQUIPMENT:

10 numbered capacitors for identification (Capacitor Kit #1), 10 numbered capacitors for checking (Capacitor Kit #2), 10 numbered capacitors for using capacitor rules (Capacitor Kit #3), digital multimeter (DMM) with a capacitor tester feature, 2 W, 15,000 Ω resistor, insulated needle-nose pliers, and screwdrivers.

⚠️ **SAFETY PRECAUTIONS:** Make certain that all capacitors have been discharged with a 2 W, 15,000 Ω resistor before touching the terminals to prevent electrical shock or possible damage to an electric meter. Capacitors often remain charged even while sitting on the shelf. Use insulated pliers.

STEP-BY-STEP PROCEDURES

1. Obtain Capacitor Kit #1 from your instructor.

2. Using insulated pliers, position a 2 W, 15,000 Ω resistor across the terminals of the capacitors to discharge any capacitor that may be charged. (See Figure MOT-11.1.)

FIGURE **MOT-11.1**

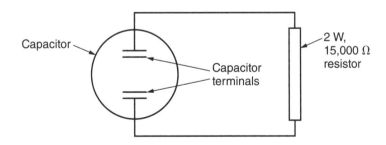

3. Record the types, voltages, and microfarad ratings of the capacitors in Capacitor Kit #1 in the following chart:

Capacitor #	Type	Voltage	Rating (µF)

4. Return Capacitor Kit #1 to its proper location.

5. Obtain Capacitor Kit #2 from your instructor.

6. Using your VOM, check the 10 capacitors in Capacitor Kit #2. Complete the following chart as each capacitor is tested:

Capacitor #	Type	CONDITION	
		Voltage	Rating (µF)

7. Have your instructor check your data sheet.

8. Return Capacitor Kit #2 to its proper location.

9. Obtain Capacitor Kit #3 from your instructor.

10. Provided here is a list of six capacitors (three run and three start) that are to be replaced using the capacitors in Capacitor Kit #3. Your instructor will provide the capacitance of each of the six capacitors. Record which of the capacitors in Capacitor Kit #3 can be used to replace each capacitor on the list. Note: This portion of the exercise could require that more than one capacitor be used.

Capacitor #	Type	Replacement Capacitor(s)
#1	Run	_____
#2	Run	_____
#3	Run	_____
#4	Start	_____
#5	Start	_____
#6	Start	_____

11. Return Capacitor Kit #3 to the proper location.

MAINTENANCE OF WORK STATION AND TOOLS: Clean and return all tools to their proper location(s). Clean up the work area.

SUMMARY STATEMENT: Explain why the total capacitance of capacitors wired in parallel with each other is equal to the sum of the individual capacitors.

TAKEAWAY: The size of the capacitor to be used on any given motor is selected by the manufacturer to help ensure maximum operating efficiency of the motor. As such, it is important that replacement capacitors be sized according to the defective component.

QUESTIONS

1. What are the major differences between a start capacitor and a run capacitor?

2. Compare the construction of a start capacitor to that of a run capacitor.

3. What is the advantage of using a bleed resistor on a start capacitor?

4. What is the purpose of a run capacitor in a motor circuit?

5. Why is it important to understand how the total capacitance is affected depending on how capacitors are wired together?

6. How would a shorted run capacitor affect the operation of a condenser fan motor?

7. How would an open run capacitor affect the operation of a motor?

8. What would be the symptoms of an open start capacitor on a capacitor-start motor?

Exercise MOT-12

Starting Relays for Single-Phase Electric Motors

Name	Date	Grade

OBJECTIVES: Upon completion of this exercise, you should be able to correctly install and troubleshoot current, potential, and solid-state relays on a single-phase motor.

INTRODUCTION: Most split-phase, capacitor-run, and capacitor-start–capacitor-run single-phase motors have some type of electrical device that removes the starting winding and/or the starting capacitor from the circuit once the motor has started and is operating close to its full speed. Current, potential, and solid-state relays are used to accomplish this in single-phase motors.

TOOLS, MATERIALS, AND EQUIPMENT:

Selection of current, solid-state, and potential relays including, some good and bad relays; power pack for PSC motor; single-phase compressors (one with a split-phase motor—one with a capacitor-start motor, one with a capacitor-start–capacitor-run motor); condensing unit with PSC hermetic compressor; power source for compressor operation; miscellaneous electrical supplies; electrical meters; and miscellaneous electrical hand tools.

⚠ **SAFETY PRECAUTIONS:** Make certain that the electrical source is disconnected when making electrical connections. In addition:

- Make sure all connections are tight.
- Make sure no bare ungrounded conductors are touching metal surfaces.
- Make sure the correct voltage is being supplied to the compressor.
- Make sure body parts do not come in contact with live electrical conductors.
- Keep hands and materials away from moving parts.

STEP-BY-STEP PROCEDURES

1. Obtain a selection of current-type and potential relays from your instructor.

2. Troubleshoot five current-type starting relays, and record the condition of the relays here:

Relay	Condition of Relay's Coil and Contacts
#1	_____
#2	_____
#3	_____
#4	_____
#5	_____

3. Troubleshoot five potential-type starting relays, and record the condition of the relays here:

Relay	Condition of Relay's Coil and Contacts
#1	_____
#2	_____
#3	_____
#4	_____
#5	_____

4. Explain what else must be done to completely check the potential relays to determine whether they are good or not.

_____

_____

_____

5. Clean up the work area, and return all tools and supplies to their correct location(s).

6. Obtain from your instructor a fractional horsepower, 115 V hermetic compressor with a current relay that has external connections between the run and start terminals. A schematic of the wiring for a motor without a start capacitor is provided in Figure MOT-12.1, and the wiring for a motor with a start capacitor is provided in Figure MOT-12.2.

FIGURE **MOT-12.1**

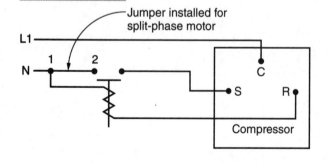

FIGURE **MOT-12.2**

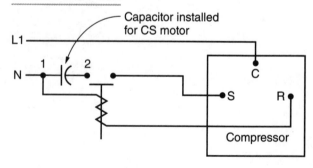

7. Make the necessary electrical connections for the compressor to operate as a split-phase motor. Most current-type starting relays plug directly into the start and run terminals of the hermetic compressor. The common terminal of a single-phase hermetic compressor will be connected to one leg of power and in series with the overload on fractional horsepower compressors. If you have any problems in making the electrical connections, consult a wiring diagram, or ask your instructor for assistance. Terminals 1 and 2 of the current relay must be connected for the motor to operate as a split-phase motor.

8. Have your instructor check the wiring of the hermetic compressor and the starting relay before supplying power to the hermetic compressor.

9. After your instructor has checked the wiring of the compressor, operate the compressor, and record the following information.

    Compressor model number: _____

    Voltage supplied to the compressor: _____ V

    Amperage draw of the compressor: _____ A

10. Disconnect power to the compressor.

11. Remove the wiring from the hermetic motor.

12. Using the same fractional horsepower compressor, obtain the correct size starting capacitor according to the manufacturer's specifications.

13. Make the necessary electrical connections for the compressor to operate as a capacitor-start motor. The capacitor should be connected between terminals 1 and 2. The jumper on terminals 1 and 2 should be removed.

14. Have your instructor check the wiring of the hermetic compressor, current relay, and starting capacitor before supplying power to the compressor.

15. After your instructor has checked the wiring of the compressor, operate the compressor, and record the following information.

    Compressor model number: _____

    Voltage supplied to the compressor: _____ V

    Amperage draw of the compressor: _____ A

16. Disconnect power to the compressor.

17. Properly discharge any capacitors.

18. Remove the wiring of the compressor, and return all equipment to the proper location.

19. Obtain from your instructor a small 115 V capacitor-start motor that operates with a potential relay. A schematic of a potential relay is shown in Figure MOT-12.3.

FIGURE **MOT-12.3**

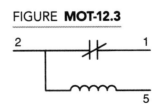

20. Using the manufacturer's information, select the correct potential relay, and start capacitor for the compressor. Obtain a potential relay of the correct coil group and calibration number and capacitor for the compressor.

21. Correctly make the necessary connections for the compressor, potential relay, and capacitor.

22. Have your instructor check the wiring of the compressor and the starting components before supplying power to the compressor.

23. After your instructor has checked the wiring of the compressor, operate the compressor, and record the following information.

    Compressor model number: _____

    Potential relay coil group: _____

    Voltage supplied to the compressor: _____ V

    Amperage draw of the compressor: _____ A

24. Disconnect power to the compressor.

25. Properly discharge any capacitors.

26. Remove the wiring from the compressor, and return all components to their correct location.

27. Obtain a 230 V hermetic compressor from your instructor that can be operated as a PSC and CSR motor.

28. Using the manufacturer's information, select the correct potential relay, and start and run capacitors for your compressor. Many hermetic compressor motors can be wired as PSC motors, and if the need arises for a high starting torque motor, a start capacitor and potential relay can be added to the PSC motor to make it a CSR motor. A schematic of a hard-start kit is shown in Figure MOT-12.4.

29. Wire the compressor as a PSC motor using the correct run capacitor. Connect the power supply to the motor.

30. Have your instructor inspect your work.

31. Operate the compressor, and record the following information.

    Compressor model number: _____

    Voltage supplied to the compressor: _____ V

    Amperage draw of the compressor: _____ A

FIGURE **MOT-12.4**

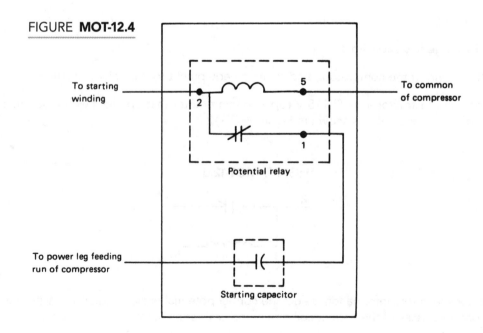

32. Disconnect power to the compressor.

33. Properly discharge any capacitors.

34. Add the necessary components to the PSC motor to convert the motor to a CSR motor. This will be the same procedure as adding a hard-start kit to the motor.

35. Have your instructor check your wiring.

36. Operate the compressor, and record the following information.

Compressor model number: _____

Potential relay coil group: _____

Voltage supplied to the compressor: _____ V

Amperage draw of the compressor: _____ A

37. Disconnect power to the compressor.

38. Properly discharge any capacitors.

39. Remove the wiring, and return all components to their proper location.

40. Obtain from your instructor a fractional horsepower split-phase compressor motor and a solid-state starting relay.

41. Make the necessary electrical connections following the wiring diagram and/or instructions that came with the relay.

42. Have your instructor check the wiring of the compressor and starting relay before supplying power to the hermetic compressor.

43. After your instructor has checked the wiring of the compressor, operate the compressor, and record the following information.

Compressor model number: _____

Relay model number: _____

Voltage supplied to the compressor: _____ V

Amperage draw of the compressor: _____ A

44. Disconnect power to the compressor.

45. Remove the wiring, and return all components to their proper location.

46. Obtain a residential air-conditioning condensing unit assignment from your instructor. The unit should be operable and equipped with a PSC hermetic compressor motor.

47. Obtain a power pack from your instructor of the correct size for the assigned condensing unit.

48. Following the instructions included with the power pack, make the necessary electrical connections to install the power pack on the hermetic compressor motor.

49. Have your instructor check your wiring.

50. Start the condensing unit, and record the following information.

Condensing unit model number: _____

Model number of power pack: _____

Voltage supplied to the condensing unit: _____ V

Amperage draw of the compressor: _____ A

51. Disconnect power to the compressor.

52. Properly discharge any capacitors.

53. Remove the power pack, and return all components to their proper location.

54. Replace all covers on the equipment used for the lab project.

MAINTENANCE OF WORK STATION AND TOOLS: Clean and return all tools and supplies to their proper location(s). Replace all equipment covers. Clean up the work area.

SUMMARY STATEMENT: Why are relays needed to drop the start winding and/or start capacitor out of the starting circuit? What would be the results if a start winding or start capacitor remained in the circuit?

TAKEAWAY: Properly selected and installed starting components will help ensure continued satisfactory motor operation. This exercise provided numerous opportunities to wire these devices as well as observe their operation.

QUESTIONS

1. Explain the operation of a current relay.

2. On what applications are current-type starting relays likely to be found?

3. What is the advantage of using a PSC motor that can be converted to a CSR motor?

4. What is the proper procedure for troubleshooting a current relay?

5. Why are potential relay contacts normally closed?

6. What is the advantage of using solid-state starting components?

Exercise MOT-13

Overloads

Name	Date	Grade

OBJECTIVES: Upon completion of this exercise, you should be able to identify, understand, and install the types of overloads used in the HVACR industry.

INTRODUCTION: Most electrical loads are protected by some type of overload device. Resistive loads are easier to protect than inductive loads, as they have a constant current draw. The current draw of inductive loads is not constant and generally starts at a higher level, diminishing as the load continues to operate. In most cases, resistive loads use fuses for protection, while inductive loads use more complicated overloads, such as the bimetal, magnetic, current, and electronic. The service technician must be able to identify and replace the type of overload used. The technician must know what overload will provide the best protection in a particular application.

TOOLS, MATERIALS, AND EQUIPMENT:

Eight overloads for student identification, inline overloads, small hermetic compressor, pilot duty current-type overloads, small electric motor, clamp-on ammeter, digital multimeter (DMM), electrical hand tools, miscellaneous wiring supplies, and external heat source to test overloads.

 SAFETY PRECAUTIONS: Make certain that the electrical source is disconnected when making electrical connections. In addition:

- Make sure all connections are tight.
- Make sure no bare current-carrying conductors are touching metal surfaces except the grounding conductor.
- Make sure the correct voltage is being supplied to the circuits.
- Make sure body parts do not come in contact with live electrical conductors.
- Keep hands and materials away from moving parts.

STEP-BY-STEP PROCEDURES

1. Obtain eight overloads from your instructor.

2. In the following table, record the type of each overload and whether it is inline or pilot duty:

Overload	Type	Inline or Pilot Duty
#1	_____	_____
#2	_____	_____
#3	_____	_____
#4	_____	_____
#5	_____	_____
#6	_____	_____
#7	_____	_____
#8	_____	_____

3. Obtain an inline overload from your instructor, and install it on a small hermetic compressor. Note: You will have to install the correct starting relay on the compressor.

4. Connect the hermetic compressor to a power source.

5. Have your instructor check your wiring.

6. Operate the motor and record the current draw.

Current draw of compressor: _____ A

7. Disconnect the circuit from the power source.

8. Remove the components from the compressor, and return them to their proper location.

9. Using the component symbols provided in Figure MOT-13.1, determine which of the following wiring diagrams accurately represents a circuit that will protect the motor.

FIGURE **MOT-13.1**

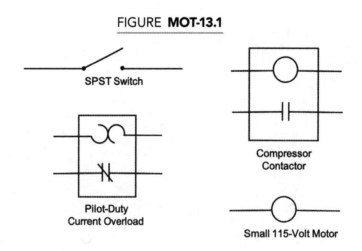

SPST Switch

Pilot-Duty
Current Overload

Compressor
Contactor

Small 115-Volt Motor

(continued)

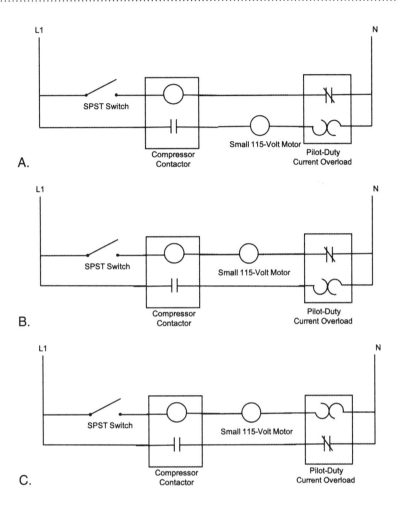

A.

B.

C.

10. Have your instructor check to ensure you have selected the correct diagram.

11. Wire the circuit on an electrical practice board.

12. Before connecting to a power source, have your instructor check your work.

13. Once your circuit has been checked, connect the circuit to a power source.

14. Operate the motor and record the amp draw.

Current draw: _____ A

15. Apply an artificial source of heat to the element of the current-type thermal overload until the pilot duty contacts open.

16. What happens to the electric motor?

17. Disconnect the circuit from the power supply.

18. Remove the components from the practice board, and return them to their proper location.

MAINTENANCE OF WORK STATION AND TOOLS: Clean and return all tools to their proper location(s). Replace all equipment covers. Clean up the work area.

SUMMARY STATEMENT: Explain the differences between an inline and a pilot overload. Provide examples of applications where each one is desirable.

TAKEAWAY: Overloads are safety devices that are intended to protect motors. They should never be jumped out, and in the event one fails, it is the job of the service technician to determine the cause of failure. Overloads, for the most part, do not fail for no reason.

QUESTIONS

1. Why are fuses used for protection of resistive loads?

2. What is the purpose of an overload?

3. Explain why inductive loads are not typically protected by traditional fast-blow-type fuses.

4. What is an internal compressor overload?

Electric Motor Bearings and Drives

Name	Date	Grade

OBJECTIVES: Upon completion of this exercise, you should be able to identify, determine the condition, and lubricate the two types of motor bearings commonly used in the industry. You will be able to correctly align a V-belt application and adjust a variable pitch pulley for a different speed without overloading the motor.

INTRODUCTION: All electric motors have some type of bearings that reduce friction and allow for free rotation of the motor and other types of rotating devices. The two types of bearings commonly used in the HVACR industry are sleeve bearings and ball bearings. Devices that require rotation must have some means of connecting the device to a source of rotating motion. The source of rotating power can be connected to the device by a direct drive or a belt-driven assembly.

TOOLS, MATERIALS, AND EQUIPMENT:

Electric motor with sleeve bearings, electric motor with ball bearings, HVACR equipment with direct-driven and belt-driven assemblies, miscellaneous hand tools for motor disassembly and pulley adjustments, tachometer, and tape measure.

 SAFETY PRECAUTIONS: Make sure that the electrical source is disconnected when examining live equipment. In addition:

- Make sure all connections are tight.
- Make sure no bare current-carrying conductors are touching metal surfaces except the grounding conductor.
- Make sure the correct voltage is being supplied to the circuits.
- Make sure no body parts come in contact with live electrical conductors.
- Keep hands and materials away from moving parts.
- Do not attempt to stop a rotating shaft, blower, or fan with your hands.
- Wait until all blowers, fans, and shafts have come to a COMPLETE STOP before attempting to make any adjustments to the assembly.
- Keep all objects, including fingers, away from the point where a V-belt meets a pulley. Severe personal injury can result from not following this precaution.

STEP-BY-STEP PROCEDURES

1. Obtain two electric motors from your instructor, one with sleeve bearings and one with ball bearings.

2. Mark the end bells and stators of the motors in order to determine their original location. This will be helpful during reassembly.

3. Disassemble the electric motors, being careful to keep the parts of each motor separate.

4. Identify the type of bearing and lubrication method in each type of motor, and record the following information.

Motor	Type of Bearing	Lubrication Method
#1	_____	_____
#2	_____	_____

5. Describe the bearings found on each of the motors.

6. Reassemble the two motors, being careful to use the marks made in step 2. Once the motors have been reassembled, make certain that each turns freely.

7. Have your instructor check your identification and motor assembly.

8. Return the motors to their proper location.

9. In the laboratory, locate at least five direct-drive applications and five V-belt applications.

10. Record the following information.

Unit	Type of Drive	Unit Location	Unit Model Number	Unit Serial Number
#1	_____	_____	_____	_____
#2	_____	_____	_____	_____
#3	_____	_____	_____	_____
#4	_____	_____	_____	_____
#5	_____	_____	_____	_____

11. Replace all equipment covers.

12. As per your instructor's instructions, position yourself at a unit that has a variable pitch, belt-driven blower assembly, on a fan motor.

13. Remove the cover from the motor/blower compartment.

14. Locate the fan motor and blower.

15. Measure the effective diameter of the motor's drive pulley and the blower's driven pulley, and record this information here:

Drive pulley effective diameter: _____ in.
Driven pulley effective diameter: _____ in.

16. Operate the motor, and record the following information.

Voltage supplied to the fan motor: _____ V
Current draw of the fan motor: _____ A
Speed of the motor: _____ rpm
Speed of the blower: _____ rpm

17. Turn off the power to the unit, making certain to lock out and tag the disconnect switch.

18. Using the formula **Motor Speed x Drive Pulley Diameter = Blower Speed x Driven Pulley Diameter,** verify that the diameter and speed values that were previously obtained are accurate.

_____

_____

_____

19. Provide explanations for any discrepancies or inaccuracies in the step 18 calculations.

_____

_____

_____

_____

20. Carefully remove the V-belt from the belt-driven assembly.

21. Inspect the variable-pitch pulley. The variable-pitch pulley has two Allen set screws. One set screw attaches the pulley to the motor shaft, and the other set screw allows for one side of the pulley to be rotated. This rotation can make the pulley smaller or larger depending upon the direction of rotation. Locate the two set screws, and determine which is responsible for the two operations.

22. Loosen the set screw holding the movable side of the pulley, and turn the pulley counterclockwise until the gap between the pulley sections has been widened. This causes the belt to ride deeper into the pulley, thus making the pulley's effective diameter smaller and ultimately decreasing the speed of the blower. Continue to turn the pulley for five or six complete counterclockwise rotations, or as many as your instructor tells you to.

23. Make a note as to how many counterclockwise rotations were made to the pulley: _____ rotations

24. Tighten the set screw on the variable-pitch pulley.

25. Replace the V-belt on the pulleys.

26. Adjust the belt tension as instructed to ensure proper operation when the motor is once again energized.

27. Measure the effective diameter of the drive pulley as it is presently configured, and record this measurement here:

Drive pulley effective diameter: _____ in.

28. Restore electrical power to the unit.

29. Operate the motor, and record the following information.

Voltage supplied to the motor: _____ V
Current draw of the fan motor: _____ A
Speed of the motor: _____ rpm
Speed of the blower: _____ rpm

30. Turn off the power to the unit, making certain to lock out and tag the disconnect switch.

31. Using the formula **Motor Speed x Drive Pulley Diameter = Blower Speed x Driven Pulley Diameter,** verify that the diameter and speed values that were previously obtained are accurate. Note: Be sure to use the correct measurement for the drive pulley.

_____

_____

_____

32. Provide explanations for any discrepancies or inaccuracies in the step 31 calculations.

_____

_____

_____

_____

33. Carefully remove the V-belt from the belt-driven assembly.

34. Loosen the set screw holding the movable side of the pulley, and turn the pulley clockwise (counting the number of turns as you proceed) until the belt will ride right at the outer edge of the pulley. This will increase the diameter of the drive pulley, increasing the speed of the blower.

35. Make a note as to how many clockwise rotations were made to the pulley: _____ rotations.

36. Tighten the set screw on the variable-pitch pulley.

37. Replace the V-belt on the pulleys.

38. Adjust the belt tension as instructed to ensure proper operation when the motor is once again energized.

39. Measure the effective diameter of the drive pulley as it is presently configured, and record this measurement here:

Drive pulley effective diameter: _____ in.

40. Restore electrical power to the unit.

41. Operate the motor, and record the following information.

Voltage supplied to the motor: _____ V
Current draw of the fan motor: _____ A
Speed of the motor: _____ rpm
Speed of the blower: _____ rpm

42. Turn off the power to the unit, making certain to lock out and tag the disconnect switch.

43. Using the formula **Motor Speed x Drive Pulley Diameter = Blower Speed x Driven Pulley Diameter,** verify that the diameter and speed values that were previously obtained are accurate. Note: Be sure to use the correct measurement for the drive pulley.

_____

_____

_____

44. Provide explanations for any discrepancies or inaccuracies in the step 43 calculations.

_____

_____

_____

_____

45. Carefully remove the V-belt from the belt-driven assembly.

46. Return the pulley pitch to its original location. The number of turns required to bring the pulley back to its original position should be equal to the number of turns recorded in step 35 minus the number of turns recorded in step 23.

47. Check the alignment of the pulley.

48. Correct alignment, if necessary.

49. Replace the V-belt on the fan and motor.

50. Check belt tension.

51. Correct belt tension, if necessary.

52. Restore electrical power to the unit.

53. Operate the motor, and record the following information.
Voltage supplied to the motor _____ V
Current draw of the fan motor _____ A
Speed of the motor: _____ rpm
Speed of the blower: _____ rpm

54. Turn off the power to the unit, making certain to lock out and tag the disconnect switch.

55. Using the formula **Motor Speed x Drive Pulley Diameter = Blower Speed x Driven Pulley Diameter,** verify that the diameter and speed values that were previously obtained are accurate. Note: Be sure to use the correct measurement for the drive pulley.

_____

_____

_____

56. Provide explanations for any discrepancies or inaccuracies in the step 43 calculations.

_____

_____

_____

_____

57. Replace the covers on the fan compartment.

MAINTENANCE OF WORK STATION: Clean and return all tools to their proper location(s). Replace all equipment covers. Clean up the work area.

SUMMARY STATEMENT: Why is belt tension and alignment important? Give the advantages of direct-drive appliances and belt-drive appliances.

TAKEAWAY: Improperly adjusted and aligned pulleys can affect system airflow, resulting in reduced system performance and an improperly conditioned space. Setting pulley size and ensuring proper alignment are tasks that should not be taken lightly.

QUESTIONS

1. Explain how a variable-pitch pulley can be used to change the speed of a blower.

2. How can the rpm be changed in a direct-drive application?

3. Why are matched sets of V-belts used?

4. Describe the differences between "A" belts and "B" belts.

Belt-Driven Blower Assemblies

Name	Date	Grade

OBJECTIVES: Upon completion of this exercise, you should be able to disassemble and reassemble a belt-driven blower assembly and align the pulleys.

INTRODUCTION: You will remove a belt-driven blower assembly from an air handler, take the motor and pulley off, and remove the pulley from the blower shaft. You will then reassemble the fan section, align the pulleys, and set the belt tension.

TOOLS, MATERIALS AND EQUIPMENT: A digital multimeter (DMM), pulley puller, two adjustable wrenches, ⅜"–½" and ⁹⁄₁₆"–⅝" open or box-end wrenches, oil can with light oil, straight-slot and Phillips-head screwdrivers, straightedge, Allen wrench set, emery cloth (300 grit), soft-face hammer, and ¼" and ⁵⁄₁₆" nut drivers.

⚠ **SAFETY PRECAUTIONS:** Turn off the power to the unit, and lock and tag the panel before you start to disassemble it. Properly discharge all capacitors. Keep hands and fingers away from rotating machinery. Always allow rotating parts to come to a complete stop before working on them.

STEP-BY-STEP PROCEDURES

1. You will be assigned an air handler or furnace with a belt-driven blower assembly. Turn the power off. Lock and tag the panel, and keep the single key in your possession. Check the voltage with a meter to ensure that the power is off. Properly discharge all capacitors.

2. Take off enough panels to remove the blower assembly.

3. Remove the blower assembly. Label any wiring that is disconnected.

4. Release the tension on the belt by loosening the motor tension adjustment.

5. Remove the belt, and then the pulley from the motor shaft. DO NOT FORCE IT OR HAMMER THE SHAFT.

6. Remove the pulley from the blower shaft. DO NOT DISTORT THE SHAFT BY FORCING IT OFF.

7. Clean the blower and motor shafts with fine emery cloth, about 300 grit.

8. Wipe both shafts and then apply a very light coat of oil to them.

9. Replace the pulleys on the motor and blower shafts, making certain that the pulleys are on the correct shafts.

10. Place the motor back in its bracket, and tighten it finger tight.

FIGURE **MOT-15.1**

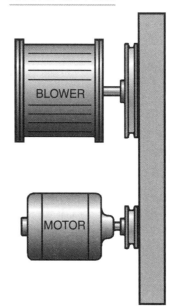

Straightedge (Such as a carpenter's level)

11. Align the pulleys as shown in Figure MOT-15.1.

12. Tighten the motor adjustment for tightening the belt. Then tighten the motor onto its base.

13. Place the blower assembly back into the air handler, and connect the motor wires.

14. Give the compartment a visual inspection to make sure no wires or obstructions are in the way.

15. After your instructor has inspected and approved your work, shut the fan compartment door, turn the power on, and start the fan. Listen for any problems.

MAINTENANCE OF WORKSTATION AND TOOLS: Wipe up any oil, clean the workstation, and return all tools to their correct places.

SUMMARY STATEMENT: Describe your experience of removing the fan shaft pulley from its shaft. Use an additional sheet of paper if more space is necessary for your answer.

TAKEAWAY: This exercise provides you with an opportunity to completely remove a belt-driven blower assembly from an air handler, disassemble the drive components, and then reassemble the blower. This exercise represents a series of tasks that are often performed in the field in tight quarters. Practicing this procedure in the lab setting will help you master the process.

QUESTIONS

1. Describe the effects of a belt that is too tight.

2. Why must pulleys be aligned accurately?

3. Why must a fan or motor shaft not be hammered?

4. State the advantages of a belt-drive motor over a direct-drive motor.

5. How do you adjust a pulley for a faster driven pulley speed?

6. You are working on a belt-driven blower assembly that has a motor that is turning at 1600 rpm. The blower is turning at a speed of 1200 rpm. If the drive pulley is 6 in. in diameter, calculate the diameter of the pulley that is on the blower shaft.

Troubleshooting Contactors and Relays

Name	Date	Grade

OBJECTIVES: Upon completion of this exercise, you should be able to correctly determine the condition of contactors and relays.

INTRODUCTION: One of the most important tasks of the HVACR technician is troubleshooting control systems. Contactors and relays are used to control loads in these control systems, and the technician must be able to correctly determine the condition of these components and replace them, if necessary.

TOOLS, MATERIALS, AND EQUIPMENT:

Contactors, relays, contactor and relay kit (five relays and five contactors), basic electrical hand tools, electrical meters, miscellaneous electrical supplies, HVACR equipment.

⚠️ **SAFETY PRECAUTIONS:** Make certain that the electrical source is disconnected when making electrical connections. In addition:

- Make sure all connections are tight.
- Make sure no bare current-carrying conductors are touching metal surfaces except the grounding conductor.
- Make sure the correct voltage is being supplied to the circuits.
- Make sure body parts do not come in contact with live electrical conductors.
- Keep hands and materials away from moving parts.

STEP-BY-STEP PROCEDURES

1. Obtain the following contactors and relays from your instructor.

    a. 24 V relay with 1 NO & 1 NC set of contacts

    b. 115 V relay with 1 NO & 1 NC set of contacts

    c. 230 V relay with 1 NO & 1 NC set of contacts

    d. 24 V, 30 A, two-pole contactor

    e. 115 V, 30 A, two-pole contactor

    f. 230 V, 30 A, two-pole contactor

    g. 230 V, 60 A, three-pole contactor

2. Measure and record the resistance of the coils of the relays and contactors obtained in step 1.

24 V relay coil: _____ Ω

115 V relay coil: _____ Ω

230 V relay coil: _____ Ω

24 V contactor coil (two-pole): _____ Ω

115 V contactor coil (two-pole): _____ Ω

230 V contactor coil (two-pole): _____ Ω

230 V contactor coil (three-pole): _____ Ω

3. Connect the coils of the relays and contactors to the appropriate voltage. Measure and record the resistance of the normally open contacts when the component is de-energized and energized.

Relay NO contacts when relay is energized: _____ Ω

Relay NO contacts when relay is de-energized: _____ Ω

Relay NC contacts when relay is energized: _____ Ω

Relay NC contacts when relay is de-energized: _____ Ω

Contactor contacts when contactor is energized: _____ Ω

Contactor contacts when contactor is de-energized: _____ Ω

4. Obtain a contactor and relay kit from your instructor. Using the appropriate piece of electrical test instrumentation, evaluate the relays and contactors. Record the condition of the five contactors and five relays here:

Device	Condition of the Device
Relay #1	_____
Relay #2	_____
Relay #3	_____
Relay #4	_____
Relay #5	_____
Contactor #1	_____
Contactor #2	_____
Contactor #3	_____
Contactor #4	_____
Contactor #5	_____

5. Return all contactors, relays, and contactor and relay kit to their appropriate location(s).

6. Evaluate various contactors or relays on live equipment as instructed by your instructor.

Device	Device Description	Device Location	Condition of Device
Device #1	_____	_____	_____
Device #2	_____	_____	_____
Device #3	_____	_____	_____

MAINTENANCE OF WORK STATION AND TOOLS: Clean and return all tools to their proper location(s). Replace all equipment covers. Clean up the work area.

SUMMARY STATEMENT: Why is it important for the technician to be able to correctly troubleshoot contactors and relays?

TAKEAWAY: Contactors and relays are often misdiagnosed, leading to costly and often unneeded repairs. Knowing how to properly evaluate these devices is an important aspect of effective system troubleshooting and repair.

QUESTIONS

1. What is the approximate resistance of a 24, 115, and 230 V relay coil?

2. What is the resistance of a set of NO contacts when the relay coil is de-energized?

3. What is the resistance of a set of NO contacts when the relay coil is energized?

4. What is likely to be the problem with a contactor if the contactor coil is good and is receiving the proper voltage but the contacts do not close?

5. How can a service technician determine the condition of the contacts of a contactor?

6. What damage could a shorted contactor coil cause in a 24 V control circuit?

7. If a technician reads 30 V across a set of contacts (L1–T1) on a contactor, what is the most likely problem?

Exercise MOT-17

Troubleshooting Overloads

Name	Date	Grade

OBJECTIVES: Upon completion of this exercise, you should be able to correctly determine the condition of overload devices.

INTRODUCTION: All major loads in an HVACR system will have some type of overload protection. It is the responsibility of the service technician to make certain that the overload protection is operating properly and, if not, to replace it with the correct replacement.

TOOLS, MATERIALS, AND EQUIPMENT:
Line break overloads, pilot duty current-type overloads, compressor with internal overload, HVACR equipment, electrical meters, miscellaneous electrical supplies, basic electrical hand tools, overload kit consisting of three line break overloads and three pilot duty overloads.

⚠ **SAFETY PRECAUTIONS:** Make certain that the electrical source is disconnected when making electrical connections. In addition:

- Make sure all connections are tight.
- Make sure no bare current-carrying conductors are touching metal surfaces except the grounding conductor.
- Make sure the correct voltage is being supplied to the circuits.
- Make sure body parts do not come in contact with live electrical conductors.
- Keep hands and materials away from moving parts.

STEP-BY-STEP PROCEDURES

1. Obtain the following items from your instructor.

   a. Line break current overload

   b. Pilot duty current overload

   c. Compressor equipped with an internal overload

2. Measure and record the resistance of the line break overload: _____ Ω

3. Measure and record the resistance of the control contacts and the controlling element of the pilot duty overload.
   Control contacts: _____ Ω
   Controlling element: _____ Ω

4. Measure and record the resistance of the compressor motor windings to determine the condition of the internal overload.

> **Compressor Resistances**
> Common to start: _____ Ω
> Common to run: _____ Ω
> Start to run: _____ Ω

5. Obtain an overload kit from your instructor. Evaluate and record the condition of the overloads in the kit.

Line break overload #1 _____
Line break overload #2 _____
Line break overload #3 _____
Pilot duty overload #1 _____
Pilot duty overload #2 _____
Pilot duty overload #3 _____

6. Return all overloads, compressors, and the overload kit to their appropriate location(s).

7. Evaluate various overloads on live equipment as assigned by your instructor.

Overload	Condition of Overload	Location of Overload
Line break overload	_____	_____
Pilot duty overload	_____	_____
Internal overload	_____	_____

MAINTENANCE OF WORK STATION AND TOOLS: Clean and return all tools to their proper location(s). Replace all equipment covers. Clean up the work area.

SUMMARY STATEMENT: Explain why pilot duty overloads are used on large loads instead of line break overloads.

TAKEAWAY: Overloads are safety devices that are intended to protect motors. They should never be jumped out, and in the event one fails, it is the job of the service technician to determine the cause of failure. Overloads, for the most part, do not fail for no reason. Proper evaluation of these devices is extremely important and a task that should not be overlooked.

QUESTIONS

1. Why is the trip-out current draw important in overload applications?

2. What is the simplest overload used in the HVACR industry?

3. What is the resistance of a good fuse?

4. What precautions should be taken when a technician suspects that an internal overload is open?

5. What procedure would a service technician use to correctly diagnose the condition of an internal overload in a hermetic compressor?

6. What is the difference between an internal overload and an internal thermostat?

7. If a hermetic compressor reads 0 Ω between start and run, what is wrong with the compressor?

8. Why can't a fuse adequately protect a compressor motor?

Exercise MOT-18

Troubleshooting Electric Motors

Name	Date	Grade

OBJECTIVES: Upon completion of this exercise, you should be able to correctly troubleshoot the common types of motors used in the HVACR industry.

INTRODUCTION: The service technician's responsibility is to diagnose and repair problems in HVACR equipment or control systems. Many times, these problems are motors that are not operating properly. The technician will have to determine whether the electric motor or the control responsible for operating the motor is the problem. Once the determination has been made that the motor is faulty, the technician must replace the motor with a proper replacement.

TOOLS, MATERIALS, AND EQUIPMENT:

Selection of loose motors including two each of the following motors: shaded-pole, split-phase, capacitor-start, PSC, and three-phase; power supply to operate motors; digital multimeter (DMM); clamp-on ammeter; and miscellaneous electrical hand tools.

SAFETY PRECAUTIONS: Make certain that the electrical source is disconnected when making electrical connections. In addition:

- Make sure all connections are tight.
- Make sure no bare current-carrying conductors are touching metal surfaces except the grounding conductor.
- Make sure the correct voltage is supplied to the motor.
- Make sure body parts do not come in contact with live electrical conductors.

STEP-BY-STEP PROCEDURES

1. There will be 10 stations set up by your instructor. You will move from station to station, evaluating each of the motors as you and your classmates work your way around the room.

2. Evaluate the first motor by checking the windings as well as the condition of the bearings.

3. Record the condition of the first motor on the appropriate line:

Motor	Condition of Motor	Current Draw If Motor Is Good
#1	_____	_____
#2	_____	_____
#3	_____	_____
#4	_____	_____
#5	_____	_____

#6 _____   _____

#7 _____   _____

#8 _____   _____

#9 _____   _____

#10 _____   _____

4. If the motor is okay, operate the motor, and record the current draw in the appropriate place.

MAINTENANCE OF WORK STATION: Clean and return all tools to their proper location(s). Clean up the work area.

SUMMARY STATEMENT: Explain the proper procedure for troubleshooting electric motors.

TAKEAWAY: Electric motors are the driving force behind fans, blowers, pumps, compressors, and other HVACR system components. It is, therefore, important to be able to properly evaluate motors, as this makes up a significant portion of what HVACR service technicians do on a daily basis.

QUESTIONS

1. How can a service technician determine the condition of the bearings in an electric motor?

2. How can a service technician determine the condition of the windings in an electric motor?

3. Explain the procedure a technician would use to check the resistance of a multispeed shaded-pole motor.

Exercise MOT-19

Megohmmeter Usage

Name	Date	Grade

OBJECTIVES: Upon completion of this exercise, you should be able to, with the aid of a megohmmeter, electrically check a compressor from the motor windings to ground for motor winding insulation breakdown, moisture accumulation, and acid formations.

TOOLS, MATERIALS, AND EQUIPMENT: Textbook, megohmmeter, refrigeration, or air-conditioning system containing a hermetic motor.

⚠ SAFETY PRECAUTIONS: Wear safety glasses and the proper attire to test electrical devices. Make sure the power to the refrigeration or air-conditioning system is off and all wires are disconnected from the compressor terminals before using a megohmmeter. Megohmmeters, Figure MOT-19.1, generate a high DC voltage (over 500 V DC) and should only be used by experienced technicians or under supervision of your instructor.

STEP-BY-STEP PROCEDURES

1. Let the compressor run as long as possible to get the windings at a steady-state temperature.

2. Carefully remove the protective cover from the compressor's motor terminal box shown in Figure MOT-19.2.

FIGURE **MOT-19.1**

Courtesy Ferris State University. Photo by John Tomczyk

FIGURE **MOT-19.2**

3. Disconnect all wires from the motor terminals. A motor terminal is shown in Figure MOT-19.3.

FIGURE **MOT-19.3**

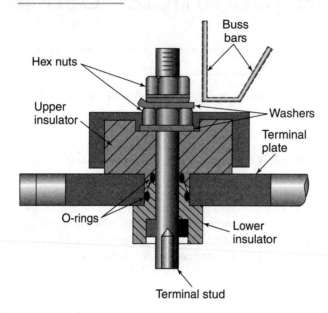

Buss bars

Hex nuts

Upper insulator

Washers

Terminal plate

O-rings

Lower insulator

Terminal stud

4. Connect one probe of the megohmmeter (megger) to one of the motor terminals and the other probe to the shell of the compressor (ground). Make sure metal is exposed at the shell of the compressor where the probe is attached so that the compressor's shell paint is not acting as an insulator to ground.

5. Activate the megohmmeter according to the instructions by pushing a button that will apply a high DC voltage between its probes, and measure all electrical paths to ground. Be careful, as activating the meter for too long can damage motor windings.

Record your reading here: _____ megohms.

6. Good motor windings should have a resistance value of a minimum of 100 MΩ relative to ground. In fact, good motor windings should have a resistance between 100 MΩ and infinity.

7. Compare your reading with the readings in Figure MOT-19.4, and determine the condition of the motor windings. Record your answer here:_____.

FIGURE **MOT-19.4**

Required Reading (Megohms)	Condition Indicated	Required Preventive Maintenance	Percent of Winding In Field
Over 100	Excellent	None	30%
100–50	Some moisture present	Change filter drier.	35%
50–20	Severe moisture and/or contamination	Several filter drier changes; change oil if acid is present	20%
20–0	Severe contamination	Check entire system and make corrections. Consider an oversized filter drier, refrigerant and oil change, and reevacuation. System burnout and clean-up procedures required.	15%

MAINTENANCE OF WORKSTATION: Return all compressors, tools, and instruments to their proper location. Be sure to leave the work area clean.

SUMMARY STATEMENT: Explain why megohmmeters are used to check for shorts to ground as opposed to simply using traditional VOM meters.

TAKEAWAY: Successful service technicians have a wide range of specialized tools and instrumentation to properly and effectively evaluate and troubleshoot the components of air-conditioning and refrigeration systems. The megohmmeter is one such piece of instrumentation.

QUESTIONS:

1. Explain why a traditional VOM might not accurately indicate that an electrical path is present between a motor terminal and ground when, in fact, there actually is one.

2. Explain how shorts to ground can affect the operation of a motor.

3. What symptoms might be observed if there is an electrical short to ground in a motor.

Replacing an Electric Motor on a Furnace or Air Handler

Name	Date	Grade

OBJECTIVES: Upon completion of this exercise, you should be able to select the appropriate replacement blower for the system being worked on and properly replace the motor, ensuring that the motor is correctly mounted and wired.

INTRODUCTION: Replacing defective motors on HVACR systems is a major part of the service technician's job. In order to help ensure that the process runs smoothly, the proper replacement must be obtained and all mounting and wiring must be correct. This exercise will provide practice in identifying the proper replacement motor, properly removing the defective motor as well as mounting and wiring the new motor.

TOOLS, MATERIALS, AND EQUIPMENT: Screwdrivers, nut drivers, Allen keys, blower wheel puller, adjustable wrench, penetrating oil, steel wool, rags, operating air handler, replacement blower motor, replacement motor relay, replacement capacitor, digital multimeter (DMM), clamp-on ammeter, and safety glasses.

⚠ **SAFETY PRECAUTIONS:** Make certain that all power to the unit is disconnected and that all capacitors have been safely discharged. Before working on any blower or fan, make certain that the fan or blower has completely stopped. Before starting any blower or fan, make certain that the fan blade or blower spins freely and does not come in contact with any wires or other blower or fan housings.

STEP-BY-STEP PROCEDURES

1. Position yourself at an air handler, as assigned by your instructor.

2. Make certain that the power to the unit is off.

3. Carefully remove the service panel(s) from the air handler that is (are) covering the blower compartment.

4. Inspect the blower and housing assembly, and determine if the blower housing is permanently secured to the air handler or if it is a slide-out module. Indicate your findings here: _____

5. If the motor has one or more capacitors wired to it, make certain that it is (they are) properly discharged.

6. List each of the motor's wires here, indicating if each is connected electrically and, if so, where the wire is connected.

   Wire 1: Color _____

       Connected electrically? YES   NO

       If connected, where is the wire connected? _____

   Wire 2: Color _____

       Connected electrically? YES   NO

       If connected, where is the wire connected? _____

Wire 3: Color _____

    Connected electrically? YES NO

    If connected, where is the wire connected? _____

Wire 4: Color _____

    Connected electrically? YES NO

    If connected, where is the wire connected? _____

Wire 5: Color _____

    Connected electrically? YES NO

    If connected, where is the wire connected? _____

Wire 6: Color _____

    Connected electrically? YES NO

    If connected, where is the wire connected? _____

7. If the motor has a capacitor, provide the device's information here:

Capacitor 1 type: START   RUN   DUAL

Voltage rating: _____ V

Microfarad rating _____ μF

Capacitor 2 type: START   RUN   DUAL

Voltage rating: _____ V

Microfarad rating _____ μF

8. Identify the terminals on the capacitor(s), and identify the wires connected to each of the terminals:

Capacitor 1   Terminal 1 _____

             Terminal 2 _____

Capacitor 2   Terminal 1 _____

             Terminal 2 _____

Capacitor 3 (if dual)   Terminal HERM _____

                        Terminal FAN _____

                        Terminal C _____

9. Remove the capacitor(s) from the unit.

10. Label the wire/terminal connections on the capacitor(s).

11. Label the wires on the capacitors, if needed.

12. Carefully remove the wires from the capacitor(s).

13. Label the wire/terminal connections on the equipment.

14. Label the wires on the motor itself, if needed.

15. Carefully remove the motor wires from their connecting points in the air handler. At this point, the motor should be electrically disconnected from the air handler.

16. Access the blower wheel, and loosen the set screw that is holding the blower wheel on the motor shaft.

17. If needed, spray penetrating oil on the shaft to help loosen the blower wheel from the shaft. Be sure to allow a few minutes for the penetrating oil to do its job. If the end of the shaft is rusted, it may need to be cleaned using steel wool.

18. Loosen the motor mounts on the blower assembly, making certain that the motor is not permitted to hang loose, as this can cause the blower wheel to become imbalanced.

19. At this point, the blower wheel can be removed from the motor shaft. It may be necessary to use a blower wheel puller to assist with the wheel's removal.

20. Mark the motor's position in the motor mount, and if desired, take a picture of the motor's position in the mount.

21. Remove the motor from the motor mounts.

22. Remove the motor's starting relay from the air handler.

23. Provide information about the motor's starting relay here:
Relay type: _____ Part number: _____
Coil voltage rating:_____ Contact amperage rating: _____
How many sets of normally closed contacts does the relay have? _____
How many sets of normally open contacts does the relay have? _____

24. Provide the motor's information here:
Make: _____ Part number: _____
Voltage ratings: _____ Phase: _____ Amperage ratings: _____
Horsepower rating: _____ Rotation: _____
Capacitor rating: _____ Shaft diameter: _____
Motor speed(s): _____ Shaft length: _____
Motor mount: _____
Motor diameter: _____ Motor length: _____

25. Provide the blower wheel's information here:
Inlet type: SINGLE   DOUBLE
Rotation: _____ Bore size: _____
Blower wheel diameter: _____ Blower wheel length: _____

26. Using the information gathered and documented in step 22, identify the proper replacement motor from either a supply house catalog or the laboratory's motor stock.

27. Provide the information for the replacement motor here:
Make: _____ Part number: _____
Voltage ratings: _____ Phase: _____ Amperage ratings: _____
Horsepower rating: _____ Rotation: _____
Capacitor rating: _____ Shaft diameter: _____
Motor speed(s): _____ Shaft length: _____
Motor mount: _____

28. Explain why this replacement motor was selected. Discuss any differences that exist between the original motor and its replacement.

_____

_____

_____

29. Make certain that the motor shaft rotates freely.

30. Position the new motor in the mount and position the blower wheel on the motor shaft. Do not tighten the blower wheel onto the motor shaft at this time.

31. Tighten the motor into the mount, and secure the mounting hardware that may have been loosened to remove the motor. The motor should now be securely mounted.

32. Tighten the blower wheel onto the motor shaft, making certain that the blower wheel is not making contact with the motor housing.

33. Manually spin the blower wheel, again checking to ensure that no contact or rubbing occurs between the motor, the blower, and the housing.

34. Mount the new capacitor(s) in the unit, making certain that it is (they are) secure and the devices cannot slip.

35. Mount the new motor's starting relay in the unit.

36. Carefully route the motor wires to their connection points, making certain that the wires are secured and cannot come in contact with the rotating blower wheel.

37. Carefully make the appropriate motor wiring connections. Double-check your work; motor damage and/or personal injury can result if wiring connections are not correct.

38. With your instructor's permission, energize the air handler, and set the thermostat so the blower motor operates.

39. Ensure that the motor is operating and that there is no unusual noise or vibrations.

40. Replace the service panels on the air handler.

41. Using your clamp-on ammeter, measure the amperage draw of the motor, and record it here:
_____ A

42. Compare the amperage reading obtained in step 41 to the amperage rating obtained from the nameplate in step 28. Discuss any differences between these two values.

_____

_____

_____

MAINTENANCE OF WORKSTATION AND TOOLS: Make certain that all pieces of test equipment are turned off. Return all tools to their proper location. Clean the work area.

SUMMARY STATEMENT: Explain the importance of comparing the specifications of a defective motor with the replacement motor. What can happen if the replacement motor is not an acceptable match for the motor that has failed? Provide as many possible scenarios as possible.

TAKEAWAY: Replacing defective motors and other parts on HVACR systems is a major part of the service technician's job. This exercise provided practice in identifying the proper replacement motor, properly removing the defective motor, and mounting and wiring the new motor. While replacing a motor, it is good field practice to replace all of the motor's starting components as well.

QUESTIONS

1. Explain why it is good field practice to replace a motor's starting components whenever a motor is replaced.

2. Explain what can happen if the blower wheel becomes imbalanced as a result of removing the wheel from, or reinstalling it on, the motor shaft.

3. Explain the importance of concentrating on your work and avoiding distractions while replacing a motor, especially when making the electrical connections on the motor.

4. Explain why taking the amperage reading of a newly installed motor is important.

5. Explain what will happen to the amperage draw of a blower motor if the service panel is removed from the air handler while the motor is operating.

SECTION 8

Heat Pump Systems (HPS)

SUBJECT MATTER OVERVIEW

Heat pumps are reverse-cycle refrigeration systems that are used to provide both heating and cooling by moving heat either from indoors to outdoors or from outdoors to indoors. Heat pump systems have both indoor and outdoor components. The coil that serves the inside of the structure is called the indoor coil, whereas the coil located outside the structure is called the outdoor coil. The indoor coil operates as the condenser in the heating mode and as the evaporator in the cooling mode. The outdoor coil operates as the condenser in the cooling mode and as the evaporator in the heating mode.

Most heat pump systems rely on a four-way reversing valve to direct the compressor's hot discharge gas to either the indoor or outdoor coil to reject the heat to the desired location. In the cooling mode, the hot gas is directed to the outdoor coil, and in the heating mode, the hot gas is directed to the indoor coil.

Heat pumps are typically classified as air-source, water-source, geothermal, or earth-coupled systems. In residential applications, the most common type of heat pump system is the air-to-air variety. Air-source heat pumps (air-to-air and air-to-water) rely on air as the source of heat while operating in the heating mode. The second part of the classification refers to the medium that is ultimately treated. So an air-to-water heat pump system, for example, uses air as the heat source to heat water. In the winter months, the air-to-air heat pump removes heat from the outside air and rejects it to the air in the conditioned space. In the summer, the heat pump acts like a conventional air-conditioning system, absorbing heat from the air inside the house and rejecting it to the outside air. Water-source (water-to-air and water-to-water), geothermal, and earth-coupled heat pump systems use water or the earth as the heat source when operating in the heating mode.

Water-source heat pumps can be of the open-loop or closed-loop variety. Open-loop systems take water from either a well in the ground or a lake. The water is circulated through a coaxial heat exchanger that operates as either an evaporator or condenser, depending on the mode of operation. The circulated water either adds heat to or removes heat from the refrigerant in the heat exchanger. Closed-loop systems use buried piping arrangements. Heat is transferred between the fluid inside the pipe and its surroundings to either heat or cool the conditioned space. The waterless, earth-coupled, closed-loop geothermal heat pump system consists of a series of buried, plastic-coated copper refrigerant lines.

Split-type heat pump systems have two field-installed refrigerant lines. The larger-diameter pipe is called the gas line because it contains gas in both modes of system operation: cool suction gas in the cooling mode and hot discharge gas in the heating mode. The smaller line is called the liquid line because it contains liquid during both the heating and cooling modes of operation.

Auxiliary, or second-stage, heat is provided on air-to-air systems because the heat pump system may not have the capacity to heat a structure in colder temperatures. The balance point is the lowest temperature at which the heat pump can meet the heating needs of the structure without assistance from a secondary heat source. When operating at the balance point, the heat pump will heat

(continued)

the building but will run continuously. Auxiliary heat is required when the temperature is below the balance point to provide the additional heat needed to keep the structure at the desired temperature. The source of auxiliary heat may be electric, oil, or gas. Water-source heat pumps do not typically need auxiliary heat because the heat source (the water) can be maintained at a relatively constant temperature year round. Because the surface of the evaporator coil must be colder than the medium passing through it, it is very common for the outdoor coil of a heat pump system operating in the heating mode to be at a temperature that is below freezing.

A defrost cycle is used to defrost the ice from the outside coil during winter operation. Defrost is accomplished with air-to-air systems by changing the position of the four-way reversing, allowing the outdoor coil to function as the condenser, and stopping the outdoor fan. During defrost, the indoor coil acts as an evaporator, so cold air might be introduced to the occupied space, creating an uncomfortable draft for the occupants. To prevent this from happening, auxiliary heat is typically energized. This air-tempering process is not efficient and is used only when necessary. There are various ways to initiate and terminate the defrost cycle.

KEY TERMS

Air-source heat pumps
Balance point
Check valves
Closed loop
Coefficient of
  performance (COP)
Defrost cycle

Direct burial
Dry well
Ductless split-system
Earth-coupled
Four-way reversing valve
Geothermal heat pumps
Heat pump

Heating seasonal
  performance factor
  (HSPF)
Horizontal loops
Open loop
Parallel loops
Pressure tank

Return well
Series loops
Split system
Supply well
Vertical loops
Water loops
Water-source heat pumps

REVIEW TEST

Name	Date	Grade

Circle the letter that indicates the correct answer.

1. A heat pump that absorbs heat from surrounding air and rejects it into air in a different place is called _____ heat pump.
   A. a water-source
   B. an air-to-air
   C. a water-to-air
   D. an air-to-water

2. A heat pump has about the same components as an electric air-conditioning system except it has:
   A. an indoor coil.
   B. an outdoor coil.
   C. a four-way reversing valve.
   D. a capillary tube.

3. The larger-diameter interconnecting tubing in a heat pump system is called the _____ line.
   A. liquid
   B. suction
   C. discharge
   D. gas

4. The indoor coil operates as the _____ in the heating cycle.
   A. condenser
   B. evaporator
   C. compressor
   D. accumulator

5. Metering devices in most heat pumps are piped in parallel with:
   A. a check valve.
   B. the evaporator.
   C. the outdoor coil.
   D. the compressor.

6. Filter-driers installed in heat pump systems should be of the biflow type or installed:
   A. with thermistors.
   B. with an electrical disconnect.
   C. in parallel with the condensate line.
   D. with check valves.

7. When a heat pump is pumping in exactly as much heat as the building is leaking out and it is running continuously, it is said to be at the:
   A. auxiliary heat capacity.
   B. coefficient of performance.
   C. beginning of the defrost cycle.
   D. balance point.

8. The coefficient of performance for electric resistance heat is:
   A. 1 to 1.
   B. 2 to 1.
   C. 3 to 1.
   D. 4 to 1.

9. The outdoor coil operates as the _____ in the heating cycle.
   A. evaporator
   B. condenser
   C. accumulator
   D. auxiliary heat

10. For the greatest efficiency, the heated air from a heat pump operating in the heating mode should be directed through the registers to the:
    A. inside walls.
    B. outside walls.
    C. ceiling.
    D. attic.

11. The larger-diameter interconnecting pipe may reach temperatures as high as _____ in the winter.
    A. 120°F
    B. 150°F
    C. 170°F
    D. 200°F

12. The defrost cycle in a heat pump is used to melt the ice that forms on the:
    A. indoor coil.
    B. outdoor coil.
    C. compressor.
    D. both the indoor and outdoor coil.

13. The geothermal heat pump uses which of the following as a heat source?
    A. The earth
    B. Chemicals
    C. Air
    D. Coal

14. The liquid heat exchanger in a geothermal heat pump is often referred to as:
    A. coaxial.
    B. radial.
    C. linear.
    D. reverse-acting.

15. Water from wells or lakes is used in:
    A. open-loop systems.
    B. closed-loop horizontal systems.
    C. slinky systems.
    D. air-to-air systems.

16. Water for open-loop systems is most likely to contain which of the following?
    A. Chlorine
    B. Fluoride
    C. Sand and clay
    D. Algae

17. When there is a potential of fouling by minerals, the heat exchanger tubes should be made of:
    A. steel.
    B. cupronickel.
    C. copper.
    D. brass.

18. A closed-loop geothermal heat pump system may use which of the following circulating fluids?
    A. Antifreeze
    B. Sugar solution
    C. Vinegar solution
    D. Fluoride water

19. When a coaxial heat exchanger becomes coated with minerals, which of the following is used to clean it?
    A. Chemical circulation
    B. Scrubbing

C. Brushes
D. Airflow

20. When a coaxial heat exchanger becomes coated with minerals, the cooling-cycle symptom would be:
    A. low suction pressure.
    B. low head pressure.
    C. low amperage.
    D. high head pressure.

21. When a coaxial heat exchanger becomes coated with minerals, the heating-cycle symptom would be:
    A. low suction pressure.
    B. low voltage.
    C. high amperage.
    D. high head pressure.

22. In a waterless, earth-coupled, closed-loop geothermal heat pump system, what is the fluid flowing in the polyethylene plastic–coated copper pipes buried in the ground?
    A. Glycol
    B. Water
    C. Refrigerant
    D. Antifreeze

23. The waterless, earth-coupled, closed-loop geothermal heat pump system extracts heat directly from the earth and is often referred to as a(n) _____ heat exchange with the earth.
    A. direct
    B. indirect
    C. exact
    D. differential

24. The purpose of a buffer tank on a water-to-water heat pump is to:
    A. keep the head pressure of the system as high as possible.
    B. reduce contamination in the water.
    C. act as the water supply tank for a radiant heating system.
    D. ensure that the system's water pump operates continuously.

Familiarization with Air-to-Air Heat Pump System Components

Name	Date	Grade

OBJECTIVES: Upon completion of this exercise, you should be able to identify and describe the components in a typical air-to-air heat pump system.

INTRODUCTION: You will inspect the components listed in what follows and record the specifications called for. By referring to your text, you should be able to recognize all components listed.

TOOLS, MATERIALS, AND EQUIPMENT: Complete split-type heat pump system, straight-slot and Phillips-head screwdrivers, ¼" and ⁵⁄₁₆" nut drivers, combination wrench set, flashlight, rags, and digital multimeter (DMM).

⚠ **SAFETY PRECAUTIONS:** Ensure that power to both indoor and outdoor units is off. To make sure, check the voltage with the digital multimeter before starting this exercise. Lock and tag the electrical power panel, and keep the key with you.

STEP-BY-STEP PROCEDURES

1. Remove the service panels on the indoor unit and record the following:

   Motor: _____ hp, _____ V, _____ A
   Motor type: _____ Shaft size: _____
   Shaft rotation: _____ Number of speeds: _____
   Electric heat: _____ kW, _____ A, _____ V, _____ rated voltage
   Resistance of heaters: _____
   Number of heat stages: _____ Factory wire size: _____
   Refrigerant coil material: _____ Number of refrigerant circuits in coil: _____ Type of metering device: _____
   Size of gas line: _____ Size of liquid line: _____
   Type of coil (A, H, N, or slant): _____ Number of tube passes: _____ Size of tubes: _____

2. Make a sketch of the refrigerant piping circuit of the indoor unit. Include any system components connected close to, but not necessarily contained within, the actual indoor unit.

3. Remove the panels on the outdoor unit and record the following:

   Fan motor: _____ hp _____ V _____ A
   Motor type: _____
   Shaft size: _____ Type of blade: _____
   Shaft rotation: _____
   Compressor motor type: _____ A _____ V
   Suction-line size: _____
   Discharge line size: _____ Number of motor terminals: _____
   Coil material: _____ Number of refrigerant circuits in coil: _____ Coil tubing size: _____

Number of tube passes: _____
Four-way valve coil voltage: _____ Line sizes: _____

4. Make a sketch of the refrigerant piping circuit of the outdoor unit. Include any system components connected close to, but not necessarily contained within, the actual outdoor unit.

MAINTENANCE OF WORKSTATION AND TOOLS: Replace any components you may have removed or disturbed. Have your instructor inspect your system to ensure it is left in proper working order. Replace all panels with appropriate fasteners. Replace all tools, equipment, and materials to their proper places. Clean your work area.

SUMMARY STATEMENT: Describe in detail the air and refrigerant flow in the air-to-air heat pump when it is in the heating cycle. Use an additional sheet of paper if more space is necessary for your answer. As part of your response, be sure to identify any components that were identified on the heat pump system that are not part of a traditional, cooling-only air-conditioning system.

TAKEAWAY: The refrigerant piping on heat pump systems can be somewhat intimidating, as there are more components than on traditional air-conditioning systems. Becoming familiar with the piping arrangements and additional components will help you better understand heat pump systems and become a better service technician.

QUESTIONS

1. What are typical condenser and evaporator tubes made of?

2. When is the solenoid coil on the four-way reversing valve energized in a typical heat pump, the heating mode or the cooling mode? Explain why this is so.

3. What does kW stand for, and how does this term relate to Btu/h?

4. What component is used to change the heat pump from cooling to heating?

5. What does "air-to-air" mean when referring to heat pumps?

6. How is the frost buildup removed? Describe this process in detail.

Using Manufacturer's Information to Check a Heat Pump System's Performance

Name	Date	Grade

OBJECTIVES: Upon completion of this exercise, you should be able to check system performance and compare it to the system performance chart furnished in the manufacturer's literature.

INTRODUCTION: You will check the indoor and outdoor wet-bulb and dry-bulb temperatures for the existing weather conditions for the season of the year, as called for in the manufacturer's performance chart. Then chart the suction and discharge pressures, and compare with the pressures indicated on the chart.

TOOLS, MATERIALS, AND EQUIPMENT: A thermometer with leads set up to check wet-bulb and dry-bulb temperatures, a gauge manifold, straight-slot and Phillips-head screwdrivers, ¼" and ⁵⁄₁₆" nut drivers, light gloves, goggles, and operating heat pump system with accompanying manufacturer's performance chart.

⚠ **SAFETY PRECAUTIONS:** Wear gloves and goggles when connecting gauges to the heat pump system. Do not allow liquid refrigerant to get on your bare skin.

STEP-BY-STEP PROCEDURES

1. Check the manufacturer's literature to determine on which lines to install the gauges of the gauge manifold. With the system off and while wearing gloves and goggles, install the gauges as indicated.

2. Start the system in the cooling mode.

3. Allow the unit to run until the conditions are stable and pressures and temperatures are not changing.

4. Measure and record the temperatures and pressures, as called for here.

   For summer or cooling: Indoor WB _____°F, DB _____°F, Outdoor DB _____°F

   Suction pressure: _____ psig, Discharge pressure: _____ psig

   Return air temperature to the indoor coil: _____ F

   Supply air temperature from the indoor coil: _____ F

   Suction-line temperature at the evaporator coil outlet: _____ F

   Liquid-line temperature at the condenser coil outlet: _____ F

   For winter or heating: Indoor DB _____°F, Outdoor WB _____°F, DB _____°F

   Suction pressure: _____ psig

   Discharge pressure: _____ psig

   Return air temperature to the indoor coil: _____ F

   Supply air temperature from the indoor coil: _____ F

   Suction-line temperature at the evaporator coil outlet: _____ F

   Liquid-line temperature at the condenser coil outlet: _____ F

5. Compare the obtained information with the manufacturer's performance chart.

6. Turn the system off.

7. Discuss any discrepancies that might exist between the obtained readings and the information in the manufacturer's system performance literature.

_____

_____

_____

_____

8. Switch the system over to the heating mode.

9. Allow the unit to run until the conditions are stable and pressures and temperatures are not changing.

10. Measure and record the temperatures and pressures, as called for here.

For summer or cooling: Indoor WB _____°F, DB _____°F, Outdoor DB _____°F

Suction pressure: _____ psig, Discharge pressure: _____ psig

Return air temperature to the indoor coil: _____ F

Supply air temperature from the indoor coil: _____ F

Suction-line temperature at the evaporator coil outlet: _____ F

Liquid-line temperature at the condenser coil outlet: _____ F

For winter or heating: Indoor DB _____°F, Outdoor WB _____°F, DB _____°F

Suction pressure: _____ psig

Discharge pressure: _____ psig

Return air temperature to the indoor coil: _____ F

Supply air temperature from the indoor coil: _____ F

Suction-line temperature at the evaporator coil outlet: _____ F

Liquid-line temperature at the condenser coil outlet: _____ F

11. Compare the obtained information with the manufacturer's performance chart.

12. Turn the system off.

13. Discuss any discrepancies that might exist between the obtained readings and the information in the manufacturer's system performance literature.

_____

_____

_____

_____

14. Wearing gloves and goggles, remove the gauges.

MAINTENANCE OF WORKSTATION AND TOOLS: With the system off, replace all panels with proper fasteners as instructed. Return all tools and equipment to their proper places. Leave your workstation in order.

SUMMARY STATEMENT: Describe, in detail, the differences/similarities between the set of readings obtained during the cooling operation and the set of readings obtained during the heating operation.

TAKEAWAY: Heat pump systems are designed to provide both heating and cooling by reversing the direction of refrigerant flow through the system and by changing the function of the heat transfer surfaces. This exercise provides an opportunity for the student to observe heat pump system operation in both modes while obtaining and evaluating valuable data from the system.

QUESTIONS

1. Why is the wet-bulb reading necessary when one is checking the charge?

2. Where is the wet-bulb temperature measured (indoors or outdoors) when one is checking the charge in the winter or heating cycle? Explain why.

3. Where is the wet-bulb temperature measured (indoors or outdoors) when one is checking the charge in the summer or cooling cycle? Explain why.

4. Explain the difference between a wet-bulb temperature and a dry-bulb temperature.

Heat Pump Performance Check in Cooling Mode

Name	Date	Grade

OBJECTIVES: Upon completion of this exercise, you should be able to make a heat pump performance check on a heat pump with fixed-bore metering devices (capillary tube, fixed-bore, piston) while it is in the cooling mode and adjust refrigerant charge for desired performance.

INTRODUCTION: You will be calculating the superheat under a standard condition. You will install gauges on the gas and liquid refrigerant lines to check pressures and to add or remove refrigerant. The head pressure will be increased to 275 psig (for an R-22 system) in the heat pump and used as a standard condition. Refrigerant will be added to or removed from the system until the desired superheat is achieved.

Note: This method of adjusting the charge is recommended only when manufacturer's literature is not available. It may be performed in the winter while operating in the summer mode.

TOOLS, MATERIALS, AND EQUIPMENT: A high-quality thermocouple digital thermometer, a gauge manifold, straight-slot and Phillips screwdrivers, ¼" and ⁵⁄₁₆" nut drivers, refrigerant, light gloves, goggles, and an operational heat pump system with gauge ports and fixed-bore metering devices.

 SAFETY PRECAUTIONS: Wear goggles and light gloves when connecting gauges to or removing them from system service ports. Liquid refrigerant can cause frostbite. Watch the low-side gauge to ensure that the system does not go into a vacuum. Turn the system off if it appears that this is happening. If the compressor makes any unusual sound or if it acts like it is under a strain, shut the system off, and check with your instructor.

STEP-BY-STEP PROCEDURES

1. With the unit off, fasten one lead of the thermometer to the gas line just before the outdoor unit service valve. Insulate the thermometer's sensor.

2. Carefully connect the low-pressure gauge line onto the gas (large line) port and the high-pressure gauge line onto the liquid (small line) port. Wear gloves and goggles, and keep your fingers to the side.

3. Switch the thermostat to the cooling cycle, and start the heat pump. Allow the unit to operate for about 15 minutes to stabilize.

4. After the system has stabilized, use a plastic bag, a piece of cardboard, or heavy paper to partially block the airflow through the outdoor coil. Adjust the restriction until the head pressure has stabilized at 275 psig (for R-22) for 15 minutes.

5. Record the temperature reading on the gas line: _____ °F. Determine the superheat and record: _____ °F.

6. The superheat should be 12°F ± 2°F. If the superheat is above this range, add refrigerant until it is within range. If it is below this range, recover refrigerant until the superheat is within range.

7. After any refrigerant charge adjustment, let the unit operate for at least 15 minutes with a stable 10°F to 14°F superheat before concluding that the charge is correct.

8. When this condition is met, turn the unit off. Carefully remove the gauge lines and the thermometer leads.

MAINTENANCE OF WORKSTATION AND TOOLS: Remove the plastic bag or cardboard from around the condenser area. Replace all panels on the heat pump unit with proper fasteners. Return tools, equipment, and materials to their proper places.

SUMMARY STATEMENT: Explain why manufacturer's literature, when available, should always be consulted for system set points as opposed to using general industry guidelines.

TAKEAWAY: Adjusting the refrigerant charge on an operating air-conditioning or heat pump system is something that service technicians do on a daily basis. Practicing these techniques in a controlled lab setting helps make mastering this skill much easier, ultimately leading to the creation of a highly skilled technician.

QUESTIONS

1. Name three metering devices that can be found on typical heat pumps.

2. What type of drier is used in the liquid line on heat pumps?

3. Where is the only true suction line?

4. Where is the only true discharge line?

5. Where is the suction-line accumulator located?

6. Why did you simulate summer conditions when checking the charge?

7. Could the superheat method of charge adjustment be used on a heat pump with a thermostatic expansion device? Explain why or why not.

8. Why should you shut the system off to remove the gauge manifold hose from the liquid-line service valve?

Exercise HPS-4

Checking the Control Circuit in the Emergency and Auxiliary Heat Modes

Name	Date	Grade

OBJECTIVES: Upon completion of this exercise, you should be able to follow the control circuits in the auxiliary and emergency heat modes in a heat pump system and measure the voltage at each terminal.

INTRODUCTION: You will study the wiring diagrams for the low-voltage control circuits in the auxiliary and emergency heat modes to help you become familiar with these circuits. After tracing each circuit on the diagram, you will locate each terminal on the heat pump system and check the voltage at each terminal.

TOOLS, MATERIALS, AND EQUIPMENT: An operating heat pump system, note pad, pencils, straight-slot and Phillips-head screwdrivers, flashlight, ¼" and ⁵⁄₁₆" nut drivers, and digital multimeter (DMM).

⚠ **SAFETY PRECAUTIONS:** Turn the power off, and lock and tag the disconnect box while becoming familiar with the control circuits and components for the auxiliary and emergency heat systems. Use the digital multimeter to verify that the power is off. The line voltage should remain off at the outdoor unit while you are taking voltage readings. Be careful not to come in contact with the high voltage at the indoor unit.

STEP-BY-STEP PROCEDURES

1. Study the wiring diagrams shown in Figures HPS-4.1 to HPS-4.11.

2. Locate each component and terminal in the actual system.

3. Turn the power to the indoor unit on.

4. Set the room thermostat to call for heat.

5. Check and record the voltage for each circuit at the indoor unit terminal block. Place one meter lead on the common terminal, and record the voltage at the terminal, controlling the following:

   - Sequencer coil voltage: _____ V
   - Fan relay coil voltage: _____ V
   - Compressor contactor coil voltage: _____ V

6. Switch the thermostat to the emergency heat mode at the thermostat, and record the following:

   - Sequencer coil voltage: _____ V
   - Fan relay coil voltage: _____ V
   - Compressor contactor coil voltage: _____ V

   Note: The difference between the two circuits is caused by the switching action from auxiliary to emergency heat in the thermostat.

MAINTENANCE OF WORKSTATION AND TOOLS: Replace all panels on the heat pump system. Return all tools and equipment to their proper places. Leave your workstation in a neat and orderly condition.

SUMMARY STATEMENT: Describe the difference between auxiliary and emergency heat.

TAKEAWAY: The control wiring on a heat pump system is more complicated than the wiring of a traditional cooling-only air-conditioning system. It will take a fair amount of exposure to heat pump wiring to become comfortable with it. By continuing to practice and work with heat pump system wiring strategies, you will become more confident in your ability to work on these systems in the field.

FIGURE **HPS-4.1**

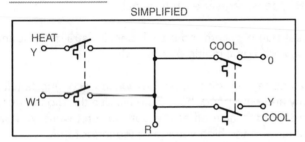

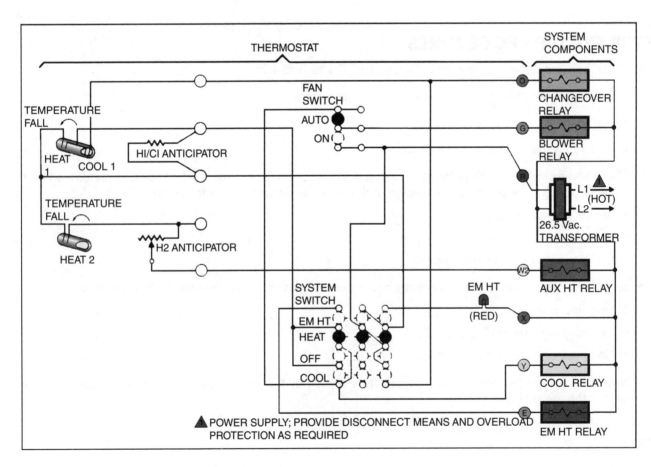

FIGURE **HPS-4.2**

1ST STAGE OF COOLING

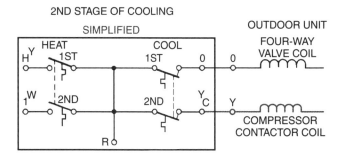

FIGURE **HPS-4.3**

2ND STAGE OF COOLING

FIGURE **HPS-4.4**

1ST STAGE OF HEAT

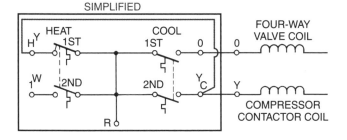

FIGURE **HPS-4.5**

2ND STAGE OF HEAT

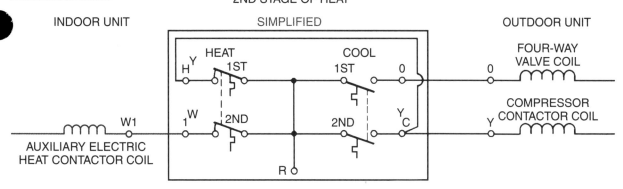

FIGURE **HPS-4.6**

2ND STAGE OF HEAT WITH
OUTDOOR THERMOSTATS

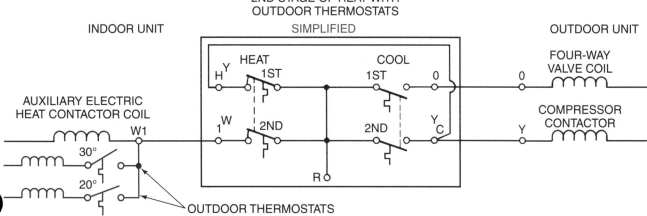

FIGURE **HPS-4.7**

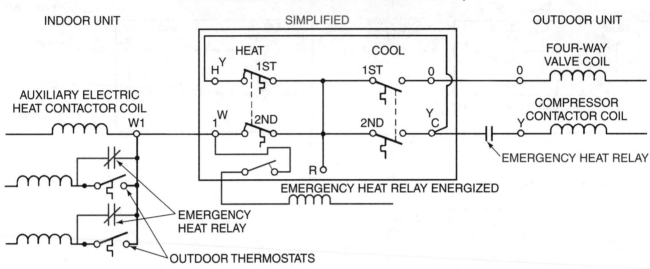

EMERGENCY HEAT MODE (THE COMPRESSOR STOPS AND
THE W TERMINAL CONTROLS ALL ELECTRIC HEAT)

FIGURE **HPS-4.8**

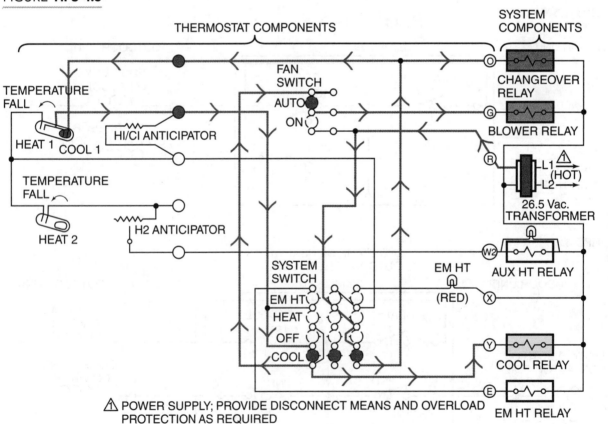

⚠ POWER SUPPLY; PROVIDE DISCONNECT MEANS AND OVERLOAD
PROTECTION AS REQUIRED

FIGURE **HPS-4.9**

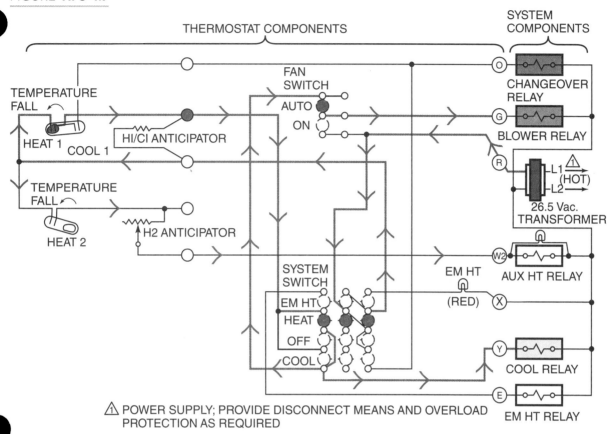

⚠ POWER SUPPLY; PROVIDE DISCONNECT MEANS AND OVERLOAD
PROTECTION AS REQUIRED

FIGURE **HPS-4.10**

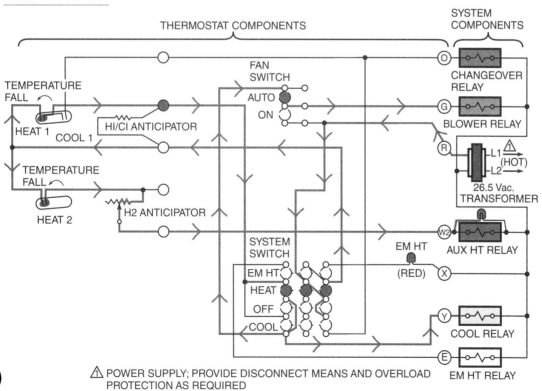

⚠ POWER SUPPLY; PROVIDE DISCONNECT MEANS AND OVERLOAD
PROTECTION AS REQUIRED

FIGURE **HPS-4.11**

⚠ POWER SUPPLY; PROVIDE DISCONNECT MEANS AND OVERLOAD PROTECTION AS REQUIRED

QUESTIONS

1. What is the COP of the emergency heat mode?

2. What is the COP of the auxiliary heat?

3. When should emergency heat be used?

4. When would auxiliary heat normally operate?

5. Where are the electric heating elements located in relation to the heat pump refrigerant coil in the airstream?

6. Where does the heat come from for defrost?

7. What keeps the unit from blowing cold air into the house during defrost?

Exercise HPS-5

Checking the Four-Way Valve During the Heating and Cooling Cycles

Name	Date	Grade

OBJECTIVES: Upon completion of this exercise, you should be able to check the four-way valve for internal leaking of gas during the cooling mode.

INTRODUCTION: Using a high-quality, digital thermocouple thermometer, you will measure the temperature of the permanent suction line and the indoor coil line at the four-way valve while the unit is in the cooling mode. You will compare the two readings to determine if the reversing valve is leaking, causing hot gas to pass into the suction line from the discharge line.

TOOLS, MATERIALS, AND EQUIPMENT: A gauge manifold; a high-quality, digital thermocouple thermometer; straight-slot and Phillips-head screwdrivers; ¼" and ⁵⁄₁₆" nut drivers; insulation material for attaching thermometer sensors to tubing; gloves; and goggles.

⚠ **SAFETY PRECAUTIONS:** If there is an indication the heat pump may go into a vacuum, stop the unit and check with your instructor. Wear gloves and goggles when connecting gauge lines to and removing them from gauge ports. Shut the power off to the outdoor unit while fastening thermometer sensors to refrigerant lines to avoid electrical contact. Lock and tag the panel, and keep the key with you.

STEP-BY-STEP PROCEDURES

1. With the power to the outdoor unit off and locked out, fasten one thermometer sensor to the permanent suction line and one to the indoor coil line. Fasten them at least 4 in. from the four-way valve. Insulate the sensors.

2. With goggles and gloves on, connect the high- and low-side gauges to the gauge ports, and start the unit in the cooling mode. Check to ensure that the low side does not go into a vacuum.

3. Let the unit run for 15 minutes, and then record the following:

   - Suction pressure: _____ psig

   - Discharge pressure: _____ psig

   - Permanent suction-line temperature: _____ °F

   - Indoor coil-line temperature: _____ °F

4. Stop the unit. Turn the power off to the outdoor unit, and lock the disconnect box. Remove the thermometer sensors. With your gloves and goggles on, remove the gauge lines.

   Note: For a complete test of the four-way valve, the same test may be performed in the heating mode by moving the lead from the indoor coil line to the outdoor coil line.

   Caution: Do not run the unit for more than 15 minutes in the heating cycle if the outdoor temperature is above 65°F.

MAINTENANCE OF WORKSTATION AND TOOLS: Replace all panels on the system as you are instructed. Return all tools, equipment, and materials to their respective places.

SUMMARY STATEMENT: Explain what the symptoms would be if the four-way valve were leaking across the various ports.

TAKEAWAY: Reversing valves are among the most difficult components to replace on a heat pump system. It is, therefore, important to be able to effectively and accurately evaluate the valve to make absolutely certain that the valve is defective before making the replacement.

QUESTIONS

1. What temperature difference should be expected between the indoor line and the permanent suction-line ports on the reversing valve during the cooling cycle?

2. What temperature difference should be expected between the outdoor line and permanent suction-line ports on the reversing valve during the heating cycle?

3. If the permanent suction line were 10°F warmer than the indoor coil line at the four-way valve while the system is operating in the cooling mode, where could the heat come from?

4. Why should you place the thermometer sensor at least 4 in. from the four-way valve when checking the refrigerant-line temperature?

5. What precautions should be taken when soldering a new four-way valve into the line?

6. What force actually changes the four-way valve over from one position to another?

Wiring a Split-Type Heat Pump System

Name	Date	Grade

OBJECTIVES: Upon completion of this exercise, you should be able:

- Correctly size line voltage conductors for the outdoor unit, indoor unit, and supplementary electrical resistance heat of a split-system heat pump.
- Create a list of supplies that will be needed to install the outdoor unit, indoor unit, and supplementary electrical resistance heat of a split-system heat pump.
- Make the electrical connection for the outdoor unit, indoor unit, and supplementary electrical resistance heat of a split-type heat pump system.
- Operate and evaluate the operating system.
- Complete a check, test, and start-up checklist on the installation.

INTRODUCTION: The installation instructions furnished with new equipment, along with local and state codes, will be the guide for the installation of new air-conditioning and heat pump equipment. The technician should be able to correctly size the line voltage conductors using the National Electrical Code or manufacturer's installation instructions. A list of materials that will be needed to complete the installation should be compiled by the installation technician. The technician will be responsible for installing and checking the operation of the newly installed equipment. Once equipment is installed and operating, the technician should complete a check test and start-up form.

TOOLS, MATERIALS, AND EQUIPMENT:

National Electrical Code, manufacturer's installation instructions for unit assigned by instructor, outdoor unit of a split-type heat pump system, indoor unit of a split-type heat pump system, resistance heat for heat pump, low-voltage heat pump thermostat, electrical supplies (technician's list), tools needed to make installation.

⚠ **SAFETY PRECAUTIONS:** Make certain that the electrical source is disconnected when making electrical connections. In addition,

- Make sure all electrical connections are tight.
- Make sure no current-carrying conductors are touching metal surfaces except the grounding conductor.
- Make sure the correct voltage is being supplied to the equipment.
- Make sure all equipment covers are replaced.
- Make sure body parts do not come in contact with live electrical conductors.
- Keep hand and materials away from moving parts.

STEP-BY-STEP PROCEDURES

1. Obtain the installation instructions for the unit being installed from your instructor.

2. Read the section in the installation instructions on wiring the unit.

3. Determine the correct wire size for the assigned unit components (outdoor unit and indoor unit). Take note of the distance between the equipment and the electrical supply panel.

4. Record the system data here:

   Outdoor unit model number: _____

   Outdoor unit serial number: _____

   Voltage rating of outdoor unit: _____

   Amperage rating of outdoor unit: _____

   Distance from power supply to condensing unit: _____

   Size of wire to be installed for condensing unit: _____

   Indoor unit model number: _____

   Indoor unit serial number: _____

   Resistance heat model number: _____

   Voltage rating of indoor unit: _____

   Amperage rating of indoor unit: _____

   Voltage rating of electric heaters: _____

   Amperage rating of electric heaters: _____

   Distance from power supply to indoor unit: _____

   Size of wire to be installed for indoor unit with supplementary heat: _____

5. Create a complete material list, including electrical supplies, that will be required to properly complete the system installation.

   _____

   _____

   _____

   _____

   _____

   _____

6. Have your instructor check your wire sizing and material list for the installation of the assigned piece of equipment.

7. Make the necessary electrical connections to the system.

8. Operate the system and complete the CHECK, TEST, AND START-UP LIST.

   Voltage being supplied to outdoor unit: _____

   Amperage draw of compressor: _____

   Amperage draw of outdoor fan motor: _____

   Voltage being supplied to the indoor unit: _____

   Amperage draw of blower motor: _____

   Amperage draw of resistance heaters: _____

   Supply air temperature (w/o supplementary heat): _____ °F

Supply air temperature (with supplementary heat): _____ °F

Return air temperature: _____ °F

CHECKLIST

All electrical connections tight	☐ YES	☐ NO
Electrical circuits properly labeled	☐ YES	☐ NO
Equipment properly grounded	☐ YES	☐ NO
All equipment level	☐ YES	☐ NO
All equipment covers in place and properly attached	☐ YES	☐ NO
Installation area clean	☐ YES	☐ NO
Homeowner instructed on operation of system	☐ YES	☐ NO

Note: The technician should also check the mechanical refrigeration cycle characteristics and gas heating operation when performing check, test, and start procedures.

9. Have instructor check your installation.

10. Clean up the work area, and return all tools and supplies to their correct locations.

MAINTENANCE OF WORKSTATION AND TOOLS: Clean and return all tools to their proper locations.

SUMMARY STATEMENT: Why is supplementary heat necessary on an air-to-air heat pump? Why type of low-voltage thermostat is used on the heat pump in this lab exercise?

TAKEAWAY: The control wiring on a heat pump system is more complicated than the wiring of a traditional cooling-only air-conditioning system. It will take a fair amount of exposure to heat pump wiring to become comfortable with it. By continuing to practice and work with heat pump system wiring strategies, you will become more confident in your ability to work on these systems in the field.

QUESTIONS

1. Why is supplementary heat required on most air-to-air heat pumps?

2. What is the difference between the mechanical refrigeration cycle of a heat pump and a regular air conditioner?

3. What two sources of supplementary heat could be used on heat pumps?

4. What is the purpose of the reversing valve in the mechanical refrigeration cycle of a heat pump?

5. What control voltage connections must be made when installing an air-to-air heat pump using electrical resistance heaters as the supplementary heat?

6. How are electrical resistance heaters mounted in most indoor blower units in a split-system heat pump installation?

Exercise HPS-7

Checking Low-Voltage Field Control Circuits at the Individual Components

Name	Date	Grade

OBJECTIVES: Upon completion of this exercise, you should be able to identify and take low-voltage readings at each individual component.

INTRODUCTION: To become familiar with the control wiring for the heat pump you are working with, you will study the manufacturer's wiring diagram and trace the individual circuits in the unit. You will then operate the system in the cooling cycle, the first- and second-stage heat cycles, and the emergency heat cycle while measuring the current flow and checking the voltage at the various components.

TOOLS, MATERIALS, AND EQUIPMENT: Straight-slot and Phillips-head screwdrivers, a flashlight, ¼" and ⁵⁄₁₆" nut drivers, a digital multimeter (DMM), and a split-system heat pump system and wiring diagram.

⚠ **SAFETY PRECAUTIONS:** Turn the power off to the outdoor unit when working with electrical wiring and components in the unit. Lock and tag the disconnect box, and keep the key with you. Always take all precautions when working around electricity.

STEP-BY-STEP PROCEDURES

1. Study the manufacturer's wiring diagram until you feel certain you understand the sequence of operations. Locate the circuits on the diagram in the unit to become familiar with their location.

2. Turn off and lock out the power to the outdoor unit. Verify with the digital multimeter that the power is off. Remove the common wire that runs from the load side of the contactor to the common terminal on the compressor. Tape it to insulate it. This will allow you to turn the system on and off without starting and stopping the compressor.

3. Turn the power to the outdoor unit on. Switch the room thermostat to the cool cycle, and adjust to call for cooling.

   Note: The outdoor fan will start and stop each time the compressor would normally start and stop. Check to see whether the indoor and outdoor fans are on.

   Trace the control circuitry for cooling on the manufacturer's diagram. With your digital multimeter, check the voltage at the following components:

   • Fan relay coil voltage: _____ V

   • Compressor contactor coil voltage: _____ V

   • Sequencer coil voltage: _____ V

   • Four-way valve coil voltage: _____ V

4. Set the thermostat to call for first-stage heat. Check the indoor and outdoor units to ensure that they are operating. Trace the control wiring for this circuit on the manufacturer's diagram. With your digital multimeter, check the voltage at the following components:

   - Fan relay coil voltage: _____ V

   - Compressor contactor coil voltage: _____ V

   - Sequencer coil voltage: _____ V

   - Four-way valve coil voltage: _____ V

5. Set the thermostat to call for second-stage heat. Check the indoor and outdoor units to ensure that they are operating. Follow the control wiring for this circuit on the manufacturer's diagram. With your digital multimeter, check the voltage at the following components:

   - Fan relay coil voltage: _____ V

   - Compressor contactor coil voltage: _____ V

   - Sequencer coil voltage: _____ V

   - Four-way valve coil voltage: _____ V

6. Turn the thermostat to the emergency heat cycle. Check to see that the indoor unit is on. Follow the control wiring for this circuit on the manufacturer's wiring diagram. With your digital multimeter, check the voltage at the following components:

   - Fan relay coil voltage: _____ V

   - Compressor contactor coil voltage: _____ V

   - Sequencer coil voltage: _____ V

   - Four-way valve coil voltage: _____ V

7. Turn off and lock out the power to the outdoor unit. Replace the wire that you previously removed from the common terminal to the contactor. Turn the power back on.

MAINTENANCE OF WORKSTATION AND TOOLS: Replace the panels with proper fasteners as instructed. Return all tools and equipment to their proper places.

SUMMARY STATEMENTS:

Describe all functions of the control system from the time:

   - The thermostat calls for cooling until the thermostat is satisfied.

   - The thermostat calls for heat, including second-stage heat, until the thermostat is satisfied.

   - The system is switched to emergency heat until the thermostat is satisfied.

TAKEAWAY: The control wiring on a heat pump system is more complicated than the wiring of a traditional cooling-only air-conditioning system. It will take a fair amount of exposure to heat pump wiring to become comfortable with it. This exercise provides you with the opportunity to identify and test the electrical control components of the heat pump system.

QUESTIONS

1. What thermostat terminals are energized in the cooling cycle?

2. What thermostat terminals are energized in the heating cycle, first stage?

3. What thermostat terminals are energized in the heating cycle, second stage?

4. What thermostat terminals are energized in the emergency heat mode?

5. How would the system's operating pressures change (increase, decrease, remain the same) if the system is operated under the following conditions?

A. Dirty filter in the heating cycle

B. Dirty outdoor coil in the cooling cycle

C. Dirty indoor coil in the cooling cycle

D. Overcharge in the heating cycle

E. Overcharge in the cooling cycle

F. The four-way valve stuck in the midposition

Exercise HPS-8

Troubleshooting Exercise: Heat Pump

Name	Date	Grade

OBJECTIVES: Upon completion of this exercise, you should be able to troubleshoot basic defrost problems in a heat pump.

INTRODUCTION: This unit should have a fault that was added by your instructor. You will locate the problem using a logical approach and instruments. THE INTENT IS NOT FOR YOU TO MAKE THE REPAIR BUT TO LOCATE THE PROBLEM AND ASK YOUR INSTRUCTOR FOR DIRECTION BEFORE COMPLETION.

TOOLS, MATERIALS, AND EQUIPMENT: An operating heat pump system, straight-slot and Phillips-head screwdrivers, ¼" and ⁵⁄₁₆" nut drivers, a digital multimeter (DMM), and needle-nose pliers.

⚠️ **SAFETY PRECAUTIONS:** Use care while checking live electrical circuits.

STEP-BY-STEP PROCEDURES

1. Start the unit in the heating mode, and allow it to run for 15 minutes. If the outdoor air temperature is above 60°F, you should either block the airflow to the outdoor coil or disconnect the outdoor fan to keep from overloading the compressor. Frost should begin to form on the outdoor coil in a short period of time.

2. With the unit running, remove the control compartment door, and locate the method for simulating defrost. If the unit has a circuit board, the manufacturer's directions will have to be followed. If the unit has time and temperature defrost, you may follow the methods outlined in the text.

3. Describe the type of defrost this unit has.

4. Wait for the coil to cover with frost before trying to defrost using the time and temperature method. REMEMBER, THE SENSING ELEMENT IS LARGE AND REQUIRES SOME TIME FOR IT TO COOL DOWN TO COIL TEMPERATURE.

5. With the coil covered with frost, force the defrost.

   • Did the reversing valve reverse to cooling? _____

   • Did the outdoor fan motor stop? _____

   • Did the first stage of strip heat energize? _____

6. Using the digital multimeter, is the defrost relay coil energized? _____ Is it supposed to be energized during defrost? _____

7. Is the defrost relay coil responding as it should? _____

8. Describe what you think the problem may be.

9. Describe any materials you think would be required to repair the equipment. _____

10. Ask your instructor whether the unit should be repaired or shut down and used for the next student.

MAINTENANCE OF WORKSTATION AND TOOLS: Leave the workstation as your instructor directs you and return all tools to their places.

SUMMARY STATEMENT: Describe the entire defrost cycle for this unit. Use an additional sheet of paper if more space is necessary for your answer.

TAKEAWAY: Troubleshooting HVACR equipment is an important part of what the service technician does on a daily basis. Developing good troubleshooting skills takes time, but the result is well worth the effort.

QUESTIONS

1. Why must an air-source heat pump have a defrost cycle?

2. Why is the electric heat energized during the defrost cycle on any heat pump?

3. Do all heat pumps have a defrost cycle? If not, which ones do not?

4. What is the purpose of the air switch used for defrost on some heat pumps?

5. How could a heat pump be safely defrosted if the technician were to arrive and find that the coil has frozen solid and resembles a block of ice?

6. What is the purpose of emergency heat on a heat pump system?

Exercise HPS-9

Troubleshooting Exercise: Heat Pump

Name	Date	Grade

OBJECTIVES: Upon completion of this exercise, you should be able to troubleshoot an electrical problem with the change from cool to heat.

INTRODUCTION: This unit should have a specified problem introduced by your instructor. You will locate the problem using a logical approach and instruments. THE INTENT IS NOT FOR YOU TO MAKE THE REPAIR BUT TO LOCATE THE PROBLEM AND ASK YOUR INSTRUCTOR FOR DIRECTION BEFORE COMPLETION.

TOOLS, MATERIALS, AND EQUIPMENT: Straight-slot and Phillips-head screwdrivers, ¼" and ⁵⁄₁₆" nut drivers, needle-nose pliers, and a digital multimeter (DMM).

⚠ **SAFETY PRECAUTIONS:** Use care while checking energized electrical connections.

STEP-BY-STEP PROCEDURES

1. Study the wiring diagram, and determine whether the four-way valve should be energized in the cooling or heating mode.

2. Start the unit in the cooling mode, and allow it to run for 15 minutes. Is it cooling or heating the conditioned space? _____ If it is heating and is supposed to be cooling, block the evaporator coil to reduce the airflow to the outdoor unit or it may overload the compressor.

3. Locate the low-voltage terminal board at the unit, and check the voltage from the common terminal to the terminal that energizes the four-way valve.

   Note: This is the "O" terminal for many manufacturers.

   Record the voltage here: _____ V

4. Is the correct voltage supplied? _____

5. Check the voltage at the four-way valve coil.

   Note: This may be full line voltage on some units. If so, the voltage is usually switched at the defrost relay.

6. Is the correct voltage being supplied to the four-way valve coil? _____

7. You should have discovered the problem with the unit by this time. Describe what the problem may be.

8. Describe what the repair should be, including parts.

9. Consult your instructor before making a repair.

MAINTENANCE OF WORKSTATION AND TOOLS: Leave the unit as your instructor directs you, and return all tools to their respective places.

SUMMARY STATEMENT: Using the four-way valve as an example, describe how a pilot-operated valve works.

TAKEAWAY: Troubleshooting HVACR equipment is an important part of what the service technician does on a daily basis. Developing good troubleshooting skills takes time, but the result is well worth the effort.

QUESTIONS

1. Why are some four-way reversing valves referred to as pilot-operated solenoid valves?

2. Name the four refrigerant-line connections on the four-way valve.

3. What is the purpose of stopping the fan during the defrost cycle?

4. What would happen if the fan on the outdoor did not stop during defrost?

5. What would the complaint be if the electric heat were not energized during the defrost cycle?

Exercise HPS-10

Changing a Four-Way Reversing Valve on a Heat Pump

Name	Date	Grade

OBJECTIVES: Upon completion of this exercise, you should be able to change a four-way reversing valve on a heat pump.

INTRODUCTION: You will recover the refrigerant, change a four-way reversing valve, pressure-check and leak-check the system, charge the system, and put it back into service. You will then check the system in the heating and cooling modes.

TOOLS, MATERIALS, AND EQUIPMENT: An operating heat pump system, four-way reversing valve, refrigerant recovery unit, recovery tank, extra refrigerant hoses, vacuum pump, soap bubble leak detection solution, straight-slot and Phillips-head screwdrivers, two pairs of slip-joint pliers, ¼" and ⁵⁄₁₆" nut drivers, brazing torch setup, solder and flux (ask your instructor what is preferred), some strips of cloth and water, leak detector, gloves, goggles, bidirectional liquid line filter drier, dry nitrogen, and refrigerant.

⚠ **SAFETY PRECAUTIONS:** Wear goggles and gloves while fastening gauges and transferring liquid refrigerant. Remove the four-way valve solenoid coil before soldering.

STEP-BY-STEP PROCEDURES

1. Put on goggles and gloves. Properly connect the gauges to the heat pump system, purging the gauge hoses of any air that might be contained in them.

2. Recover the refrigerant from the unit with an approved recovery system.

3. With all of the refrigerant recovered, raise the system pressure to the atmosphere, using the gauge manifold and dry nitrogen.

4. Turn the power off. Lock and tag the power panel, and keep the key with you. Remove enough panels to comfortably access and work on the four-way valve. Remove the control wires from the solenoid on the reversing valve.

5. It is recommended that you cut the old valve out of the system, leaving stubs for fastening the new valve. Remove the valve according to your instructor's directions. Prevent filings from entering the system.

6. With the old valve out, prepare the new valve and piping stubs for installation. Clean all connections thoroughly.

7. With all connections clean, apply flux as recommended and put the new valve in place. WRAP THE VALVE BODY WITH THE STRIPS OF CLOTH, AND DAMPEN THEM. HAVE SOME WATER IN RESERVE. Make certain that no water gets into the system piping.

8. Braze the new valve into the system, using care not to overheat the valve body. Point the torch tip away from the body. If the body begins to heat, water can be added from the reserve water.

9. When the valve body cools, pressure-test the system with nitrogen and a refrigerant trace. Leak-check the valve installation.

10. When you are satisfied with the valve installation, remove the nitrogen and refrigerant from the system, and install a bidirectional liquid-line filter drier.

11. Pressurize the system, and leak-check the drier connections.

12. When you have established that the system is leak-free, evacuate the system, and prepare to charge it.

13. Remount the solenoid coil back onto the reversing valve.

14. After evacuation, introduce a holding charge into the system, making certain to weigh the refrigerant that is being introduced to the system.

15. Replace all panels, and turn the power on.

16. Start the system in the cooling mode and observe the pressures.

17. Continue to add refrigerant to the system until the proper charge level has been reached.

18. Take a complete set of temperature, system, and pressure readings from the system, and record them here:

    Return air temperature: _____ °F
    Supply air temperature: _____ °F
    Outside air temperature: _____ °F
    High-side pressure: _____ psig
    Low-side pressure: _____ psig
    Evaporator saturation temperature: _____ °F
    Condenser saturation temperature: _____ °F
    Evaporator outlet temperature: _____ °F
    Condenser outlet temperature: _____ °F
    Evaporator superheat: _____ °F
    Condenser subcooling: _____ °F
    System refrigerant: _____
    Total refrigerant charge: _____ lb, _____ oz

19. Remove the gauges from the system

20. Turn the unit off.

MAINTENANCE OF WORKSTATION AND TOOLS: Leave the unit as your instructor directs you, and return all tools to their places.

SUMMARY STATEMENT: Describe how you could tell by touching the lines when the heat pump was in the cooling mode versus the heating mode.

TAKEAWAY: This is an exciting exercise in that it provides the opportunity to practice not only your brazing skills, but also your system-charging techniques. Brazing and charging are two skills that the successful service technician must master.

QUESTIONS

1. How can the four-way reversing valve become damaged if it is overheated?

2. What is the seat in the four-way valve made of?

3. Why are some four-way reversing valves classified as pilot-operated valves?

4. What was the coil voltage of this system's four-way reversing valve?

5. Describe the permanent discharge line.

6. On which line would a suction-line filter drier be installed if it were needed?

7. What is the purpose of a suction-line accumulator?

Wiring a Split-System Heat Pump, Low-Voltage Control Wiring

Name	Date	Grade

OBJECTIVES: Upon completion of this exercise, you should be able to follow a wiring diagram and wire the field control wiring for a split-system heat pump.

INTRODUCTION: You will connect the low-voltage wires on a live heat pump system, start the system, and then perform a control checkout.

TOOLS, MATERIALS, AND EQUIPMENT: A split-type heat pump system with control wiring connections removed, Straight-slot and Phillips-head screwdrivers, ¼" and ⁵⁄₁₆" nut drivers, a digital multimeter (DMM), a pencil and paper, and a flashlight.

⚠ **SAFETY PRECAUTIONS:** Turn all power off before starting this exercise. Lock and tag the panel where you turned off the power. Keep the key in your possession.

STEP-BY-STEP PROCEDURES

1. Turn off and lock out the power to the unit, and remove all panels. With a split system, this should be the indoor unit, the outdoor unit, and the thermostat (remove the thermostat from the subbase).

2. Study the manufacturer's low-voltage wiring diagram.

3. Connect the wires to the indoor coil using the diagram.

4. Connect the wires to the outdoor coil using the diagram.

5. Connect the wires to the thermostat subbase using the diagram.

6. Ask your instructor to look over the connections, and get permission to start the system.

7. Replace the panels on the units, and start the system in the cooling mode. Allow it to run for about 15 minutes. Verify that it is in the cooling mode. If it is not, remove the panels to the indoor unit, and check for the correct voltage at the proper terminals.

8. After you are satisfied with cooling, shut the unit off, and leave it off for 5 minutes, allowing the pressures to equalize; then start it in the heating mode. Verify that it is heating; then shut the system off.

9. Start the system in the emergency heat mode, and verify that it is operating correctly.

MAINTENANCE OF WORKSTATION AND TOOLS: Leave the unit as your instructor directs you, and return all tools to their places.

SUMMARY STATEMENT: Describe the complete function of the thermostat. Use an additional sheet of paper if more space is necessary for your answer.

TAKEAWAY: Developing strong electrical skills is a must for all successful service technicians in the HVACR industry. This exercise provides the opportunity to use wiring diagrams to make electrical connections on a heat pump system. Be sure to trace out wires one at a time, to avoid getting confused.

QUESTIONS

1. Is a heat pump compressor the same as an air-conditioning compressor? In what ways are they different? In what ways are they the same?

2. Did this heat pump have more than one stage of auxiliary heat? How do you know?

3. Did this heat pump have outdoor thermostats for the auxiliary heat? How do you know?

4. What was the terminal designation for emergency heat on the system you worked on?

5. What was the thermostat terminal designation for the four-way valve?

6. What was the coil voltage for the four-way valve?

7. Did the thermostat have any indicator lights, for example, for auxiliary heat? If so, name them all.

8. What gauge wire was the field low-voltage wire?

9. How many wires were in the cable from the outdoor unit to the indoor unit? List all the colors, including any spare wires that were run.

10. How many wires were in the cable from the indoor unit to the room thermostat? List all the colors, including any spare wires that were run.

Exercise HPS-12

Familiarization with Geothermal Heat Pumps

Name	Date	Grade

OBJECTIVES: Upon completion of this exercise, you should be able to observe and record the various components of the geothermal heat pump.

INTRODUCTION: You will remove the necessary panels to study and record the various components of the geothermal heat pump.

TOOLS, MATERIALS, AND EQUIPMENT: An operating geothermal heat pump system, tape measure, straight-slot and Phillips-head screwdrivers, and ¼" and ⁵⁄₁₆" nut drivers.

⚠ **SAFETY PRECAUTIONS:** Turn the power off, and lock it out before removing the electrical panels.

STEP-BY-STEP PROCEDURES

1. After turning the power off, remove all of the panels to expose the fan, compressor, controls, and heat exchanger. Record the following:

Fan motor information:

- Fan motor rated-load amperage (RLA): _____ A

- Type of fan motor: _____

- Diameter of fan motor: _____ in.

- Diameter of fan motor shaft: _____ in.

- Motor rotation (looking at the fan shaft, if the motor has two shafts, look at the motor lead end): _____

- Fan wheel diameter: _____ in.

- Width: _____ in.

- Number of motor speeds: _____

- High-speed rpm: _____; Low-speed rpm: _____

- Location of the return-air inlet: _____

- Location of the supply-air discharge: _____

Nameplate information:

- Manufacturer: _____

- Model number: _____

- Serial number: _____

- Capacity in tons of cooling: _____

- Heating capacity: _____ Btu/h
- Voltage: _____; Amperage: _____
- Type of pressure controls: _____ high; _____ low

2. Identify the coaxial heat exchanger:

- Inlet pipe size: _____ in.
- Outlet pipe size: _____ in.
- If available, what is the water-side pipe made of, copper or cupronickel? _____
- Gas-line size at the heat exchanger: _____ in.
- Liquid-line size at the heat exchanger: _____ in.
- Type of metering device used for the indoor coil: _____
- Type of metering device used on the coaxial heat exchanger coil: _____

3. Control identification:

- Line voltage to the unit: _____ V
- Control voltage to the unit: _____ V
- Is there a control voltage terminal block? _____
- When is the four-way valve energized, in cooling or heating? _____

MAINTENANCE OF WORKSTATION AND TOOLS: Replace all panels with the correct fasteners, and return all tools to their places.

SUMMARY STATEMENT: Describe how the four-way valve directs the refrigerant gas flow to determine how the pump heats or cools.

TAKEAWAY: Geothermal heat pump systems are growing in popularity, so gaining knowledge about them is extremely useful. Identifying the parts of these systems is a great first step in gaining a working knowledge of these systems.

QUESTIONS

1. Why is there no defrost cycle on a geothermal heat pump?

2. List and describe some possible disadvantages of the open-loop geothermal heat pump system.

3. Why are cupronickel heat exchangers often used for geothermal heat pumps?

4. What type of refrigerant was used in the geothermal heat pump you worked on?

5. What was the total refrigerant charge for this system?

6. Why is there no auxiliary heat for geothermal heat pumps?

7. Name three different fluids that may be circulated in a closed-loop system.

8. Describe a dry well.

9. Describe a waterless, earth-coupled, closed-loop geothermal heat pump system.

Exercise HPS-13

Checking the Water Flow on a Geothermal Heat Pump

Name	Date	Grade

OBJECTIVES: Upon completion of this exercise, you should be able to check for the correct water flow through a coaxial heat exchanger on a geothermal heat pump.

INTRODUCTION: You will use pressure gauges, thermometers or thermistors, and manufacturer's data to check for the correct water flow through a coaxial heat exchanger on a geothermal heat pump.

TOOLS, MATERIALS, AND EQUIPMENT: A geothermal heat pump with Pete's test ports, a thermometer or thermistors and pressure gauge for Pete's plug-type access ports, Figures HPS-13.1 and HPS-13.2, a high-quality digital thermocouple thermometer, and manufacturer's literature with pressure drop charts for the coaxial heat exchanger coil.

⚠ **SAFETY PRECAUTIONS:** None.

FIGURE **HPS-13.1**

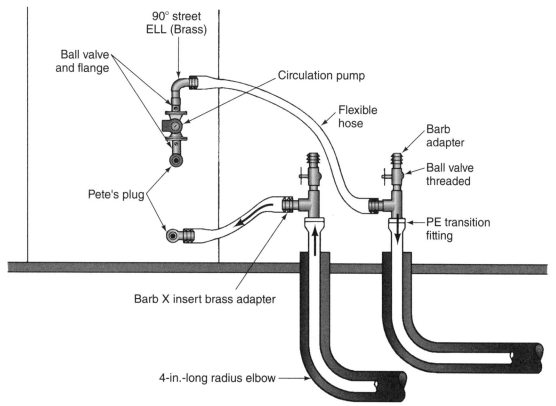

Courtesy of Oklahoma State University

FIGURE **HPS-13.2**

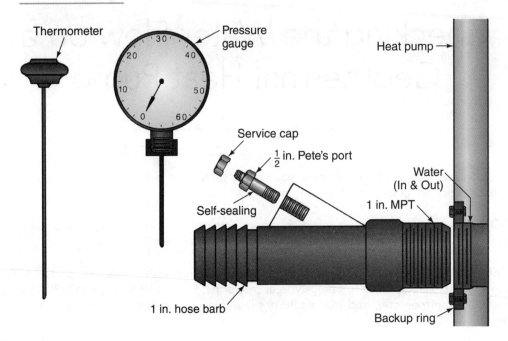

STEP-BY-STEP PROCEDURES

1. Place one lead of the temperature tester in the return-air inlet of the air handler and the other lead in the supply-air outlet.

2. Start the unit cooling cycle, and wait for the air temperatures to settle down until there is no change, at least 30 minutes.

3. Measure the pressure drop across the coaxial heat exchanger using the pressure gauge at the Pete's plug connection, first at the inlet and then at the outlet.

Note: Using the same gauge gives a more accurate reading.

- Pressure in: _____ psig
- Pressure out: _____ psig
- Pressure difference: _____ psig

4. Measure the temperature difference (delta-T) across the coaxial heat exchanger using the Pete's plug connection, first at the inlet and then at the outlet. Again, using the same thermometer or thermistor for temperature difference will give the best results.

- Temperature in: _____°F
- Temperature out: _____°F
- Temperature difference: _____°F

5. Record the air temperatures:

- Temperature in: _____°F
- Temperature out: _____°F
- Temperature difference: _____°F

6. Compare the pressure and temperature differences to the manufacturer's pressure chart, and record the estimated flow rate here: _____ gallons per minute (gpm).

MAINTENANCE OF WORKSTATION AND TOOLS: Remove all gauges or temperature probes, and cap the Pete's plug ports. Replace all panels with the correct fasteners, and return all tools to their correct places.

SUMMARY STATEMENT: Describe how the pressure drop can give an accurate measure of the amount of water flowing through a liquid heat exchanger. Use an additional sheet of paper if more space is necessary for your answer.

TAKEAWAY: This exercise provides the opportunity to practice taking readings from the water side of a heat pump system and utilizing manufacturer's charts to determine system water flow rates. It is important to obtain the literature for the specific heat exchanger being evaluated, as the specifications vary from one manufacturer to the next.

QUESTIONS

1. What type of fluid flowed in the system you worked on, water or antifreeze? How were you able to determine this?

2. Where is water ordinarily exhausted to in an open-loop geothermal heat pump system?

3. What material is used for the coaxial heat exchanger when there may be excess minerals present?

4. How can a coaxial heat exchanger be cleaned should it become fouled with minerals or scum?

5. What would the temperature rise across a coaxial heat exchanger do in the cooling cycle if the water flow were to be reduced?

6. How would the condition described in question 5 affect the head pressure?

7. What would the temperature drop across a coaxial heat exchanger do in the heating cycle if the water flow were reduced?

8. How would the condition described in question 7 affect the suction pressure?

Exercise HPS-14

Situational Service Ticket

Name	Date	Grade

Note: Your instructor must have placed the correct service problem in the system for you to successfully complete this exercise.

Technician's Name: _____ Date: _____

CUSTOMER COMPLAINT: No heat with heat pump.

CUSTOMER COMMENTS: Indoor fan runs and emergency heat works, but heat pump will not run.

TYPE OF SYSTEM:

OUTDOOR UNIT:

    MANUFACTURER _____   MODEL NUMBER _____   SERIAL NUMBER _____

INDOOR UNIT:

    MANUFACTURER _____   MODEL NUMBER _____   SERIAL NUMBER _____

COMPRESSOR (where applicable):

    MANUFACTURER _____   MODEL NUMBER _____   SERIAL NUMBER _____

TECHNICIAN'S REPORT

    1. SYMPTOMS: _____

    2. DIAGNOSIS: _____

    3. ESTIMATED MATERIALS FOR REPAIR: _____

    4. ESTIMATED TIME TO COMPLETE THE REPAIR: _____

SERVICE TIP FOR THIS CALL

Use a digital multimeter to check all components for heat cycles that pertain to the heat pump.

Exercise HPS-15

Situational Service Ticket

Name	Date	Grade

Note: Your instructor must have placed the correct service problem in the system for you to successfully complete this exercise.

Technician's Name: _____ Date: _____

CUSTOMER COMPLAINT: Outdoor coil is frosting all the time.

CUSTOMER COMMENTS: Unit cannot be operating efficiently with outdoor coil frosting all the time.

TYPE OF SYSTEM:

OUTDOOR UNIT:

     MANUFACTURER _____   MODEL NUMBER _____   SERIAL NUMBER _____

INDOOR UNIT:

     MANUFACTURER _____   MODEL NUMBER _____   SERIAL NUMBER _____

COMPRESSOR (where applicable):

     MANUFACTURER _____   MODEL NUMBER _____   SERIAL NUMBER _____

TECHNICIAN'S REPORT

1. SYMPTOMS: _____

2. DIAGNOSIS: _____

3. ESTIMATED MATERIALS FOR REPAIR: _____

4. ESTIMATED TIME TO COMPLETE THE REPAIR: _____

SERVICE TIP FOR THIS CALL

Fasten gauges to the system, and check pressures and superheat on the outdoor coil in heating mode.

SECTION 9

Heating (HTG)

SUBJECT AREA OVERVIEW

Due to the many different climate conditions across the country, the type of heating system and the fuel used varies greatly as we move from east to west and from north to south. The source of heat is also an important factor in determining the type of heating system used. In some areas, certain fuel types are more readily available than others and, as such, are typically less expensive than other options. Commonly used heat sources are electricity, gas, and oil. This section will discuss these three fuel types as well as hydronic heating, which relies on water (or steam) to act as the heat transfer medium.

Electric heat is produced when current flows through nickel chromium, or nichrome wire. This wire creates a resistance to electron flow and, as a result, produces heat. Many electric heaters use fans or blowers to move air from the conditioned space over the heating elements in order to heat the air. Central forced-air furnaces equipped with electric heating elements use ductwork to distribute heated air to remote rooms or spaces. This type of heating system often has central air conditioning and humidification added to it.

Natural gas and liquefied petroleum (LP) are the fuels most commonly used in gas furnaces. Natural gas consists mostly of methane. Liquefied petroleum is liquefied propane, butane, or a combination of both. In order to release the heat energy contained in the gas, it must be burned through a process called combustion. The combustion process requires fuel, oxygen, and an ignition source. When mixed properly, this combination will cause a chemical reaction known as rapid oxidation, commonly referred to as burning.

In order for combustion to occur, fuel is fed to the appliance's gas valve, which regulates the flow of gas into the manifold. From the manifold, the gas is forced through an orifice into a burner, where it mixes with air and is then ignited. The fuel/air mixture may be ignited by a number of different methods including a standing pilot, high-voltage spark, or hot surface igniter. The fuel and air must be mixed in the proper proportions to start and continue the combustion process.

The combustion process causes hot flue gases to be formed in the heat exchanger, heating the air both in and around the heat exchanger. The furnace blower moves air from the space being conditioned across the heat exchanger, where it absorbs heat. The heated air is then distributed through the ductwork to the areas where the heat is wanted. The by-products of combustion must be removed from the appliance and the structure. This process can be accomplished by natural convection currents or by using blower-assisted methods.

Most modern heat exchangers are serpentine-shaped and are made of aluminized steel. The combustion gases travel a winding path through the exchangers before entering the vent system. Serpentine heat exchangers use inshot burners instead of the ribbon or slotted-port burners used in natural draft furnaces.

High-efficiency furnaces circulate the combustion gases through additional heat exchangers to capture more of the heat energy before it is vented from the appliance. Some modern condensing furnaces have two or even three sets of heat exchangers and are designed to condense the flue gases. Such systems are designed to handle this condensate and provide a means for the water to be drained from the furnace to an external drain.

Most modern gas furnaces use electronic ignition modules or integrated furnace controllers (IFCs) to control the operation of the furnace.

(continued)

Controls can be integrated, controlling all of the functions of the furnace, or nonintegrated, controlling only the burner functions. These controls often have dual-in-line pair (DIP) switches, which can be used to change certain functions of the furnace to make it more versatile.

Ignition control modules can be 100% shut-off, non-100% shutoff, or continuous retry with 100% shutoff, depending on the manufacturer's requirements. The control can cycle the combustion blower for a prepurge, interpurge, and/or postpurge of the heat exchangers, depending on the module's control program. They can also utilize a hard or soft lockout in the event of a flame detection failure.

Furnace efficiency ratings are determined by the amount of heat transferred to the space compared to the total amount of heat produced by the appliance. Factors that can affect the efficiency of a furnace are the type of draft, the amount of excess air, the temperature difference across the heat exchanger, and the flue stack temperature.

To increase the efficiency of the appliance, modern modulating gas furnaces are controlled to match their heating output to the heat loss of a structure. They use a modulating gas valve instead of a staged valve to vary the furnace's Btu output. They also come with variable-speed combustion blowers and warm-air blowers to vary the speed and amount of warm air. Variable-output thermostats, which send a proportional signal instead of an on–off signal to the furnace control, are used with modulating furnaces.

Oil is a popular fuel choice when natural gas is not readily available. Just as with natural gas and LP, oil must be mixed with air (oxygen) in order to be ignited. Since oil is a liquid, it must first be atomized, or broken into very small droplets, to allow it to vaporize more effectively. Once ignited, the heat energy from the fuel is released, heating the appliance's heat exchanger, where the heat is transferred to the heated space.

There are six grades of fuel oil used as heating oils. Fuel oil No. 2 is the one most commonly used in residential and light commercial oil heating. One gallon of No. 2 fuel oil produces about 140,000 Btu of heat energy when burned. Oil can be stored either aboveground or underground.

In hydronic heating systems, water is heated and pumped to terminal units such as radiators or finned-tube baseboard units. In some systems, the water is heated to form steam, which is then piped to the terminal units. The source of heat can be any one of those just mentioned in this introduction.

The heated water is circulated through the system with one or more circulator pumps. Thermostatically controlled zone valves can also be used to direct the heated water to the spaces requiring heat. Most residential installations use finned-tube baseboard terminal units to transfer heat from the water to the air. Many residential systems use a series loop system, where all of the hot water flows through all of the terminal units. This is not the most efficient design, but it is the least expensive from an installation standpoint. There are other, more expensive but more efficient design strategies that can be used.

KEY TERMS

Animal fuel utilization efficiency (AFUE)
Atomization
Boiler
Cad cell
Combustion
Condensing boiler
Condensing oil furnace
Conventional boiler
Diverter tee
Direct ignition
Downflow furnace
Draft diverter
Dry-base boiler

Electronic module
Electronic timer
Gas valve
Heat anticipator
Heat exchange
Hot surface igniter
Hydronic heating
Ignition module
Impeller
Integrated furnace control (IFC)
Intermittent ignition
Ladder diagram
Light-emitting diode (LED)

Line wiring diagram
Liquified petroleum (LP)
Motor protection module
Multi-position furnace
Natural gas
Nichrome
PEX tubing
Pilot
Primary air
Primary control
Rapid oxidation
Rectifier
Schematic wiring diagram
Secondary air

Semiconductor
Sequencer
Solid state
Spark igniter
Temperature rise
Terminal unit
Thermistor
Thermocouple
Thermostat
Upflow furnace
Venturi
Wet-base boiler

REVIEW TEST ELECTRIC HEAT

Name	Date	Grade

Circle the letter that indicates the correct answer.

1. The resistance wire often used in an electric heating element is made of:
A. copper.
B. stainless steel.
C. nickel chromium.
D. magnesium.

2. Radiant heat is transferred by:
A. ultraviolet rays.
B. infrared rays.
C. ionization.
D. diffraction.

3. Most electric baseboard units are _____ heaters.
A. natural-draft convection
B. forced-draft convection
C. forced-draft conduction
D. radiant

4. Central forced-air electric furnaces frequently use _____ to start and stop multiple heating circuits.
A. capacitors
B. sequencers
C. limit controls
D. cool anticipators

5. A sequencer often utilizes a _____ to initiate its operation.
A. bimetal heat motor
B. bimetal snap disc
C. capacitor
D. nichrome resistor

6. The component located in the thermostat that shuts the electric furnace down prematurely so that the space will not be overheated is the:
A. sequencer.
B. cool anticipator.
C. heat anticipator.
D. mercury bulb.

7. The disadvantage in using a contactor to open and close the circuit to the electric heating elements is:
A. large current draw when the contactor coil is energized.
B. low-control voltage.
C. heating elements all starting at the same time unless separate time delay relays were used.
D. both A and C.

8. A limit switch or control is used to:
A. start the furnace if the thermostat malfunctions.
B. ensure that the fan stops when the furnace cools down.
C. start the furnace prematurely on a cold day.
D. open the circuit to the heating elements if the furnace overheats.

9. Electric unit heaters are usually installed:
A. on ceilings or walls.
B. as baseboard heat.
C. as forced-air furnaces.
D. with hot water systems.

10. Electric ceiling panel heat:
A. easily adapts to central air conditioning.
B. is radiant-type heat.
C. uses circulating hot water.
D. can easily have humidity equipment added to the system.

11. The _____ thermostat is commonly used for forced-air electric furnace installations.
A. low-voltage
B. dual-voltage
C. high-wattage
D. high-voltage

12. The _____ thermostat is commonly used for baseboard electric heat installations.
A. low-voltage
B. dual-voltage
C. no-voltage
D. high-voltage

13. The control voltage for the controls on a sequencer is typically:
    A. 24 V.
    B. 208 V.
    C. 460 V.
    D. 12 V.

14. An electric furnace is operating on 225 V and is drawing 83 A. How much power is being consumed by the furnace?
    A. 18,675 W
    B. 18.675 kW
    C. 18,675 kW
    D. Both A and B are correct.

15. Using the information from review question 14, what heat capacity is this furnace putting out?
    A. 55,787 Btu/h
    B. 33,467 Btu/h
    C. 98,953 Btu/h
    D. 63,738 Btu/h

16. What electrical control device is used in an electric furnace to bring on electric resistance heaters in stages?
    A. Contactor
    B. Line voltage thermostats
    C. Sequencers
    D. None of the above

REVIEW TEST GAS HEAT

Name	Date	Grade

Circle the letter that indicates the correct answer.

1. Which of the following is a common energy source used for heating?
   - A. Gas
   - B. Oil
   - C. Electricity
   - D. All of the above

2. Which of the following components is required in a fossil fuel, warm air furnace but not required in an electric furnace?
   - A. Fan
   - B. Limit switches
   - C. Combustion chamber
   - D. Fan control

3. Natural gas is composed of 90% to 95%:
   - A. oxygen.
   - B. ethanol.
   - C. methane.
   - D. propane.

4. Liquid petroleum (LP) is composed of liquefied:
   - A. natural gas.
   - B. propane or butane.
   - C. methane.
   - D. sulfur compounds.

5. Natural gas:
   - A. is lighter than air.
   - B. has approximately the same specific gravity as air.
   - C. is heavier than air.
   - D. has the same heat content as fuel oil #2.

6. Combustion is sometimes expressed as:
   - A. a venturi effect.
   - B. hydrocarbonization.
   - C. rapid oxidation.
   - D. the mixing of propane and butane.

7. The natural gas flame should be:
   - A. yellow.
   - B. yellow with some orange tips.
   - C. orange with some yellow tips.
   - D. blue with some orange tips.

8. It takes approximately _____ f³ of air to produce 2 f³ of oxygen.
   - A. 10
   - B. 20
   - C. 30
   - D. 40

9. The purpose of the fan switch in a warm air furnace is to:
   - A. control the fan motor in order to deliver warm air to the structure at the correct temperature.
   - B. control the combustion fan.
   - C. control the primary ignition control.
   - D. do all of the above.

10. A standing pilot in a conventional gas furnace:
    - A. is ignited by an electric spark when the thermostat calls for heat.
    - B. is used as a flame-proving device.
    - C. burns continuously.
    - D. burns only when the main burner is lit.

11. A thermocouple is used in a conventional gas furnace as a:
    - A. flame-proving device.
    - B. standing pilot.
    - C. limit switch.
    - D. spark igniter.

12. The orifice is a precisely sized hole in the:
    - A. manifold.
    - B. spud.
    - C. burner tube.
    - D. combustion chamber.

13. The limit switch is a safety device that:
    - A. shuts the gas off at the pressure regulator when a leak is detected.
    - B. shuts the gas off if there is a power failure.
    - C. closes the gas valve if the furnace overheats.
    - D. closes the gas valve if the pilot is blown out.

14. In a conventional gas furnace, the hot flue gases are:
    - A. forced out of a PVC vent with a blower.
    - B. recirculated through a second heat exchanger and then vented through the draft diverter.
    - C. forced through the type B vent by a blower.
    - D. vented by natural convection.

15. A direct-spark ignition (DSI) system is designed to:
    A. reignite a pilot light if it is blown out.
    B. ignite the pilot when the thermostat calls for heat.
    C. provide ignition to the main burners.
    D. ignite the pilot following a furnace shutdown over the summer.

16. A condensing gas furnace uses _____ from a condenser-type heat exchanger to improve the furnace efficiency.
    A. sensible heat
    B. latent heat
    C. superheat
    D. subcooling

17. Where are auxiliary limit switches always found in heating applications?
    A. Upflow furnaces
    B. Downflow furnaces
    C. Horizontal furnaces
    D. None of the above

18. A small microprocessor with integrated circuits mounted on a board to control the logic or sequence of events of the furnace is a(n):
    A. microboard.
    B. integrated furnace controller (IFC).
    C. transducer.
    D. transformer.

19. When a draft blower is located at the outlet of the heat exchanger and causes a negative pressure in the heat exchanger, it is called:
    A. induced draft.
    B. forced draft.
    C. inlet draft.
    D. outlet draft.

20. Flame current in a flame rectification system is measured in _____ units.
    A. milliamp
    B. millivolt
    C. microvolt
    D. microamp

21. A time-controlled fan switch is nothing more than a _____.
    A. time clock
    B. time-delay relay
    C. bimetal element
    D. mechanical timer

22. Which of the following components is used to interrupt the power source to a load when an unsafe condition occurs?
    A. Fan switch
    B. Gas valve
    C. Time clock
    D. Limit switch

23. What are the three basic types of gas ignition systems?
    A. Standing pilot, intermittent pilot, and direct ignition
    B. Standing pilot, intermittent pilot, and automatic ignition
    C. Direct-spark ignition, standing pilot, and bimetal ignition
    D. Intermittent pilot, direct-spark ignition, and subsurface ignition

24. What component in a standing pilot ignition system produces the signal that a pilot is available?
    A. Thermocouple
    B. Flame rod
    C. Bimetal element
    D. Hot surface igniter

25. Which of the following types of flame sensors could not be used with an intermittent pilot ignition system?
    A. Flame rod
    B. Liquid-filled pilot sensor
    C. Temperature sensor
    D. Thermocouple

26. What type of gas valve is used with an intermittent pilot ignition system?
    A. Combination
    B. Direct
    C. Redundant
    D. Combustion

27. A direct ignition system:
    A. lights the pilot.
    B. lights the main burner.
    C. does both A and B.
    D. does none of the above.

28. An upflow furnace:
    A. is generally used where there is little headroom.
    B. has the return air at the bottom.
    C. has the return air at the top.
    D. has the air discharge at the bottom.

REVIEW TEST OIL HEAT

Name	Date	Grade

Circle the letter that indicates the correct answer.

1. What ignites the atomized fuel oil in a residential oil burner?
   A. Pilot
   B. High-voltage spark
   C. Hot surface igniter
   D. None of the above

2. What would be the results if the oil burner were allowed to run without ignition?
   A. Explosion
   B. Delayed ignition
   C. Accumulation of fuel oil in the combustion chamber
   D. None of the above

3. Which of the following is a light-sensitive device that changes its resistance according to the intensity of the light?
   A. Thermocouple
   B. Fuel limiter cell
   C. Subradiant cell
   D. Cad cell

4. What type of primary control used on an oil burner uses temperature to prove ignition?
   A. Stack switch
   B. Fan switch
   C. Limit switch
   D. Photo cell

5. The resistance of the cad cell should be below _____ ohms to continue operation of an oil burner.
   A. 35,000
   B. 2500
   C. 1600
   D. 300

6. What is the proper location of a stack switch for proper supervision of the oil burner?
   A. Combustion chamber
   B. Oil burner stack
   C. Oil burner flame
   D. Blower compartment

7. A limit switch used on an oil- or gas-fired warm air furnace would usually _____ on a temperature rise.
   A. open
   B. close

8. Of the six regular grades of fuel oil, the grade most commonly used in residential and light commercial heating systems is:
   A. No. 1.
   B. No. 2.
   C. No. 4.
   D. No. 6.

9. How much heat is produced when 2 gallons of No. 2 fuel oil are burned?
   A. 10,000 Btu
   B. 50,000 Btu
   C. 140,000 Btu
   D. 280,000 Btu

10. When hydrocarbons unite rapidly with oxygen, it is called:
    A. vaporization.
    B. condensation.
    C. atomization.
    D. combustion.

11. Assuming perfect combustion, how many pounds of oxygen are required to burn 1 lb of fuel oil?
    A. 1.44 lb
    B. 3.00 lb
    C. 14.4 lb
    D. 30.00 lb

12. Assuming 50% excess air is being provided, how many cubic feet of air will be supplied to an oil burner to burn 1 lb of fuel oil?
    A. 3.0 ft^3
    B. 14.4 ft^3
    C. 144 ft^3
    D. 288 ft^3

13. When fuel oil is broken up to form tiny droplets, it is called:
  A. heat of fusion.
  B. condensation.
  C. atomization.
  D. combustion.

14. The motor on a residential gun-type oil burner is usually a _____ motor.
  A. split-phase multi-horsepower
  B. split-phase fractional horsepower
  C. three-phase
  D. shaded pole

15. The pump on a residential gun-type oil burner is a _____ pump.
  A. centrifugal
  B. single-stage or two-stage
  C. scroll-type
  D. air

16. The fuel oil enters the orifice of the nozzle through the:
  A. hollow cone.
  B. squirrel cage blower.
  C. swirl chamber.
  D. flame retention head.

17. The ignition transformer steps up the 120 V residential voltage to approximately _____ V.
  A. 208
  B. 440
  C. 1000
  D. 10,000

18. The stack switch or relay:
  A. has cold contacts that are open when the flue pipe is cold.
  B. has hot contacts that are closed when the flue pipe is cold.
  C. has hot contacts that close when the temperature of the flue pipe rises.
  D. Both A and B are correct.

19. The cad cell is a safety device that:
  A. senses heat in the flue.
  B. senses light in the flue.
  C. senses light in the combustion chamber.
  D. senses heat in the heat exchanger.

20. The cad cell is often coupled with a:
  A. diode.
  B. triac.
  C. transistor.
  D. rectifier.

21. A booster pump is often needed when the oil supply tank is _____ below the oil burner.
  A. any distance
  B. 5 ft or more
  C. 10 ft or more
  D. 15 ft or more

22. An oil deaerator:
  A. increases the pressure of the oil.
  B. allows a one-pipe oil delivery system to operate as a two-pipe oil delivery system.
  C. causes the temperature of the oil entering the fuel pump to drop.
  D. eliminates the need for an oil filter.

REVIEW TEST HYDRONIC HEAT

Name	Date	Grade

Circle the letter that indicates the correct answer.

1. A terminal unit in a hydronic heating system may be a _____ unit.
 A. gas- or oil-fired
 B. finned-tube baseboard
 C. one-pipe fitting
 D. Any of the above

2. Boilers that have efficiencies over 90% are commonly referred to as:
 A. noncondensing boilers.
 B. conventional boilers.
 C. condensing boilers.
 D. inefficient boilers.

3. In a wet-base boiler, the water being heating is located:
 A. only above the combustion area.
 B. only below the combustion area.
 C. both above and below the combustion area.
 D. nowhere near the combustion area.

4. All of the following are characteristics of a cast iron boiler except:
 A. they heat up very quickly.
 B. they are classified as high-mass boilers.
 C. they have long run times and long off cycles.
 D. they typically contain between 15 and 30 gallons of water.

5. Copper water-tube boilers:
 A. contain more water than cast-iron boilers.
 B. are low-mass boilers.
 C. heat water very slowly.
 D. All of the above are correct.

6. A limit control:
 A. regulates the amount of water to each unit.
 B. regulates the water pressure from the supply line.
 C. is used to balance the flow rate.
 D. shuts the heat source down if the temperature becomes too high.

7. The pressure relief valve:
 A. regulates the pressure of the water in the supply line.
 B. is set to relieve at or below the maximum allowable working pressure of the boiler.
 C. shuts the boiler down if the temperature becomes too high.
 D. relieves the pressure at the expansion tank.

8. The expansion tank provides a place for air initially trapped in the system and:
 A. space for excess water from the supply line.
 B. water for the domestic hot water coil.
 C. space for the expanded water in the system.
 D. All of the above.

9. If the height of a hot water hydronic piping arrangement extends 23 ft above the inlet of the expansion tank, the pressure in the expansion tank should be about:
 A. 10 psig.
 B. 14 psig.
 C. 16 psig.
 D. 20 psig.

10. The part of the centrifugal pump that spins and forces water through the system is the:
 A. heat motor.
 B. impeller.
 C. flow control.
 D. propeller fan.

11. Centrifugal pumps are positive displacement pumps.
 A. True
 B. False

12. It is desirable to have the inlet of the circulator connected to the:
 A. return side of the boiler.
 B. point of no pressure change.
 C. inlet of the water pressure regulating valve.
 D. inlet of the terminal units.

13. Which of the following components would be used in the control system of a steam boiler?
   A. Pressure switch
   B. Limit switches
   C. Low water cut-off
   D. All of the above

14. Air in a hot water hydronic system can cause which of the following?
   A. Noise
   B. Corrosion
   C. Flow blockage
   D. All of the above

Exercise HTG-1

Electric Furnace Familiarization

Name	Date	Grade

OBJECTIVES: Upon completion of this exercise, you should be able to describe the components in an electric heating system and state the functions of each one.

INTRODUCTION: You will inspect all the components in an electric furnace and read specifications listed in the component nameplates or available literature.

TOOLS, MATERIALS, AND EQUIPMENT: Straight-slot and Phillips-head screwdrivers, ¼" and ⁵⁄₁₆" nut drivers, digital multimeter (DMM), thermometer, calculator, wire size chart, and complete electric furnace.

⚠ **SAFETY PRECAUTIONS:** Turn the power off, and lock out the power to the heating unit before beginning your inspection. You should keep the only key to the locked panel in your possession. After turning the power off, check the line voltage with your digital multimeter to make sure that it is turned off.

STEP-BY-STEP PROCEDURES

1. Locate each of the following components. Place a checkmark after the name when you have located it.

   - Thermostat _____
   - Heat anticipator _____
   - Control device(s) for energizing heating elements (contactor or sequencer) _____
   - Heating elements _____
   - Fan motor and blower wheel _____

2. Provide the following information from available literature or component nameplates:

   - Fan motor horsepower: _____
   - Fan motor shaft size: _____
   - Capacitor rating of fan motor (if any): _____
   - Fan motor FLA: _____
   - Fan motor LRA: _____
   - Fan motor operating voltage: _____
   - Blower wheel diameter: _____
   - Number of heating elements: _____
   - Amperage draw of each heating element: _____
   - kW rating of each heating element: _____
   - Resistance of each heating element: _____
   - Current draw in the control circuit: _____
   - Heat anticipator setting: _____
   - Minimum wire size of each heating element: _____
   - Btu capacity of each heating element: _____
   - Minimum wire size to furnace: _____
   - Minimum wire size to fan motor: _____

MAINTENANCE OF WORKSTATION AND TOOLS: Replace all panels on the furnace, and leave the work area clean. Return all tools, the DMM, and other materials to their proper locations.

SUMMARY STATEMENT: Describe the heating operation of the furnace that you were assigned. Include both the low- and high-voltage circuits. Use an additional sheet of paper if more space is necessary for your answer.

TAKEAWAY: Electric furnaces utilize different components than gas-burning furnaces. Becoming familiar with the components, their location, and function can help you to fully understand how the unit operates.

QUESTIONS

1. Electric heating elements are rated in which of the following terms: kW, Btu/h, or gal/h?

2. The individual heater wire size is dependent on which of the following: voltage, current, kilowatt, or horsepower?

3. Electric heat is supposed to glow red hot in the airstream. True or false?

4. When a fuse blows several times, a larger fuse should be installed. True or false?

5. What control component starts the fan on a typical electric heat furnace?

6. When the furnace air filter is clogged, which of the following controls would de-energize the electric heat first: the limit switch or the fusible link? Why?

7. Which function of a digital multimeter is used to check the current draw in the electric heat circuit: ohms, voltage, or amperage?

8. Is electric heat normally less expensive or more expensive to operate than gas heat?

Exercise HTG-2

Determining Airflow (CFM) by Using the Air Temperature Rise Method: Electric Furnace

Name	Date	Grade

OBJECTIVES: Upon completion of this exercise, you should be able to calculate the quantity of airflow in cubic feet per minute (cfm) in an electric furnace by measuring the air temperature rise, amperage, and voltage.

INTRODUCTION: You will measure the air temperature at the electric furnace supply and return ducts and the voltage and amperage at the main electrical supply. By using the formulas furnished, you will calculate the cubic feet per minute being delivered by the furnace.

Note: The power supply must be single phase.

TOOLS, MATERIALS, AND EQUIPMENT: A digital multimeter with an amp clamp, electronic thermometer, calculator, Phillips-head and straight-slot screwdrivers, ¼" and ⁵⁄₁₆" nut drivers, and operating electric furnace.

⚠ **SAFETY PRECAUTIONS:** Use proper procedures when making electrical measurements. Your instructor should check all measurement setups before you turn the power on. Use digital multimeter leads with alligator clips. Keep your hands away from all moving or rotating parts. Lock and tag the panel where you turn the power off. Keep the single key in your possession.

STEP-BY-STEP PROCEDURES

1. Turn the power to the furnace off, and lock out the panel. Under the supervision of your instructor and with the power off, connect the digital multimeter so that it will measure both the voltage and total amperage supplied to the electric furnace. See Figure HTG-2.1 as an example problem.

FIGURE **HTG-2.1**

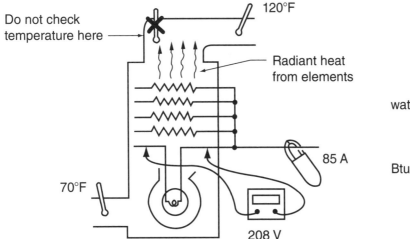

watts = amperes × volts
= 85 × 208
= 17,680 W

Btu/h = watts × 3.413
= 17,680 × 3.413
= 60,341.8 Btu/h

2. Turn the unit on, and call for heat. Wait for all elements to be energized. Record the current and voltage here:

Current _____ A (Amperes × Volts = Watts)

Voltage _____ V (_____ × _____ = _____ Watts)

3. Convert watts to Btu per hour, Btu/h. This is the furnace input. Watts × 3.413 = Btu/h.

_____ × 3.413 = _____ Btu/h

4. Find the change in temperature (ΔT) from the return duct to the supply duct. Place a temperature sensor in the supply duct (around the first bend in the duct to prevent radiant heat from the elements from hitting the probe) and a temperature sensor in the return duct. Record the temperatures here:

Supply _____°F   (Supply − Return = Change in temperature or ΔT)

Return duct _____°F   (_____ − _____ = _____°F ΔT)

5. To calculate the cubic feet per minute (cfm) of airflow, use the following formula:

$$\text{cfm airflow} = \frac{\text{Total heat input (step 3)}}{1.08 \times \Delta T \text{ (step 4)}} = \frac{\underline{\quad} \text{ Btu/h}}{1.08 \times \underline{\quad} °F} = \underline{\qquad}$$

Note: 1.08 is a constant used for standard air conditions. Its derivation is Specific Heat of air (0.24 btu/lb) × Density of air (0.075 lb/ft³) × 60 minutes = 1.08.

6. Allow the furnace to cool down and then turn it off. Replace all panels on the unit with the correct fasteners.

MAINTENANCE OF WORKSTATION AND TOOLS: Return all tools and instruments to their proper places. Leave the furnace as you are instructed.

SUMMARY STATEMENT: Describe why it is necessary for a furnace to provide the correct cfm. Use an additional sheet of paper if more space is necessary for your answer.

TAKEAWAY: During service, it may become necessary to verify the airflow (cfm) being delivered by a furnace. Taking accurate measurements and applying the appropriate formulae are critical to obtaining an accurate air volume flow.

QUESTIONS

1. The kW rating of an electric heater is the _____ (temperature range, power rating, voltage rating, or name of the power company).

2. The Btu rating is a rating of the _____ (time, wire size, basis for power company charges, or heating capacity).

3. The factor for converting watts to Btu is _____.

4. A kW is equal to (1000 W, 1200 W, 500 W, or 10,000 W) _____.

5. The typical air temperature rise across an electric furnace is normally considered (high or low) compared to a gas furnace.

6. The air temperature at the outlet grilles of an electric heating system is considered _____ (warmer, cooler) compared to a gas furnace.

7. The circulating blower motor normally runs _____ on (high or low) speed in the heating mode.

8. What is the current rating of a 7.5 kW heater operating at 230 V?

9. A 12 kW heater will put out _____ Btu/h.

Exercise HTG-3

Setting the Heat Anticipator for an Electric Furnace

Name	Date	Grade

OBJECTIVES: Upon completion of this exercise, you should be able to take an amperage reading and properly adjust a heat anticipator in an electric furnace low-voltage control circuit.

INTRODUCTION: You will be working with a single-stage and a multistage electric heating furnace, taking amperage readings in the low-voltage circuit, and checking or adjusting the heat anticipator.

TOOLS, MATERIALS, AND EQUIPMENT: Phillips-head and straight-slot screwdrivers, ¼" and ⁵⁄₁₆" nut drivers, digital multimeter with an amp clamp, single strand of thermostat wire, a single-stage electric furnace, and a multistage electric furnace.

⚠ **SAFETY PRECAUTIONS:** Make sure that the power is turned off, the panel is locked, and you have the only key before removing panels on the electric furnace. Turn the power on only after your instructor approves your setup.

STEP-BY-STEP PROCEDURES

1. Turn the power off, and lock and tag the panel. With the power off on a single-stage electric heating unit, remove the panels. Use the small piece of thermostat wire to create a jumper from the "R" terminal to the "W" terminal on the thermostat subbase. Connect the amp clamp of the multimeter around the jumper wire.

2. Turn the power on. With the jumper wire in place on the thermostat, the furnace will begin a call for heat. Check and record the amperage of the circuit: _____ A

3. Adjust the heat anticipator under the thermostat cover to reflect the actual amperage if it is not set correctly.

4. With the power on a multistage electric heating unit off and the panels locked and tagged, remove the panels. Connect the amp clamp of the multimeter to the low-voltage heating circuit as in step 1.

5. Turn the power on. With the jumper wire in place on the thermostat, the furnace will begin a call for heat. Check and record the amperage: _____ A

6. Adjust the heat anticipator to reflect the actual amperage if it is not set correctly.

MAINTENANCE OF WORKSTATION AND TOOLS: Replace all panels. Return all tools and instruments, leaving your work area neat and orderly.

SUMMARY STATEMENT: Write a description of how the heat anticipator works, telling what its function is. Describe why the amperage is greater when more than one sequencer is in the circuit. Use an additional sheet of paper if more space is necessary for your answer.

TAKEAWAY: Thermostats that have adjustable heat anticipators require a technician to take a current reading of the low-voltage heating circuit in order to set them properly. When working on multistage electric heating units, technicians need to wait until all sequencers are energized for the final reading.

QUESTIONS

1. Most heat anticipators are _____ (fixed resistance or variable resistance).

2. A heat anticipator that is set wrong may cause _____ (no heat, excessive temperature swings, or noise in the system).

3. If the service technician momentarily shorted across the sequencer heater coil, it would cause which of the following?
   A. Blown fuse
   B. Burned heat anticipator
   C. The heat turned on
   D. The cooling turned on

4. All heating thermostats have heat anticipators. True or false?

5. When a system has several sequencers that would cause too much current draw through the heat anticipator, which of the following could be done to prevent the anticipator from burning up?
   A. Installing a higher-amperage-rated thermostat
   B. Wiring the sequencer so that all of the coils did not go through the anticipator
   C. Changing to a package sequencer
   D. B or C

6. The heat anticipator differs from the cooling anticipator. True or false?

7. When the furnace is two-stage, how many heat anticipators are there?

8. The heat anticipator is located in which component, the thermostat or the subbase?

9. If the heat anticipator was burned so that it was in two parts, the symptoms would be which of the following?
   A. No heat
   B. Too much heat
   C. The cooling coming on
   D. The residence overheating

Low-Voltage Control Circuits Used in Electric Heat

Name	Date	Grade

OBJECTIVES: Upon completion of this exercise, you should be able to interpret wiring diagrams illustrating the low-voltage control circuits in single- and multistage electric furnaces.

INTRODUCTION: You will study the low-voltage wiring in a single- and a multistage electric furnace. You will then select the wiring diagram that most accurately illustrates the operation of each furnace.

TOOLS, MATERIALS, AND EQUIPMENT: Phillips-head and straight-blade screwdrivers, ¼" and ⁵⁄₁₆" nut drivers, and single- and multistage electric furnaces.

⚠ **SAFETY PRECAUTIONS:** Ensure that the power is turned off, that the panel is locked and tagged, and that you have the only key before removing panels on the electric furnaces. Turn the power on only after your instructor approves your setup.

STEP-BY-STEP PROCEDURES

1. Turn the power off, lock and tag the electrical panel, and remove the panels from an electric heater that has one stage of heat. The electric circuits should be exposed.

2. Study the wiring in the low-voltage circuit until you understand how it controls the heat strip.

3. Which of the following most accurately illustrates the operation of an electric heater with one stage of heat?

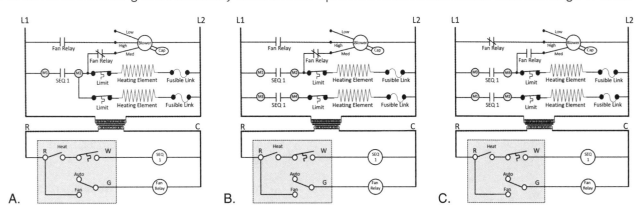

A.   B.   C.

4. Make sure that the power is off and the electrical panel is locked out, and remove the panels from an electric furnace that has more than one stage of heat. The electric circuits should be exposed.

5. Study the wiring in the low-voltage circuit until you understand how it controls the strip heaters.

6. Which of the following most accurately illustrates the operation of an electric heater with multiple stages of heat?

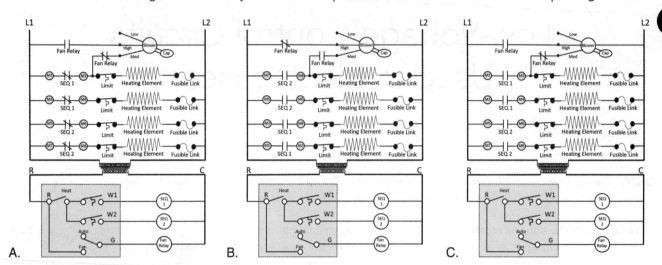

A.   B.   C.

MAINTENANCE OF WORKSTATION AND TOOLS: Replace all panels. Return tools to their respective places, and leave the workstation as instructed.

SUMMARY STATEMENT: Describe the sequence of events in the complete low-voltage circuit for each system studied. Use an additional sheet of paper if more space is necessary for your answer.

TAKEAWAY: Electric furnaces use sequencers to energize loads one at a time. These sequencers have a low voltage coil that is part of the 24 V heating circuit. Multistage units may have more than one sequencer to accomplish this. Studying the system's schematic can help you understand how the furnace should operate under normal conditions.

QUESTIONS

1. Low-voltage controls are used for which of the following reasons?
   A. Less expensive
   B. Easier to troubleshoot
   C. Safety
   D. Both A and C

2. Low-voltage wire has an insulation value for which of the following?
   A. 50 V
   B. 100 V
   C. 500 V
   D. 24 V

3. Wire size is determined by which of the following?
   A. Voltage
   B. Amperage
   C. Ohms
   D. Watts

4. The gauge of wire that is normally used in residential low-voltage control circuits is which of the following?
   A. No. 12
   B. No. 18
   C. No. 14
   D. No. 22

5. Low-voltage wire always has to be run in conduit. True or false?

6. The nominal low voltage for control circuits is which of the following?
   A. 50 V
   B. 24 V
   C. 115 V
   D. 230 V

7. All units have only one low-voltage power supply. True or false?

Exercise HTG-5

Checking a Package Sequencer

Name	Date	Grade

OBJECTIVES: Upon completion of this exercise, you should be able to check a package sequencer by energizing the operating coil and checking the contacts of the sequencer with a digital multimeter (DMM).

INTRODUCTION: You will energize the operating coil of a package sequencer for electric heat with a low-voltage power source to cause the contacts to close. You will then check the contacts with a meter to make sure that they close as they are designed to do, similar to those in Figure HTG-5.1.

TOOLS, MATERIALS, AND EQUIPMENT: Low-voltage power supply, a digital multimeter (DMM), and a package sequencer used in an electric furnace.

⚠ **SAFETY PRECAUTIONS:** You will be working with low voltage (24 V). Do not allow the power supply leads to arc together.

FIGURE **HTG-5.1** Package sequencer.

Photo by Jason Obrzut

STEP-BY-STEP PROCEDURES

1. Turn the DMM function switch to AC voltage, and, if applicable turn the, the range selector switch to the 50 V or higher scale.

2. Turn the power off, and lock out the panel. Fasten one meter lead to one of the 24 V power supply leads and the other meter lead to the other supply lead.

3. Turn the power on, and record the voltage: _____ V. It should be very close to 24 V.

4. Turn the power off.

5. Fasten the two leads from the 24 V power supply to the terminals on the sequencer that energize the contacts. They are probably labeled heater terminals.

6. Use either the ohms or continuity function on the multimeter. Fasten the meter leads on the terminals for one of the electric heater contacts. See Figure HTG-5.1 for an example of how the contacts may be oriented.

7. Turn the 24 V power supply on. You should hear a faint audible click each time a set of contacts makes or breaks.

8. After about 3 minutes, check each set of contacts with the meter to see that all sets are made. You may want to disconnect the power supply and allow all contacts to open and then repeat the test to determine the sequence in which the contacts close. You can also repeat the test to determine the time delay between each set of contacts.

9. Turn the power off and disconnect the sequencer.

MAINTENANCE OF WORKSTATION AND TOOLS: Return the meter to its proper storage place. You may need the sequencer while preparing the Summary. Put it away in its proper place when you are through with it.

SUMMARY STATEMENT: Describe the operation of the package sequencer, including all components. Use an additional sheet of paper if more space is necessary for your answer.

TAKEAWAY: A sequencer is an integral component of an electric furnace. There are different models of sequencers that use different time delays between contacts being made. They can also differ in the number of contacts they contain. A typical multimeter can be used to learn how a sequencer operates and to test whether it is good or bad.

QUESTIONS

1. What type of device is inside the sequencer to cause the contacts to close when voltage is applied to the coil?

2. What type of coil is inside the sequencer, magnetic or resistive?

3. What is the advantage of a sequencer over a contactor?

4. How many heaters can a typical package sequencer energize?

5. How is the fan motor started with a typical package sequencer?

6. What is the current-carrying capacity of each contact on the sequencer you worked with?

7. Do all electric furnaces use package sequencers?

8. If the No. 2 contact failed to close on a package sequencer, would the No. 3 contact close?

9. How should you determine the heat anticipator setting on the room thermostat when using a package sequencer?

10. How should you determine the heat anticipator setting on the room thermostat when using individual sequencers?

Exercise HTG-6

Checking Electric Heating Elements Using an Ammeter

Name	Date	Grade

OBJECTIVES: Upon completion of this exercise, you should be able to check an electric heating system by using a digital multimeter with an amp clamp to make sure that all heating elements are drawing power.

INTRODUCTION: You will identify all electric heating elements and then start an electric heating system. You will then use your digital multimeter to prove that each element is drawing power.

TOOLS, MATERIALS, AND EQUIPMENT: A digital multimeter with an amp clamp, goggles, straight-slot and Phillips-head screwdrivers, a flashlight, ¼" and ⁵⁄₁₆" nut drivers, and an electric heating system.

⚠ **SAFETY PRECAUTIONS:** You will be placing the amp clamp of the digital multimeter around the conductors going to the electric heat elements. Wear goggles. Do not pull on the wires or they may become loose at the terminals.

STEP-BY-STEP PROCEDURES

1. Turn the power to an electric furnace off and remove the door to the control compartment. Lock and tag the panel when the power is off and keep the key in your possession.

2. Locate the conductor going to each electric heating element, Figure HTG-6.1. Consult the diagram for the furnace you are working with. Make sure that you can fit the amp clamp around each conductor.

FIGURE **HTG-6.1**

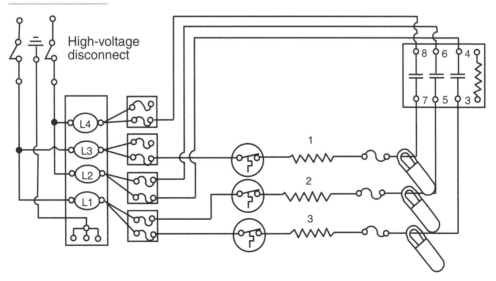

3. Turn the power on, and listen for the sequencers to start each element. Wait for about 3 minutes, the fan should have started.

4. Place the amp clamp around one conductor to each heating element, and record the readings here. When the element is drawing current, it is heating.

- Element 1: _____ A   Element 2: _____ A
- Element 3: _____ A   Element 4: _____ A
- Element 5: _____ A   Element 6: _____ A

Note: Your furnace may have only two or three elements.

5. Turn the room thermostat off and, using the meter observe which elements are de-energized first.

6. Replace all panels with the correct fasteners.

MAINTENANCE OF WORKSTATION AND TOOLS: Return all tools to their respective places. Replace all panels, if appropriate, on the electric furnace.

SUMMARY STATEMENT: Describe the function of an electric heating element, telling how the current relates to the resistance in the actual heating element wire. Use an additional sheet of paper if more space is necessary for your answer.

TAKEAWAY: Electric furnaces use one or more resistance heating elements to generate heat. Verifying operation should be done with a meter, as the elements typically cannot be seen once installed in a unit. A multimeter with an amp clamp can be used to verify current draw, proving that the element was energized and is producing heat. The elements can also be checked for resistance when they are de-energized.

QUESTIONS

1. What are the actual heating element wires made of?

2. If the resistance in a heater wire is reduced, what will happen to the current if the voltage remains constant?

3. How much heat would a 5 kW heater produce in Btus per hour?

4. What started the fan on the unit you were working on?

5. What is the difference between a contactor and a sequencer?

6. Why are sequencers preferred over contactors in ductwork applications?

7. How much current would a 5 kW heater draw with 230 V as the applied voltage?

8. What is the advantage of a package sequencer over a system with individual sequencers?

9. If sequencer coil 1 were to fail, would all the heat be off?

Exercise HTG-7

Troubleshooting Exercise: Electric Furnace

Name	Date	Grade

OBJECTIVES: Upon completion of this exercise, you should be able to troubleshoot electrical problems in an electric furnace.

INTRODUCTION: This unit should have a specified problem introduced to it by your instructor. YOU WILL LOCATE THE PROBLEM USING A LOGICAL APPROACH AND INSTRUMENTS. THE INTENT IS NOT FOR YOU TO MAKE THE REPAIR, BUT TO LOCATE THE PROBLEM AND ASK YOUR INSTRUCTOR FOR DIRECTION BEFORE COMPLETION.

TOOLS, MATERIALS, AND EQUIPMENT: Straight-slot and Phillips-head screwdrivers, ¼" and ⁵⁄₁₆" nut drivers, and a digital multimeter with an amp clamp.

⚠ **SAFETY PRECAUTIONS:** You will be working with live voltage. Use only insulated meter leads, and be very careful not to touch any energized terminals.

STEP-BY-STEP PROCEDURES

1. With the power off and the electrical panel locked out, remove the panel to the electric heat controls. WHILE THE POWER IS OFF, IDENTIFY HOW MANY STAGES OF STRIP HEAT ARE IN THIS UNIT. IDENTIFY BOTH SIDES OF LINE VOLTAGE FOR EACH STAGE OF STRIP HEAT. YOU ARE GOING TO ATTACH AN AMMETER AND VOLTMETER TO EACH LATER. THE TIME TO IDENTIFY THEM IS WHILE THEY ARE NOT ENERGIZED.

2. Set the room thermostat to call for heat, and turn the unit on.

3. Give the unit a few minutes to get up to temperature. If the unit is controlled by sequencers and has several stages of heat, it takes from 20 to 45 seconds per stage of heat to be fully energized.

4. Check the current to each stage of strip heat and record here:

   1. _____ A   2. _____ A   3. _____ A

   If there are more than three stages, add them to the end.

5. Check and record the voltage for each heater here:

   1. _____ V   2. _____ V   3. _____ V

   If there are more than three stages, add them to the end.

6. Was the blower motor operating?

7. Were all stages drawing current?

8. Were all stages energized with line power?

9. Describe the problem and your recommended repair.

MAINTENANCE OF WORKSTATION AND TOOLS: Return the unit and tools per the instructor's directions.

SUMMARY STATEMENT: Describe Ohm's law and how it applies to electric heat.

TAKEAWAY: Troubleshooting a furnace requires knowledge of its sequence of operations, the ability to read a schematic, skilled use of a multimeter, and the power of observation. Each of these can give the technician a piece of the puzzle that, together, can show the whole picture and help make a proper diagnosis.

QUESTIONS

1. What material is an electric heating element made of?

2. Why doesn't the wire leading up to the electric furnace get hot?

3. How is the heat concentrated at the electric heating element?

4. What would happen to the electric heating element if the heat were not dissipated by the fan?

5. What is the advantage of a sequencer over a contactor?

6. Why is electric heat not used exclusively instead of gas and oil?

7. What is the efficiency of electric heat?

8. Describe how the ohms function of a digital multimeter may have been used in the this exercise.

9. What voltage does the typical room thermostat operate on?

10. How would reduced airflow affect the operation of an electric furnace?

Exercise HTG-8

Changing a Sequencer

Name	Date	Grade

OBJECTIVES: Upon completion of this exercise, you should be able to change a sequencer in an electric furnace.

INTRODUCTION: You will remove a sequencer in an electric furnace and replace and rewire it just as though you have installed a new one.

TOOLS, MATERIALS, AND EQUIPMENT: Straight-blade and Phillips-head screwdrivers, ¼" and ⁵⁄₁₆" nut drivers, a digital multimeter with an amp clamp, and needle-nose pliers.

⚠ **SAFETY PRECAUTIONS:** Make sure the power is off and the panel is locked and tagged before you start this project. Use a digital multimeter to verify that the power is off.

STEP-BY-STEP PROCEDURES

1. Turn the power off and lock out the electrical panel.

2. Check to make sure the power is off using the multimeter.

3. Study the system wiring diagram that shows all of the wires on the sequencer to be changed. If some of the wires are the same color, make a tag and mark them for easy replacement.

4. Remove the wires from the sequencer one at a time. Use care when removing wires from spade or push-on terminals, and do not pull the wire out of the connector. Check for properly crimped terminals.

5. Remove the sequencer to the outside of the unit, and reuse it as if it were a new one.

6. Now remount the sequencer.

7. Replace the wiring one at a time. Be very careful that each connection is tight. Electric heat pulls a great deal of current, and any loose connection will cause a problem.

8. When the sequencer is installed to your satisfaction, ask the instructor to approve the installation before you turn the power on.

9. Turn the power on, and set the thermostat to call for heat.

10. After the unit has been on long enough to allow the sequencers to energize the heaters, use the ammeter to verify that the heaters are operating.

11. Turn the unit off when you are satisfied that it is working properly.

MAINTENANCE OF WORKSTATIONS AND TOOLS: Return all tools to their places, and leave the unit as your instructor directs you.

SUMMARY STATEMENT: Describe the internal operation of a sequencer. Use an additional sheet of paper if more space is necessary for your answer.

TAKEAWAY: When replacing a sequencer, it is important that all of the electrical connections are made properly and are tight. Loose connections can result in damage to the wire, sequencer, or heating element. Making a sketch of the wires can help with wiring the new one.

QUESTIONS

1. What would low voltage do to the heating capacity of an electric heater?

2. What would high voltage do to the heating capacity of an electric heater?

3. An electric heater has a resistance of 11.5 ohms and is operating at 230 V. What would the amperage reading be for this heater? Show your work.

4. What would the power be for the heater in question 3? Show your work.

5. What would the Btu/h rating be on the heater in question 3? Show your work.

6. If the voltage for the heater in question 3 were to be increased to 245 V, what would the wattage and Btu/h be? Show your work.

7. If the voltage were reduced to 208 V, what would the wattage and Btu/h be for the same unit? Show your work.

8. Using the voltages in questions 6 and 7, would the heater be operating within the ±10% recommended voltage?

9. What is the unit of measure the power company charges for electricity?

10. What starts the fan motor in the unit you worked on?

Exercise HTG-9

Gas Furnace Familiarization

Name	Date	Grade

OBJECTIVES: Upon completion of this exercise, you should be able to recognize and state the various components of a typical gas furnace.

INTRODUCTION: You will remove the front panels and possibly the motor and blower housing to visually inspect and identify all of the components of the furnace.

TOOLS, MATERIALS, AND EQUIPMENT: Straight-slot and Phillips screwdrivers, ¼" and 5/16" nut drivers, a 6" adjustable wrench, Allen wrenches, a flashlight, and a gas furnace.

⚠ **SAFETY PRECAUTIONS:** Turn the gas and power to the furnace off before beginning this exercise. Lock and tag the power at the disconnect panel. There should be only one key. Keep this key on your person while performing this exercise.

STEP-BY-STEP PROCEDURES

1. With the power off and the electrical panel locked, remove the front burner and blower compartment panels.

2. Fan information:
   - Motor full-load current: _____ A
   - Type of motor: _____
   - Diameter of motor: _____ in.
   - Shaft diameter: _____ in.
   - Motor rotation (looking at motor shaft): _____
   - Fan wheel diameter: _____ in.; width: _____ in.
   - Number of motor speeds: _____; high rpm: _____; low rpm: _____

3. Burner information:
   - Type of ignition: _____
   - Type of burner: _____
   - Number of burners: _____
   - Type of pilot or flame safety: _____
   - Gas valve voltage: _____ V
   - Gas valve current: _____ A

4. Unit nameplate information:
   - Manufacturer: _____
   - Model number: _____
   - Serial number: _____

- Type of gas: _____
- Input capacity: _____ Btu/h
- Output capacity: _____ Btu/h
- Voltage: _____ V
- Temperature rise: _____ ° (degrees)

5. Heat exchanger information:

What type of heat exchanger was used? _____ Type of metal: _____
Number of burner passages: _____ Flue size: _____ in.
Furnace design: _____ (upflow, downflow, or horizontal)

6. Replace all panels with the correct fasteners.

MAINTENANCE OF WORKSTATION AND TOOLS: Return all tools to their respective places, and be sure the work area is clean.

SUMMARY STATEMENT: Describe the combustion process for the furnace that you worked on. Use an additional sheet of paper if more space is necessary for your answer.

TAKEAWAY: There are numerous different types of forced-air gas furnaces. Becoming familiar with the various components, where they are located, and what they do can help to determine how the furnace operates.

QUESTIONS

1. What component transfers the heat from the products of combustion to the room air?

2. What is the typical manifold pressure for a natural gas furnace?

3. Which of the following gases requires a 100% gas shutoff, natural or propane?

4. Why does this gas require a 100% shutoff?

5. What is the typical control voltage for a gas furnace?

6. What types of safety switches are used on gas furnaces?

7. What is the purpose of the drip leg in the gas piping located just before a gas appliance?

8. What is the typical line voltage for gas furnaces?

9. What is the purpose of the flue pipe on a gas furnace?

10. What is the purpose of the draft diverter on a gas furnace?

Identification of the Flame or Pilot Safety Feature of a Gas Furnace

Name	Date	Grade

OBJECTIVES: Upon completion of this exercise, you should be able to look at the controls of a typical gas furnace and determine the type of flame safety features used.

INTRODUCTION: You will remove the panels from a typical gas-burning furnace, examine the control arrangement, and describe the type of ignition system and flame safety control.

TOOLS, MATERIALS, AND EQUIPMENT: A straight-slot and Phillips-head screwdrivers, a flashlight, ¼" and 5⁄16" nut drivers, a box of matches, and a modern gas-burning furnace.

⚠ **SAFETY PRECAUTIONS:** Turn the power off. Lock and tag the electrical power panel, and keep the key in your possession. Use caution while working with any gas-burning appliance. If there is any question, ask your instructor.

STEP-BY-STEP PROCEDURES

1. Select a gas-burning furnace. Shut off and lock out the power.

2. Remove the panel to the burner compartment.

3. Remove the cover from the burner section if there is one.

4. Using the flashlight, examine the ignition components as well as any flame safety devices.

5. Determine what type of ignition is used in this furnace and what type of flame safety controls are used.

6. If the furnace has a pilot light and the pilot light is lit, blow it out for the purpose of learning how to relight it.

7. Turn on the power and adjust the room thermostat to call for heat. Make sure that the burner ignites.

8. Allow the furnace to run until the fan starts. Turn the room thermostat to off, and make sure the burner goes out.

9. Allow the fan to run to ensure that the furnace cools down.

10. Replace all panels with the correct fasteners.

MAINTENANCE OF WORKSTATION AND TOOLS: Return all tools to their respective places. Leave the area around the furnace clean.

SUMMARY STATEMENT: Describe the exact sequence used for the pilot or other ignition system and flame safety control on this furnace. Use an additional sheet of paper if more space is necessary for your answer.

TAKEAWAY: All gas-burning furnaces use a safety mechanism to ensure the presence of a flame. These controls are a safety measure that, in the absence of a flame, will shut down or prevent the ignition sequence and stop the continued flow of gas into the furnace. Some components are used to prove a pilot flame, and others are used to prove ignition of the main burners.

QUESTIONS

1. Describe the principle of flame rectification.

2. Describe the ignition sequence of a hot surface ignition (HSI) system.

3. Describe the ignition sequence in an intermittent pilot (IP) system.

4. What is meant by the term direct ignition?

5. What is the typical heat content of 1 ft^3 of natural gas?

6. Where does natural gas come from?

7. What is the recommended manifold pressure for a natural gas furnace?

8. What component reduces the inlet gas pressure for a typical gas furnace?

9. What is the recommended manifold pressure in an LP gas furnace?

10. What is the difference between an intermittent ignition system and a direct ignition system?

Exercise HTG-11

Gas Valve, Blower Motor, Blower Assembly, and Draft Analysis

Name	Date	Grade

OBJECTIVES: Upon completion of this exercise, you should be able to understand and analyze a gas furnace gas valve, blower motor, blower assembly, and draft and be able to perform airflow calculations.

INTRODUCTION: You will remove the panels of the furnace for the purpose of identification and analysis of many components within the furnace. Some components may need to be removed for analysis. The furnace will then be reassembled and fired up to perform calculations based on its operation. Reference Figures HTG-11.1 and HTG-11.2.

TOOLS, MATERIALS, AND EQUIPMENT: A gas furnace, straight-slot and Phillips-head screwdrivers, ¼" and ⁵⁄₁₆" nut drivers, a 6" adjustable wrench, velometer and/or anemometer, flashlight, calculator, mirror, digital multimeter (DMM), manometer, two thermometers, and pencil with eraser.

⚠ **SAFETY PRECAUTIONS:** Make certain the furnace is properly grounded and that you have safety glasses and the proper protective clothing for making these observations and tests.

FIGURE **HTG-11.1**

STEP-BY-STEP PROCEDURES

Gas Valve Information

FIGURE **HTG-11.2**

1. What is the amp draw of the gas valve? _____ A

2. What type of ignition system is being used?

3. Is the gas valve a redundant gas valve?

4. Are there any safety switches wired in series with the valve?

5. Is the gas valve single-stage or two-stage?

6. Is the pressure regulator part of the gas valve or a separate component?

Fan Motor and Blower Assembly Information

Note: For the following tests and observations, run the furnace in "Fan On" mode.

1. What is the rated amperage of the fan motor? _____ A

2. What is the actual amperage draw of the fan motor with the access panels in place? _____ A

Photo by Jason Obrzut

3. With the fan running, remove the door panel to observe the change in amp draw. Remember to close the door interlock switch using tape or a clamp. Record the amperage reading: _____ A

4. With the fan motor running, block the return air inlet. Describe what happened to the amp draw and why.

5. With the fan motor running, block the discharge air. Describe what happened to the amp draw and why.

6. With the information you obtained from questions 4 and 5, explain how you would determine if an air blockage was in the return or discharge duct.

7. Is the fan motor a multispeed or variable-speed motor?

8. How many speeds can the fan motor be wired for? If the fan motor has more than one speed, which speed is used for heating purposes? Which speed is being used for the fan relay circuit?

9. What type of fan blades are used in blower assembly?

10. Is the fan motor rotation reversible?

Draft Information

1. List the type of draft used on the unit (atmospheric, forced, induced).

2. If the unit is induced or forced draft, what type of component is used to prove that the draft has been established? Also explain how this type of component proves that the draft has been established.

General Checks and Test Data

1. Level the thermostat if it is a mercury bulb type.

2. Check the thermostat for proper calibration. If out of calibration, calibrate it.

3. If the thermostat has an adjustable anticipator, make sure it is set properly.

4. Adjust primary air for proper flame if there is an adjustment.

5. Hook the manometer up to the unit, and check the gas pressure before and after the regulator. Adjust if necessary.

6. Properly adjust the fan control and high-limit setting if the controls are adjustable.

7. Check to make sure electrical components are tight.

8. Check all safety switches on the unit to make sure they will turn off the unit in the event of a component failure.

9. Check the temperature rise across the heat exchanger and record it: _____ °F

10. Using a velometer and/or anemometer, check the velocity of the air being moved through the unit during the heating sequence. Determine the cfm of air through the unit. Show all work and units. Use an additional sheet of paper if more space is necessary for your answer.

11. Cover up ¼ of the discharge air opening and repeat step 10. Show all work and units. Use an additional sheet of paper if more space is necessary for your answer.

12. Now cover up ½ of the discharge air opening, and repeat step 10. Show all work and units. Use an additional sheet of paper if more space is necessary for your answer.

13. Compare the velocity and the cfm in steps 11 and 12. Explain what happened to both the velocity and the cfm and why. Use an additional sheet of paper if more space is necessary for your answer.

14. Compare the temperature rise data collected in step 9 to the required rise listed on the furnace data plate. Was this furnace operating within its specified range? Show your calculations. Use an additional sheet of paper if more space is necessary for your answer.

MAINTENANCE OF WORKSTATION AND TOOLS: Return all tools to their respective places. Leave the area around the furnace clean.

SUMMARY STATEMENT: Describe the effects that airflow can have both on the heat output of a furnace and the fan motor itself. Use a separate sheet of paper if necessary.

TAKEAWAY: Components in a gas-burning furnace can vary greatly from one unit to another. However, they all utilize similar safety components and require proper airflow to operate as intended.

QUESTIONS

1. What effects would a dirty air filter have on the amp draw of the fan motor?

2. In a furnace that utilizes atmospheric draft, what is the function of the draft hood?

3. What adjustments can be made to the furnace if it is not operating with the correct temperature rise?

4. Why must a thermostat with mercury contacts be hung perfectly level?

5. What is the purpose of the door interlock switch in the blower compartment?

Exercise HTG-12

Wiring for a Gas Furnace

Name	Date	Grade

OBJECTIVES: Upon completion of this exercise, you should be able to describe the sequence of operations for the furnace as well as identify the wiring configurations for the various electrical components.

INTRODUCTION: You will trace the wiring on a gas furnace to answer questions about its operation. For this exercise, the furnace should be equipped with an integrated furnace control (IFC) and utilize a flame sensor as a flame proving device. Refer to Figure HTG-12.1.

TOOLS, MATERIALS, AND EQUIPMENT: A typical gas furnace with an integrated furnace controller (IFC) that uses flame rectification for flame safety, a flashlight, ¼" and ⁵⁄₁₆" nut drivers, and straight-blade and Phillips-head screwdrivers.

 SAFETY PRECAUTIONS: Before starting this exercise, turn the power to the furnace off, lock and tag the electrical panel, and keep the key on your person.

STEP-BY-STEP PROCEDURES

1. Make sure the power is off and the electrical panel is locked. Remove the front cover to the furnace.

2. Remove the cover to the junction box where the power enters the furnace.

3. Study the wiring of the furnace. After you feel you understand the wiring, answer the following questions using the furnace and the diagram provided (Figure HTG-12.1) as references.

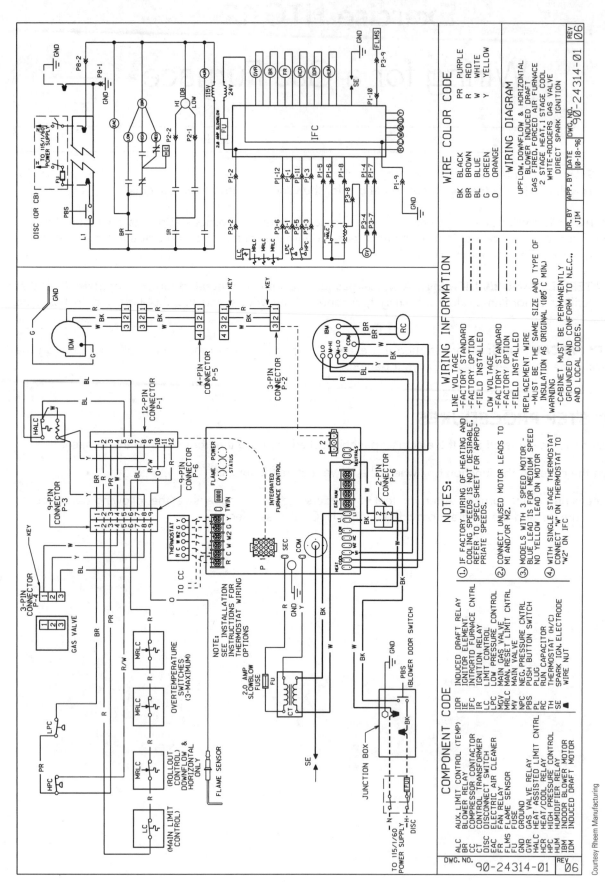

FIGURE **HTG-12.1**

Courtesy Rheem Manufacturing

4. Which of the following most accurately illustrates the gas and safety circuits of a single-stage furnace equipped with and integrated furnace control (IFC)?

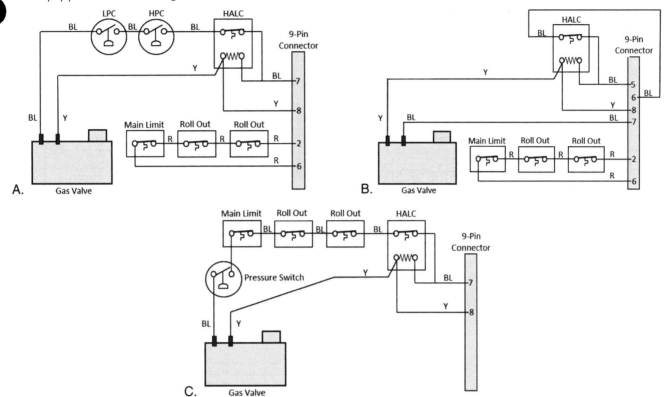

5. Which of the following most accurately illustrates the vent safety circuit of a two-stage furnace equipped with and integrated furnace control (IFC)?

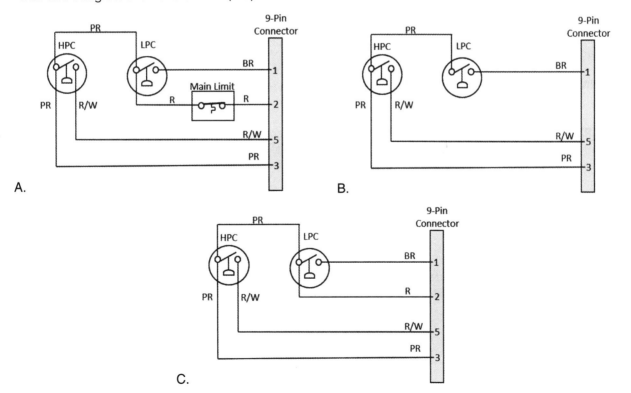

6. Which of the following most accurately illustrates the gas and safety circuits of a two-stage furnace equipped with and integrated furnace control (IFC)?

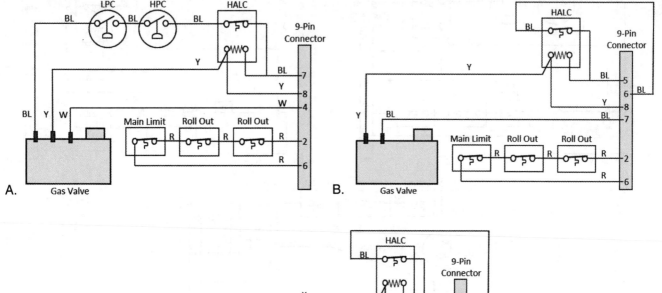

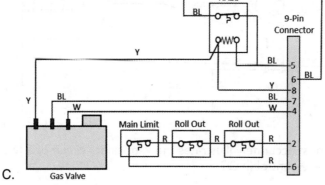

7. Replace all covers and panels.

MAINTENANCE OF WORKSTATION AND TOOLS: Replace all covers and panels on the furnace properly. Return all tools and clean the work area.

SUMMARY STATEMENT: Using the manufacturer's wiring diagram, describe the complete sequence of operations when a call for heat is initiated starting with the thermostat. Explain what happens with each control. Use an additional sheet of paper if more space is necessary for your answer.

TAKEAWAY: Schematics are important tools to both understand the sequence of operations of a furnace and service it. Technicians must be able to relate what they see on the schematic to what they see in the unit.

QUESTIONS

1. What is the job of the flame rectification system?

2. Explain flame rectification.

3. Why do modern furnaces use integrated furnace controllers (IFCs)?

4. Explain how the limit switch functions in the furnace used in this exercise. In the event the furnace overheats, what components would shut off? Which ones would stay on?

5. Were there any other safety switches wired in series with limit switch? If so, which ones?

6. Did the furnace used in this exercise have a door switch?

7. Where is the door switch physically located in a furnace? Why?

8. At how many speeds will the fan motor in the provided diagram operate?

9. What is the typical control voltage on a residential gas furnace?

10. What types of furnaces typically use auxiliary limit switches?

Exercise HTG-13

Maintenance of Gas Burners

Name	Date	Grade

OBJECTIVES: Upon completion of this exercise, you should be able to perform routine maintenance on a gas burner.

INTRODUCTION: You will remove the burners from a gas furnace, perform routine maintenance on the burners, place them back in the furnace, and put the furnace back into operation.

TOOLS, MATERIALS, AND EQUIPMENT: A gas furnace, straight-slot and Phillips-head screwdrivers, a flashlight, two 6" adjustable wrenches, two 10" pipe wrenches, ¼" and ⁵⁄₁₆" nut drivers, soap bubbles for leak testing, a heat exchanger brush, compressed air, goggles, and a vacuum cleaner.

⚠ **SAFETY PRECAUTIONS:** Turn the power and gas to the furnace off before beginning. Lock and tag the electrical panel and keep the key with you. Wear safety glasses when using compressed air.

STEP-BY-STEP PROCEDURES

1. With the power and gas off and the electrical panel locked, remove the front panel of the furnace.

2. Disconnect the burner manifold from the gas line.

3. Remove any screws or fasteners that hold the burners in place, and remove the burners from the furnace.

4. Examine the burner ports for rust and clean as needed, Figure HTG-13.1. Check the primary air openings for obstructions, dust, or debris. Put on safety goggles, and use compressed air to blow the burners out.

5. Vacuum the burner compartment and the manifold area.

6. If there is a draft hood, remove it to access the heat exchanger outlets. Inspect all visible areas for cracks or weak spots. Insert a heat exchanger brush into each opening to clean any soot or rust.

7. Using the flashlight, examine the heat exchanger openings in the burner area for cracks, soot, and rust.

8. Shake any rust or soot down to the burner compartment, and vacuum again. Heat exchanger brushes may be used to break loose any debris. You can see how much debris came from the heat exchanger area by vacuuming it first and then again after cleaning the heat exchanger.

9. Replace the burners and align them properly. Connect the manifold, gas valve, and any other gas components.

10. Turn the gas on, and use soap bubbles for leak testing any connection you loosened or are concerned about.

11. Turn the power on, and start the furnace. Wait for the burner to settle down and burn correctly. The burner may burn orange for a few minutes because of the dust and particles that were broken loose during the cleaning.

FIGURE **HTG-13.1**

Photo by Jason Obrzut

12. Observe the flames to determine if there is enough primary air. If the burners are equipped with air shutters, adjust them to produce a proper flame.

13. When you are satisfied that the burner is operating correctly, turn off the burners, allow the furnace to cool down, and then replace all of the panels.

MAINTENANCE OF WORKSTATION AND TOOLS: Return all tools to their places. Make sure the work area is clean and that the furnace is left as instructed.

SUMMARY STATEMENT: Describe the complete combustion process, including the characteristics of a properly burning flame. Use an additional sheet of paper if more space is necessary for your answer.

TAKEAWAY: Combustion can leave components of the furnace dirty, full of soot, damaged, or rusted. This can lower the efficiency of the furnace and cause improper operation. Cleaning these components regularly is required for the furnace to operate as designed and to maximize the life of the unit.

QUESTIONS

1. What three things are needed for combustion?

2. How many cubic feet of air are necessary to properly burn 1 ft^3 of gas?

3. What is the difference between primary air and secondary air?

4. Name the three typical types drafts used in gas furnaces.

5. What is the difference between an orange gas flame and a yellow gas flame?

6. What is the part of the burner that increases the air velocity to induce the primary air?

7. Where is secondary air induced into the furnace?

8. What is the purpose of a draft hood?

9. What type of air adjustment did the furnace you worked with have?

10. What are the symptoms of a shortage of primary air?

Testing the Flue on a Natural Draft Furnace

Name	Date	Grade

OBJECTIVES: Upon completion of this exercise, you should be able to check a typical natural draft gas furnace to make sure that it is venting correctly.

INTRODUCTION: You will start up a natural draft gas furnace and use a match, candle, or smoke generator to determine that the products of combustion are rising up the flue properly.

TOOLS, MATERIALS, AND EQUIPMENT: Straight-slot and Phillips-head screwdrivers, ¼" and ⁵⁄₁₆" nut drivers, a flashlight, matches, and an operating natural draft gas furnace.

⚠ **SAFETY PRECAUTIONS:** You will be working around live electricity and gas lines. Take care not to burn yourself on any components that become hot during operation. Wear gloves and safety goggles.

STEP-BY-STEP PROCEDURES

1. Select any gas-burning appliance that is naturally vented through a flue pipe.

2. Remove the cover to the burner.

3. Turn the thermostat up to the point that the main burner lights.

4. While the furnace is heating up to temperature, you may CAREFULLY touch the flue pipe to see whether it is becoming hot.

5. Locate the draft hood on the furnace.

6. Place a lit match, candle, or smoke-generating device at the entrance to the draft hood where the room dilution air enters the hood. The flame will follow the airflow and should be drawn toward the furnace and up the flue pipe, Figure HTG-14.1.

FIGURE **HTG-14.1**

Photo by Jason Obrzut

7. Shut the furnace off with the thermostat, and allow the fan to run and cool the furnace.

8. Replace all panels with the correct fasteners.

MAINTENANCE OF WORKSTATION AND TOOLS: Return all tools to their respective places, and make sure the area is clean.

SUMMARY STATEMENT: Describe the function of the draft hood. Use an additional sheet of paper if more space is necessary for your answer.

TAKEAWAY: Natural draft furnaces use natural convection as means of transferring the products of combustion through the heat exchangers and up the flue pipe to the outside. The draft hood allows dilution air to be pulled into the flue to mix with the flue gases. Obstructions in the flue pipe can cause these flue gases to flow back into the conditioned space through the opening in the draft hood.

QUESTIONS

1. What is the air that is drawn into the draft hood called?

2. What gases are contained in the products of complete combustion?

3. What gases are contained in the products of incomplete combustion?

4. Name the component that transfers heat from the products of combustion to the circulated air.

5. Explain the difference between natural draft and induced draft.

6. Describe a gas flame that is starved for oxygen.

7. When a gas flame is blowing and lifting off the burners, what is the most likely cause?

8. What is the most dangerous by-product of incomplete combustion?

9. Why does the flue pipe have to be pitched?

10. What type of switch will prevent a natural draft furnace from operating with an obstructed flue pipe?

Exercise HTG-15

Combustion Analysis of a Gas Furnace

Name	Date	Grade

OBJECTIVES: Upon completion of this exercise, you should be able to perform a combustion analysis on a gas furnace.

INTRODUCTION: You will use a combustion analyzer to analyze the combustion gases in the flue pipe during operation.

TOOLS, MATERIALS, AND EQUIPMENT: Straight-slot and Phillips-head screwdrivers, ¼" and ⁵⁄₁₆" nut drivers, a flue-gas analysis kit including thermometer, and an operating natural gas furnace.

⚠ **SAFETY PRECAUTIONS:** Be very careful working around live electricity. You will be working on a live furnace during operation. Some components will be very hot. Wear gloves and safety goggles.

STEP-BY-STEP PROCEDURES

1. Remove the panel to the burner section of the furnace.

2. Identify the type of ignition system and flame safety control.

3. Initiate a call for heat, and watch for ignition.

4. Let the furnace run for 10 minutes, and then examine the flame. Observe the flame color and characteristics; if the flame needs adjusting, make the adjustments.

5. Following the instructions provided with the combustion analyzer that you have, perform a flue-gas analysis. Record the results here:

   - Percent carbon dioxide content: _____ %

   - Flue-gas temperature: _____ °F

   - Efficiency of the burner: _____ %

6. Adjust the air shutter for minimum air to the burner, and wait 5 minutes.

7. Perform another combustion analysis, and record the results here:

   - Percent carbon dioxide content: _____ %

   - Flue-gas temperature: _____ °F

   - Efficiency of the burner: _____ %

8. Adjust the burner air back to the original setting, and wait 5 minutes. Perform one last test, and record the results here:

   - Percent carbon dioxide content: _____ %

   - Flue-gas temperature: _____ °F

   - Efficiency of the burner: _____ %

9. Allow the furnace to cool; then turn it off and replace any panels with the correct fasteners.

MAINTENANCE OF WORKSTATION AND TOOLS: Return all tools to their respective places.

SUMMARY STATEMENT: Describe the difference between perfect combustion and "acceptable" combustion. Use an additional sheet of paper if more space is necessary for your answer.

TAKEAWAY: Combustion analysis can provide valuable information about the operation of a gas furnace. Adjustments can be made to the furnace to increase efficiency, which can be seen in the analysis. Each analyzer works a little differently, and technicians should become familiar with the proper operation of the tool prior to use.

QUESTIONS

1. What is CO_2?

2. What is CO?

3. What are the products of complete combustion?

4. What is excess air?

5. Why is perfect combustion never obtained?

6. What is the ideal CO_2 content of flue gas for a natural gas furnace?

7. How much air is consumed to burn 1 ft³ of gas in a natural gas furnace?

8. Why must flue-gas temperatures in older, standard-efficiency, noncondensing gas furnaces be kept high?

9. What do yellow tips on a flame indicate?

10. If the gas burner has an air adjustment, should it be adjusted?

Determining Airflow (cfm) Using the Air Temperature Rise Method: Gas Furnace

Name	Date	Grade

OBJECTIVES: Upon completion of this exercise, you should be able to determine, using the air temperature rise method, the air volume, in cfm, moving through a gas furnace.

INTRODUCTION: You will use a thermometer or thermistor and the gas input for a standard-efficiency gas furnace to determine the airflow of a system. A manometer will also be used to verify the correct gas pressure.

TOOLS, MATERIALS, AND EQUIPMENT: A standard-efficiency gas furnace, a thermometer, and a manometer.

⚠ **SAFETY PRECAUTIONS:** Shut the gas off before connecting the manometer to the furnace.

STEP-BY-STEP PROCEDURES

1. Make sure the gas is off, and hook the manometer up to the outlet side of the gas valve.

2. Turn the gas on, start the furnace, and record the outlet or manifold gas pressure: _____ in. wc. This should typically be 3.5 in. wc for natural gas. If you have a gas other than natural gas, your instructor will help you determine the correct pressure and Btu content of the gas. If the pressure is not correct, adjust the regulator until it is correct.

3. When you have established the correct gas pressure at the manifold, stop the furnace, shut the gas off, and remove the manometer. Remember to replace the pressure tap.

4. Turn the gas on; start the furnace; and, while it is getting up to temperature, install a thermometer or thermistor lead in the supply and return ducts. You will achieve the best reading by placing the supply duct lead several feet from the furnace, where the air leaving the furnace will have some distance to mix, Figure HTG-16.1.

FIGURE **HTG-16.1**

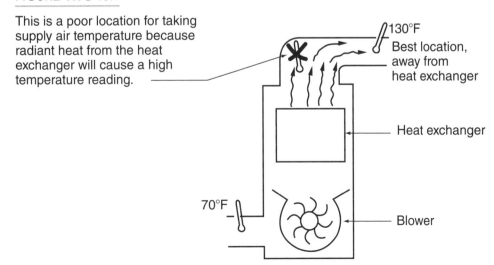

This is a poor location for taking supply air temperature because radiant heat from the heat exchanger will cause a high temperature reading.

130°F
Best location, away from heat exchanger

Heat exchanger

70°F

Blower

5. Record the change in temperature after the furnace has operated for 10 minutes: _____ °F ΔT.

6. If the Btu output is not listed on the furnace nameplate, calculate it by multiplying the input Btu rating by the furnace efficiency. For example, a 100,000 Btu/h furnace has a typical input of 100,000 × 0.80 (80%) = 80,000 Btu/h output. The remaining heat goes up the flue of the furnace as products of combustion.

7. Use the following formula to determine the airflow.

$$\text{cfm} = \frac{\text{Total Hourly Btu Output}}{1.08 \times \Delta T \text{ (change in temperature)}} = \underline{\hspace{2cm}} \text{ cfm}$$

Note: 1.08 is a constant used for standard air conditions. Its derivation is Specific Heat of air (0.24 btu/lb) × Density of air (0.075 lb/ft³) × 60 minutes = 1.08.

8. Allow the furnace to cool; then turn it off and replace all panels.

MAINTENANCE OF WORKSTATION AND TOOLS: Return all tools to their places. Make sure the work area is clean and that the furnace is left as you are instructed.

SUMMARY STATEMENT: Describe how heat is transferred from the burning gas to the air leaving the furnace. Use an additional sheet of paper if more space is necessary for your answer.

TAKEAWAY: The amount of air moving through a furnace will directly impact how much heat it can deliver to the conditioned space. The actual cfm can be determined by using the temperature rise method. When performing this test, be sure to use accurate thermometers and place them in the appropriate locations in the ducts.

QUESTIONS

1. What would happen to the temperature of the supply air if the air filter became clogged?

2. What would the flue-gas symptoms be if there were too much airflow?

3. What is the Btu heat content of 1 ft³ of natural gas?

4. What is the typical temperature rise for a standard-efficiency gas furnace?_____ to _____ °F

5. What could cause a high temperature rise?

6. What would cause a low temperature rise?

7. What is the efficiency of a standard-efficiency gas furnace?

8. What is the efficiency of a high-efficiency gas furnace?

9. How is high efficiency accomplished?

10. What is the manifold pressure of a typical natural gas furnace?

Exercise HTG-17

Main Burner, Ignition System, Flame Sensing System, Fan and Limit Control, Transformer, and Thermostat Analysis

Name	Date	Grade

OBJECTIVES: Upon completion of this exercise, you should be able to understand and analyze the gas furnace's main burner, ignition and flame sensing system, fan and limit control, and transformer.

INTRODUCTION: You will remove the panels of the furnace for the purpose of identification and analysis of many components within the furnace.

TOOLS, MATERIALS, AND EQUIPMENT: A gas furnace, straight-slot and Phillips-head screwdrivers, ¼" and 5⁄16" nut drivers, a 6" adjustable wrench, flashlight, mirror, and pencil with eraser.

⚠ **SAFETY PRECAUTIONS:** Make certain the furnace is properly grounded and that you have goggles and the proper protective clothing for making these observations and tests.

STEP-BY-STEP PROCEDURES

General Information

1. What is the Btu input of the unit?

2. What is the recommended temperature rise across the heat exchanger?

3. What are the recommended static pressure requirements for the unit?

4. What type of ignition system is used?

5. Is the unit an upflow, counterflow, low-boy, horizontal, or multipositional furnace?

Burner Information

1. How many burners are in the unit?

2. What type of burners are they (inshot, ribbon, slotted)?

3. What sizes are the orifices in the unit?

4. Is the furnace natural or mechanical draft?

Pilot, Ignition, Flame Sensing, and Flame Rectification System Information

1. If the system uses a pilot, is it manual freestanding, automatic freestanding, or automatic recycling?

2. If the pilot is automatic, what is the ignition source for it?

3. If the system is direct ignition, what is used to light the main burners?

4. What type of flame-proving device is used on the unit? Is it millivolt, bimetallic, liquid-filled tube, or flame rectification?

5. What controls the flow of the gas?

Fan Control and High-Limit Information

1. How many limits are used in the unit? If more than one limit is used, explain the purpose for each one.

2. What types of safety switches are used on the heating unit: helix, thermal disc, or fuse?

3. If the safeties are thermal disc or fuse, at what temperature will they open the circuit?

4. What type of fan control is used on the unit: helix, thermal disc, time delay, or an IFC?

5. If the fan control is a thermal disc, at what temperature will it close the circuit and energize the fan?

6. If the fan control is a time delay or an IFC, list how long it takes to close the fan circuit after it has been energized.

Transformer Information

1. What is the VA rating of the control transformer?

2. Determine the maximum amperage used by the control circuit when the maximum loads are energized at any given time.

3. What is the maximum amount of current draw that could be added to the control circuit without overloading the control transformer?

Thermostat Information

1. Is the thermostat a heating thermostat or combination heating/cooling thermostat?

2. Does the thermostat have an anticipator?

3. If the thermostat has an anticipator, is it fixed or adjustable?

4. If the anticipator is adjustable, take an amp draw reading of the control circuit, and set the anticipator to the correct setting. What is this correct setting?

5. What components in the unit determine the anticipator setting?

MAINTENANCE OF WORKSTATION AND TOOLS: Allow the furnace to cool down. Then replace all of the panels on the furnace, and make sure you leave the area neat and clean.

SUMMARY STATEMENT: Explain what may happen with the operation of the unit if the anticipator were set at a higher or lower amperage setting than it is supposed to be. Also explain why. Use an additional sheet of paper if more space is necessary for your answer.

TAKEAWAY: Gas-burning furnaces utilize numerous different components from one ignition system to the next. They can be standing pilot, intermittent ignition or direct ignition and can use different methods for controlling the fan. It is important to understand and recognize these different ignition systems, their components, and their typical operation.

QUESTIONS

1. Explain the principle of flame rectification.

2. What is the purpose of the heat anticipator?

3. What is the difference between intermittent ignition and direct ignition?

4. In a system that utilizes a standing pilot, how does the thermocouple function?

5. Explain how changing blower speeds can change the system's temperature rise.

6. What is the maximum load that can be placed on a 70 VA low-voltage transformer?

Troubleshooting Exercise: Gas Furnace

Name	Date	Grade

OBJECTIVES: Upon completion of this exercise, you should be able to troubleshoot a fault in a gas furnace with spark ignition.

INTRODUCTION: This unit has a specified problem introduced by your instructor. You will locate the problem using a logical approach and instruments. The manufacturer's troubleshooting procedures would help. THE INTENT IS NOT FOR YOU TO MAKE THE REPAIR BUT TO LOCATE THE PROBLEM AND ASK YOUR INSTRUCTOR FOR DIRECTION BEFORE COMPLETION.

TOOLS, MATERIALS, AND EQUIPMENT: Straight-slot and Phillips-head screwdrivers, ¼" and ⁵⁄₁₆" nut drivers, and a digital multimeter (DMM).

⚠ **SAFETY PRECAUTIONS:** Use care while working with electricity. Always use a multimeter to verify electrical circuits.

STEP-BY-STEP PROCEDURES

Note: Your instructor will have placed a non-operative component into the furnace prior to this exercise.

1. Turn the power on, and set the thermostat to call for heat.

2. Give the furnace a few minutes to start. Describe the sequence of events while the furnace is trying to start.

3. If the system uses a pilot, did it ignite?

4. Did the burners ignite?

5. Did the fan start?

6. Use a multimeter to make sure the thermostat is calling for heat. Check for low voltage at the gas valve.

7. If there is an induced draft motor, did it energize?

8. If applicable, use a multimeter to verify that the contacts in the negative pressure switch closed when the induced draft motor energized.

9. Describe your findings and give your diagnosis.

10. If you have made the correct diagnosis, with your instructor's permission, make the repair and verify that the furnace operates properly.

MAINTENANCE OF WORKSTATION AND TOOLS: Allow the furnace to cool off; then replace all the panels. Return all tools to their proper location, and leave the area clean.

SUMMARY STATEMENT: Describe the operation of a correctly working spark or hot surface ignition system. Use an additional sheet of paper if more space is necessary for your answer.

TAKEAWAY: Troubleshooting a furnace requires a thorough understanding of the ignition system and the sequence of operations. It also requires skilled use of a digital multimeter and the power of observation. Each of these can give the technician a piece of the puzzle that, when put together, can show the whole picture and help make a proper diagnosis.

QUESTIONS

1. What is the difference between intermittent-spark ignition and direct-spark ignition?

2. Why is spark ignition important for rooftop or outdoor equipment?

3. How does a hot surface igniter function?

4. What is the approximate voltage for the spark-to-pilot ignition system?

5. Did the ignition system on this furnace have a circuit board or ignition module?

6. Were you able to follow the wiring diagram furnished with the furnace?

7. What would you suggest to make the wiring diagram better?

8. Was there a troubleshooting procedure furnished with the furnace?

9. If so, was it clear and did it direct you to the problem?

10. What type of flame-proving device was used on this unit?

Changing a Gas Valve on a Gas Furnace

Name	Date	Grade

OBJECTIVES: Upon completion of this exercise, you should be able to change a gas valve on a typical gas furnace using the correct tools and procedures.

INTRODUCTION: You will be using the appropriate tools to remove the gas valve from a furnace and then reinstall it to simulate changing a valve.

TOOLS, MATERIALS, AND EQUIPMENT: Straight-slot and Phillips-head screwdrivers, needle-nose pliers, two 10" adjustable wrenches, ⅜" and ⁷⁄₁₆" open-end wrenches, two 14" pipe wrenches, soap bubbles for leak checking, a digital multimeter (DMM), and thread seal compatible with natural gas.

⚠ **SAFETY PRECAUTIONS:** Make sure the power and the gas are off before starting this exercise. The electrical panel should be locked, and you should have the key.

STEP-BY-STEP PROCEDURES

1. Turn the power off, and check it with the multimeter. Lock and tag the electrical panel.

2. Turn the gas off at the main valve before the unit is serviced.

3. Remove the furnace door. Observe how it comes off so you will know how to replace it.

4. The wiring can be tagged or labeled for proper identification. Remove the wires from the gas valve. Use needle-nose pliers on the connectors if they are the spade type. Do not pull the wires out of their connectors.

5. Look at the gas piping, and decide where to take it apart. There may be a pipe union or a flare nut connection to the inlet of the system. Either one will work.

6. Remove the pilot line connections from the gas valve. USE THE CORRECT-SIZE END WRENCH, AND DO NOT DAMAGE THE FITTINGS. THE PILOT LINE IS ALUMINUM AND CAN KINK OR CRACK EASILY.

7. Use the adjustable wrenches for flare fitting connections and pipe wrenches for gas piping connections, and disassemble the piping to the gas valve. LOOK FOR SQUARE SHOULDERS ON THE GAS VALVE, AND BE SURE TO KEEP THE WRENCHES ON THE SAME SIDE OF THE GAS VALVE. TOO MUCH STRESS CAN EASILY BE APPLIED TO A GAS VALVE BY HAVING ONE WRENCH ON ONE SIDE OF THE VALVE AND THE OTHER ON THE PIPING ON THE OTHER SIDE OF THE VALVE.

8. Remove the valve from the gas manifold. USE CARE NOT TO STRESS THE VALVE OR THE MANIFOLD.

9. While you have the valve out of the system, examine the valve for some of its features, such as where the pilot tubing or electrical connections are made. Look for damage.

10. After completely removing the gas valve, treat it like a new one, and fasten it back to the gas manifold. BE SURE TO USE THREAD SEAL ON ALL EXTERNAL PIPE THREADS. DO NOT OVERUSE AND DO NOT USE ON FLARE FITTINGS.

11. Fasten the inlet piping back to the gas valve.

12. Fasten the pilot line, if there is one, back using the correct wrench.

13. Fasten all wiring back using the manufacturer's wiring diagram.

14. Ask your instructor to look at the job and approve; then turn the gas on. IMMEDIATELY USE SOAP BUBBLES OR A LEAK DETECTOR TO CHECK THE GAS LINE UP TO THE GAS VALVE FOR LEAKS.

15. If there is a standing pilot, ignite the pilot flame and test the pilot tubing for leaks.

16. Turn the thermostat to call for heat. WHEN THE BURNER LIGHTS, LEAK CHECK THE VALVE OUTLET IMMEDIATELY.

17. When you are satisfied there are no leaks, turn the burners off, allow the furnace to cool down, and then turn the furnace off.

18. Turn the power off, and remove the lock and tag from the electrical panel.

19. Use a wet cloth to clean any soap bubble residue off the fittings.

MAINTENANCE OF WORKSTATION AND TOOLS: Return the workstation to proper condition as the instructor directs you. Return the tools to their respective places.

SUMMARY STATEMENT: Describe what may happen if too much thread seal is used on the inlet fittings to the gas valve and the outlet to the gas valve. Use an additional sheet of paper if more space is necessary for your answer.

TAKEAWAY: When changing a gas valve, it is important to understand the requirements for the type of gas line used, for example, using thread seal on black iron pipe but not on flare fittings. It is also important to use the appropriate tools for each task. When the job is completed, a thorough leak check of all gas lines and components is required.

QUESTIONS

1. Did the gas valve on this system have a square shoulder for holding with a wrench?

2. Did you have any gas leaks after the repair?

3. Why is thread seal used on threaded pipe fittings?

4. Why is thread seal not used on flared fittings?

5. Is thread seal used to connect the two halves of a pipe union together?

Exercise HTG-20

Changing a Fan Motor on a Gas Furnace

Name	Date	Grade

OBJECTIVES: Upon completion of this exercise, you should be able to change a fan motor on a gas furnace and then restart the furnace.

INTRODUCTION: You will pull the blower assembly to gain access to the blower fan motor. You will then remove the motor, replace it, reassemble the blower assembly, and replace it in the furnace.

TOOLS, MATERIALS, AND EQUIPMENT: Straight-slot and Phillips-head screwdrivers, ¼" and ⁵⁄₁₆" nut drivers, Allen wrenches, a small wheel puller, needle-nose pliers, a digital multimeter with an amp clamp, a 6" adjustable wrench, and slip-joint pliers.

⚠ **SAFETY PRECAUTIONS:** Make sure the power is off and the electrical panel is locked and tagged before starting this exercise. Be sure to properly discharge any fan capacitors. Wear gloves and safety goggles during this exercise.

STEP-BY-STEP PROCEDURES

1. Turn the power off, and check it with a multimeter. Lock and tag the electrical panel.

2. Remove the panel from the blower compartment.

3. Tag or label the motor wires to ensure that the correct speeds are used when the furnace is put back into operation.

4. Discharge any fan capacitors, and disconnect the motor wires.

5. If the motor is direct-drive, remove the blower from the compartment. If the motor is a belt-drive, loosen the tension on the belt.

6. If the motor is direct-drive, remove the motor from the blower wheel. If the motor is a belt-drive, remove it from the mounting bracket, and remove the pulley with a puller, if needed.

7. Treat this motor as a new motor, and replace it in the reverse order in which you removed it. Be sure to select the correct motor speed. Take note of the amperage listed on the motor data plate prior to installing it.

8. Double-check the wiring using the manufacturer's wiring diagram.

9. Once everything is reassembled, ask your instructor to check the installation.

10. When the installation has been inspected, start the motor and check the amperage. REMEMBER TO CHECK THE AMPERAGE WITH THE FAN DOOR IN PLACE. YOU MAY NEED TO CLAMP THE MULTIMETER AROUND THE WIRE ENTERING THE FURNACE.

11. Turn the furnace off.

MAINTENANCE OF WORKSTATION AND TOOLS: Leave the furnace as specified by your instructor. Return all tools to their appropriate places.

SUMMARY STATEMENT: Describe the airflow through a gas furnace during a typical heating cycle. Use a separate sheet of paper if you need more space.

TAKEAWAY: Removing the blower assembly is a routine maintenance task that should be performed regularly. The blades on the blower wheel need to be kept clean to ensure proper air movement through the furnace. The various fan motors that can be found in furnaces today may have different methods of removal for access. Sketching a small wiring diagram is good way to ensure proper wiring when the unit is reassembled.

QUESTIONS

1. What type of fan motor did this furnace have: shaded-pole, permanent split capacitor (PSC), electronically commutated (ECM), or variable frequency drive (VFD)?

2. In this exercise, which motor speed was used for heating? For cooling?

3. When the motor and the blower wheel are difficult to separate, what tool can be used to make the job easier?

4. What type of motor mount did the motor in this exercise have: rigid or rubber?

5. What type of bearings did the motor have, sleeve or ball-bearing?

6. Under what circumstances would the other speeds be used on a multispeed motor?

7. How many capacitor leads were there, and what color were they (if applicable)?

8. What happens to the motor amperage if the leaving airflow is restricted?

9. What happens to the motor amperage if the entering airflow is restricted?

10. Did the motor in this exercise have oil ports?

Testing Ignition Controls and Measuring Flame Rectification Signals

Name	Date	Grade

OBJECTIVES: Upon completion of this exercise, you should be able to identify and test ignition controls. You should also be able to measure the flame sensing signal (flame rectification) in microamps.

INTRODUCTION: You will be testing the flame current by placing a microammeter in series with the flame sensing circuit in various ignition systems.

TOOLS, MATERIALS, AND EQUIPMENT: Operating gas-fired appliances with various ignition systems, digital multimeter (DMM), microammeter, flashlight or droplight, screwdrivers, and nut drivers.

⚠️ **SAFETY PRECAUTIONS:** If a microammeter is not connected properly, damage to the furnace and/or meter may result. Check all electrical connections, and wear the proper personal protective equipment.

STEP-BY-STEP PROCEDURES

1. With safety glasses on and appliance power off, remove the service panels from the appliance as necessary.

2. Identify the type of ignition system being used: intermittent pilot (IP), Figure HTG-21.1; direct-spark ignition (DSI), Figure HTG-21.2; intermittent-hot-surface ignition, Figure HTG-21.3; or direct-hot-surface ignition, Figure HTG-21.4.

FIGURE **HTG-21.1**

Photo by Jason Obrzut

FIGURE **HTG-21.2**

Photo by Jason Obrzut

FIGURE **HTG-21.3**

Photo by Jason Obrzut

FIGURE **HTG-21.4**

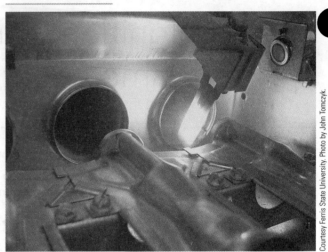

Courtesy Ferris State University. Photo by John Tomczyk.

3. If there is a pilot flame that ignites the main burner, what ignites the pilot flame?

4. If the unit uses hot surface ignition, record the igniter resistance at room temperature. _____

5. If the unit uses hot surface ignition, measure and record the amperage of the HSI when it is energized. _____ A

6. If the ignition type is HSI, what is the supply voltage to the HSI when it is energized? _____ V

7. If the ignition system is spark, locate where the high voltage comes from to generate the spark. List the voltage used for the spark igniter. _____ V

8. Use a microammeter to measure and record the flame current. This measurement requires a meter capable of measuring DC microamps. Connect the meter leads in series with the flame rod. Record you reading here: _____ µA. Refer to Figures HTG-21.5 and HTG-21.6.

FIGURE **HTG-21.5**

Flame rectification pilot and probe

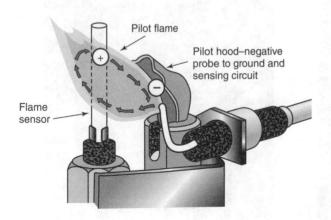

Pilot flame

Pilot hood–negative probe to ground and sensing circuit

Flame sensor

Rectification circuit

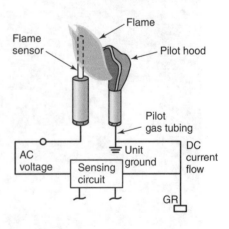

Flame

Flame sensor

Pilot hood

Pilot gas tubing

AC voltage

Unit ground

Sensing circuit

DC current flow

GR

FIGURE **HTG-21.6**

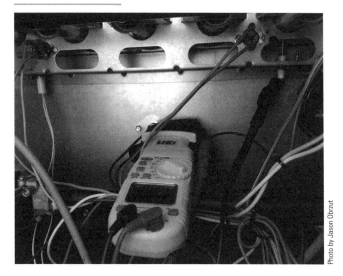

Photo by Jason Obrzut

9. If the system is direct-spark ignition, is it a single-rod (local sensing) or two-rod (remote sensing) system? _____

10. Flame safety control test: Disconnect the HSI or spark igniter. Observe the operation of the ignition module or integrated furnace control.

    A. In detail, explain the operation of the control when a pilot or main flame is not established.

    B. Does the safety control have troubleshooting indicator lights?

    C. What do the lights indicate?

11. Does the furnace have an interpurge?   Yes   No

12. How long is the timing between trials for ignition? _____

13. Does the control have a soft lockout, hard lockout, or both?

14. Reconnect the HSI or spark igniter. Disconnect the flame rod wire, and call for heat. Observe the operation of the ignition module or integrated furnace control.

    Were the results similar to those from the flame safety control test? Explain why or why not.

Once finished, have the instructor check your lab report, and then select another furnace that utilizes a different type of ignition.

MAINTENANCE OF WORKSTATION AND TOOLS: Be sure to properly reconnect all wires, and have your instructor inspect them. Replace all panels on the furnace, and store any tools in the appropriate location.

SUMMARY STATEMENT: Explain flame rectification and how it is used in a gas-burning furnace. Use a separate sheet of paper if more space is necessary.

TAKEAWAY: All ignition systems utilize some sort of flame safety control to prevent operation when a hazardous condition exists in the furnace. A technician must be able to identify what type of ignition system is being used and what type of flame proving device it is equipped with. Testing the flame current can be helpful when troubleshooting a furnace or when performing routine maintenance.

QUESTIONS

1. What is the typical flame current range for a properly operating furnace?

2. What symptoms would a furnace with a dirty flame sensor exhibit?

3. Why does a hot surface igniter have to be at room temperature to be tested for resistance?

4. What is the difference between direct-spark ignition and intermittent-spark ignition?

5. What other conditions in a furnace can have an effect on the flame current?

Exercise HTG-22

Testing Furnace Safety Controls

Name	Date	Grade

OBJECTIVES: Upon completion of this exercise, you should be able to identify and test furnace safety controls. You should also be able to properly use a digital multimeter to determine if safety control contacts are "open" or "closed."

INTRODUCTION: Safety devices are intended to help ensure that an appliance only operates when conditions to do so are correct. If an unsafe condition is present, a properly mounted, wired, and maintained safety device will prevent the system from operating. It is, therefore, extremely important for the service technician to fully understand these devices and how to properly evaluated them. Under no circumstances should a safety device be jumped out or otherwise removed from the circuit.

TOOLS, MATERIALS, AND EQUIPMENT: Manometer, digital multimeter (DMM), thermistor or thermometer, insulated pliers, insulated screwdrivers, insulated nut drivers, pencil, and paper.

⚠ **SAFETY PRECAUTIONS:** Make sure you are wearing the appropriate clothing. Make sure all electrical devices you are testing or working with are chassis grounded.

STEP-BY-STEP PROCEDURES

1. With safety glasses on and the power to the appliance off, remove service panels as needed.

2. Connect a manometer, and measure the air pressure that the combustion air-proving switch, Figure HTG-22.1, is monitoring. Turn the power back on, and set the thermostat to call for heat. Observe the gauge reading during normal heating operations.

FIGURE **HTG-22.1**

Combustion blower motor

Pressure switch and diaphragm for proving draft

Blower housing

Courtesy Ferris State University. Photo by John Tomczyk

3. Air-proving switch test: With the manometer connected, slowly restrict the vent until the pressure switch contacts open. Use a voltmeter to monitor the switch contacts. Write down the air pressure when the safety contacts open. _____

4. Shut the furnace off, and disconnect one tube from the air-proving switch. Turn the furnace back on, and initiate a call for heat.

   • Observe the furnace operation, and give a description of what occurs when the pressure switch contacts do not close during the call for heat.

   • Use your multimeter to check the position of the switch contacts. Set the meter to the alternating current (AC) voltage scale, and put the test leads on the switch terminals (one test lead on each terminal). What voltage do you read when the contacts are open? _____ V

   • Use your meter to check the voltage for each wire terminal of the air pressure switch to the "C" terminal of the transformer (transformer common). What are the results? _____

5. Does the furnace have more than one air pressure safety switch?   Yes   No
   If there is more than one, what is the purpose of the second switch? _____ Refer to Figure HTG-22.2.

FIGURE **HTG-22.2**

6. Return the furnace to its normal operating conditions. Use a manometer to measure the pressure observed during the heating sequence, and record it here: _____

7. High-limit test: Make sure that the furnace has been returned to its original operating conditions. Disconnect the fan motor "white" wire from the integrated furnace controller (IFC) so that the fan will not

start. Put a thermistor, thermometer, or some kind of temperature sensor as close to the high limit sensing element as possible. Make sure you can read the temperature during the test. Set the thermostat to call for heat. The burner should start and operate until the high limit temperature is reached. Record the temperature at the moment the high limit is reached and the burners cycle off: _____°F.

Note: Shut the furnace power off if the burner does not shut off when the temperature reads 280°F.

- Use your meter to check voltage from each wire terminal of the limit to the "C" terminal (transformer common). What are the results? _____. Observe the furnace operations, and give a description of what occurs when the limit switch contacts open during the call for heat.

8. Locate all of the safety controls in the furnace. Check the furnace's service literature and schematic for information if needed. You are very likely to encounter a fan limit switch—Figures HTG-22.3, HTG-22.4, and HTG-22.5—and a high-limit control—Figure HTG-22.6. Notice that the high limit control is positioned very close to the appliance's heat exchanger. You will also likely come across a rollout switch, Figure HTG-22.7. Figure HTG-22.8 shows the location of the rollout switch.

9. Flame rollout safety control failure simulation: Turn power to furnace off, and use a meter to ensure the power is off. Disconnect one of the two wires connected to the flame rollout safety, and secure (cover with electrical tape) the wire so that the wire connector does not touch anything. Turn the power back on, and make sure the thermostat is calling for heat. Pay close attention to how the furnace operates, and record your observations. Example: fan runs continuously, light-emitting diodes (LEDs) on the control board flash to indicate a _____ code.

FIGURE **HTG-22.3**

FIGURE **HTG-22.4**

Photo by Jason Obrzut

FIGURE **HTG-22.5**

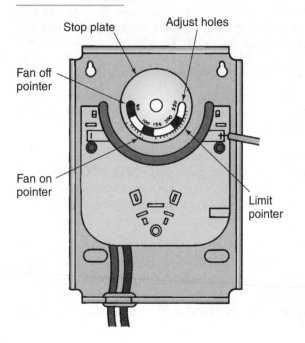

Stop plate
Adjust holes
Fan off pointer
Fan on pointer
Limit pointer

FIGURE **HTG-22.6**

by Jason Obrzut

FIGURE **HTG-22.7**

Photo by Jason Obrzut

FIGURE **HTG-22.8**

MANUALLY RESETTABLE
FLAME ROLLOUT SAFETY

Courtesy Ferris State University. Photo by John Tomczyk.

MAINTENANCE OF WORKSTATION AND TOOLS: Make sure that all wires have been reconnected and that all safety controls are functioning normal. The furnace should be left clean with its panels in place. Return all tools to their appropriate locations.

SUMMARY STATEMENT: Study the wiring diagram, and follow the path through all of the safety circuits. Describe how all the safety controls are wired to the main control and how they function. Use a separate sheet of paper if more space is needed.

TAKEAWAY: Gas-burning furnaces can become very dangerous under certain circumstances. Safety controls are a very important component to the safe operation of these units. Technicians must be aware of the different types of safety controls used, how they operate, why they are used, and what causes them to terminate the heating sequence and shut the furnace down.

QUESTIONS

1. What conditions can cause a pressure switch failure to close fault?

2. What conditions can cause the limit switch to open during furnace operation?

3. Where in the furnace are rollout switches located, and why are they used?

4. What is meant by the term "manual reset" when referring to a safety control?

5. When there is a second pressure switch in the furnace, what is it typically used for?

Familiarization with Oil Furnace Components

Name	Date	Grade

OBJECTIVES: Upon completion of this exercise, you should be able to identify the various components of an oil furnace and describe their function.

INTRODUCTION: You will inspect the components of an oil furnace and record their characteristics to help you become familiar with the appliance's configuration and operation.

TOOLS, MATERIALS, AND EQUIPMENT: Straight-slot and Phillips-head screwdrivers, two 6" adjustable wrenches, an Allen wrench set, an open-end wrench set, a nozzle wrench, rags, an oil burner (Figure HTG-23.1), safety goggles, and a flashlight.

FIGURE **HTG-23.1**

Courtesy R. W. Beckett Corporation

⚠ **SAFETY PRECAUTIONS:** Turn the power and oil supply line off before starting this exercise. Lock and tag the electrical panel, and keep the key with you. Be sure to wear safety goggles.

STEP-BY-STEP PROCEDURES

After turning the power and oil supply off, and locking the electrical panel, remove the front burner and blower compartment panels. Record the following:

1. Furnace blower information (you may have to remove or slide out the blower assembly):

   • Motor full-load amperage: _____ A

   • Type of motor: _____

- Diameter of motor: _____ in.
- Shaft diameter: _____ in.
- Motor rotation (looking at motor shaft end): _____
- Fan wheel diameter: _____ in.
- Width: _____ in.
- Number of motor speeds: _____ high rpm _____ low rpm _____

2. Burner information (burner may have to be removed):

- Nozzle size: _____ gpm
- Nozzle angle: _____ degrees
- Pump speed: _____ rpm
- Pump motor amperage: _____ A
- Number of pump stages: _____
- One- or two-pipe system: _____
- Type of primary control: cad cell or stack switch: _____

3. Furnace nameplate information:

- Manufacturer: _____
- Model number: _____
- Serial number: _____
- Capacity: _____ Btu/h
- Furnace voltage: _____ V
- Control voltage: _____ V
- Recommended temperature rise: low: _____, high: _____

4. Is the oil tank above or below the furnace? _____

5. Is there an oil filter in the line leading to the furnace? _____ What is the oil-line size? _____

6. Reinstall all components. Have your instructor inspect to ensure that the furnace is in working order.

MAINTENANCE OF WORKSTATION AND TOOLS: Replace all furnace panels, and leave the oil burner as you are instructed. Return all tools and materials to their respective places. Clean your work area.

SUMMARY STATEMENT: Describe the relationship between the gpm rating of the nozzle and the Btu/h capacity of the furnace. Provide as much support information regarding the properties of fuel oil and system efficiency to describe this relationship.

TAKEAWAY: In order to effectively service and troubleshoot any piece of air-conditioning or heating equipment, it is important to become familiar with the basic system components, where they are located, and what functions they are intended to perform. This exercise serves to aid in the process of component familiarization.

QUESTIONS

1. What is the heat content of a gallon of No. 2 fuel oil?

2. In what unit of heat is an oil furnace capacity expressed?

3. Are oil furnace capacities expressed in output or input?

4. What is the purpose of a two-pipe oil supply system?

5. How is the blower motor started in a typical oil-fired furnace?

6. What is the typical voltage for an oil-fired furnace?

7. What is the typical control voltage for an oil-fired furnace?

8. What is the function of the limit control for an oil-fired furnace?

9. What is the function of the stack switch on an oil-fired furnace?

10. What is the function of a cad cell on an oil-fired furnace?

Exercise HTG-24

Oil Burner Maintenance

Name	Date	Grade

OBJECTIVES: Upon completion of this exercise, you should be able to perform routine maintenance on a typical gun-type oil burner. You will also be able to describe the difference in resistance in a cad cell when it "sees" or "does not see" light.

INTRODUCTION: You will remove a gun-type oil burner from a furnace and remove and inspect the oil nozzle. You will also check and set the electrodes.

TOOLS, MATERIALS, AND EQUIPMENT: Straight-slot and Phillips-head screwdrivers, a set of open-end wrenches (⅜" through ¾"), two 6" adjustable wrenches, rags, a small shallow pan, a digital multimeter (DMM), goggles, a flashlight, and a nozzle wrench.

 SAFETY PRECAUTIONS: Shut the power and oil supply off, and lock and tag the electrical panel before starting this exercise.

STEP-BY-STEP PROCEDURES

1. With the electrical power and oil supply off and the electrical panel locked, disconnect the oil supply line (and return if any). Let any oil droplets drip in the pan.

2. Disconnect the electrical connections to the burner. There should be both high- (120 V) and low-voltage connections.

3. Remove the burner from the furnace, and place the burner assembly on a workbench. Refer to your textbook for an example of a burner assembly.

4. Loosen the fastener that holds the transformer in place. The transformer is hinged and will usually open to the back and stop. This exposes the electrode and nozzle assembly.

5. Loosen the oil-line connection on the back of the burner. This will allow the nozzle assembly to be removed from the burner.

   Note: This assembly can be removed with the burner in place on the furnace, but we are doing it on a workbench for better visibility.

6. Examine the electrodes and their insulators. The insulators may be carefully removed and cleaned in solvent if needed. They are ceramic and very fragile, so be sure to handle them very carefully.

7. Remove the oil nozzle using the nozzle wrench. Inspect and replace or reinstall the old nozzle. Have your instructor inspect the installation.

8. Set the electrodes in the correct position using an electrode gauge, use Figure HTG-24.1, as a guide.

9. Replace the nozzle assembly in the burner assembly.

10. Remove the cad cell (if this unit has a cad cell).

11. Connect an ohmmeter to the cad cell. Turn the cell eye toward a light, and record the ohm reading here: _____ Ω.

12. Cover the cell eye, and record the ohm reading here: _____ Ω.

FIGURE **HTG-24.1**

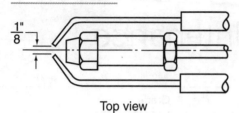

Top view

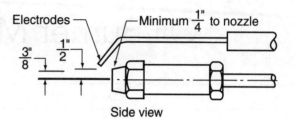

Side view

Height adjustment

$\frac{1"}{2}$ Residential installation

$\frac{3"}{8}$ Commercial installation

Position of electrodes in front of nozzle is determined by spray angle of nozzle.

Electrodes cannot be closer than $\frac{1"}{4}$ to any metal part.

13. Replace the cad cell in the burner assembly. Make sure the burner is reassembled correctly before replacing it on the furnace. Check to see that all connections are tight.

14. Return the burner assembly to the furnace, and fasten the oil line or lines. Make all electrical connections. Have your instructor inspect your work.

MAINTENANCE OF WORKSTATION AND TOOLS: Return all tools to their correct places. Make sure there is no oil left on the floor or furnace.

SUMMARY STATEMENT: Describe the complete burner assembly, including the static disc in the burner's air tube and the cad cell. Use an additional sheet of paper if more space is necessary for your answer.

TAKEAWAY: There are many steps involved in the combustion process in an oil-fired heating appliance. Two important steps involve atomizing the fuel, which is then mixed with air to help ensure proper combustion, and the actual igniting of the fuel/air mixture. The nozzle and the electrodes are responsible for these important tasks. Without atomization and ignition, the appliance cannot function as intended. It is, therefore, extremely important to fully understand these system components.

QUESTIONS

1. Why should you be sure that no oil is left on any part, in addition to its being a possible fire hazard?

2. What is the color of No. 2 fuel oil? Why is it this color?

3. What must be done to oil to prepare it to burn?

4. Why is the typical oil burner called a "gun-type" oil burner?

5. What is the typical pressure at which an oil burner operates?

6. Is there a strainer in an oil burner nozzle?

7. What is the purpose of the tangential slots in an oil burner nozzle?

Exercise HTG-25

Electronic Controls for Oil Burners

Name	Date	Grade

OBJECTIVES: Upon completion of this exercise, you should be able to follow the electrical circuit in an oil-fired furnace and check a cad cell for correct resistance.

INTRODUCTION: You will use a multimeter to follow the line-voltage circuit through a primary control and a cad cell on a typical oil-fired furnace.

TOOLS, MATERIALS, AND EQUIPMENT: A digital multimeter (DMM) having one lead with an alligator clip and the other lead with a probe, a screwdriver, electrical tape, and an oil-fired furnace that has a cad cell primary control configured for intermittent, or constant, ignition.

⚠ **SAFETY PRECAUTIONS:** Be sure to turn the power off while removing the primary control from its junction box and performing the cad cell check. Lock and tag the electrical panel, and keep the key with you. Be very careful while making electrical measurements. Make these measurements only under the supervision of your instructor.

STEP-BY-STEP PROCEDURES

1. Turn the power off, lock the electrical panel, and remove the primary control from its junction box. Move it to the side so voltage readings may be taken.

2. Set the DMM to read voltage.

3. Use a lead with an alligator clip, and place it on the white wire (neutral).

4. Start the furnace and make sure that ignition takes place—that is, the furnace is fired.

5. Prepare to take voltage readings on the black and orange wires. Remember the black wire is the hot lead feeding the primary and the orange wire is the hot lead leaving the primary to the burner motor and transformer.

6. Using the pointed end of the second probe, check for power at the black terminal, and record here: _____ V.

7. Check the voltage at the orange terminal, and record here: _____ V.

8. Turn the furnace power supply off, and lock the panel.

9. Remove the cad cell. This is usually accomplished by removing one screw that holds the transformer and raising the transformer to one side.

10. Using the multimeter set to ohms, check the resistance of the cad cell with the eye pointed at the room light source, and record it here: _____ Ω. With the eye covered with electrical tape: _____ Ω.

11. With the electrical tape still covering the eye, replace the cad cell in its mounting in the furnace, and prepare the furnace for firing.

12. Place the alligator clip lead on the white wire. Start the furnace while measuring the voltage at the orange wire. The burner should operate for about 90 seconds before shutting off.

13. Record the running time here: _____ seconds. If you miss the timing the first time, allow 5 minutes for the control to cool, and reset the primary. Caution: Do not reset the primary more than three times or else excess oil may accumulate in the combustion chamber.

14. Turn the power off, remove the tape from the cad cell, start the furnace, and run it for 10 minutes to clear any excess oil from the combustion chamber.

15. Remove one lead from the cad cell, and tape it to keep it from touching another circuit. This produces an open circuit in the cad cell.

16. Start the furnace while measuring the voltage from the white to the orange wire. Record the voltage here: _____ V.

17. Remove the meter leads, and place all wiring panels and screws back in their correct order.

MAINTENANCE OF WORKSTATION AND TOOLS: Return all tools to their respective places.

SUMMARY STATEMENT: Describe how the cad cell and the primary control work together to protect the oil furnace from accumulating too much oil in the combustion chamber. Use an additional sheet of paper if more space is necessary for your answer.

TAKEAWAY: The primary control is often considered to be the brain of an oil burner, as it helps ensure the starting and stopping of the burner, as well as helping to ensure that the burner operates safely. Properly installed and maintained primary controls ensure that the oil burner will not operate if unsafe conditions are present.

QUESTIONS

1. What material is in a cad cell that allows it to respond to light?

2. What is the response that a cad cell makes to change in light?

3. What would the symptoms be of a dirty cad cell?

4. What would the result of excess oil in the combustion chamber be if it were ignited?

5. What would the symptom of an open cad cell be?

6. What would cause a cad cell to become dirty?

Exercise HTG-26

Checking the Fan and Limit Control

Name	Date	Grade

OBJECTIVES: Upon completion of this exercise, you should be able to check a fan and limit switch on an oil-fired furnace for correct operation.

INTRODUCTION: You will use a thermometer probe in the area of the fan and limit control to determine the temperature and then observe the control action.

TOOLS, MATERIALS, AND EQUIPMENT: Straight-slot and Phillips-head screwdrivers, an electronic thermometer, ¼" and 5⁄16" nut drivers, and a flashlight.

⚠️ **SAFETY PRECAUTIONS:** Turn the power off, and lock out before installing the thermometer sensor. Be careful not to get burned on hot surfaces.

STEP-BY-STEP PROCEDURES

1. With the power off, locked, and tagged, remove the front cover of the furnace, and locate the fan/limit switch.

2. Remove the cover from the fan limit control to see whether it has a circular dial that turns as it is heated. If so, it can be observed during the test.

3. Slide the thermometer sensor in above the heat exchanger, close to the fan limit sensor location.

4. Start the furnace and watch the temperature climb. Record the temperature every 2 minutes.

   2 minutes _____ °F   4 minutes _____ °F

   6 minutes _____ °F   8 minutes _____ °F

   10 minutes _____ °F   12 minutes _____ °F

5. Record the temperature when the furnace blower starts: _____ °F

6. When the blower starts, shut the furnace off using the system switch. The furnace will not have time to cool.

7. Carefully disconnect the neutral wire from the fan motor so the blower will not start.

8. Start the furnace again f, and record the temperature when the limit switch stops the burner pump and blower: _____ °F. Note: Turn the furnace off manually if the temperature rises above 280°F.

9. Shut the power off and reconnect the fan motor wire.

10. Turn the power on; the blower should start and cool the furnace's heat exchanger.

11. Wait until the heat exchanger has cooled to the point that the blower stops; then remove the temperature lead from the area of the fan limit control.

12. Start the furnace again to make sure that you have left it operational.

MAINTENANCE OF WORKSTATION AND TOOLS: Return all tools to their proper places, and make sure your work area is clean.

SUMMARY STATEMENT: Describe what happens within the fan limit control from the time the stack starts to heat until it cools to the point where the fan shuts off. Use an additional sheet of paper if more space is necessary for your answer.

TAKEAWAY: The fan limit control is a unique control device in that it has the ability to operate as an operational device as well as a safety device. In addition, it can be connected so that one portion of the device controls a low-voltage circuit, while the other controls a line-voltage circuit. Gaining a solid understanding of this device and its various configurations is an important step in becoming a successful service technician.

QUESTIONS

1. State a common cause for a limit switch to stop a forced-air furnace.

2. Describe the sensor on the fan limit control that senses heat.

3. What would happen if a technician wired around a limit switch and the fan became defective and would not run?

4. What would the symptoms be on an oil furnace if the fan setting on the fan limit control were set too low?

5. Describe the location of the fan limit control on the furnace you worked with.

6. How many wires are attached to the fan limit control on the furnace you worked with?

7. What circuit does the limit control break when there is an overheat condition?

8. Is the limit switch normally in the high- or low-voltage circuit?

9. Can a fan limit control with a 5-inch long bimetal shaft be replaced with a switch that has a longer one? Why or why not?

10. What would be the symptom for a defective limit switch in which the contacts remained open?

Adjusting an Oil Burner Pump Pressure

Name	Date	Grade

OBJECTIVES: Upon completion of this exercise, you should be able to install the gauges on an oil pump and adjust it to the correct pressure.

INTRODUCTION: You will install a pressure gauge on the oil pump discharge and a compound gauge (that will read into a vacuum) on the suction side of the oil pump. You will then start the pump and adjust the discharge pressure to the correct pressure.

TOOLS, MATERIALS, AND EQUIPMENT: Straight-slot and Phillips-head screwdrivers, a vacuum or compound gauge, a pressure gauge (150 psig), any adapters needed to adapt gauges to an oil pump, a shallow pan to catch oil drippings, goggles, two adjustable wrenches, thread sealer, rags, a set of Allen wrenches, and an operating oil-fired furnace.

SAFETY PRECAUTIONS: Do not spill any fuel oil on the floor; use a pan under the pump while connecting and disconnecting gauges. Turn the power to the furnace off when you are not calling for it to run. Lock and tag the electrical panel, and keep the only key in your possession. Remember to wear safety goggles.

STEP-BY-STEP PROCEDURES

1. Shut off and lock out the power to the furnace.

2. Shut the fuel supply valve off if the supply tank is above the oil pump.

3. Remove the gauge plug for the discharge pressure. This is usually the one closest to the small oil line going to the nozzle. Connect the high-pressure gauge here. Refer to your textbook for an example.

4. Remove the gauge plug for the suction side of the pump. This is usually located next to the inlet piping on the pump.

5. Remove the pressure-regulating screw protective cap, and insert the correct Allen wrench in the slot.

6. Open the fuel valve, start the furnace, and quickly make sure that there are no oil leaks. If there are, shut the furnace off and repair them. IF THERE ARE NO LEAKS, MAKE SURE THAT THE OIL HAS IGNITED AND IS BURNING.

7. Record the pressures on the gauges here:
   - Inlet pressure: _____ psig or in. Hg
   - Outlet pressure: _____ psig

8. Adjust the outlet pressure to 90 psig (this is 10 psig below the recommended pressure) and then back to 100 psig.

9. Adjust the outlet pressure to 110 psig (this is 10 psig above the recommended pressure) and then back to 100 psig. Count the turns per 10 psig.

10. Shut the furnace off and remove the gauges. Replace the plugs in the appropriate openings.

11. Wipe any oil away from the fittings, and dispose of any oil in the pan as your instructor advises.

12. Replace all panels with the correct fasteners.

MAINTENANCE OF WORKSTATION AND TOOLS: Make sure the workstation is clean, and return all tools to their respective places.

SUMMARY STATEMENT: Describe how the oil burner pump responded to adjustment—approximately how many turns per 10 psig change? Use an additional sheet of paper if more space is necessary for your answer.

TAKEAWAY: As part of effective oil burner service, it is important to check the pressure and vacuum produced by the fuel pump. If the pump is not producing the desired oil pressure, the appliance cannot function properly. It is not a good field practice to simply assume that the pump is operating properly.

QUESTIONS

1. What is the pressure at which oil burner nozzles are rated?

2. What is the recommended outlet or discharge oil pressure?

3. What was the size of the pressure tap in the oil pump for the high-pressure gauge?

4. What was the size of the pressure tap in the oil pump for the low-pressure gauge?

5. On average, how many Btus are there in a gallon of fuel oil?

6. What is the typical efficiency of a conventional oil-fired furnace?

7. When is a two-pipe oil delivery system recommended for an oil burner?

8. How far can an oil burner pump vertically without an auxiliary pump?

9. Name the two places that oil is filtered before it is burned.

Exercise HTG-28

Combustion Analysis of an Oil Burner

Name	Date	Grade

OBJECTIVES: Upon completion of this exercise, you should be able to perform a draft check and a smoke test and analyze an oil burner for combustion efficiency.

INTRODUCTION: You will use a draft gauge to measure the draft over the fire, a smoke tester to measure the smoke in the flue, and a combustion analyzer to perform a combustion analysis on an oil-burning system.

TOOLS, MATERIALS, AND EQUIPMENT: A portable electric drill, ¼" drill bit, ¼" and ⁵⁄₁₆" nut drivers, straight-slot and Phillips-head screwdrivers, a draft gauge, a smoke test kit, goggles, a combustion analysis kit, and an operating oil-fired furnace.

⚠ **SAFETY PRECAUTIONS:** The flue pipe of an oil-fired furnace can be very hot, even when the appliance is not running; do not touch it.

STEP-BY-STEP PROCEDURES

1. Wear safety goggles. If holes are not already available, drill a ¼" hole in the flue pipe at least 6" before the draft regulator and one in the burner inspection door. Get advice from your instructor.

2. Insert the thermometer from the analysis kit in the hole in the flue pipe.

3. Start the furnace and determine that the oil burner fires.

4. Wait for 10 minutes or until the thermometer in the flue stops rising; then use the draft gauge and measure the draft above the fire, at the inspection door, Figure HTG-28.1. Record the reading here: _____ in. wc.

5. Remove the thermometer from the flue pipe, and perform a smoke test following the instructions in the kit, Figure HTG-28.2. Record the smoke test number here: _____.

FIGURE **HTG-28.1**

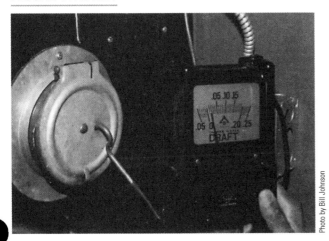

Photo by Bill Johnson

FIGURE **HTG-28.2**

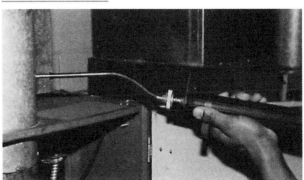

Photo by Bill Johnson

6. Perform a combustion analysis at the ¼" hole in the flue pipe. Record the CO_2 reading here: _____.

7. Record the combustion efficiency here: _____% efficient.

8. Turn the furnace off, and remove the instruments.

9. Replace all panels with the correct fasteners.

10. Fill the hole with a ¼" cap, if available.

MAINTENANCE OF WORKSTATION AND TOOLS: Return all tools to their respective places.

SUMMARY STATEMENT: Describe the efficiency of this oil furnace. If it was running below 70% efficiency, explain the probable causes. Use an additional sheet of paper if more space is necessary for your answer.

TAKEAWAY: Performing smoke, draft, and combustion efficiency tests on an oil-fired heating appliance are not optional. These tests are extremely important to ensure that the appliance is operating safely and efficiently. It is IMPOSSIBLE to properly set up or determine the efficiency of an oil-fired heating appliance without performing these tests.

QUESTIONS

1. What could be the cause of a high stack temperature?

2. How would you prove a furnace was stopped up with soot in the heat exchanger?

3. What is the function of the heat exchanger?

4. What is the function of the combustion chamber?

5. What is the purpose of the restrictor in the blast tube?

6. What is the difference between a hollow-core and a solid-core oil nozzle?

7. What is the purpose of the tangential slots in the nozzle?

8. What must be done to oil to prepare it to burn?

9. What would be the result of a situation where a customer reset the primary control many times before the service person arrived?

10. What are the symptoms of an oil forced-air furnace with a hole in the heat exchanger?

11. Comment on the smoke number obtained in step 5.

12. What should be done if the initial smoke reading is zero?

Exercise HTG-29

Troubleshooting Exercise: Oil Furnace

Name	Date	Grade

OBJECTIVES: Upon completion of this exercise, you should be able to troubleshoot a problem in an oil furnace.

INTRODUCTION: This furnace should have a specific problem introduced by your instructor. You will locate the problem using a logical approach and instruments. THE INTENT IS NOT FOR YOU TO MAKE THE REPAIR BUT TO LOCATE THE PROBLEM AND ASK YOUR INSTRUCTOR FOR DIRECTION BEFORE COMPLETION.

TOOLS, MATERIALS, AND EQUIPMENT: Straight-slot and Phillips screwdrivers, Allen wrenches, two 6" adjustable wrenches, a set of socket wrenches, a flashlight, and an oil gauge.

⚠ **SAFETY PRECAUTIONS:** Make sure all power to the furnace is off. Lock and tag the panel before removing any parts.

STEP-BY-STEP PROCEDURES

1. Set the thermostat to call for heat, and start the furnace.

2. Did the burner ignite?

3. If not, did the furnace blower start?

4. Did the burner motor start?

5. Turn the power off, and lock out. Install a gauge on the burner outlet.

6. With the gauge in place, restart the furnace. Record the pressure reading here: _____ psig.

7. If the burner is running and there is no pressure, what could the problem be?

8. Use the flashlight to view the pump coupling. Is the coupling rotating? Is the oil pump turning?

9. Describe what you think the problem could be.

10. What is your recommended next step?

MAINTENANCE OF WORKSTATION AND TOOLS: Return the furnace to the condition recommended by your instructor. Return all tools to their places.

SUMMARY STATEMENT: Describe the difference between a single-stage and a two-stage oil pump. As part of your response, be sure to include the intended application for each one. Use an additional sheet of paper if more space is necessary for your answer.

TAKEAWAY: Troubleshooting a furnace requires knowledge of its sequence of operations, the ability to read a schematic, a firm grasp of tool use, and the power of observation. Each of these can give the technician a piece of the puzzle that, together, can show the whole picture and help make a proper diagnosis.

QUESTIONS

1. Does fuel oil burn as a liquid or a vapor?

2. What is the purpose of the oil pump and nozzle?

3. What is the typical oil pressure for a gun-type oil burner?

4. What is the purpose of the cutoff inside the oil pump?

5. At what pressure does the cutoff function?

6. What is the purpose of the coupling between the pump and motor?

7. How long can an oil furnace run after start-up without firing?

8. What happens if the burner does not fire in the prescribed time?

9. If the burner does not fire in the prescribed time, how do you restart it?

10. Describe the operation of a cad cell.

Exercise HTG-30

Filling a Hot Water Heating System

Name	Date	Grade

OBJECTIVES: Upon completion of this exercise, you should be able to fill an empty hot water system with fresh-water and put it back into operation.

INTRODUCTION: You will start with an empty boiler and hot water system and fill it with water, bleed the air, and put it back into operation.

TOOLS, MATERIALS, AND EQUIPMENT: Two pipe wrenches (12"), straight-slot and Phillips-head screwdrivers, a water hose if needed, and a hot water heating system.

⚠️ **SAFETY PRECAUTIONS:** NEVER ALLOW COLD WATER TO ENTER A HOT BOILER. IF YOUR BOILER IS HOT, SHUT IT OFF THE DAY BEFORE THIS EXERCISE, OR DO NOT DRAIN THE BOILER AS A PART OF THE EXERCISE. Shut the power off to the boiler when it is not in operation. Lock and tag the electrical panel, and keep the single key in your possession.

STEP-BY-STEP PROCEDURES

1. Turn the power off, and lock out the boiler and water pumps. Close the water inlet valve. Drain the entire system at the boiler drain valve if the boiler is at the bottom of the circuit. If the boiler is at the top of the system, drain from the system valve at the lowest point.

2. Make sure the boiler is cool to the touch of your hand.

3. Close all valves in the system, and start to fill the system with water. CHECK TO MAKE SURE THE SYSTEM HAS A WATER-REGULATING OR PRESSURE-REDUCING VALVE, such as that in Figure HTG-30.1. With this valve in the system, you cannot apply too much pressure to the system. IF THE SYSTEM DOES NOT HAVE AN AUTOMATIC FILL VALVE, WATCH THE PRESSURE GAUGE, AND DO NOT ALLOW THE PRESSURE TO RISE ABOVE THE RECOMMENDED LIMIT, USUALLY 12 PSIG.

FIGURE **HTG-30.1**

Courtesy Bell and Gossett

FIGURE **HTG-30.2**

Courtesy Bell and Gossett

FIGURE **HTG-30.3**

Courtesy Bell and Gossett

4. Fill the system until the automatic valve stops feeding or the correct pressure is reached (in which case, shut the water off).

5. Go to the air bleed valves, and make sure that they are allowing any air to escape. Bleed valves can be manual, Figure HTG-30.2, or automatic, Figure HTG-30.3. If they are manual valves, bleed them until a small amount of water is bleeding out.

6. Start the pump and check the air bleed valves again.

7. Start the boiler and wait until it becomes hot to the touch.

8. Carefully feel the water line leaving the boiler. It should become hot, and the hot water should start leaving the boiler and move toward the system. The pump should be moving the water through the system. You should now be able to trace the path of the hot water throughout the system, with the cooler return water going back to the boiler.

9. You should now have a system that is operating properly.

10. Clean up the work area; do not leave water on the floor. LEAVE THE BOILER AS YOUR INSTRUCTOR INDICATES.

MAINTENANCE OF WORKSTATION AND TOOLS: Return all tools to their respective places.

SUMMARY STATEMENT: Describe the distribution piping system in the unit you were working with.

TAKEAWAY: It may become necessary to completely drain a boiler system to make a repair. When refilling the boiler, it is important to monitor the boiler water pressure, taking care not to over-pressurize the system. Also, it is just as important that all air be removed from the system to ensure proper operation.

QUESTIONS

1. How does air get into a hot water heating system?

2. How can you get air out of a hot water heating system?

3. What are the symptoms of air in a hot water heating system?

4. Why should cold water not be introduced into a hot boiler?

5. What is inside an automatic air bleed valve that allows the air to get out but not the water?

6. What type of pump is typically used on a small hot water heating system?

7. What is the purpose of balancing valves in a hot water heating system?

8. How is air separated from the water at the boiler in some systems?

9. Once air has been separated from the water, how is it ultimately removed from the system?

10. What is the purpose of a system pressure relief valve?

Exercise HTG-31

• ## Determining Expansion Tank Pressure

Name	Date	Grade

OBJECTIVES: Upon completion of this exercise, you should be able to determine the proper pressure for a bladder-type expansion tank.

INTRODUCTION: Most of the time, bladder-type expansion tanks come from the factory with a preset pressure of 12 psig. For most applications, this is sufficient, but the actual pressure that is needed is determined by the height of the piping circuit with respect to the inlet of the expansion tank.

TOOLS, MATERIALS, AND EQUIPMENT: A tape measure and a hot water hydronic heating system with terminal units on multiple floors, if possible.

⚠ **SAFETY PRECAUTIONS:** Always be aware that the pipes in a hydronic heating system may be very hot. Be sure to protect yourself by wearing proper personal protection equipment.

STEP-BY-STEP PROCEDURES

1. At the boiler, measure the distance from the inlet of the expansion tank to the floor, and record that distance here: _____ ft _____ in.

2. Determine the number of floors in the structure that have baseboard heating (above the floor on which the boiler is located), and record that number here: _____.

3. Go to the highest floor on which there is a baseboard heating unit.

4. Measure the distance from the top of the highest pipe to the floor, and record that distance here: _____ ft _____ in.

5. Go to the next-highest floor on which there is a baseboard heating unit. If the next-highest floor is the level on which the boiler is located, go to step 10.

6. Measure the height of the ceiling, and record it here: _____ ft _____ in.

7. Go to the next-highest floor on which there is a baseboard heating unit. If the next-highest floor is the level on which the boiler is located, go to step 10.

8. Measure the height of the ceiling, and record it here: _____ ft _____ in.

9. Measure the height of the ceiling, of the boiler room, and record it here: _____ ft _____ in.

10. Add the measurements obtained in steps 4, 6, 8, and 9, and record that total here: _____ ft _____ in.

11. Multiply the number in step 2 by 0.50, and enter that number here: _____ ft _____ in. If the number from step 2 is even, the entry will be only in feet. If the number from step 2 is odd, the entry will be the number of whole feet and 6". For example, if the number in step 2 is 3, the result will be 1.5, or 1 ft, 6 in. If the number in step 2 is 2, the result will be 1.0, or 1 ft, 0 in.

12. Add the numbers in steps 10 and 11 and record that value here: _____ ft _____ in.

13. Subtract the measurement from step 1 from the value in step 12, and record that value here: _____ ft _____ in.

14. Convert the value in step 13 to the decimal form of feet, and record the result here: _____ ft. For example, 20 ft, 6 in. is 20.5 ft.

15. Divide the value in step 14 by 2.31, and record that value here: _____ psi.

16. Add 5 to the result from step 15, and record the result here: _____ psig.

17. The result in step 16 is the required expansion tank pressure for that system.

MAINTENANCE OF WORKSTATION AND TOOLS: Make certain that all tools are put back in their proper places and that the hot water hydronic heating system is left in good working order.

SUMMARY STATEMENT: Compare the value obtained in this exercise to the "normal" factory preset pressure, and explain why the manufacturer of expansion tanks typically uses this pressure. Use an additional sheet of paper if more space is necessary for your answer.

TAKEAWAY: Expansion tanks allow room for the water to expand as it is heated. If the pressure in the tank is too high, the system pressure may rise, causing the pressure relief valve to open. If the tank pressure is too low, the tank may overfill with water and become water-logged. It is important that the pressure in the tank matches the requirements of the system that it is installed on.

QUESTIONS

1. Explain where the 2.31 conversion factor in step 15 comes from.

2. Explain why we add 5 psig to the end result.

3. What prediction can be made regarding the required expansion tank pressure on systems that are used on structures with more floors?

4. Why is it that only the height of the piping arrangement is factored into the expansion tank pressure calculation?

5. How can the pressure of the tank be checked?

6. What is the recommended installation location of the expansion tank?

Exercise HTG-32

Near-Boiler Piping Component Identification

Name	Date	Grade

OBJECTIVES: Upon completion of this exercise, you should be able to identify the components typically found on and around a typical hot water hydronic heating system.

INTRODUCTION: There are a number of system components that are located on or very close to the boiler. These components enhance system operation and allow it to function in an efficient manner. This exercise will allow you to become familiar with these system components.

TOOLS, MATERIALS, AND EQUIPMENT: Hot water hydronic heating system, pad, pencil, and flashlight.

⚠ **SAFETY PRECAUTIONS:** Always be aware that the pipes on a hot water hydronic heating system may be very hot. Protect yourself by wearing proper personal protection equipment.

STEP-BY-STEP PROCEDURES

1. Study the near-boiler piping and system components for the unit that you were assigned. Note the pipe sizes and the fittings that were used to complete the piping of the unit.

2. Identify all of the system components that are found in the piping arrangement on the sketch.

3. Obtain information from the following system components:

   Water pressure reducing valve:

       Manufacturer: _____ Model: _____

   Expansion tank:

       Manufacturer: _____ Model: _____

       Circle one: Steel tank   Bladder type

   Air separator:

       Manufacturer: _____ Model: _____

   Air vent:

       Manufacturer: _____ Model: _____

   Pressure relief valve:

       Manufacturer: _____ Model: _____

   Circulator pump 1:

       Manufacturer: _____ Model: _____

   Circulator pump 2:

       Manufacturer: _____ Model: _____

Circulator pump 3:

    Manufacturer: _____ Model: _____

Zone valve:

    Manufacturer: _____ Model: _____

Other _____:

    Manufacturer: _____ Model: _____

Other _____:

    Manufacturer: _____ Model: _____

4. Identify the point of no pressure change on the system.

5. Determine the direction of water flow as it makes its way through the system.

MAINTENANCE OF WORKSTATION AND TOOLS: Make certain that all tools are put back in their proper places and that the hot water hydronic heating system is left in good working order.

SUMMARY STATEMENT: Explain the importance of properly piping in a boiler and knowing how all of the near-boiler system components operate. Use an additional sheet of paper if more space is necessary for your answer.

TAKEWAY: The components located on or near the boiler should be considered part of the boiler. Each performs a task that is essential to the proper operation of the boiler and the transfer of heat to the structure.

QUESTIONS

1. What is the typical opening pressure on the boiler's pressure relief valve?

2. Explain why the pressure relief valve must be installed directly on the boiler with as few fittings as possible and with no valves in between the component and the boiler.

3. If an expansion tank has a static (resting) pressure of 12 psig, what is the desired fill pressure (water pressure) in the hot water hydronic heating system prior to boiler operation?

4. Of the near-boiler piping components that are found on a typical hot water hydronic heating system, determine which are intended to be operational controls and which are intended to operate specifically as safety devices.

5. What would happen to a hot water hydronic heating system if the bladder on the expansion tank ruptured?

6. What is the difference between a bladder-style expansion tank and a conventional steel tank?

Exercise HTG-33

System Configuration Identification

Name	Date	Grade

OBJECTIVES: Upon completion of this exercise, you should be able to identify various types of hot water hydronic heating systems.

INTRODUCTION: There are many different configurations for hot water hydronic heating systems. To properly troubleshoot and evaluate a hydronic heating system, the technician must first understand what type of system he is dealing with. By evaluating the near-boiler piping and the components that are present, the technician will be able to determine the piping configuration that is being used.

TOOLS, MATERIALS, AND EQUIPMENT: Hot water hydronic heating system, pad, pencil, and flashlight.

SAFETY PRECAUTIONS: Always be aware that the pipes in a hydronic heating system may be very hot. Protect yourself by wearing proper personal protection equipment.

STEP-BY-STEP PROCEDURES

1. Determine which system piping components are installed on the system. Enter information on the following lines:

   Circulator pump   Yes   No

   How many circulator pumps are installed on this system? _____

   List the different models of the pumps here:

   Zone valves   Yes   No

   Diverter tees   Yes   No

   Mixing valves   Yes   No

   Thermostatic radiator valves   Yes   No

   Balancing valves   Yes   No

2. Based on the information obtained in step 1, compare the system being evaluated to the following:

	Circulator (single or multiple)	Zone Valves	Diverter Tees	Mixing Valves	Thermostatic Radiator Valves	Balancing Valves
Series loop	Single	No	No	No	No	No
1 Pipe	Single	Maybe	Yes	No	Maybe	No
2 Pipe direct	Single or multiple	Maybe	No	No	Maybe	Yes
2 Pipe reverse	Single or multiple	Maybe	No	No	Maybe	No
High-temperature primary-secondary	Multiple	Maybe	No	Maybe	Maybe	No
Multi-temperature primary-secondary	Multiple	Maybe	No	Yes	Maybe	No

3. From the information obtained in step 1 and the table in step 2, the system being evaluated is most likely a _____ system.

4. Study the boiler piping until are confident that you understand the water flow through the system.

MAINTENANCE OF WORKSTATION AND TOOLS: Make certain that all tools are put back in their proper places and that the hot water hydronic heating system is left in good working order.

SUMMARY STATEMENT: Explain the importance of properly piping in a boiler and knowing the type of system that is being worked on. Use an additional sheet of paper if more space is necessary for your answer.

TAKEAWAY: There are numerous different piping arrangements that can be found on hydronic heating systems. Each configuration may require more or less piping components for proper operation. By observing the near-boiler piping and components, a technician can determine what type of piping arrangement the system utilizes.

QUESTIONS

1. Why are zone valves, thermostatic radiator valves, and other temperature-controlling valves not found on series loop hydronic heating systems?

2. Why are series loop hydronic heating systems equipped with only one circulator?

3. Explain the function/purpose of the diverter tee.

4. Explain how a thermostatic mixing valve is used to provide different heating zones or areas with water at different temperatures.

SECTION 10

Mechanical System Troubleshooting (MST)

SUBJECT AREA OVERVIEW

The servicing of any air-conditioning or heating system is important because sooner or later the equipment will malfunction and need attention. Service technicians must develop a systematic approach to diagnosing faulty equipment and making the necessary repairs. They must also use a certain amount of common sense and observational skills to become a good troubleshooter. A firm grasp of the operating conditions that are typical of the unit being worked on is just as important as the tools being used to service it.

Air-conditioning equipment is designed to operate at its rated capacity and efficiency at a specified design condition. This condition is generally accepted to be at an outside temperature of 95°F and an inside temperature of 80°F (though some homeowners prefer 75°F) with a relative humidity of 50%. System design conditions have been established by the Air-Conditioning, Heating, and Refrigeration Institute (AHRI). A system's operating temperatures and pressures will change in response to the conditions acting on the system, as well as the amount of refrigerant contained in the system.

There are normally three grades of equipment: economy, standard efficiency, and high efficiency. Economy and standard grades have about the same efficiency, while the high-efficiency grade has different operating characteristics. Higher-efficiency equipment has heat transfer surfaces (evaporators and condensers) that are physically larger than those on standard efficiency and, as a result, have the ability to transfer more heat. This system characteristic allows for the refrigerant in high-efficiency appliances to condense at lower temperatures, thereby reducing the energy usage of the system.

There are also typical electrical operating conditions. Before determining whether the cause of failure is equipment-related, it is important for the technician to determine that the voltage being supplied to the equipment is within an acceptable range, which is typically ±10% of the nameplate rating. If the voltage being supplied to the equipment is not correct, this condition needs to be resolved first.

The gauge manifold indicates the low- and high-side pressures while the air-conditioning system is operating. These pressures can then be converted to the system refrigerant's boiling and condensing temperatures. In addition to the pressure readings, a service technician will need to take various temperature readings in order to effectively troubleshoot a system. These temperature readings include the evaporator inlet air, evaporator outlet air, suction line, discharge line, liquid line, and outside ambient air.

A multimeter that is capable of measuring ohms, amps, voltage, and temperature is the main tool used for electrical troubleshooting. For a split system, there are usually separate circuit breakers in the main panel for the indoor unit and the outdoor unit. For a package system, there is usually one breaker for the unit. Before performing any troubleshooting, be sure that you are aware of all safety practices, and follow them. Not all safety practices are repeated in this manual.

KEY TERMS

Approach temperature
Condenser saturation
  temperature (CST)
Current-sensing lockout
  relay

Electronic self-diagnostic
  feature
Evaporator saturation
  temperature (EST)
Fault isolation diagram
Hopscotching

Installation and service
  instructions
Nominal ratings
Schrader valve
Service valve
Subcooling

Superheat
Temperature difference
  (TD)
Troubleshooting tree

REVIEW TEST

Name	Date	Grade

1. What design conditions are most air-conditioning systems designed for?
   A. Outside temperature 90°F, inside temperature 70°F, relative humidity 50%
   B. Outside temperature 95°F, inside temperature 75°F, relative humidity 50%
   C. Outside temperature 95°F, inside temperature 80°F, relative humidity 50%
   D. Outside temperature 100°F, inside temperature 80°F, relative humidity 60%

2. A standard evaporator will operate with a _____ refrigerant boiling temperature with a 75°F inside air temperature and other conditions being standard.
   A. 40°F
   B. 45°F
   C. 50°F
   D. 55°F

3. If the head pressure increases at the condenser, the suction pressure will:
   A. decrease.
   B. increase.
   C. not be affected.

4. The refrigerant condensing temperature in a standard-efficiency condenser will be _____ above the outside ambient temperature.
   A. 20°F
   B. 25°F
   C. 30°F
   D. 35°F

5. A normal superheat in a standard efficiency system operating under design conditions will be:
   A. 10°F.
   B. 15°F.
   C. 20°F.
   D. 25°F.

6. If the head pressure at the condenser is 297 psig, the condensing temperature will be _____ if the system uses R-22.
   A. 115°F
   B. 120°F

C. 125°F
D. 130°F

7. The head pressure for a high-efficiency condenser will be _____ for a standard condenser under the same conditions.
   A. lower than
   B. higher than
   C. about the same as

8. If the space temperature is higher than standard conditions, the suction pressure will be _____ normal.
   A. lower than
   B. higher than
   C. about the same as

9. If the head pressure at a high-efficiency condenser is 243 psig and the system is using R-22, how high will the condensing temperature be above the ambient of 95°F?
   A. 20°F
   B. 25°F
   C. 30°F
   D. 35°F

10. The more accurate rating system for air-conditioning equipment that takes into consideration operation during the complete cycle is the:
    A. EER.
    B. SEER.
    C. AHRI.
    D. NEC.

11. Which of the following would cause a lower condensing temperature?
    A. Insulating the liquid line
    B. More airflow through the evaporator
    C. A higher ambient temperature
    D. More airflow through the condenser

12. Higher efficiency for a condenser may be accomplished by using:
    A. a smaller condenser.
    B. larger size condenser tubing.
    C. a larger condenser.
    D. a faster compressor.

13. The approximate full load for a 5 hp single-phase motor operating on 230 V would be:
    A. 28 A.   C. 10 A.
    B. 15.2 A.   D. 18 A.

14. If the head pressure at the condenser is 475 psig, the condensing temperature will be _____ if the system uses R-410A.
    A. 115°F   C. 125°F
    B. 120°F   D. 130°F

15. The low-side pressure of an air-conditioning system reads 118 psig at the evaporator. If the air-conditioning system is using R-410A as its refrigerant, what will the evaporating temperature be?
    A. 30°F   C. 40°F
    B. 35°F   D. 45°F

16. The suction or low-side gauge is located on the _____ side of the gauge manifold.
    A. left   B. right

17. Which refrigerant blend is replacing R-22 in air-conditioning applications for new equipment?
    A. R-410A
    B. R-401A
    C. R-402B
    D. R-438A

18. A service valve can be used to:
    A. check the electrical service to the compressor.
    B. isolate the refrigerant so the system can be serviced.
    C. check the airflow across the evaporator.
    D. check the ambient temperature at the condenser.

19. The liquid-line temperature may be used to check the _____ efficiency of the condenser.
    A. superheat
    B. low-side pressure
    C. suction pressure
    D. subcooling

20. A capillary tube is _____ metering device.
    A. an adjustable
    B. a fixed-bore
    C. a thermostatic
    D. a nonpressure-equalizing

21. To accurately add refrigerant to a system, a _____ ambient temperature should be simulated.
    A. 75°F
    B. 85°F
    C. 95°F
    D. 105°F

22. If a short in a compressor motor winding is suspected, which of the following instruments would be used to check it?
    A. Ohmmeter
    B. Ammeter
    C. Voltmeter
    D. Milliammeter

23. Gauge lines should not be connected to a system unless necessary, as:
    A. the Schrader valve may become worn.
    B. the service valve is used only to isolate refrigerant.
    C. most gauge lines leak.
    D. a small amount of refrigerant will leak from the system when the gauge lines are disconnected.

24. A Schrader valve is a:
    A. pressure check port.
    B. low-side valve only.
    C. high-side valve only.
    D. high-efficiency expansion device.

25. A short service connection may be used:
    A. to make quick service calls.
    B. to eliminate the loss of large amounts of liquid while checking the liquid-line pressure.
    C. for heat pumps only.
    D. only in summer.

26. Electronic thermometers are normally used for checking superheat and subcooling because:
    A. they are cheap.
    B. they are easy to use and accurate.
    C. they are all that is available.
    D. none of the above.

27. The common meter for checking the current flow in the compressor electrical circuit is the:
    A. VOM.
    B. clamp-on ammeter.
    C. electronic thermometer.
    D. ohmmeter.

Exercise MST-1

Gauge Installation and Removal on Systems Equipped with Service Valves

Name	Date	Grade

OBJECTIVES: Upon completion of this exercise, you should be able to properly install and remove gauges from an operating air-conditioning or refrigeration system equipped with service valves without contaminating the system and with only minimal loss of system refrigerant.

INTRODUCTION: While servicing air-conditioning and refrigeration equipment, it may be necessary to install a set of gauges on the system in order to obtain the system's operating pressures. During the process of installing or removing gauges, it is important that system refrigerant not be released from the system and that contaminants are not permitted to enter the system. By following a detailed procedure for installing and removing gauges, the chances of releasing refrigerant or contaminating the system are greatly reduced.

TOOLS, MATERIALS, AND EQUIPMENT: Gauge manifold, refrigeration service wrench, adjustable wrench, safety glasses, gloves, and an operating air-conditioning or refrigeration system that is equipped with service valves.

SAFETY PRECAUTIONS: Be sure to protect yourself from refrigerant releases. Contact with released liquid refrigerant can cause frostbite, so wear appropriate personal protection equipment. The system's discharge line can be very hot, so avoid touching it. Check to make certain that the refrigerant hoses are rated appropriately for the system that you will be working on.

STEP-BY-STEP PROCEDURES:

1. Position yourself at the unit as assigned by your instructor.

2. Energize the system, and allow it to operate for 5 minutes.

3. Remove the stem caps from the suction- and liquid-line service valves, and make certain that the valves are fully backseated (stem is turned completely counterclockwise when looking at the end of the stem). If the stem caps are too tight, use an adjustable wrench to loosen the caps.

4. Remove the high- and low-side hoses from the blank ports on the gauge manifold.

5. Inspect the gauges for proper calibration. Calibrate as needed.

6. Make certain that the center hose on the gauge manifold is in place and that the hose connections on the manifold are thumb tight.

7. Remove the service port caps from the suction- and liquid-line service valves.

8. Place the service port caps on the blank ports on the gauge manifold.

9. Connect the high-side hose from the gauge manifold to the liquid-line service port, and connect the low-side hose from the gauge manifold to the suction service port.

10. Using a refrigeration service wrench, crack the liquid service valve off the backseat (half to one turn clockwise from the fully backseated position).

11. Open the high-side and low-side valves on the gauge manifold.

12. Slightly loosen the low-side hose connection on the suction service valve to purge any air from the hoses and gauge manifold. This should take no more than 2 or 3 seconds.

13. Tighten the low-side hose connection on the suction service valve.

14. Briefly loosen the center hose connection at the manifold's blank port to purge any air from the center hose. Retighten the center hose's blank port connection.

15. Close the high-side and low-side valves on the gauge manifold.

16. Crack the suction-line service valve off the backseat.

17. Replace the stem caps (thumb tight) on the liquid-line and suction-line service valves.

18. The gauges have now been properly installed on the system.

19. Record the operating pressures of the system here:

   High-side pressure: _____ psig; Low-side pressure: _____ psig.

20. Have your instructor check your work and verify that the service valves and manifold valves are in the proper position.

21. Remove the stem caps from the liquid-line and suction-line service valves.

22. Backseat the liquid-line service valve.

23. Replace the stem cap (thumb tight) on the liquid-line service valve.

24. Open the high-side and low-side valves on the gauge manifold. The refrigerant in the high-side hose will flow through the manifold and enter the system through the low-side hose connection on the system. Wait until the high-side pressure and the low-side pressure have equalized before moving on to the next step.

   Note: Do not disconnect the hoses from a system if the suction pressure is at or below atmospheric pressure (0 PSIG). De-energize the system, and allow the lines to equalize.

25. Close the high-side and low-side valves on the gauge manifold.

26. Backseat the suction service valve.

27. Replace the stem cap (thumb tight) on the suction-line service valve.

28. If the gauge manifold's hoses are equipped with manually operated low-loss fittings, close the valves on both the high- and low-side hoses. If the low-loss fittings on the hoses are self-sealing, continue to the next step.

29. Remove the high- and low-side hoses from the liquid- and suction-line service valves.

30. Remove the service port caps from the blank ports on the gauge manifold.

31. Place the service port caps on the suction-line and liquid-line service ports.

32. Connect the high-side and low-side hoses to the blank ports on the gauge manifold.

33. The gauges have now been properly removed from the system.

34. Have your instructor check your work and verify that the service valves and manifold valves are in the proper position.

MAINTENANCE OF WORKSTATION AND TOOLS: Make certain that all service valves are fully backseated and that all stem caps are placed back (thumb tight) on the service valves. Make certain that both ends of all refrigerant hoses are connected to the manifold ports to prevent manifold and hose contamination.

SUMMARY STATEMENT: Explain how improperly installing gauges on an air-conditioning or refrigeration system can lead to system contamination. Provide as many examples of system contamination as you can think of as well as its effects on system operation and performance.

TAKEAWAY: The process of installing gauges on and removing gauges from an air-conditioning or refrigeration system should not have a negative effect on the system. Keeping the refrigerant contained in the system, as well as keeping the system contaminant free, should be a top priority for the HVACR service technician.

QUESTIONS

1. Explain why the backseated position is the only service valve position that will result in no refrigerant loss when removing a refrigerant hose from the service port.

2. Explain why it is strongly recommended that the service port caps be placed on the manifold's blank ports when refrigerant hoses are connected to the system's service valves.

3. Explain what can happen if both manifold valves are in the open position and both (liquid-line and suction-line) service valves are cracked off the backseat on an operating air-conditioning or refrigeration system.

4. Explain the importance of replacing the stem caps on the service valves.

5. Explain why it is important to use only a refrigeration service wrench on the stems of service valves.

6. Explain why gauge calibration should only be done when the hoses are disconnected from the blank ports on the gauge manifold.

Exercise MST-2

Evaporator Operating Conditions

Name	Date	Grade

OBJECTIVES: Upon completion of this exercise, you should be able to state typical evaporator pressures and describe the results of reduced airflow across an evaporator.

INTRODUCTION: You will use a gauge manifold and thermometer to measure and then record pressures and temperatures under normal conditions in an evaporator in an air-conditioning system. You will then reduce the airflow and measure and record pressures and temperatures again.

TOOLS, MATERIALS, AND EQUIPMENT: A gauge manifold, thermometer or thermistor, gauge, goggles, gloves, cardboard for blocking the evaporator coil, service valve wrench, straight-slot and Phillips-head screwdrivers, ¼" and ⁵⁄₁₆" nut drivers, and operating air-conditioning system.

⚠️ **SAFETY PRECAUTIONS:** Wear gloves and goggles while connecting and removing gauges. Be careful not to come in contact with any electrical connections when cabinet panels are off.

STEP-BY-STEP PROCEDURES

1. Place a thermometer or thermistor sensor in the supply and return airstreams.

2. With gloves and goggles on, fasten the low- and high-side gauges to the gauge ports, and purge any air from the hoses.

3. Start the unit, and record the following after the unit and conditions have stabilized:

- Suction pressure: _____ psig

- Evaporator saturation temperature (EST): _____ °F

- Inlet air temperature: _____ °F

- Outlet air temperature: _____ °F

- Discharge pressure: _____ psig

4. Reduce the airflow across the evaporator to approximately one-fourth by sliding the cardboard into the filter rack. If there is a single return inlet, the cardboard may be placed across three-fourths of the opening.

5. When the unit stabilizes, record the following:

- Suction pressure: _____ psig

- Evaporator saturation temperature (EST): _____ °F

- Inlet air temperature: _____ °F

- Outlet air temperature: _____ °F

- Discharge pressure: _____ psig

6. Remove the air blockage , disconnect the gauges, replace the service port caps, and return the unit to normal operation.

MAINTENANCE OF WORKSTATION AND TOOLS: Replace all panels with proper fasteners. Return all tools to their correct places.

SUMMARY STATEMENT: Describe how the unit suction pressure and air temperature differences responded. Use an additional sheet of paper if more space is necessary for your answer.

TAKEAWAY: Technicians should be familiar with the typical operating conditions of a system that they are charged with troubleshooting. It is just as important to recognize proper operation as it is to recognize improper operation.

QUESTIONS

1. How does a fixed-bore expansion device respond to a reduction in load, such as when the filter is restricted?

2. How does a thermostatic expansion valve respond to reduction in load, such as when the filter is restricted?

3. What are the results of small amounts of liquid reaching a hermetic compressor?

4. How does the temperature drop across an evaporator respond to a reduced airflow?

5. What happens to the head pressure when a reduced airflow is experienced at the evaporator?

6. List four things that can happen to reduce the airflow across an evaporator.

7. What procedure may be used to thaw a frozen evaporator?

8. What would happen if an air conditioner were operated for a long period of time without air filters?

9. How may an evaporator coil be cleaned?

Exercise MST-3

Condenser Operating Conditions

Name	Date	Grade

OBJECTIVES: Upon completion of this exercise, you should be able to describe how a condenser responds to reduced airflow.

INTRODUCTION: You will measure and record a typical condenser's refrigerant pressure and temperature, the condenser entering airstream temperature, and the compressor amperage under normal operating conditions. You will then reduce the airflow across the condenser, measure and record the same conditions, and compare the results.

TOOLS, MATERIALS, AND EQUIPMENT: A thermometer or thermistor, a clamp-on ammeter, a gauge manifold, goggles, gloves, sheet plastic or cardboard to reduce the airflow through the condenser, a service valve wrench, straight-slot and Phillips-head screwdrivers, ¼" and ⁵⁄₁₆" nut drivers, and an operating air-conditioning system.

⚠ **SAFETY PRECAUTIONS:** Wear goggles and gloves while attaching and removing gauge lines. Be very careful not to come in contact with any electrical terminals or moving fan blades.

STEP-BY-STEP PROCEDURES

1. With goggles and gloves on, fasten the gauge lines to the gauge ports on an air-conditioning system, and purge the air from the hoses.

2. With the unit off, fasten one thermometer or thermistor lead to the discharge line on the compressor, and place the other in the airstream entering the condenser. Replace the compressor compartment panel if the compressor compartment is not isolated from the condenser.

3. Clamp the ammeter on the compressor common terminal wire.

4. Start the unit and allow it to run until stabilized (when the temperature of the discharge line does not change). This may take 10 minutes or more. Record the following information:

   - Suction pressure: _____ psig

   - Evaporator saturation temperature (EST): _____ °F

   - Discharge pressure: _____ psig

   - Condenser saturation temperature (CST): _____ °F

   - Compressor current: _____ A

5. Cover part of the condenser coil with the plastic or cardboard until the condenser saturation temperature rises to about 130°F, 300 psig for R-22 or 478 psig for R410A. Allow the unit to run until it stabilizes, about 10 minutes. Then record the following information:

   - Suction pressure: _____ psig

   - Evaporator saturation temperature (EST): _____ °F

   - Discharge pressure: _____ psig

   - Condenser saturation temperature (CST): _____ °F

   - Compressor current: _____ A

6. Remove the cover and allow the unit to run for about 5 minutes to reduce the head pressure.

7. Shut the unit off, and remove the gauges.

8. Replace all panels with the correct fasteners. Return the unit to its normal condition.

MAINTENANCE OF WORKSTATION AND TOOLS: Return all tools to their places.

SUMMARY STATEMENT: Describe how and why the pressures and temperatures responded to the reduced airflow across the condenser. Use an additional sheet of paper if more space is necessary for your answer.

TAKEAWAY: Technicians should be familiar with the typical operating conditions of a system that they are charged with troubleshooting. Low condenser airflow will result in a higher head pressure and condensing temperature. It is just as important to recognize proper operation as it is to recognize improper operation.

QUESTIONS

1. Does the compressor work more or less efficiently with an increased head pressure?

2. What does the discharge-line temperature do when an increase in head pressure is experienced?

3. What would a very high discharge-line temperature do to the oil circulating in the system?

4. Name four things that would cause a high discharge pressure.

5. How does the compressor amperage react to an increase in discharge pressure?

6. Would the discharge pressure be more or less for a high-efficiency unit when compared to a standard-efficiency unit?

7. How is high efficiency accomplished?

8. Is a high-efficiency unit more or less expensive to purchase?

9. Why do many manufacturers offer both standard- and high-efficiency units?

Manufacturer's Charging Procedure

Name	Date	Grade

OBJECTIVES: Upon completion of this exercise, you should be able to use a manufacturer's charging chart to check for the correct operating charge for a central air-conditioning system.

INTRODUCTION: You will use a manufacturer's charging chart to determine a suggested charge and charging procedures for maximum efficiency in a central air-conditioning system in outside temperatures above 65°F. This unit may have any type of metering device.

TOOLS, MATERIALS, AND EQUIPMENT: A thermometer or thermistor, a psychrometer, a gauge manifold, two 8" adjustable wrenches, a service valve wrench, straight-slot and Phillips-head screwdrivers, ¼" and ⁵⁄₁₆" nut drivers, light gloves, goggles, tape, insulation (to hold the temperature sensor and insulate it), and an air-conditioning system with the manufacturer's charging chart.

Note: If the manufacturer's literature is not available, a superheat/subcooling calculator or charging app on a smart device can be used.

⚠ **SAFETY PRECAUTIONS:** Wear gloves and goggles while fastening gauge lines and removing them from the gauge ports. Be very careful around rotating components.

STEP-BY-STEP PROCEDURES

1. Review the manufacturer's procedures for charging. They may call for a superheat check using charts in the provided literature or on the unit.

   Note: All manufacturers' literature specifies that the indoor airflow be correct for their charging procedures to work. If the readings you get are off more than 10% and cannot be corrected by adjusting the refrigerant charge, suspect the airflow.

2. With goggles and gloves on, fasten the gauge lines to the gauge ports (Schrader ports or service valve ports), and purge the air from the hoses. Fasten thermometers to both the suction and discharge lines, and insulate properly.

3. Start the unit and allow it to run for at least 10 minutes.

4. While the unit is stabilizing, prepare to take and record the following information:

   - Suction pressure reading: _____ psig
   - Suction line temperature: _____ °F
   - Evaporator saturation temperature, EST (converted from suction pressure): _____ °F
   - Superheat (suction line temperature–evaporator saturation temperature): _____ °F
   - Return-air temperature: _____ °F
   - Evaporator split (return-air temperature–evaporator saturation temperature): _____ °F
   - Discharge pressure: _____ psig
   - Discharge line temperature: _____ °F
   - Condenser saturation temperature, CST (converted from discharge pressure): _____ °F

- Subcooling (condenser saturation temperature–discharge line temperature): _____ °F
- Condenser entering air temperature: _____ °F
- Condenser split (condenser saturation temperature–outdoor air temperature): _____ °F
- Indoor wet-bulb: _____ °F, dry-bulb: _____ °F
- Outdoor dry-bulb: _____ °F

5. If the unit has the correct charge because it corresponds to the manufacturer's chart, you may shut the unit off and disconnect the gauges.

6. If it does not, under your instructor's supervision, adjust the charge to correspond to the chart. Then shut the unit off, and disconnect the gauges.

7. Replace all panels with the correct fasteners.

MAINTENANCE OF WORKSTATION AND TOOLS: Return all tools to their respective places, and be sure your workstation is in order.

SUMMARY STATEMENT: Describe how the wet-bulb temperature reading affects the load on the evaporator.

TAKEAWAY: When troubleshooting a system that is functioning electrically but not cooling, check the charge. An over- or undercharged system may cool somewhat, but not properly. Always refer to the manufacturer's recommended charging procedures when available.

QUESTIONS

1. What would a typical superheat reading be at the outlet of an evaporator on a design temperature day for a central air-conditioning system?

2. How does the outdoor ambient temperature affect the head pressure?

3. What is the recommended airflow per ton of air conditioning for the unit you worked on?

4. What would the result of too little airflow be on the readings you took?

5. What type of metering device did the unit you worked with have?

6. Did you have to adjust the charge on the unit you worked with?

7. How would an overcharge affect a unit with a capillary tube?

8. Would the overcharge in question 7 be evident in mild weather?

Exercise MST-5

Charging a System in the Field Using Typical Conditions (Fixed-Bore Metering Device)

Name	Date	Grade

OBJECTIVES: Upon completion of this exercise, you should be able to correctly charge a system using a fixed-bore metering device when no manufacturer's literature is available.

INTRODUCTION: You will check the charge in a central air-conditioning system under field conditions when no manufacturer's literature is available. This system must have a fixed-bore metering device. Always use manufacturer's literature when it is available. The following procedure may be done in any outdoor ambient temperature; however, the colder it is, the more condenser cover you will need.

TOOLS, MATERIALS, AND EQUIPMENT: A thermometer or thermistor, digital psychrometer, gauge manifold, tape, insulation, gloves, goggles, two 8" adjustable wrenches, service valve wrench, plastic or cardboard to restrict airflow to the condenser, straight-slot and Phillips-head screwdrivers, ¼" and ⁵⁄₁₆" nut drivers, cylinder of refrigerant that is used in the system, superheat calculator or charging app on a smart device, and central air-conditioning system with a fixed-bore metering device.

⚠ **SAFETY PRECAUTIONS:** Wear gloves and goggles when attaching and removing gauges. Do not let the condenser saturation temperature rise above 140°F, about 350 psig for R-22 and 540 psig for R-410A. Be careful of rotating parts and electrical terminals and connections.

STEP-BY-STEP PROCEDURES

1. With goggles and gloves on, fasten the gauges to the gauge ports.

2. Start the unit, and allow it to run for at least 30 minutes to stabilize. During this time, fasten the thermometer or thermistor to the suction line at the condensing unit, and insulate it.

3. Observe the pressures, and cover the condenser to increase the condenser saturation temperature to 125°F, about 275 psig for R-22 and 448 psig for R-410A, if necessary. Keep track of it, and do not let it rise above 125°F saturation.

   Note: If it will not rise to a saturation temperature of 125°F, the unit may not have enough charge. Check with your instructor.

   The total superheat, or the superheat the compressor sees, should typically be 10°F to 15°F at the condensing unit, with a correct charge and an average-length suction line (see example, Figure MST-5.1. Notice that 5°F superheat was gained between the air handler and the condensing unit in this figure). A superheat calculator or charging app on a smart device should be used to verify the recommended superheat for your conditions.

FIGURE **MST-5.1**

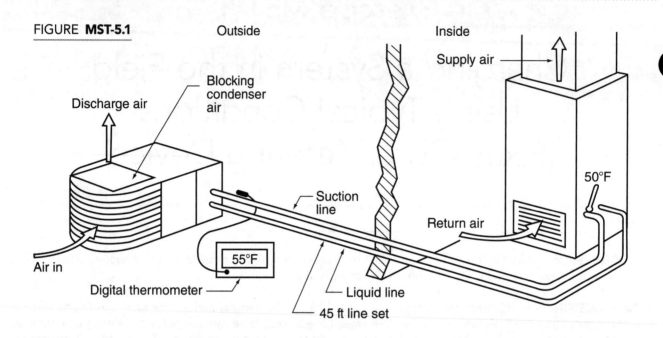

4. If the charge is incorrect, add refrigerant to lower the superheat, or remove refrigerant to raise it. Do not add refrigerant too fast or you will overcharge the system. It is easier to adjust the charge while adding refrigerant than while removing it.

5. When the charge is correct, record the following:

   • Suction pressure: _____ psig

   • Suction-line temperature: _____ °F

   • Evaporator saturation temperature, EST (converted from suction pressure): _____ °F

   • Superheat (suction-line temperature–evaporator saturation temperature): _____ °F

   • Recommended superheat (from superheat calculator or charging app): _____ °F

   • Discharge pressure: _____ psig

   • Condenser saturation temperature, CST (converted from discharge pressure): _____ °F

6. When you have adjusted the charge, remove the head pressure control (cardboard or plastic), and allow the unit to stabilize. Then record the following information:

   • Suction pressure: _____ psig

   • Suction-line temperature: _____ °F

   • Evaporator saturation temperature, EST (converted from suction pressure): _____ °F

   • Superheat (suction-line temperature–evaporator saturation temperature): _____ °F

   • Discharge pressure: _____ psig

   • Condenser saturation temperature, CST (converted from discharge pressure): _____ °F

7. Shut the unit off, remove the temperature sensors and gauges, and replace all panels with the correct fasteners.

MAINTENANCE OF WORKSTATION AND TOOLS: Return all tools, equipment, and materials to their storage places. Remember to put the caps back on the service valves.

SUMMARY STATEMENT: Describe how head pressure affects superheat with a fixed-bore metering device. Use an additional sheet of paper if more space is necessary for your answer.

TAKEAWAY: When servicing an air-conditioning system, the manufacturer's literature may not always be available. Technicians should be able to field-charge a system in a variety of conditions. Fixed-bore systems should be charged using the superheat method. Technicians should be able to use a superheat calculator or charging app on a smart device to determine the correct charge.

QUESTIONS

1. How does line length affect refrigerant charge?

2. Why is a 125°F condensing temperature used instead of some other value?

3. What are some typical refrigerants used for central air conditioning?

4. In a troubleshooting scenario, what symptoms would an undercharged system exhibit?

5. In a troubleshooting scenario, what symptoms would an overcharged system exhibit?

Exercise MST-6

Charging a System in the Field Using Typical Conditions (Thermostatic Expansion Valve)

Name	Date	Grade

OBJECTIVES: Upon completion of this exercise, you should be able to correctly charge a system using a thermostatic expansion valve metering device when no manufacturer's literature is available.

INTRODUCTION: You will use the subcooling method to ensure the correct charge for the unit. This procedure is to be used when there is no manufacturer's literature available and may be performed in any outdoor ambient temperature. A unit with R-22 is suggested because it is the most common refrigerant; however, a unit using another refrigerant may be used.

TOOLS, MATERIALS, AND EQUIPMENT: A thermometer or thermistor, a gauge manifold, a tank of refrigerant that is used in the system, goggles, light gloves, plastic or cardboard (to restrict airflow to the condenser), two 8" adjustable wrenches, a service valve wrench, straight-slot and Phillips-head screwdrivers, ¼" and ⁵⁄₁₆" nut drivers, tape, insulation for the temperature sensor, a subcooling calculator or charging app on a smart device, and an operating air conditioner equipped with a TXV as a metering device.

⚠ **SAFETY PRECAUTIONS:** Wear gloves and goggles while fastening and removing gauges. Do not let the condenser saturation temperature rise above 140°F. Be careful of rotating components and electrical terminals and connections.

STEP-BY-STEP PROCEDURES

1. Fasten the gauge lines to the test ports (Schrader or service valve ports), and purge any air from the hoses. Locate the sight glass in the liquid line if there is one.

2. Start the unit. While it is stabilizing, fasten the thermometer or thermistor to the liquid line at the condensing unit, and insulate it.

3. Place the condenser airflow restrictor in place, and allow the condenser saturation temperature of 125°F, about 275 psig for R-22 and 448 psig for R-410A. Watch closely, particularly if the weather is warm, to make sure that the saturation temperature stays around 125°F. If it will not reach a 125°F saturation temperature, look to see whether there are bubbles in the sight glass. The unit is probably undercharged, Figure MST-6.1.

4. Record the following information:

   - Suction pressure: _____ psig

   - Discharge pressure: _____ psig

   - Condenser saturation temperature, CST (converted from head pressure): _____ °F

   - Liquid-line temperature: _____ °F

   - Subcooling (condenser saturation temperature–liquid-line temperature): _____ °F

FIGURE **MST-6.1**

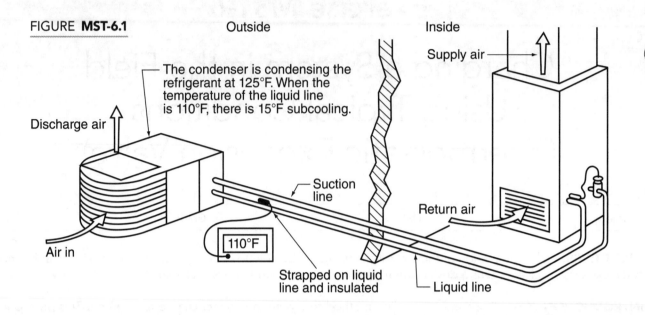

Outside   Inside

The condenser is condensing the refrigerant at 125°F. When the temperature of the liquid line is 110°F, there is 15°F subcooling.

Discharge air

Air in

Suction line

110°F

Strapped on liquid line and insulated

Supply air

Return air

Liquid line

5. Using a subcooling calculator or charging app on a smart device, determine the correct subcooling for your system and conditions. Record it here: _____ °F

6. If the charge is incorrect, adjust it to the proper level (using the recommended subcooling), and record the following:

   - Suction pressure: _____ psig

   - Discharge pressure: _____ psig

   - Condenser saturation temperature, CST (converted from head pressure): _____ °F

   - Liquid-line temperature: _____ °F

   - Subcooling (condenser saturation temperature–liquid-line temperature): _____ °F

7. When the charge is correct, remove the temperature sensor and gauges and replace all panels with the correct fasteners. Remove the restrictions from the condenser, and allow the head pressure to stabilize.

MAINTENANCE OF WORKSTATION AND TOOLS: Return all tools, equipment, and materials to their storage places, and make sure your workstation is clean and in order.

SUMMARY STATEMENT: Describe subcooling and where it takes place. Use an additional sheet of paper if more space is necessary for your answer.

TAKEAWAY: Systems that utilize a thermostatic expansion valve as a metering device should be field charged using the subcooling method. When the literature is unavailable, technicians should be able to use technology to find the relevant information.

QUESTIONS

1. If the unit you worked on had a sight glass, did it show bubbles when the charge was correct? What caused the bubbles?

2. Where is the correct location for a sight glass in a liquid line in order to check the charge?

3. How is subcooling different from superheat?

4. Does superheat ever occur in the condenser? Why or why not?

5. Why do you raise the condenser saturation temperature of 125°F to check the subcooling of the condenser in this exercise?

6. Why do you insulate the temperature lead when checking refrigerant-line temperatures?

Troubleshooting Exercise: Central Air Conditioning with Fixed-Bore Metering Device

Name	Date	Grade

OBJECTIVES: Upon completion of this exercise, you should be able to troubleshoot a basic mechanical problem in a typical central air-conditioning system.

INTRODUCTION: This exercise should have a specific problem introduced by your instructor. You will locate the problem using a logical approach and instruments. Instruments may be used to arrive at the final diagnosis; however, you should be able to determine the problem without instruments. THE INTENT IS NOT FOR YOU TO MAKE THE REPAIR BUT TO LOCATE THE PROBLEM AND ASK YOUR INSTRUCTOR FOR DIRECTION BEFORE COMPLETION.

TOOLS, MATERIALS, AND EQUIPMENT: Straight-slot and Phillips-head screwdrivers, ¼" and ⁵⁄₁₆" nut drivers, a thermometer or thermistor, two 8" adjustable wrenches, goggles, gloves, and a set of gauges.

 SAFETY PRECAUTIONS: Use care while connecting gauge lines to the liquid line. Wear goggles and gloves.

STEP-BY-STEP PROCEDURES

1. Start the system, and allow it to run for 15 minutes.

2. Touch-test the suction line leaving the evaporator. HOLD THE LINE TIGHT FOR TOUCH TESTING. Record your opinion of the suction-line condition here:

3. Touch-test the suction line entering the condensing unit, and record your impression here:

4. Remove the compressor compartment door. MAKE SURE THAT AIR IS STILL FLOWING OVER THE CONDENSER. IF AIR PULLS IN OVER THE COMPRESSOR WITHOUT A PARTITION, HIGH HEAD PRESSURE WILL OCCUR VERY QUICKLY. COMPLETE THE FOLLOWING VERY QUICKLY, AND SHUT THE DOOR. Touch-test the compressor body all over. Record the following:

   A. The temperature of the top and sides of the compressor in relation to hand temperature: _____

   B. The temperature of the compressor crankcase compared to your hand temperature: _____

5. All of these touch-test points should be much colder than hand temperature if the system has the correct problem. Record your opinion of the system symptoms so far:

6. Fasten gauges to the gauge ports, and purge any air from the hoses.

7. Fasten a thermometer or thermistor lead to the suction line at the condensing unit.

8. Record the superheat here: _____ °F

9. Place a thermometer or thermistor lead in the inlet and outlet airstreams to the evaporator and record their readings here:

- Inlet air temperature: _____ °F

- Outlet air temperature: _____ °F

- Temperature difference: _____ °F

10. What should the temperature difference be? _____ °F

11. Describe your opinion of the problem and the recommended repair.

12. Ask your instructor before making the repair; then do as instructed.

MAINTENANCE OF WORKSTATION AND TOOLS: Leave the unit as directed, and return all tools to their respective places.

SUMMARY STATEMENT: Describe how to determine the problem with this unit by the touch test.

TAKEAWAY: When a technician has a working knowledge of typical system operating conditions, observations such as the touch test can lead to a general diagnosis of the problem. Instruments should be used to verify the diagnosis.

QUESTIONS

1. How should the touch test of the evaporator suction line feel when a unit has an overcharge of refrigerant?

2. How would the compressor housing feel when a unit has an overcharge of refrigerant?

3. How would the suction line feel when a unit has a restricted airflow?

4. How would the compressor housing feel when a unit has a restricted airflow?

5. How can you tell the difference between an overcharge and a restricted airflow by the touch test?

6. Why should gauges not be fastened to the system until the touch test is completed?

Troubleshooting Exercise: Central Air Conditioning with Thermostatic Expansion Valve

Name	Date	Grade

OBJECTIVES: Upon completion of this exercise, you should be able to troubleshoot a mechanical problem in a central air-conditioning system with a thermostatic expansion valve metering device.

INTRODUCTION: This exercise should have a specific problem introduced by your instructor. You will locate the problem using a logical approach and instruments. The problem in this exercise will require instruments to properly solve and diagnose. THE INTENT IS NOT FOR YOU TO MAKE THE REPAIR BUT TO LOCATE THE PROBLEM AND ASK YOUR INSTRUCTOR FOR DIRECTION BEFORE COMPLETION.

TOOLS, MATERIALS, AND EQUIPMENT: Straight-slot and Phillips-head screwdrivers, ¼" and ⁵⁄₁₆" nut drivers, gloves, goggles, two 8" adjustable wrenches, a thermometer or thermistor, a cylinder of refrigerant, and a gauge manifold.

⚠ SAFETY PRECAUTIONS: Use care while fastening gauge lines to the liquid line. Wear gloves and goggles.

STEP-BY-STEP PROCEDURES

1. Start the system, and allow it to run for 15 minutes.

2. Touch-test the suction line leaving the evaporator. HOLD THE LINE TIGHTLY IN YOUR HAND. Record the temperature of the line in comparison to hand temperature: _____

3. Touch-test the suction line entering the condensing unit. Record the temperature of the line in comparison to hand temperature: _____

4. Remove the compressor compartment door for the purpose of touch-testing the compressor.

   Note: If the compressor compartment is not isolated, this test must be performed quickly or else the head pressure will become too high due to air not flowing over the condenser. Look the system over before leaving the door off too long.

   Touch-test the compressor and record:

   A. The top and sides of the compressor housing compared to hand temperature: _____

   B. The bottom (crankcase) of the compressor compared to hand temperature: _____

   Replace the door quickly if air is bypassing the condenser.

5. Using the touch test for diagnosis, what is your opinion of the problem with this unit so far?

6. Fasten gauges to the gauge ports, and purge any air from the hoses.

7. Place a thermometer or thermistor lead on the suction line entering the condensing unit, and record the superheat here: _____ °F.

8. Take the temperature difference across the evaporator by placing a thermometer or thermistor lead in the inlet and outlet air.

   • Inlet air temperature: _____ °F

   • Outlet air temperature: _____ °F

   • Temperature difference: _____ °F

9. What should the temperature difference be? _____ °F

10. Record your opinion of the unit problem and your recommended repair:

11. Consult your instructor as to what to do next.

MAINTENANCE OF WORKSTATION AND TOOLS: Return the unit to the condition your instructor directs you to, and return all tools to their respective places.

SUMMARY STATEMENT: Describe the typical operation of a thermostatic expansion valve and how it functions in a residential air-conditioning refrigerant circuit. Use an additional sheet of paper if more space is necessary for your answer.

TAKEAWAY: Systems that utilize a TXV metering device will respond differently to changes in operating conditions than a fixed-bore system. With a working knowledge of both types of systems, observations such as the touch test can lead to a general diagnosis of the problem. Instruments should be used to verify the diagnosis.

QUESTIONS

1. How does liquid refrigerant affect the oil in the crankcase of a compressor?

2. What is the difference between liquid slugging in a compressor and a small amount of liquid refrigerant entering the compressor crankcase?

3. Do most compressors have oil pumps?

4. How is a compressor lubricated if it does not have an oil pump?

5. Describe how a suction line feels if liquid refrigerant is moving toward the compressor.

6. Do all central air-conditioning systems use fixed-bore metering devices?

7. Describe how a TXV metering device differs from a fixed-bore metering device.

Changing a Hermetic Compressor

Name	Date	Grade

OBJECTIVES: Upon completion of this exercise, you should be able to change a compressor in a typical air-conditioning or refrigeration system.

INTRODUCTION: You will recover the refrigerant from a system; then you will completely remove a hermetic compressor from the unit and replace it as though you had a new compressor. You will then leak-check, evacuate, and charge the unit with the correct charge.

TOOLS, MATERIALS, AND EQUIPMENT: Straight-slot and Phillips-head screwdrivers, an adjustable wrench, ¼" and ⁵⁄₁₆" nut drivers, a torch setup suitable for this exercise, solder, flux, slip-joint pliers, a gauge manifold, refrigerant recovery equipment, a vacuum pump, a micron gauge, a nitrogen tank and regulator, a digital multimeter with an amp clamp, a tank of refrigerant that is used in the system, a leak detector, gloves and goggles, a set of socket wrenches, needle-nose pliers, a superheat/subcooling calculator or charging app on a smart device, and an operating air-conditioning or refrigeration system.

⚠ **SAFETY PRECAUTIONS:** Use care while working with liquid refrigerant. When using torches, take care not to overheat any connections or damage any nearby components or wiring. When brazing refrigerant lines, be sure to use a low-pressure nitrogen purge. Turn all power off before starting to remove the compressor. Lock and tag the panel or disconnect box where the power is turned off. There should be a single key; keep it on your person.

STEP-BY-STEP PROCEDURES

1. Put on your safety goggles and gloves. Recover the refrigerant from the system using an approved recovery system.

2. With the refrigerant out of the system, TURN OFF AND LOCK OUT THE POWER.

3. Remove the door to the compressor compartment.

4. Remove the compressor mounting bolts or nuts.

5. Mark all electrical connections before removing. USE CARE REMOVING THE COMPRESSOR SPADE TERMINALS, AND DO NOT PULL ON THE WIRE. USE THE NEEDLE-NOSE PLIERS, AND ONLY HOLD THE CONNECTOR.

6. Disconnect the compressor refrigerant lines. You would normally use the torch to heat the connections and remove them. Your instructor may instruct you to cut the compressor suction and discharge lines, leaving some stubs for later exercises.

7. Completely remove the compressor from the condenser section. You now have the equivalent of a new compressor, ready for replacement.

8. Place the compressor on the mounts.

9. Reconnect the compressor suction and discharge lines using whatever method that your instructor has determined to be proper for this exercise.

10. When the refrigerant lines are reconnected, pressurize the system with nitrogen, and perform a leak check.

    Note: Your instructor may want you to add a filter-drier. This would be normal practice with a real changeout.

11. Using the vacuum pump and micron gauge, evacuate the system. While the evacuation is taking place, fasten the motor terminal connectors and the motor mounts.

12. When evacuation is complete, add enough refrigerant to the system so that it can be started.

13. Place the amp clamp around the common wire of the compressor and the meter leads across the line side of the contactor, and turn the power on.

14. Ask your instructor to look at the system; then start the compressor and observe the amperage and voltage. Compare with those listed on the nameplate.

15. Let the unit run for 10 minutes, and check the charge by using a superheat/subcooling calculator or charging app on your smart device.

16. When you are satisfied all is well, turn the unit off, and remove the gauges.

MAINTENANCE OF WORKSTATION AND TOOLS: Replace all panels, place the caps back on the system service valves, and return the tools to their places.

SUMMARY STATEMENT: Describe the steps of triple evacuation.

TAKEAWAY: During a component changeout, such as a compressor, the refrigerant should always be recovered prior to opening the system. A new filter-drier should be installed and a thorough leak check should be performed upon completion of the repair. An evacuation to the recommended level (usually 500 microns) should be done prior to charging the system.

QUESTIONS

1. Why is it necessary to recover the refrigerant rather than just release it to the atmosphere?

2. Why should you always check the voltage and amperage after a compressor changeout?

3. Should you always add a filter-drier when changing a compressor?

4. What type of compressor overload did this unit have?

5. What method did you use to measure the charge back into the system?

6. Why is system evacuation necessary?

7. What solder or brazing material did you use to replace the compressor?

Exercise MST-10

Changing an Evaporator in an Air-Conditioning System

Name	Date	Grade

OBJECTIVES: Upon completion of this exercise, you should be able to change an evaporator on a central air-conditioning system.

INTRODUCTION: You will recover the refrigerant from a central air-conditioning system and change the evaporator coil. You will pressure check and evacuate the system, then add the correct charge.

TOOLS, MATERIALS, AND EQUIPMENT: Straight-slot and Phillips-head screwdrivers, refrigerant recovery system, vacuum pump, micron gauge, nitrogen tank and regulator, ¼" and ⁵⁄₁₆" nut drivers, two 8"adjustable wrenches, set of sockets, gloves, goggles, liquid-line drier, adjustable wrench, cylinder of refrigerant, gauges, superheat/subcooling calculator or charging app on a smart device, and operating air-conditioning system.

⚠ **SAFETY PRECAUTIONS:** Wear goggles and gloves while transferring liquid refrigerant. If torches are used, take care not to overheat the connections, any nearby components, or wiring. When brazing refrigerant lines, be sure to use a low-pressure nitrogen purge.

STEP-BY-STEP PROCEDURES

1. Put on your goggles and gloves, fasten gauges to the gauge ports, and purge the gauge lines.

2. Recover the refrigerant into an approved recovery system.

3. Turn the electrical power off, and lock and tag the disconnect box. Using either torches or a tubing cutter, disconnect the liquid and vapor lines to the evaporator

4. Remove the condensate drain line. Plan out how this can be done so it can be replaced leak-free.

5. Remove the evaporator door or cover.

6. Slide the evaporator out.

7. Examine the evaporator for any problems. Look at the tube turns. Clean the evaporator if needed. IF YOU CLEAN IT, BE CAREFUL NOT TO ALLOW WATER TO ENTER THE COIL. TAPE THE ENDS.

8. Examine the condensate drain pan for rust and possible leakage. You may want to clean and seal the drain pan while it is out.

9. Examine the coil housing for air leaks. Seal them from the inside if the coil housing is in a positive pressure to be sure it is leak-free.

10. When you are sure all is well, slide the coil back into its housing.

11. Reconnect the suction and liquid lines using whichever method your instructor has determined to be proper for this exercise.

12. Purge the system with nitrogen; then install the liquid-line drier.

13. Pressurize the system with nitrogen for leak-checking purposes, and leak-check all connections.

14. Using a vacuum pump and micron gauge, evacuate the system down to the proper level.

15. While evacuation is taking place, install the condensate drain line, and replace the coil cover.

16. When evacuation is complete, add enough refrigerant to the system so that it may be started.

17. With your instructor's permission, start the system.

18. Allow the system to run for at least 10 minutes; then check the charge using either a superheat/subcooling calculator or a charging app on your smart device. Add refrigerant as needed.

19. Turn the system off.

MAINTENANCE OF WORKSTATION AND TOOLS: Return the system to the condition the instructor directs you to, and return all tools to their places.

SUMMARY STATEMENT: Describe why the system was purged before adding the liquid-line drier.

TAKEAWAY: During a component changeout, such as an evaporator, the refrigerant should always be recovered prior to opening the system. A new filter-drier should be installed, and a thorough leak-check should be performed upon completion of the repair. An evacuation to the recommended level (usually 500 microns) should be done prior to charging the system.

QUESTIONS

1. What will a liquid-line filter-drier remove from the system?

2. What will the vacuum pump remove from the system?

3. Why is it recommended to measure the charge into the system?

4. Why would an evaporator ever need changing?

5. Is it necessary to remove an evaporator to clean it properly if it is really dirty?

6. Name three styles of evaporator.

7. What type of tubes did the evaporator have, copper or aluminum?

8. What material was the condensate drain line made of?

Exercise MST-11

Troubleshooting Refrigeration, Heating, or Air-Conditioning Systems

Name	Date	Grade

OBJECTIVES: Upon completion of this exercise, you should be able to correctly troubleshoot basic air-conditioning problems using a schematic diagram.

INTRODUCTION: One of the most important jobs of air-conditioning technicians is to be able to locate and repair a system problem and return the system to normal operation. Approximately 80% of the problems in air-conditioning and heating systems will be electrical. Service technicians must be able to use electrical meters and read schematic diagrams in order to troubleshoot HVACR systems.

TOOLS, MATERIALS, AND EQUIPMENT: Pencil and a temperature/pressure chart.

⚠ **SAFETY PRECAUTIONS:** There are no safety precautions for this exercise.

STEP-BY-STEP PROCEDURES
Troubleshooting Problems

1. You answer a complaint of "no cooling." The condenser fan motor and evaporator fan are operating. You check the schematic diagram (see Figure MST-11.1) and decide that the problem is:

   A. an open internal overload in the compressor.

   B. blown fuses in the disconnect.

   C. a bad transformer.

   D. a bad thermostat.

2. You answer a complaint of "not enough cooling." All unit components are operating, but after about 5 minutes, the compressor cuts out. You check the pressure of the refrigeration system and find that the discharge pressure is 160 psig and the suction pressure is 20 psig. The refrigerant in the system is R-22. You check the schematic diagram (see Figure MST-11.1) and decide that the problem is:

   A. a bad transformer.

   B. a bad thermostat.

   C. an open low-pressure switch.

   D. an open high-pressure switch.

3. You answer a complaint of "no cooling." The compressor and condenser fan motor are operating, but the indoor fan motor is not (refer to MST-11.1). The probable cause is:

   A. a bad indoor fan motor.

   B. a bad transformer.

C. a bad indoor fan relay.

D. Both A and C.

FIGURE **MST-11.1** *(Courtesy of York International Corporation)*

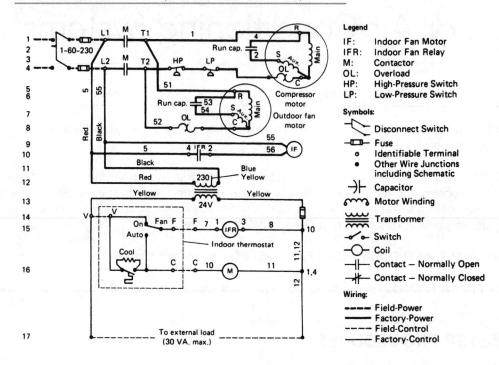

4. You answer a complaint of "no heating." The compressor and outdoor fan motor are not operating but are good. The indoor fan is operating properly. After checking the schematic diagram (see Figures MST-11.2 and MST-11.3), you determine that the problem is a bad:

A. transformer.

B. thermostat.

C. MS contactor.

D. supplementary heater.

5. You answer a complaint of "insufficient heating." The compressor, the condenser fan motor, and indoor fan motor are operating, and there is a heavy coating of ice on the outdoor unit coil. The defrost thermostat is closed. (See Figure MST-11.2 for the schematic diagram.) The probable cause is:

A. a bad defrost timer.

B. a bad transformer.

C. a bad defrost relay.

D. Both A and C.

FIGURE **MST-11.2** *(Courtesy of The Trane Company)*

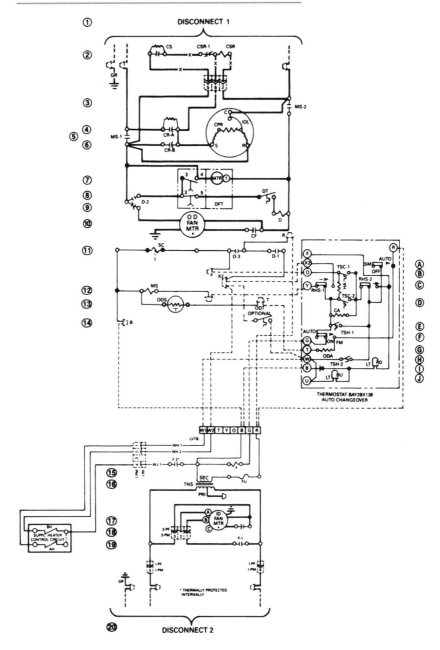

FIGURE **MST-11.3** Legend of Figure MST-11.2

Legend

AH:	Supplementary heat contactor	(19)	LT:	Light	
BH:	Supplementary heat contactor	(18)	LVTB:	Low-voltage terminal board	
CA:	Cooling anticipator		MS:	Compressor motor contactor	(5,3, & 12)
CR:	Run capacitor	(4 & 6)	MTR:	Motor	
CPR:	Compressor		ODA:	Outdoor temperature anticipator	(G)
D:	Defrost relay	(9)	ODS:	Outdoor temperature sensor	(13)
DFT:	Defrost timer	(7)	ODT:	Outdoor thermostat	
DT:	Defrost termination thermostat	(8)	RHS:	Resistance heat switch	(C)
F:	Indoor fan relay	(15)	SC:	Switchover valve solenoid	(11)
FM:	Manual fan switch	(F)	SM:	System switch	(A)
HA:	Heating anticipator		TNS:	Transformer	(16)
HTR:	Heater		TSC:	Cooling thermostat	(B & D)
IOL:	Internal overload protection		TSH:	Heating thermostat	(E & H)

6. You answer a complaint of "no heating." The combustion chamber and blower section are extremely hot, and the LC is open. The blower motor is good, and there is no restriction in the supply air distribution system. The blower relay contacts and coil are good. The schematic diagram of the unit is shown in Figure MST-11.4. The probable cause is a bad:

A. gas valve.

B. pilot relight control.

C. flame rollout switch.

D. fan cycle control.

FIGURE **MST-11.4** *(Courtesy of Rheem Air Conditioning Division, Fort Smith, AR)*

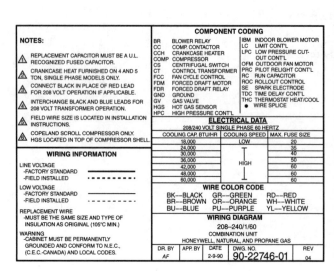
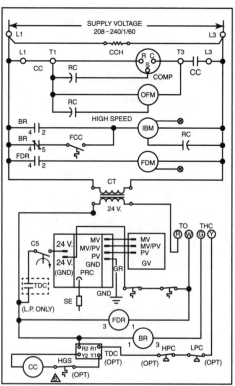

7. You answer a complaint of "no heating." The pilot ignites, but the main burner does not. You check, and 24 V are available to the PRC, but no voltage is available to the GV. The schematic diagram is shown in Figure MST-11.4. What is the probable cause?

   A. Gas valve

   B. CS

   C. ROC

   D. PRC

8. You answer a complaint of rising temperature in a walk-in freezer. The evaporator is completely covered with frost. The schematic of the unit is shown in Figure MST-11.5. The unit is being supplied the correct voltage. The DTC contacts are good, and the DTM and the DH are good. The probable cause of the problem is:

   A. LLS.

   B. open DT.

   C. closed DT.

   D. CC.

FIGURE **MST-11.5**

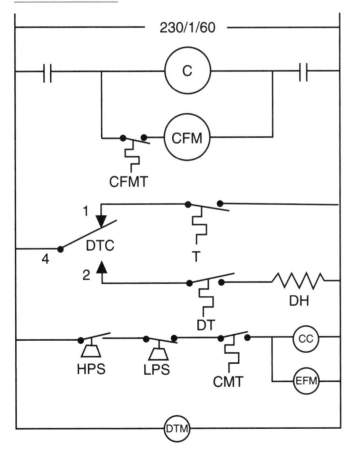

Legend

C:	Compressor
CC:	Compressor contactor
CFMT:	Condenser fan motor thermostat
CFM:	Condenser fan motor
DT:	Defrost timer motor
DTC:	Defrost timer contacts
T:	Thermostat
LLS:	Liquid line solenoid
DH:	Defrost heater
HPS:	High-pressure switch
LPS:	Low-pressure switch
CMT:	Compressor motor thermostat
EFM:	Evaporator fan motor

9. You answer a complaint of a walk-in freezer not cycling off properly, and the temperature of the freezer is −55°F. The schematic of the unit is shown in Figure MST-11.5. The thermostat is set at 0°F. The probable cause of the problem is:

   A. DTC.

   B. LPS.

   C. CMT.

   D. None of the above.

10. You answer a complaint of a walk-in-freezer temperature of 40°F. The schematic of the unit is shown in Figure MST-11.5. You discover that the discharge pressure reaches 500 psig when the HPS opens, stopping the compressor. The condenser fan motor is good. The probable cause is:

   A. CFMT.

   B. LLS.

   C. DH.

   D. CMT

MAINTENANCE OF WORK STATION AND TOOLS: There are no maintenance steps for this exercise.

SUMMARY STATEMENT: Briefly explain the procedures used when troubleshooting HVACR systems.

TAKEAWAY: A schematic is an important tool when troubleshooting a system. Along with understanding these diagrams, technicians should have a working knowledge of the typical system operations.

QUESTIONS

1. If no part of a unit is operating, what is the first check that a service technician should make?

2. What is hopscotching?

3. When a service technician arrives on the job, what are the steps that should be taken first?

4. A hermetic compressor in a small residential condensing unit is not operating, but voltage is available to the compressor terminals. What is the probable cause?

5. What would be some possible causes of a compressor motor humming when the contactor is closed?

6. What safety control would be likely to open if the unit had an extremely dirty condenser coil?

7. What are some common problems that occur with low-voltage thermostats?

8. What are some common causes of contactor failures?

9. Why are digital multimeters important to the service technician?

10. Why are schematic wiring diagrams important to the service technician?

Situational Service Ticket

Name	Date	Grade

Note: Your instructor must have placed the correct service problem in the system for you to successfully complete this exercise.

Technician's Name: _____ Date: _____

CUSTOMER COMPLAINT: No cooling.

CUSTOMER COMMENTS: The unit does not appear to be running.

TYPE OF SYSTEM:

OUTDOOR UNIT:

    MANUFACTURER _____   MODEL NUMBER _____   SERIAL NUMBER _____

INDOOR UNIT:

    MANUFACTURER _____   MODEL NUMBER _____   SERIAL NUMBER _____

COMPRESSOR (where applicable):

    MANUFACTURER _____   MODEL NUMBER _____   SERIAL NUMBER _____

TECHNICIAN'S REPORT

1. SYMPTOMS: _____

2. DIAGNOSIS: _____

3. ESTIMATED MATERIALS FOR REPAIR: _____

4. ESTIMATED TIME TO COMPLETE THE REPAIR: _____

SERVICE TIP FOR THIS CALL

Use your digital multimeter (DDM) to check all power supplies.

Exercise MST-13

Situational Service Ticket

Name	Date	Grade

Note: Your instructor must have placed the correct service problem in the system for you to successfully complete this exercise.

Technician's Name: _____ Date: _____

CUSTOMER COMPLAINT: No cooling.

CUSTOMER COMMENTS: The house will not cool down properly.

TYPE OF SYSTEM:

OUTDOOR UNIT:
    MANUFACTURER _____ MODEL NUMBER _____ SERIAL NUMBER _____

INDOOR UNIT:
    MANUFACTURER _____ MODEL NUMBER _____ SERIAL NUMBER _____

COMPRESSOR (where applicable):
    MANUFACTURER _____ MODEL NUMBER _____ SERIAL NUMBER _____

TECHNICIAN'S REPORT

1. SYMPTOMS: _____

2. DIAGNOSIS: _____

3. ESTIMATED MATERIALS FOR REPAIR: _____

4. ESTIMATED TIME TO COMPLETE THE REPAIR: _____

SERVICE TIP FOR THIS CALL

Use your digital multimeter (DDM) to check all power supplies.

SECTION 11

Controls (CON)

SUBJECT AREA OVERVIEW

The HVACR industry requires many types of automatic controls to modulate (start, stop, or adjust) the operation of equipment. These devices help provide safe and comfortable conditions for people. Although the most common controls monitor and react to changes in either temperature or pressure, other control devices are designed to react to fluid flow, fluid level, or a number of other physical properties. The changes sensed by various control devices can cause mechanically actuated devices, such as electrical contacts, to open or close or can be used in conjunction with electronic circuitry. Controls are, for the most part, power-passing devices and are normally wired in series with the power-consuming devices in a circuit. Power-passing devices are more commonly known as switches, which can be manual or automatic devices.

Common temperature-sensing controls rely on the expansion and contraction of metals or fluids. Other devices, such as thermocouples, generate small voltages to confirm the presence of a flame, as in the case of the pilot light on a heating appliance. Still other temperature-sensing control devices, such as the thermistor, are electronic in nature and vary their resistance to current flow based on the temperature that they sense. The thermistor, when used in conjunction with other electronic components, can start, stop, or modulate system components. Thermistors are designed to either increase or decrease their resistance as the sensed temperature changes.

Temperature-sensing controls are commonly referred to as thermostats. The thermostat is the primary control used to control the temperature of structures, spaces, and objects, but is also widely used as a safety device to de-energize system components if unsafe system temperatures occur. Thermostats can be either line-voltage or low-voltage devices and can be configured to control multiple stages of heating and/or cooling. When used to control space temperature, thermostats can be heating thermostats, cooling thermostats, or a combination of the two. Digital, programmable thermostats are the most common variety of primary space control, and are used to automatically raise or lower the temperature set points as desired, based on a user-established program.

In the HVACR industry, communicating control systems, including thermostats, are becoming increasingly popular. These systems allow for the system components to communicate with each other as well as the equipment owner/operator to provide the desired comfort level.

Pressure-sensing devices are normally used in conjunction with fluids such as refrigerant, air, gas, oil, and water. Pressure switches are designed to open and close based on the sensed pressure, depending upon their application. High-pressure controls, for example, open on a rise in pressure, while low-pressure controls open on a drop in pressure. Pressure controls and switches can be found on both electromechanical and electronic control systems. In order for pressure levels to be communicated to solid-state controls, pressure transducers are used. These devices translate a sensed pressure into an electronic signal that is then interpreted by a microprocessor.

As in the case of temperature-sensing controls, pressure controls can be used as either operational or safety devices. For example, the oil-pressure control, used primarily on larger compressors, ensures that there is adequate lubricating oil pressure. If the required oil pressure is not detected, the compressor will be de-energized. In this case, the pressure control is acting as a safety device. On heat pump systems, pressure controls are used to determine whether or not there is significant ice buildup on the outdoor

(continued)

coil, in an effort to control the unit's defrost cycle. In this case, the pressure control is being utilized as an operational device.

The HVACR industry has moved to incorporate electronic control circuits into most of the control systems being used today, at both the residential and commercial/industrial levels. Modern electronic control systems produce better control parameters, more efficient operation, and multifunctional control systems. Many of these solid-state controls incorporate some means to control compressor and blower speed, depending on the structural heat load.

Early-generation building control systems were configured with pneumatic devices. These systems utilized air pressure instead of electricity. Pneumatic systems were simple enough that a person with a basic understanding of mechanics could maintain them. Although there are still many pneumatic control systems in operation, the latest systems are electronic and direct digital controls (DDCs). Many of the older pneumatic systems have been successfully adapted to operate with electronic and direct digital control systems to create what are referred to as hybrid control systems.

The modern office building today may be operated by a computer in a distant city that is scanning the weather, even the long-term predicted weather, to decide how to react to keep the building comfortable. The computer can also act as a diagnostic tool to tell the technician what the likely problem is when dispatched to one of the building systems. Some high-end homes have incorporated such technologies into their HVACR and building control systems.

One of the most important tasks performed by an HVACR technician is the diagnosis of electrical components. Therefore, it is imperative for service technicians to understand the system's sequence of operations, know how the system components work, know how to use electric test instruments, and know the proper procedures for checking electric components. The service technician must be able to quickly and accurately locate the component that is not functioning properly and to perform the necessary repairs. If the operation of an electrical component is not understood, it would, therefore, be impossible to accurately diagnose the component. Manufacturers furnish troubleshooting information with each new piece of equipment. This information should be left with the equipment to aid future system service.

KEY TERMS

Algorithms
Analog
Anti-short-cycling timer
Bellows
Bimetal
Binary
Clock thermostat
Closed-loop control
Cold anticipator
Configuration
Contactor
Control loop
Controlled agent
Controlled environment
Controlled medium
Controlled output
Controlled point
Controlled system
Current-sensing lockout relay
Cut-in
Cut-out
Defrost module
Demand metering
Detent

Diaphragm
Differential
Digital
Diode
Direct digital control (DDC)
Duty cycling
Electric heat relay
Electrical resistance heater
Electronic module
Electronic self-diagnostic
Electronic timer
Element
Fan switch
Fault isolation diagram
Feedback loop
Fluid expansion
Gain
Gas valve
Heat anticipator
Hopscotching
Hot surface ignition
Humidistat
Ignition module
Ladder wiring diagram
Light-emitting diode

Line-voltage thermostat
Load shedding
Low-voltage thermostat
Memory
Modulating flow
Motor protection module
Net oil pressure
Oil safety switch
One-function solid-state device
Open-loop control
Pictorial diagram
Pilot
Pilot assembly
Pilot positioner
Pressure control
Pressure switch
Primary control
Range
Rectifier
Relay
Rod and tube
Sensitivity
Sensors
Sequencer

Set points
Signal converters
Snap action
Snap-disc
Solenoid valve
Solid state
Spark igniter
Staging thermostat
Subbase
System lag
System overshoot
Temperature swing
Thermistor
Thermocouple
Thermopile
Thermostat
Time clock
Time-delay relay
Transistor
Triac
Varistor
Volt-amperes
Voltage feedback
Voltage spike

REVIEW TEST

Name	Date	Grade

Circle the letter that indicates the correct answer.

1. Modulating controls, in general, are used to:
   A. start and stop a process.
   B. start, stop, or adjust a process.
   C. stop or adjust a process.
   D. start or adjust a process.

2. The travel of a bimetal device can be increased by:
   A. increasing the voltage.
   B. lengthening the device.
   C. decreasing the temperature.
   D. all of the above.

3. An example of a bimetal device is a:
   A. snap-disc.
   B. mercury-filled tube.
   C. thermistor.
   D. Bourdon tube.

4. The thermocouple:
   A. is constructed of several different metals.
   B. is constructed of two different metals fastened on one end.
   C. varies its resistance with temperature.
   D. uses pressure to generate an electrical current.

5. The thermocouple:
   A. generates voltage when heated at the hot junction.
   B. varies its resistance with temperature.
   C. produces an increase in pressure when heated at the hot junction.
   D. produces an increase in pressure when heated at the cold junction.

6. The thermopile:
   A. is made up of several thermocouples wired in series with each other.
   B. varies its resistance with temperature.
   C. tends to straighten out with an increase in vapor pressure.
   D. produces a pressure to activate a Bourdon tube.

7. The thermistor is an electronic solid-state semiconductor that:
   A. produces an electrical current when heated.
   B. shows an increase in pressure when heated.
   C. causes fluid expansion when heated.
   D. varies its resistance to current flow based on its temperature.

8. A common use for a thermocouple is:
   A. a household thermostat.
   B. to check pressure.
   C. to check water level.
   D. to detect a pilot light flame.

9. The heat anticipator is:
   A. used to stop a furnace prematurely.
   B. used to stop an air-conditioning system prematurely.
   C. used to start an air-conditioning system prematurely.
   D. used to start a furnace prematurely.

10. A cold anticipator is:
    A. used to stop a furnace prematurely.
    B. used to stop an air-conditioning system prematurely.
    C. used to start a furnace prematurely.
    D. used to start an air-conditioning system prematurely.

11. The heating thermostat:
    A. opens on a rise in temperature.
    B. closes on a rise in temperature.
    C. opens on a drop in temperature.
    D. both B and C are correct.

12. Detent is a term that refers to:
    A. a slow action to ensure that electrical contact is made.
    B. a snap or quick action.
    C. a part of the motor wiring.
    D. a gas used in mercury bulbs.

13. The low-voltage thermostat passes power from _____ to the loads in the control circuit.
A. a relay
B. a transformer
C. the inert gas
D. the line voltage

14. The cold anticipator is:
A. wired in series with the cooling contacts.
B. wired in series with the bimetal.
C. wired in parallel with the cooling contacts.
D. wired in parallel with the compressor relay.

15. What are two types of motor temperature protection devices?
A. Resistors and capacitors
B. Diodes and heat sinks
C. Bimetal and thermistor
D. Silicon and germanium

16. A thermistor motor temperature protection device operates by:
A. reversing the polarity when the motor overheats.
B. causing a circuit to open when the motor overheats.
C. producing excessive voltage when the motor overheats.
D. causing a bimetal to warp when the motor overheats.

17. The pressure control can be used:
A. as an operational or a safety device.
B. to measure current in a refrigeration system.
C. to detect excessive voltage.
D. instead of an electrical circuit breaker.

18. A mechanical device that detects air movement in a duct is a:
A. pressure relief valve.
B. sail switch.
C. P&T valve.
D. positive displacement valve.

19. The oil pressure safety control is:
A. a pressure relief valve.
B. a P&T valve.
C. a pressure differential control.
D. an electrically controlled valve.

20. The difference between the suction pressure and the compressor oil pump outlet pressure is:
A. the net oil pressure.
B. the compression ratio.
C. the ambient air-pressure differential.
D. the suction- and liquid-line pressure differential.

21. The air pressure control on a heat pump outdoor coil:
A. measures the pressure/temperature difference across the coil.
B. measures the high- and low-pressure refrigerant differential.
C. senses the air pressure drop across the coil to determine when it should go into defrost.
D. indicates the pressure at the heat pump thermostat.

22. An example of a mechanical control is a:
A. high-pressure switch.
B. low-pressure control.
C. water-pressure-regulating valve.
D. gas furnace limit control.

23. An example of an electromechanical control is a:
A. high-pressure switch.
B. water-pressure-regulating valve.
C. pressure relief valve.
D. boiler valve.

24. The difference between the cut-out and the cut-in of a control is the:
A. span.
B. dead band.
C. differential.
D. sum.

25. A device that senses the pressure in a system and then converts the sensed pressure signal to an electronic signal usually to be processed by a microprocessor is:
A. a pressure transducer.
B. a pressure regulator.
C. a pressure adjuster.
D. none of the above.

26. A control loop has _____ components.
  A. twelve
  B. three
  C. seven
  D. nine

27. Pneumatic controls use which of the following as the power to operate?
  A. Water
  B. Electricity
  C. Magnetism
  D. Air

28. Modulating flow means:
  A. extra fast.
  B. extra slow.
  C. variable.
  D. no flow.

29. The sensitivity of a control has to do with:
  A. the range of operation.
  B. the maximum setting on the device.
  C. how it responds to a change in conditions.
  D. how durable the device is.

30. A pilot positioner is used with which type of controls?
  A. Pneumatic
  B. Electronic
  C. Electromechanical
  D. DDC

31. The purpose of the fan switch in a warm-air furnace is to:
  A. control the blower motor.
  B. control the combustion fan.
  C. control the primary ignition control.
  D. do all of the above.

32. Which of the following components is used to interrupt the power source to a load when an unsafe condition occurs?
  A. Fan switch
  B. Gas valve
  C. Time clock
  D. Limit switch

33. What type of gas valve is used with an intermittent pilot ignition system?
  A. Combination
  B. Direct
  C. Redundant
  D. Combustion

34. A direct ignition system:
  A. lights the pilot.
  B. lights the main burner.
  C. does both a and c.
  D. does none of the above.

35. What ignites the atomized fuel oil in a residential oil burner?
  A. Pilot
  B. High-voltage spark
  C. Hot surface igniter
  D. None of the above

36. What would be the results if the oil burner were allowed to run without ignition?
  A. Reduced fuel flow into the chamber
  B. Reduced air flow into the chamber
  C. Accumulation of fuel oil in the chamber
  D. Excessive sparking of the electrodes

37. Which of the following is a light-sensitive device that changes its resistance according to the intensity of the light?
  A. Thermocouple
  B. Fuel limiter cell
  C. Subradiant cell
  D. Cad cell

38. What electrical control device is used in an electric furnace to bring in electric resistance heaters in stages?
  A. Contactor
  B. Line-voltage thermostat
  C. Sequencer
  D. None of the above

39. Which of the following components would be used in the control system of a steam boiler?
  A. Pressure switch
  B. Limit switches
  C. Low water cutoff
  D. All of the above

40. The line side of a switch is the:
    A. side the power-consuming device is connected to.
    B. neutral side of the circuit.
    C. side nearest the power source.
    D. side with the white wire.

41. The cool anticipator in a thermostat is often:
    A. a variable resistor.
    B. a fixed resistor.
    C. a capacitor.
    D. a variable relay coil.

42. Control transformers are normally rated in:
    A. AC voltage.
    B. volt-amperes.
    C. capacitance.
    D. microfarads.

43. The electrical line wiring diagram is often called:
    A. a pictorial diagram.
    B. a ladder diagram.
    C. an orthographic projection.
    D. an oblique drawing.

44. The electrical wiring diagram that shows the approximate location of components is:
    A. a pictorial diagram.
    B. a ladder diagram.
    C. an orthographic projection.
    D. an oblique drawing.

45. If the resistance reading of a 3 hp hermetic compressor motor is 220 Ω between one of the compressor motor terminals and the compressor housing, the condition of the motor is:
    A. shorted.
    B. open.
    C. grounded.
    D. good.

46. Which of the following is a procedure used when checking electrical circuits?
    A. Hopscotching
    B. Double-checking
    C. Ampere check
    D. All of the above

47. A compressor and condenser fan motor are connected in parallel to a contactor. The compressor is operating, but the condenser fan motor is not. There are no broken connections. Which of the following components could be faulty?
    A. Contactor
    B. Condenser fan motor
    C. Transformer
    D. Thermostat

48. If a service technician reads the correct voltage to a contactor coil and the contactor is not closing, what is the problem?
    A. Thermostat
    B. Transformer
    C. Contacts of contactor
    D. Contactor coil

49. A technician reads 50 V across a closed set of contactor contacts. What does this indicate?
    A. Good contactor coil
    B. Bad contactor contacts
    C. Bad transformer
    D. None of the above

50. What would the results be if two safety controls were connected in parallel?
    A. If one opened, the circuit would be de-energized.
    B. Both would have to open to de-energize the circuit.
    C. Both would have to close to energize the circuit.
    D. None of the above.

Exercise CON-1

Automatic Controls

Name	Date	Grade

OBJECTIVES: Upon completion of this exercise, you should be able to identify several temperature-sensing elements and state why they respond to temperature changes.

INTRODUCTION: You will use several types of temperature-sensing devices, subjecting them to temperature changes and recording their responses.

TOOLS, MATERIALS, AND EQUIPMENT: A straight-blade and a Phillips-head screwdriver, a digital multimeter (DMM), a source of low heat (the sun shining in an area or heat from a warm-air furnace), an air-acetylene torch kit including a striker, ice and water, a fan-limit control, a remote-bulb cooling thermostat with a temperature range close to room temperature, a thermocouple from a gas furnace, a digital thermocouple thermometer, and a vise.

⚠ **SAFETY PRECAUTIONS:** Care should be used when working with open flames. DO NOT DIRECT THE FLAME DIRECTLY AT OTHER PEOPLE OR THE REMOTE-BULB THERMOSTAT. Wear safety goggles. Have a fire extinguisher nearby, and know how to use it in the event of a fire. Only use a flint striker to light the air-acetylene torch.

STEP-BY-STEP PROCEDURES

1. Obtain a fan-limit control from your instructor, and loosely secure the device vertically in a vise.

2. Connect the two meter leads together, and set the meter to read resistance. A reading of 0 Ω, or close to 0, will establish that the test leads on the meter are good and that the meter is functioning, at least in the resistance/continuity mode.

3. Remove the cover of the fan-limit control, and fasten the meter leads to the fan terminals on the device. Your instructor may need to show you which terminals are used to control the operation of the blower. The meter should read infinity, Figure CON-1.1. This indicates an open circuit and is normal when the control senses room temperature.

FIGURE **CON-1.1**

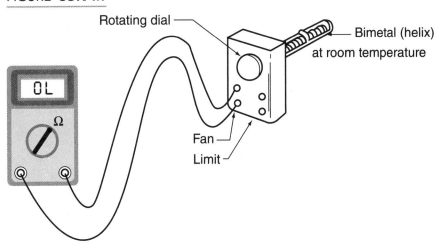

4. Observe the temperature settings on the fan-limit control, and record them here:

Fan OFF: _____
Fan ON: _____
Limit: _____

5. Light the air-acetylene torch, and pass the flame close to, but not touching, the fan control bimetal element. Pass the flame back and forth. Notice the dial on the front of the control start to turn as the element is heated. Soon the contacts on the fan switch will close, and the ohmmeter will read a resistance of zero or close to zero, Figure CON-1.2. If the digital multimeter (DMM) is set to sound an audible signal when continuity is present, the meter will do so at this point. The fan switch may also produce an audible click at this time.

FIGURE **CON-1.2**

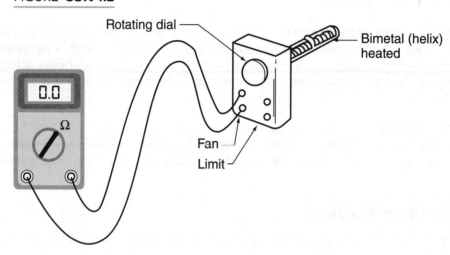

6. Remove the heat source as soon as the meter reads 0 Ω and allow the control to cool. In a few moments, the control contacts will open, and the ohmmeter will again show an open circuit by reading infinity. The closing and opening of these contacts would start and stop the blower in a furnace.

7. Fasten the meter leads to the terminals that control the high-limit portion of the device. Again, your instructor may need to help you. The meter should read a resistance of 0 Ω, indicating that there is continuity through the limit portion of the device, Figure CON-1.3. This is correct, as the limit contacts are normally closed at room temperature.

FIGURE **CON-1.3**

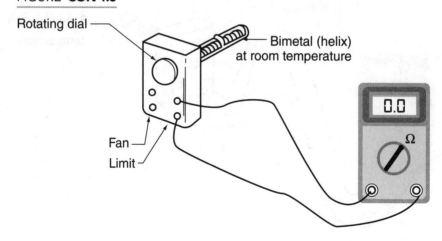

8. Again, move the torch slowly back and forth close to the sensing element. You may hear the audible click as the fan contacts on the device close. This is normal. With a little more heat applied, you should then hear the audible click of the high-limit contacts opening to stop the heat source. This will cause the meter to read infinity, indicating an open circuit, Figure CON-1.4.

FIGURE **CON-1.4**

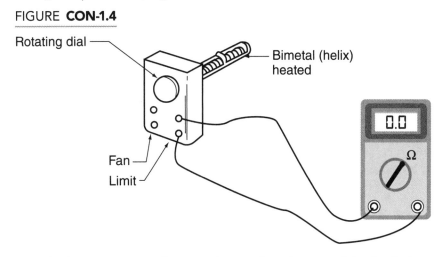

9. Place the thermocouple thermometer probe near the sensing element of the fan limit control.

10. Allow the control to cool to room temperature. The limit contacts will close first; then the fan contacts will open.

11. When the fan contacts open, note the air temperature near the sensing element, and record it here:

_____

12. Compare the air temperature obtained from step 11 with the fan OFF set point recorded in step 4. Explain the similarities and differences between these two readings, including possible reasons for any discrepancies that might have been observed.

_____

_____

13. Once the fan-limit control has completely cooled, remove it from the vise, and properly store it.

14. Obtain a remote-bulb thermostat from your instructor, and remove its cover.

15. Set the DMM to read resistance, and fasten the leads from the meter to the terminals on the remote-bulb thermostat, Figure CON-1.5.

FIGURE **CON-1.5**

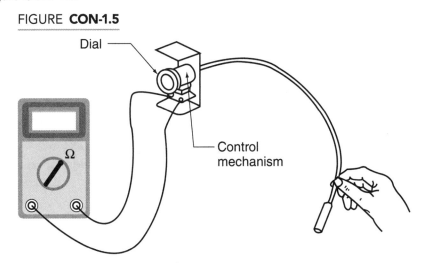

16. If the meter indicates that the circuit is closed, turn the adjustment dial until the contacts just barely open.

17. Position the control so that you can see the action inside as the temperature begins to move its mechanism. Now grasp the control bulb in your hand, and observe the control levers start to move. Soon you should hear the audible click of the control contacts closing. The meter should display 0 Ω or a reading close to 0, Figure CON-1.6.

FIGURE **CON-1.6**

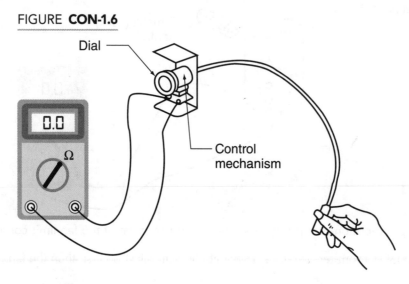

18. Now set the control at 40°F.

19. Place the control bulb in the ice-and-water mixture, and observe the contacts and their action. The contacts should open, Figure CON-1.7.

FIGURE **CON-1.7**

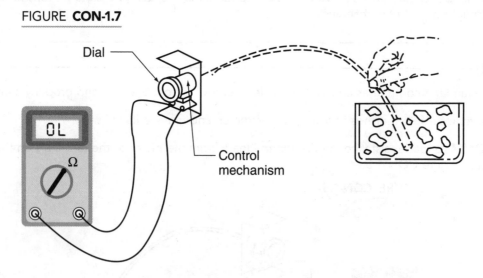

20. When the contacts open, remove the bulb from the water so that it will be exposed to room temperature. The contacts should close again in a few minutes.

21. Obtain a thermistor from your instructor, and place the thermistor on the work bench until its temperature has stabilized at room temperature. Record the resistance of the thermistor: _____ Ω. Record the temperature of the room: _____ °F.

22. Place the thermistor in the palm of your hand along with the probe from the digital thermocouple thermometer. Close your hand tightly around the thermistor and thermocouple for 30 seconds. As quickly

as possible, record the resistance of the thermistor: _____ Ω, and record the temperature reading obtained from your closed hand: _____ °F.

23. Place the thermistor along with the probe from the digital thermocouple thermometer in an ice-water mixture, and allow it to stay submerged for 30 seconds. As quickly as possible, record the resistance of the thermistor: _____ Ω, and the temperature reading obtained from the ice-water mixture: _____ °F.

24. Place the thermistor along with the probe from the digital thermocouple thermometer next to a heat source, such as close to a light bulb or in the sun. Allow the thermistor to sense this heat for 30 seconds. As quickly as possible, record the resistance of the thermistor: _____ Ω, and the temperature reading obtained from the heat source: _____ °F.

25. Record three unique temperatures on the lines below. Based on the information obtained in steps 22–24, estimate the resistance values that correspond to each recorded temperature. Record the estimated resistance values here:

Temperature 1: _____ °F   Resistance 1: _____ Ω

Temperature 2: _____ °F   Resistance 2: _____ Ω

Temperature 3: _____ °F   Resistance 3: _____ Ω

26. Fasten a thermocouple (such as a gas furnace thermocouple) lightly in a vise. Fasten meter lead to the end of the thermocouple and the other lead to the thermocouple housing, Figure CON-1.8.

FIGURE **CON-1.8**

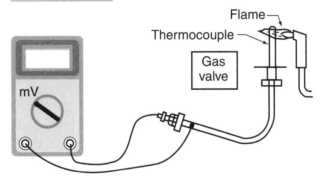

27. Light the torch and pass the flame lightly over the end of the thermocouple. If the meter displays a voltage that is below 0 V, reverse the leads. This is direct current, and the meter may be connected wrong. DO NOT OVERHEAT THE THERMOCOUPLE LEAD. Record the thermocouple voltage: _____ mV.

28. Allow the thermocouple to cool, and watch the meter reading move toward 0 mV.

MAINTENANCE OF WORKSTATION AND TOOLS: Turn off the torch at the tank valve. Bleed the torch hose. Return all tools, materials, and equipment to their respective places. Clean the work area.

SUMMARY STATEMENT: Describe the action of the four temperature-sensing devices that you checked and observed. Tell how each functions with a temperature change.

A. Bimetal device:

B. Liquid-filled bulb device:

C. Thermocouple device:

D. Thermistor device:

TAKEAWAY: Temperature-sensing devices can be configured to react differently when changes in temperature are sensed. These reactions vary from device to device and from application to application, so it is important to understand the operation of not only one, but all of the different types of temperature-sensing controls.

QUESTIONS

1. State three methods used to extend the length of a bimetal device for more accuracy.

2. How does the action of a room thermostat make and break an electrical circuit to control temperature?

3. What causes a bimetal element to change with a temperature change?

4. What causes a mechanical action in a liquid-filled bulb with a temperature change?

5. Name an application for a liquid-filled bulb sensing device.

6. What is the tube connecting the bulb to the control called on a liquid-filled remote bulb control?

7. Describe how a thermocouple reacts to temperature change.

Exercise CON-2

Bimetal Switches

Name	Date	Grade

OBJECTIVES: Upon completion of this exercise, you should be able to construct a circuit that utilizes a commonly used bimetal device, the defrost termination switch. As part of this exercise, you will observe the operation of the device as well as take amperage, voltage, and resistance readings in the circuit.

INTRODUCTION: In this exercise, you will wire, based on a sample diagram, a circuit that will control the operation of two light bulbs based solely on temperature.

TOOLS, MATERIALS, AND EQUIPMENT: Wire cutters, wire strippers, crimping tools, screwdrivers, digital multimeter (DMM), clamp-on ammeters, safety goggles, and the following materials:

    14-gauge stranded wire
    Solderless connectors (forks)
    Wire nuts
    1 blue 75 W incandescent light bulb
    1 red 75 W incandescent light bulb
    2 light sockets
    3 conductor (black, white, green) 14-gauge SJ cable
    Male plug
    1 single-pole, single-throw switch
    1 single-pole, double-throw defrost termination switch (DTS)
    Heat source

⚠ **SAFETY PRECAUTIONS:** When working on electric circuits, the following safety issues should be adhered to:
- Whenever possible, work on circuits that are de-energized.
- Avoid coming in contact with bare conductors.
- Avoid becoming part of the active electric circuit.
- Use ohmmeters only on circuits that are de-energized.
- Be sure that the meters are set to the proper scale BEFORE taking readings.
- Exercise caution when working with heat sources.
- Have a fire extinguisher nearby, and know how to use it in the event of a fire.

STEP-BY-STEP PROCEDURES

1. Create a power cord by connecting the male plug to one end of a 3 ft length of SJ cable.
   Make certain that all three conductors are securely connected to the line, neutral, and ground prongs on the plug.

2. Remove about 6 in. of the casing from the loose end of the SJ cable.

3. Strip about ¼ in. of the insulation from the ends of the three conductors.

4. Connect solderless connectors (forks) to the ends of the three conductors.

5. Connect the black wire to one end of the single-pole, single-throw switch.

6. Connect the green wire to the ground terminal on the single-pole, single-throw switch.

7. Connect the remainder of the circuit according to the sample wiring diagram in Figure CON-2.1.

FIGURE **CON-2.1**

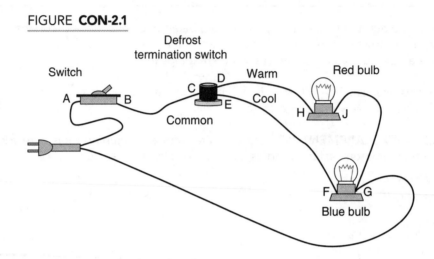

8. The defrost termination switch is a single-pole, double-throw switch that has three wires connected to it. One of the wires is a common that connects to either the "warm" wire or the "cool" wire. Contact is made between "common" and "cool" when the device senses a cold temperature. Contact is made between "common" and "warm" when the device senses a warm temperature. Check the specifications for the particular device being used to determine the actual temperatures at which the device changes its position. Enter the information for the device being used here:

Common wire color: _____

"Warm" wire color: _____

"Cool" wire color: _____

Temperature that the "common-cool" contact is made: _____ °F

Temperature that the "common-warm" contact is made: _____ °F

Device temperature differential: _____ °F

9. After the circuit has been constructed, the circuit can be energized by first turning the switch to the OFF position and then plugging the power cord into an accessible outlet.

10. Turn the switch to the ON position, and enter which bulb is energized here:

The _____ bulb is energized.

11. Explain why this bulb is energized and the other is not.

12. Using a multimeter, obtain the following readings:

Location	Voltage Reading (V)
Switch: "A" to "B"	
DTS: "C" to "D"	
DTS: "C" to "E"	
Red bulb: "H" to "J"	
Blue bulb: "F" to "G"	
"C" to "J"	
"C" to "G"	

13. Compare the voltage readings that were obtained between points "C" and "J" with the voltage reading obtained between points "C" and "G." Explain the relationship between these two voltage readings.

14. Gently apply heat to the defrost termination switch until an audible click is heard coming from the device. Once the audible click is heard, remove the heat source.

15. Describe your observations here:

16. Using a multimeter, obtain the following readings:

Location	Voltage Reading (V)
Switch: "A" to "B"	
DTS: "C" to "D"	
DTS: "C" to "E"	
Red bulb: "H" to "J"	
Blue bulb: "F" to "G"	
"C" to "J"	
"C" to "G"	

17. Compare the readings taken in step 16 with those obtained in step 12. Describe your observations here:

MAINTENANCE OF WORKSTATION AND TOOLS:

Make certain that all light bulbs are properly stored to prevent damage.

Make certain that all tools are properly stored.

Make certain that all materials (wire nuts, solderless connectors) are properly sorted and stored.

Make certain that the work area is cleaned.

Keep the power cord assembled as it will be used in other exercises.

SUMMARY STATEMENT: Explain the operation of the defrost termination switch.

TAKEAWAY: Since the interior workings of many control devices cannot be visually inspected, the HVACR service technician must rely on pieces of test equipment to determine how the component functions properly or if the device is faulty. In order to accomplish this, the technician must first understand how the component is supposed to operate. This exercise provides practice in both wiring circuits and evaluating the components that make up the circuit.

QUESTIONS

1. What set of contacts is in the closed position when the defrost termination switch is sensing a temperature that is lower than the "common-warm" temperature and the "common-cool" temperature. Explain your answer.

2. Under what conditions will the blue and red bulbs both be on? Explain your answer.

Exercise CON-3

Air-Conditioning Circuit Evaluation

Name	Date	Grade

OBJECTIVES: Upon completion of this exercise, you should be able to follow the circuit in a typical electric air-conditioning system and check the amperage in a low-voltage circuit.

INTRODUCTION: You will use a volt-ohmmeter and a clamp-on ammeter to follow the circuit characteristics of a high-voltage and a low-voltage circuit.

TOOLS, MATERIALS, AND EQUIPMENT: A digital multimeter (DMM), goggles, a screwdriver, a clamp-on ammeter, and some low-voltage wire (about 3 ft of 1-strand, 18-gauge), a packaged or split-type air-conditioning system, and a panel lock and tag.

⚠ **SAFETY PRECAUTIONS:** You will be taking voltage readings on a live circuit. Do not perform this function until your instructor has given you instruction on the use of the meter. The power must be off, locked, and tagged at all times until the instructor has approved your setup and tells you to turn it on. Make certain that all capacitors are properly discharged. Wear goggles and do not touch any of the bare terminals.

STEP-BY-STEP PROCEDURES

1. Turn the power to the air-conditioning system off. Lock and tag the panel. Remove the door to the control compartment.

2. Set the meter to read AC voltage. With your instructor's approval, turn the power on and start the unit.

   Caution: *BE SURE THAT THE CONTROL COMPARTMENT IS ISOLATED FROM THE AIRFLOW PATTERN SO AIR IS NOT BYPASSING THE CONDENSER COIL DURING THIS TEST.*

   Check the voltage at the following points and record:

   - Line voltage: _____ V
   - Line side of the contactor: _____ V
   - Load side of the contactor: _____ V
   - Compressor, common to run terminal: _____ V
   - Compressor, common to start terminals: _____ V
   - Low voltage, C to R terminals: _____ V
   - Low voltage, C to Y terminals: _____ V
   - Low voltage, R to Y terminals: _____ V
   - Low voltage, R to G terminals: _____ V
   - Low voltage, C to G terminals: _____ V

3. Turn the power off at the disconnect, lock, and tag.

4. Remove one wire from the magnetic holding coil of the compressor contactor and one wire from the coil of the indoor fan relay.

5. Record the following:
   - Resistance of the fan relay coil: _____ Ω
   - Resistance of the compressor contactor: _____ Ω

6. Label the wires connected to the compressor terminals, and then remove the wires from the terminals.

7. Measure and record the following resistances at the compressor terminals:
   - Resistance from common to run: _____ Ω
   - Resistance from common to start: _____ Ω
   - Resistance from run to start: _____ Ω

8. Reconnect any wires that were disconnected, and have your instructor inspect your work.

9. After the instructor has inspected and approved your setup, start the unit.

10. Using your clamp-on ammeter, obtain and record the following amperages:
    - Indoor fan relay: _____ A
    - Compressor contactor and fan relay: _____ A

MAINTENANCE OF WORKSTATION AND TOOLS: Replace all wires and panels, and return the tools and materials to their respective places. Clean your work area.

SUMMARY STATEMENT: Describe how the low-voltage circuit controls the power to the high-voltage circuit. Use an additional sheet of paper if more space is necessary for your answer.

TAKEAWAY: Being able to differentiate between the line-voltage power circuit and the low-voltage control is the first step in fully understanding how these two circuits work together to control the operation of the main electrical loads in an air-conditioning system. This exercise provides practice in identifying system components as well as taking resistance, voltage, and current readings at various points in the system.

QUESTIONS

1. Why is low voltage used to control high voltage?

2. What is the typical amperage draw in the low-voltage circuit? _____

3. What does the term VA mean when applied to a control transformer? _____

4. How much amperage can a 60 VA transformer provide if the secondary voltage is 24 V? _____

5. How much amperage can a typical nondigital thermostat safely handle? _____

Exercise CON-4

Controlling Temperature

Name	Date	Grade

OBJECTIVES: Upon completion of this exercise, you should be able to use a temperature-measuring device to determine the accuracy of thermostats while observing the control of the temperature of product and space.

INTRODUCTION: You will use a thermometer with its element located at a temperature-control device to determine the accuracy of a thermostat used in a refrigerated box and one used to control space temperature.

TOOLS, MATERIALS, AND EQUIPMENT: A high-quality, multi-probe thermometer (an electronic thermometer with a thermistor or thermocouple is preferred because of its response time), a digital multimeter (DMM), a small screwdriver, a medium-temperature refrigerated case that uses a remote-bulb temperature control, and a space-temperature thermostat.

⚠ **SAFETY PRECAUTIONS:** You will be using the screwdriver to make adjustments to the temperature-control setting. Be sure to make note of the control's settings prior to altering the set points so that the original settings can be restored at the end of the exercise. Remove the control's cover prior to starting the test. Your instructor must give you instruction in operating the particular refrigerated box and room thermostat before you begin. Do not touch the electrical terminals.

STEP-BY-STEP PROCEDURES

1. Locate the remote bulb for the temperature control in a refrigerated box, and fasten the thermometer sensor next to it.

2. Place a second sensor in the refrigerated space, suspended in the air.

3. Place a third sensor in the product to be cooled—for instance, a glass of water on the food shelf, Figure CON-4.1. Close the door on the case if the case is equipped with a door.

FIGURE **CON-4.1**

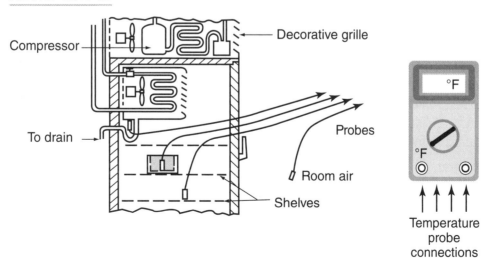

4. After 5 minutes, record the following: Probe A, by the remote bulb _____ °F; Probe B, in the air _____ °F; Probe C, in water _____ °F

5. Listen for the compressor to stop and record the same three temperatures when it stops.

   A. _____ °F   B. _____ °F   C. _____ °F

   Record every 5 minutes:

   A. _____ °F   B. _____ °F   C. _____ °F
   A. _____ °F   B. _____ °F   C. _____ °F
   A. _____ °F   B. _____ °F   C. _____ °F
   A. _____ °F   B. _____ °F   C. _____ °F
   A. _____ °F   B. _____ °F   C. _____ °F
   A. _____ °F   B. _____ °F   C. _____ °F
   A. _____ °F   B. _____ °F   C. _____ °F

6. Set the control for 5°F higher.

   Record every 5 minutes:

   A. _____ °F   B. _____ °F   C. _____ °F
   A. _____ °F   B. _____ °F   C. _____ °F
   A. _____ °F   B. _____ °F   C. _____ °F
   A. _____ °F   B. _____ °F   C. _____ °F
   A. _____ °F   B. _____ °F   C. _____ °F
   A. _____ °F   B. _____ °F   C. _____ °F
   A. _____ °F   B. _____ °F   C. _____ °F

7. Remove the cover of a room thermostat. This thermostat should not be connected to an operating system. The thermostat should be properly mounted but not connected.

8. Place the temperature probe of a thermometer next to the thermostat.

9. Set the digital multimeter (DMM) to measure resistance. Switch the thermostat to the cooling mode, and place one meter lead on "R" and the other on "Y." Set the thermostat so that the meter shows an open circuit. When the space around the thermostat warms, which is possible by bringing an illuminated light bulb closer to the thermostat, the contacts will close. By varying the amount of heat that is acting on the thermostat, the contacts should open and close from time to time. If they do not function in about 5 minutes, you may move closer to the thermostat so your body heat will cause a response. This can be verified by the use of an ohmmeter if the thermostat is on the workbench.

10. Record the time and temperature changes.

    Time: _____   Temperature: _____ °F
    Time: _____   Temperature: _____ °F
    Time: _____   Temperature: _____ °F
    Time: _____   Temperature: _____ °F
    Time: _____   Temperature: _____ °F
    Time: _____   Temperature: _____ °F
    Time: _____   Temperature: _____ °F
    Time: _____   Temperature: _____ °F

MAINTENANCE OF WORKSTATION AND TOOLS: Clean the workstation, and return all tools and equipment to their respective places.

SUMMARY STATEMENT: Describe the action of the two thermostats in relation to the thermometer that you used. Use an additional sheet of paper if more space is necessary for your answer.

TAKEAWAY: The thermostat is one of the most common control devices found on air-conditioning and refrigeration systems. Although thermostats all sense temperature, they all do so differently and also react differently. This exercise provides two examples of how thermostats react to changes in temperature and how fast they respond.

QUESTIONS

1. How does a bimetal sensing element make and break an electrical circuit?

2. What type of action does a bimetal element produce with a temperature change?

3. What type of action does a thermocouple produce with a change in temperature?

4. What type of action does a thermistor produce with a change in temperature?

5. Why does the product temperature lag behind the air temperature in a refrigerated box?

6. Explain the differential setting on a temperature control.

7. Describe a good location for a space temperature thermostat.

Low-Voltage Thermostats

Name	Date	Grade

OBJECTIVES: Upon completion of this exercise, you should be able to correctly install a low-voltage, digital programmable thermostat on a heating/cooling and heat pump system.

INTRODUCTION: Low-voltage thermostats are one of the most-used electrical devices in the heating, cooling, and refrigeration industry. Almost all residences use low-voltage thermostats, so the technician must be able to correctly install them.

TOOLS, MATERIALS, AND EQUIPMENT:

Single-stage heating and cooling low-voltage digital programmable thermostats with instructions; two-stage heating and two-stage cooling digital programmable thermostats with instructions; other electrical components as determined by the diagrams; air-conditioning system using single-stage heating and cooling thermostat; heat pump system; digital multimeter (DMM); miscellaneous wiring supplies and hardware; basic electric hand tools; colored light bulbs; light sockets; two double-pole, 24 V contactor; three single-pole, 24 V general purpose relays; control transformer; power cord; and plywood test boards.

⚠ **SAFETY PRECAUTIONS:** Make certain that the electrical source is disconnected when making electrical connections. In addition:

- Make sure all electrical connections are tight.
- Make sure no bare current-carrying conductors are touching metal surfaces except ungrounded conductors.
- Make sure the correct voltage is being supplied to the circuit or equipment.
- Make sure body parts do not come in contact with live electrical conductors.
- Keep hands and materials away from moving parts.

STEP-BY-STEP PROCEDURES

1. Study the following pictorial diagram.

FIGURE **CON-5.1**

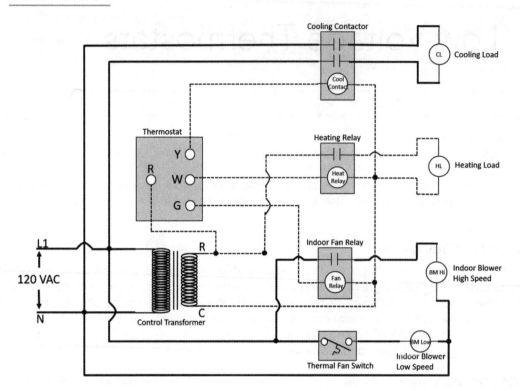

2. Evaluate the following four ladder diagrams.

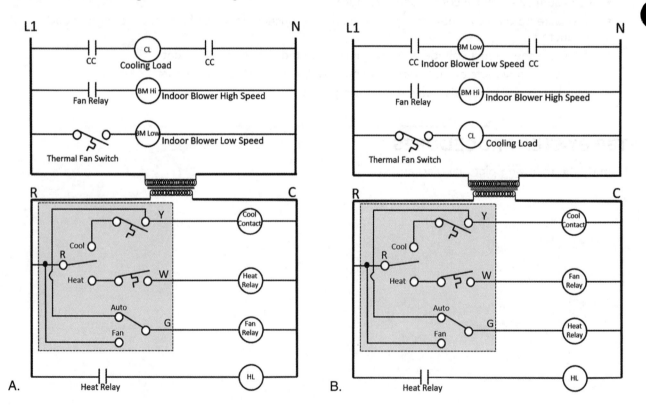

3. Identify the ladder diagram that best matches the pictorial diagram, and record your answer here: _____

4. Using your diagram, wire the control system on an electrical practice board. Read the installation instructions provided with the thermostat. Use colored bulbs to represent the cooling load, the heating load, and the indoor blower.

5. Connect your system to a 115 V power supply.

6. Have your instructor check your circuit.

7. Operate the control system by setting the thermostat for the desired function; check cooling, heating, and blower operation. In each mode, be sure to operate the system with the thermostat set to temperatures both above and below the space temperature.

8. Record what happens when the thermostat is set for each function here:

Mode	Thermostat Setting	Fan Setting	Observations
Off	Above room temp	Auto	_____
Off	Below room temp	Auto	_____
Off	Above room temp	On	_____
Off	Below room temp	On	_____
Cool	Above room temp	Auto	_____
Cool	Below room temp	Auto	_____
Cool	Above room temp	On	_____
Cool	Below room temp	On	_____

Heat	Above room temp	Auto	_____
Heat	Below room temp	Auto	_____
Heat	Above room temp	On	_____
Heat	Below room temp	On	_____

9. Disconnect the power source from the circuit, and remove the components from the board.

10. Position yourself at the unit that is assigned to you in preparation for the installation of a single-stage heating/cooling thermostat.

11. Read the installation instructions for the assigned equipment and thermostat.

12. Disconnect the electrical power source from each piece of equipment.

13. Remove the necessary covers from the equipment.

14. Make the low-voltage connections between the equipment and the thermostat for proper operation.

15. Have your instructor check your wiring.

16. Restore electrical power to the equipment.

17. If the blower compartment cover was removed from a fossil fuel furnace, this cover must be replaced before operating the system.

18. Operate the system by setting the thermostat for the desired function; check cooling, heating, and blower operation.

19. Record your observations here:

Mode	Stage	Thermostat Setting	Fan Setting	Observations
Off	N/A	Above room temp	Auto	_____
Off	N/A	Below room temp	Auto	_____
Off	N/A	Above room temp	On	_____
Off	N/A	Below room temp	On	_____
Cool	N/A	Above room temp	Auto	_____
Cool	First	Below room temp	Auto	_____
Cool	Second	Below room temp	Auto	_____
Cool	N/A	Above room temp	On	_____
Cool	First	Below room temp	On	_____
Cool	Second	Below room temp	On	_____
Heat	FIRST	Above room temp	Auto	_____
Heat	Second	Above room temp	Auto	_____
Heat	N/A	Below room temp	Auto	_____
Heat	First	Above room temp	On	_____
Heat	Second	Above room temp	On	_____
Heat	N/A	Below room temp	On	_____

20. Replace all covers on equipment.

21. Disconnect the electrical power supply from the equipment.

22. Study the following pictorial diagram.

FIGURE **CON-5.2**

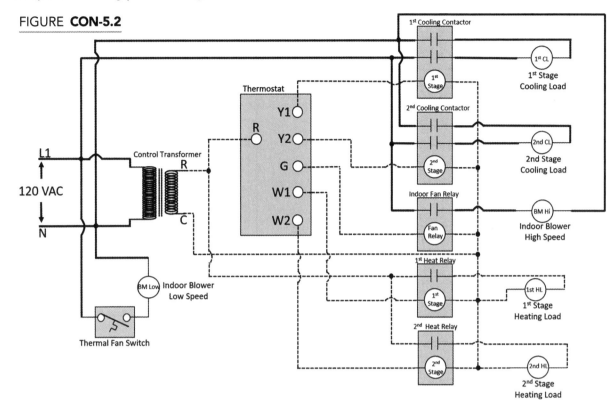

23. Evaluate the following four ladder diagrams.

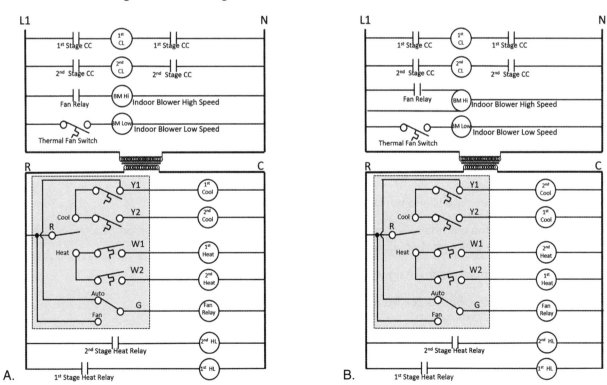

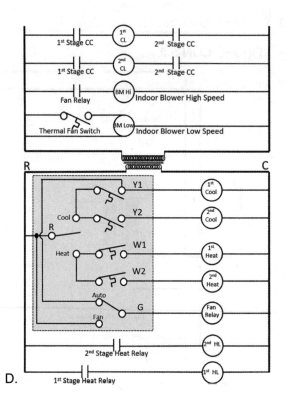

24. Identify the ladder diagram that best matches the pictorial diagram, and record your answer here: _____

25. Using your diagram, wire the control system on an electrical practice board. Read the installation instructions provided with the thermostat. Use colored bulbs to represent the cooling loads, the heating loads, and the indoor blower.

26. Connect your system to a 115 V power supply.

27. Have your instructor check your circuits.

28. Operate the control system by setting the thermostat for the desired function; check cooling, heating, and fan operation.

29. Record what happens when the thermostat is set to each function. Note the staging of the thermostat.

30. Disconnect the power supply, and remove the components from the board.

31. Position yourself at the unit that is assigned to you in preparation for the installation of a two-stage heating, single-stage cooling thermostat on a heat pump system.

32. Read installation instructions for assigned equipment and thermostat.

33. Disconnect electrical power source from each piece of equipment.

34. Remove necessary covers from equipment.

35. Make the necessary low-voltage connections between the equipment and the thermostat for proper operation.

36. Have your instructor check your wiring.

37. Restore electrical power to the equipment.

38. Replace indoor unit covers.

39. Operate the system by setting the thermostat for the desired function: cooling, heating, second-stage heat, and fan operation.

40. Write a brief paragraph explaining the operation of the heat pump system.

41. Replace all equipment covers.

42. Disconnect the electrical power supply from the equipment.

MAINTENANCE OF WORKSTATION: Clean all tools, and return them to their proper location(s). Replace all equipment covers. Clean up the work area.

SUMMARY STATEMENT: Why are low-voltage thermostats used for the control of most residential heating and cooling systems?

TAKEAWAY: One of the most desirable traits of a successful HVACR technician is a solid understanding of system controls. This exercise provided numerous opportunities to practice working with wiring diagrams, evaluating actual control circuits, and wiring live equipment.

QUESTIONS

1. How do heating and cooling thermostats differ?

2. What is the purpose of a heat anticipator in a thermostat?

3. Why is it important that thermostat contacts are snap acting?

4. What is the purpose of the letter designations on low-voltage thermostats?

5. Briefly explain proper installation procedures for installing thermostats.

6. What is a staging thermostat?

7. Why are staging thermostats used on heat pumps?

8. What is the purpose of using an electronic programmable thermostat?

Exercise CON-6

Line-Voltage Thermostats

Name	Date	Grade

OBJECTIVES: Upon completion of this exercise, you should able to correctly install a line-voltage thermostat on a unit.

INTRODUCTION: Line-voltage thermostats are widely used in the refrigeration, heating, and air-conditioning industry. They are used as operating controls on domestic refrigerators and freezers, in commercial refrigeration appliances to maintain the temperature in walk-in-coolers and walk-in-freezers, and in other applications where temperature is the element that must be controlled, such as limit switches and fan switches on fossil fuel furnaces. The technician must know the function of the line-voltage thermostat in the control system.

TOOLS, MATERIALS, AND EQUIPMENT:

Line-voltage thermostats, other components as determined by diagrams, digital multimeter, miscellaneous wiring supplies, basic electrical hand tools, line-voltage cooling thermostat, 115 V motor, power cord, plywood test boards, and appliances equipped with line-voltage thermostats.

⚠ **SAFETY PRECAUTIONS:** Make certain that the electrical power source is disconnected when making electrical connections. In addition:

- Make sure all connections are tight.
- Make sure no bare current-carrying conductors are touching metal surfaces except the grounding conductor.
- Make sure body parts do not come in contact with live electrical conductors.
- Keep hands and materials away from moving parts.

STEP-BY-STEP PROCEDURES

1. Study the wiring diagram in Figure CON-6.1, which represents an electric circuit where the line-voltage thermostat energizes a small 115 V fan motor when the temperature exceeds 85°F.

FIGURE **CON-6.1**

2. Evaluate the following four ladder diagrams.

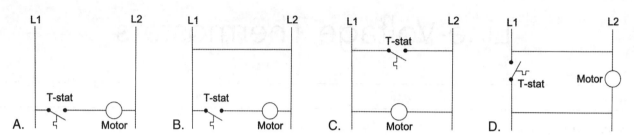

3. Select the ladder diagram that best matches the pictorial diagram, and record your answer here: _____

4. Using your diagrams, wire the circuit on an electrical practice board.

5. Connect the circuit to a 115 V power source.

6. Have your instructor check your circuit.

7. Operate the circuit by closing the thermostat.

8. Record what happens when the thermostat closes.

9. Disconnect the power source from the circuit and remove the components from the board.

10. Write a brief paragraph explaining the function of the line-voltage thermostat in the circuit. In your response, include the relationship between the amperage draw of the load and the current rating of the thermostat.

MAINTENANCE OF WORKSTATION: Clean all tools, and return them to their proper location(s). Replace all equipment covers. Clean up the work area.

SUMMARY STATEMENT: Give three applications each of line-voltage thermostats in residential air-conditioning and heating systems and commercial refrigeration systems.

TAKEAWAY: Line-voltage thermostats are used in many types of appliances and have different installation requirements than low-voltage thermostats. It is important for the HVACR technician to have experience working with both types to ensure success in the industry.

QUESTIONS

1. Why are line-voltage thermostats used to control domestic refrigerators?

2. What are some disadvantages of using a line-voltage thermostat?

3. Which of the following represents the schematic of a line-voltage thermostat that starts a condenser fan motor when the outdoor temperature reaches 90°F and de-energizes the motor when the outdoor temperature drops below 75°F?

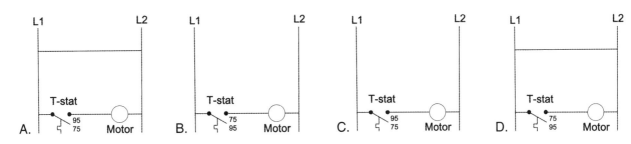

4. What is the advantage of using a line-voltage thermostat over a low-voltage thermostat?

5. Why are line-voltage thermostats less accurate than low-voltage thermostats?

6. Why are line-voltage thermostats widely used in commercial refrigeration equipment?

Exercise CON-7

Pressure Switches

Name	Date	Grade	

OBJECTIVES: Upon completion of this exercise, you should be able to correctly adjust and install a pressure switch to function as an operating and safety control.

INTRODUCTION: Pressure switches are used in control circuits as both operational and safety devices. When a pressure switch is used as a safety control, it de-energizes a control circuit, interrupting the power source to a major load when an unsafe condition exists. An operational control starts and stops a load during normal system operation. It is important that the technician knows the function of the pressure switch in the control circuit.

TOOLS, MATERIALS, AND EQUIPMENT:

Eight pressure switches for identification, high-pressure switch (opens on rise), low-pressure switch (closes on rise), dual-pressure switch, operating systems for installation of pressure switches, digital multimeter, miscellaneous wiring supplies, basic electrical hand tools, refrigeration gauge manifold, and thermocouple thermometer.

⚠ **SAFETY PRECAUTIONS:** Make certain that the electrical source is disconnected when making electrical connections. In addition:

- Make sure all electrical connections are tight.
- Make sure no bare current-carrying conductors are touching metal surfaces except ungrounded conductors.
- Make sure the correct voltage is being supplied to the circuit or equipment.
- Make sure body parts do not come in contact with live electrical conductors.
- Keep hands and materials away from moving parts.

STEP-BY-STEP PROCEDURES

1. Obtain eight pressure switches from your instructor.

2. Using component information from various sources (installation paperwork, product catalog, online search, device nameplate, etc.) record the model number, type, function, and setting for each control here:

Switch	Model Number	Type	Function	Setting
#1	_____	_____	_____	_____
#2	_____	_____	_____	_____
#3	_____	_____	_____	_____
#4	_____	_____	_____	_____
#5	_____	_____	_____	_____
#6	_____	_____	_____	_____
#7	_____	_____	_____	_____
#8	_____	_____	_____	_____

3. Determine the location of each of the following pressure switches in each of the systems to be used in this exercise:

   Adjustable high-pressure switch (safety): _____

   Fixed high-pressure switch (safety): _____

   Adjustable high-pressure switch (operating): _____

   Adjustable low-pressure switch (safety): _____

   Fixed low-pressure switch (safety): _____

   Adjustable low-pressure switch (operating): _____

   Dual-pressure switch (safety): _____

4. Obtain instructor assignment of a commercial refrigeration system to set a low-pressure switch to maintain the temperature of a medium-temperature application to instructor specifications.

5. Set the pressure switch to obtain the assigned temperature. Record the cut-in, cut-out, and differential values here:

   Cut-in: _____ psig   Cut-out: _____ psig   Differential: _____ psig

6. Install gauges on the system.

7. Position a temperature probe in the center of the box.

8. Allow the system to operate, monitoring the system pressures.

9. At the time when the system cycles off, record the following information:

   Box temperature: _____

   System low-side pressure: _____

   Time of day (hour, minutes, and seconds) the unit cycled off: _____

10. Continue to monitor the system pressures during the off cycle.

11. At the time when the system cycles back on, record the following information:

    Box temperature: _____

    System low-side pressure: _____

    Time of day (hour, minutes, and seconds) the unit cycled on: _____

12. Turn the system off.

13. Determine the length of the off-cycle: _____

14. Compare the desired box temperature to the actual box temperature (from step 11), and discuss the differences between these two values: _____

    _____

15. Compare the pressure control's cut-out set point with the pressure that the control actually de-energized the system. Discuss the differences between these two values: _____

    _____

16. Compare the pressure control's cut-in set point with the pressure that the control actually energized the system. Discuss the differences between these two values: _____

    _____

17. Properly remove the gauges from the system.

18. Obtain from your instructor a unit assignment and a high-pressure switch that closes when the temperature rises.

19. Disconnect the power supply from the equipment.

20. Install the high-pressure switch, and make all necessary connections (pressure and electrical). Make certain that you have closed all necessary valves and isolated the section of the refrigeration system where the switch is to be installed.

21. Set the pressure switch for the assigned pressure.

22. Install gauges on the system.

23. Have your instructor check your pressure connections, wiring connections, gauge installation, and pressure switch setting.

24. Connect the power supply to the system.

25. Operate the system, observe the actions of the pressure switch, and record the pressures when the condenser fan cuts off and on.

    Condenser fan on pressure: _____ psig
    Condenser fan on pressure: _____ psig
    Time interval between fan on and fan off: _____ minutes, _____ seconds

26. Compare the desired high-side pressure (control setting) to the actual high-side pressure range, and discuss the differences between these two values: _____

    _____

27. Properly remove the gauges from the system.

28. Disconnect the power supply from the equipment.

29. Obtain a dual-pressure switch and a unit assignment from your instructor.

30. Disconnect the equipment from the power supply.

31. Install the dual-pressure switch and make all necessary connections (pressure and electrical). Use valves to isolate parts of the refrigeration system in order to make pressure connections.

32. Install gauges on the system.

33. Set and record the pressure switch for the proper safety setting.

    High-pressure setting _____
    Low-pressure setting _____

34. Have your instructor check your pressure connections, gauge installation, wiring connections, and setting of the pressure switch.

35. Connect the power supply to the equipment.

36. Operate the system and observe the actions of the pressure switch.

37. Block the condenser with a piece of cardboard and operate the system. Record the pressure at which the equipment cuts off. If the discharge pressure exceeds safe limits as set by the instructor, cut the unit off using the disconnect.

    Cut-out pressure: _____

38. Compare the desired high-side pressure cut-out (control setting) to the actual high-side pressure at which the system shut down, and discuss the differences between these two values:

    _____

    _____

39. Remove the cardboard, and return the system to normal operation.

40. Close the compressor suction service valve, and operate the system. Record the pressure at which the equipment cuts off. If the suction pressure goes below safe limits as set by the instructor, cut the unit off using the disconnect.

    Cut-out pressure: _____

41. Compare the desired low-side pressure cut-out (control setting) to the actual low-side pressure at which the system shut down and discuss the differences between these two values: _____

_____

42. Open the suction service valve and return the system to normal operation.

43. Remove gauges from the system.

44. Disconnect the equipment from the power supply.

MAINTENANCE OF WORKSTATION AND TOOLS: Clean all tools, and return them to their proper location(s). Replace all equipment covers. Clean up the work area.

SUMMARY STATEMENT: Why does the pressure setting change on pressure switches when they are used as safety controls with different refrigerants?

TAKEAWAY: Pressure controls can be used as both safety and operational devices. However, to ensure proper control operation, the actual opening and closing pressures must be determined and compared to the desired settings. This exercise provides numerous opportunities for setting and verifying control settings.

QUESTIONS

1. What would be the action of a low-pressure switch used as an operational control?

2. What would be the low-pressure cut-out for a low-pressure switch used as a safety control?

3. What is the difference between fixed and adjustable high-pressure switches?

4. What factors must the technician know about a pressure switch before obtaining a replacement?

5. What is the differential of a pressure switch?

6. In what applications are high-pressure switches used as operating controls?

7. In what applications are low-pressure switches used as operating controls?

8. Why is it important for the technician to know the function of a pressure switch in a control circuit?

Pressure-Sensing Devices

Name	Date	Grade

OBJECTIVES: Upon completion of this exercise, you should be able to determine the setting of a refrigerant-type low-pressure control and reset it for low-charge protection.

INTRODUCTION: You will use dry nitrogen as a refrigerant substitute and determine the existing setting of a low-pressure control, set that control to protect the system from operating under a low-charge condition, and use a multimeter to prove the control will open and close the circuit.

TOOLS, MATERIALS, AND EQUIPMENT: An adjustable low-pressure control (such as in Figure CON-8.1), screwdriver, ¼ in. flare union, goggles, gloves, digital multimeter, gauge manifold, nitrogen regulator, and a cylinder of dry nitrogen.

⚠ SAFETY PRECAUTIONS: You will be working with dry nitrogen under pressure, and care must be taken while fastening and loosening connections. Wear goggles and gloves.

FIGURE **CON-8.1**

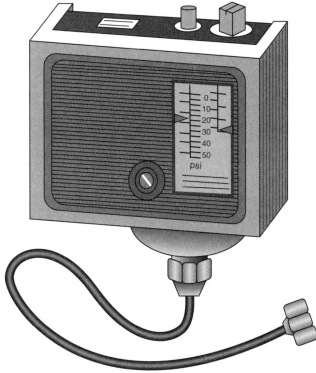

STEP-BY-STEP PROCEDURES

1. Connect the low-pressure control to the low-side hose on the gauge manifold, Figure CON-8.2.

FIGURE **CON-8.2**

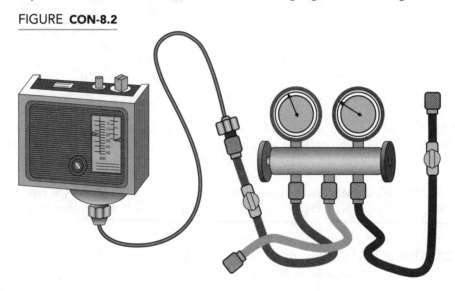

2. Typically for an R-134a system, the lowest allowable pressure for a medium temperature system is 7 psig. A typical high setting may be 38 psig. Set the control at these approximate settings.

3. Turn both valves to the off position on the gauge manifold.

4. Make certain that the nitrogen tank is properly secured.

5. Remove the protective cap from the nitrogen tank.

6. Connect the nitrogen regulator to the nitrogen tank.

7. Turn the regulator adjusting knob counterclockwise to lower the outlet pressure of the regulator to zero.

8. Connect the center hose from the gauge manifold to the outlet port on the nitrogen regulator.

9. Open the valve on the nitrogen tank.

10. Slowly turn the adjusting knob on the nitrogen regulator until the outlet pressure on the regulator reads about 50 psig.

11. Attach the meter leads to the terminals on the low-pressure control. The meter should read ∞, infinity, showing that the control contacts are open. If the contacts are closed, the control is set for operation below atmosphere and must be reset for a higher value.

12. Slowly open the low-side valve on the gauge manifold, and record the pressure at which the control contacts close.

   Note: If the control makes contact so fast that you cannot follow the action, allow a small amount of gas to bleed out around the gauge line at the flare union while slowly opening the valve.

   You may have to repeat this step several times in order to determine the exact pressure at which the control contacts close. Record the pressure reading at which the contacts close: _____ psig. (This is the cut-in pressure of the control.)

13. With the control contacts closed and pressure on the control line, shut the gauge manifold valves. Pressure is now trapped in the gauge line and control.

14. Slowly bleed the pressure out of the control by slightly loosening the connection at the flare union. Record the pressure at which the control contacts open: _____ psig. (This is the cut-out pressure of the control.) You may have to perform this operation several times to get the feel of the operation. You now know what the control is set at to cut in and out. Record here:

   • Cut-in: _____ psig

   • Cut-out: _____ psig

   • Differential (cut-in pressure − cut-out pressure): _____ psig

Procedures for Resetting the Pressure Control

1. Determine the setting you want to use and record here:

   • Cut-in: _____ psig

   • Cut-out: _____ psig

   • Differential: _____ psig

2. Set the control range indicator at the desired cut-in point.

3. Set the control's differential indicator to the desired differential.

4. Recheck the control pressure settings. Record your results here:

   • Cut-in: _____ psig

   • Cut-out: _____ psig

   • Differential: _____ psig

   You may notice that the pointer and the actual cut-in and differential are not the same. The pointer is only an indicator. The only way to properly set a control is to do just as you have done and then make changes until the control responds as desired. If the control differential is not correct, make this adjustment first. For more differential, adjust the control for a higher value. For example, if the differential is 15 psig and you want 20 psig, turn the differential dial to a higher value.

5. If the control needs further setting, make the changes, and recheck the control using pressure. Record the new values here.

   • Cut-in: _____ psig

   • Cut-out: _____ psig

   • Differential: _____ psig

MAINTENANCE OF WORKSTATION AND TOOLS: Return all tools and materials to their respective places. Clean your work area.

SUMMARY STATEMENT: Describe the range and differential functions of this control. Use an additional sheet of paper if more space is necessary for your answer.

TAKEAWAY: Pressure controls can be used as both safety and operational devices. However, to ensure proper control operation, the actual opening and closing pressures must be determined and compared to the desired settings. This exercise provides numerous opportunities for setting and verifying control settings.

QUESTIONS

1. Why should a unit have low-pressure protection when a low charge is experienced?

2. What typical pressure-sensing device may be used in a low-pressure control?

3. What is the function of the small tube connecting the pressure-sensing control device to the system?

4. If the cut-out point of a pressure control were set correctly but the differential set too high, what would the symptoms be?

5. If a control were cutting in at the correct setting and the differential were set too high, what would the symptoms be?

Exercise CON-9

Controlling the Head Pressure in a Water-Cooled Refrigeration System

Name	Date	Grade

OBJECTIVES: Upon completion of this exercise, you should be able to set a water pressure regulating valve for a water-cooled condenser and set the cut-out and cut-in point for a high-pressure control in the same system.

INTRODUCTION: You will control the head pressure in a water-cooled refrigeration (or air-conditioning) system. This system will have a wastewater condenser and a water pressure regulating valve to control the head pressure. You will set the high-pressure control for protection against a water shutoff.

TOOLS, MATERIALS, AND EQUIPMENT: A gauge manifold; a valve wrench; goggles; gloves; and a water-cooled, wastewater refrigeration system with a water regulating valve, Figure CON-9.1.

⚠ **SAFETY PRECAUTIONS:** You will be working with refrigerant, so care should be taken. Wear your goggles and gloves. Never use a jumper to take a high-pressure control out of a system because excessively high pressure may occur.

FIGURE **CON-9.1**

OPERATING VALVE

STEP-BY-STEP PROCEDURES

1. Properly install the refrigeration gauges on the system.

2. Start the system.

3. After 5 minutes of running time, record the following:

   • Suction pressure: _____ psig

   • Discharge pressure: _____ psig

4. Turn the water regulating valve adjustment clockwise two full turns, wait 5 minutes, and record the following:

   • Suction pressure: _____ psig

   • Discharge pressure: _____ psig

5. Turn the water regulating valve adjustment counterclockwise four full turns, wait 5 minutes, and record the following:

   • Suction pressure: _____ psig

   • Discharge pressure: _____ psig

6. After completion of the preceding steps, slowly turn the adjustment stem of the water pressure regulating valve in the direction that causes the head pressure to rise until the unit stops because of head pressure. Be sure to keep track of how many times the valve adjustment stem was turned. **CAUTION:** Do not let the pressure rise above the level that corresponds to a 115°F condensing temperature.

7. If the compressor stops before the previously listed settings, it is set correctly.

8. If the high-pressure control does not stop the compressor before the condensing temperature reaches 115°F, slowly turn the range adjustment on the high-pressure control to stop the compressor.

9. After completion of checking and setting the high-pressure control, reset the water pressure regulating valve to maintain a 105°F condensing temperature.

10. Turn the system off.

11. Remove the gauge manifold.

MAINTENANCE OF WORKSTATION AND TOOLS: Return all tools to their respective places. Reset the system to the manner in which you found it.

SUMMARY STATEMENT: Describe what would happen to the water flow if the condenser tubes became dirty. Use an additional sheet of paper if more space is necessary for your answer.

TAKEAWAY: Water-cooled air-conditioning and refrigeration systems are more efficient than air-cooled systems. For this reason, water-cooled systems are the system of choice whenever that option is available. It is, therefore, very important for the HVACR service technician to be able to properly adjust water regulating valves to ensure efficient system operation.

QUESTIONS

1. What is the purpose of the high-pressure control?

2. What is the purpose of the water pressure regulating valve on a wastewater system?

3. What senses the pressure of the water pressure regulating valve?

4. Where is the typical water pressure regulating valve connected to the system?

5. Where is the high-pressure control typically connected to the refrigeration system?

Exercise CON-10

Direct Digital Controls (DDC)

Name	Date	Grade

OBJECTIVES: Upon completion of this exercise, you should be able to understand direct digital control (DDC) operations and procedures and evaluate flow charts and diagrams of DDC controller operations.

TOOLS, MATERIALS, AND EQUIPMENT: There are no tools, materials, or equipment required for this exercise.

⚠ **SAFETY PRECAUTIONS:** There are no safety precautions.

STEP-BY-STEP PROCEDURES

1. Explain the three-step procedure for conditioning a signal on a DDC control system.

2. Define an open-loop control configuration.

3. Define a closed-loop control configuration.

4. Define an analog input a, and list five examples of analog inputs to a controller.

5. Define a digital input, and list four examples of digital inputs to a controller.

6. Define the following terms in sentence form:

Direct digital control:

Controlled environment:

Sensor:

Database:

Algorithm:

Control point:

Controlled output devices:

Passive sensors:

Analog input:

Control system:

Feedback loop:

Set point:

Control medium:

Load shedding:

Sensitivity (gain):

TAKEAWAY: Most later-generation HVACR systems incorporate direct digital control strategies to some degree. Unlike electromechanical controls, the internal workings of DDC systems cannot be observed. It is, therefore, very important for the service technician to understand how to interact with these control modules to ensure proper and efficient system operation.

Exercise CON-11

Single-Function Solid-State Modules

Name	Date	Grade

OBJECTIVES: Upon completion of this exercise, you should be able to identify single-function solid-state modules and install an electronic timer to delay the starting of a motor.

INTRODUCTION: Many single-function solid-state modules are used in the HVACR industry. Single-function solid-state controls are used to operate fan motors, defrost heat pumps, and perform other time functions in a controls system. Technicians should be able to install and troubleshoot single-function solid-state controls in HVACR equipment.

TOOLS, MATERIALS, AND EQUIPMENT:

Selection of single-function solid-state modules, 110 V power supply, 24 V transformer, low-voltage thermostat, contactor (24 V coil), small 110 V electric motor, 12 in. × 12 in. Plywood board, and assorted mounting hardware.

⚠ **SAFETY PRECAUTION:** Make certain that the electrical source is disconnected when making electrical connections.

- Make sure all connections are tight.
- Make sure no bare current-carrying conductors are touching metal surfaces except the grounding conductor.
- Make sure the correct voltage is being supplied to the circuit.
- Make sure body parts do not come in contact with live electrical conductors.
- Keep hands and materials away from moving parts.

STEP-BY-STEP PROCEDURES

1. Obtain a selection of six single-function solid-state modules from your instructor.

2. Record the model and function of each single-function solid-state device:

Solid-State Module	Part Number	Function
1.	_____	_____
2.	_____	_____
3.	_____	_____
4.	_____	_____
5.	_____	_____
6.	_____	_____

3. Study the wiring diagram in Figure CON-11.1, which shows how the small 120 V electric motor will start 30 seconds after the low-voltage thermostat closes on a call for heat.

FIGURE **CON-11.1**

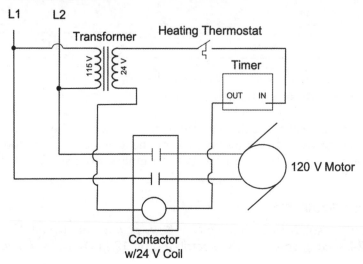

4. Identify which of the following diagrams best matches the diagram in Figure CON-11.1. Record your answer here: _____

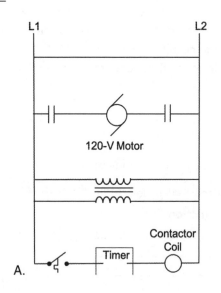

A.

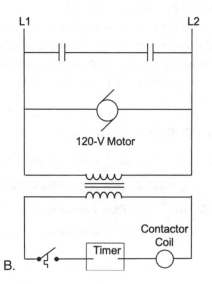

B.

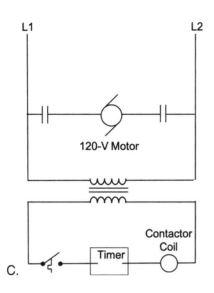

C.

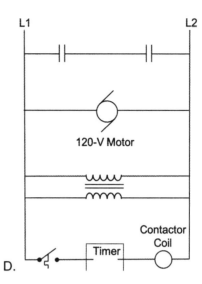

D.

5. Using your diagram, wire the circuit on a plywood board.

6. Connect the circuit to 115 V power supply.

7. Have your instructor check your circuit.

8. Close the switch for the 115 V power supply.

9. Close the thermostat, and begin timing to see how long it takes for the motor to start.

10. Record what happens when the thermostat closes. How long did it take for the motor to start? _____ seconds.

11. Compare the time entered in step 10 with the time delay set on the device. Discuss any discrepancies that exist between these two values. _____

12. Disconnect the power source from the circuit and remove the components from the board.

MAINTENANCE OF WORKSTATION AND TOOLS: Clean all tools, and return them to their proper locations. Replace all equipment covers. Clean up the workstation.

SUMMARY STATEMENT: What is the action of a solid-state timer when it is energized?

TAKEAWAY: Although solid-state modules and controls are electronic devices, the set points are often entered manually via rotating dials. Once set, the actual performance of the device must be observed and verified to ensure proper operation.

QUESTIONS

1. What would be the function in a control circuit of a solid-state timer that delays on closing?

2. What would be the function in a control circuit of a solid-state timer that delays on opening?

3. Why is a defrost control needed for a heat pump in the heating cycle?

4. Name at least five single-function solid-state controls used in the HVACR industry.

5. What procedure would a technician use to troubleshoot a single-function electronic module?

6. What are the advantages of using solid-state devices control systems?

7. What is a thermistor?

8. What are some reasons that electronic control modules have made their rapid advancement into the HVACR industry?

Multifunction Solid-State Modules

Name	Date	Grade

OBJECTIVE: Upon completion of this exercise, you should be able to identify and install a multifunction electronic module. In addition, you should be able to identify the input, ouput, and sensory ports on an electronic module.

INTRODUCTION: Many HVACR control systems utilize electronic modules to monitor the operation of heating and air-conditioning equipment. These electronic modules rely on multiple input signals that allow the module to control the operation of the main system components. The control module on a heat pump system, for example, must monitor system pressures, coil temperatures, air temperatures, and other system parameters to produce the desired system output. Gas furnaces use electronic modules to monitor pilot safety, system lockout, prepurge, postpurge, blower motor operation, heat exchanger temperature, and other system parameters. Many of these electronic modules are equipped with diagnostic functions that assist the technician with system evaluation and troubleshooting.

TOOLS, MATERIALS, AND EQUIPMENT:

Four multifunction electronic modules (two gas furnace modules and two heat pump modules), one operating heat pump with electronic module, and one operating gas furnace with electronic module.

 SAFETY PRECAUTIONS: When you inspect modules in operating heat pump and furnace, make certain that body parts do not come in contact with live electrical circuits.

STEP-BY-STEP PROCEDURES

1. Obtain four electronic modules from your instructor (two heat pump modules and two gas furnace modules).

2. Record the part number, the type of each module and the inputs and outputs:

Module	Part Number	Type	Inputs	Outputs
#1	_____	_____	_____	_____
	_____		_____	_____
	_____		_____	_____
	_____		_____	_____
#2	_____	_____	_____	_____
	_____		_____	_____
	_____		_____	_____
	_____		_____	_____
#3	_____	_____	_____	_____
	_____		_____	_____
	_____		_____	_____
	_____		_____	_____

#4

_____   _____   _____   _____

_____   _____   _____

_____   _____   _____

_____

3. Locate and observe the operation of the heat pump with electronic module.

4. Open the control panel of the unit. If necessary, turn the system off, and lock the power supply.

5. Closely examine the electronic control module, and provide the following information:

Inputs of board: _____

_____

_____

_____

Outputs of board: _____

_____

_____

_____

Function of board: _____

_____

_____

_____

6. Locate and observe the operation of a gas furnace with electronic module.

7. Closely examine the electronic control module and provide the following information:

Inputs of board: _____

_____

_____

_____

Outputs of board: _____

_____

_____

_____

Function of board: _____

_____

_____

_____

MAINTENANCE OF WORKSTATION AND TOOLS: Return electronic modules to the proper location, and clean tools and return them to their proper locations. Replace all equipment covers. Clean up work area.

SUMMARY STATEMENT: Why are electronic modules used to control the operation of heat pumps and gas furnaces?

TAKEAWAY: Upon initial inspection, solid-state control modules can be intimidating. However, upon closer inspection, it can be seen that the module simply produces electrical output signals based on the input signals it receives. Continued exposure to these modules will help reduce the intimidation factor.

QUESTIONS

1. What is the purpose of the signal light on an electronic module used on a gas furnace?

2. Why is it important for the technician to know the inputs and outputs to an electronic module used on a heat pump or gas furnace?

3. Why is it impractical for a technician to check the electronic components built into an electronic module?

4. What functions of a gas furnace does the electronic control module control?

5. What are the inputs and outputs of the heat pump electronic control module shown in Figure CON-12.1?

FIGURE **CON-12.1**

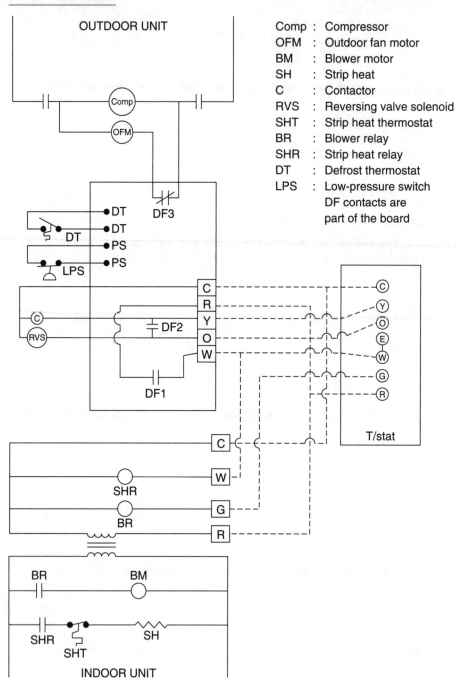

Comp	:	Compressor
OFM	:	Outdoor fan motor
BM	:	Blower motor
SH	:	Strip heat
C	:	Contactor
RVS	:	Reversing valve solenoid
SHT	:	Strip heat thermostat
BR	:	Blower relay
SHR	:	Strip heat relay
DT	:	Defrost thermostat
LPS	:	Low-pressure switch
		DF contacts are
		part of the board

6. What are the inputs and outputs of the gas furnace electronic control module shown in Figure CON-12.2?

FIGURE **CON-12.2**

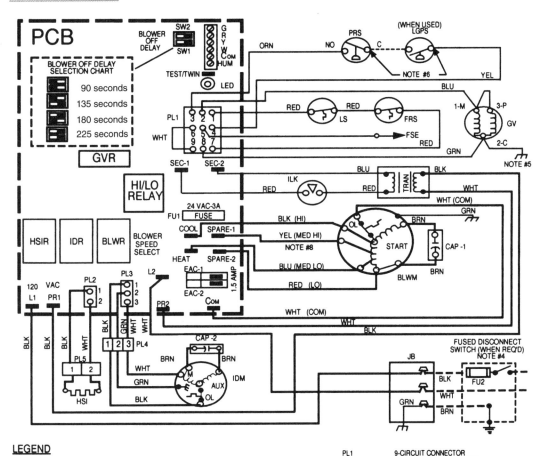

LEGEND

BLWR	BLOWER MOTOR RELAY, SPST-(N.O.)
BLWM	BLOWER MOTOR
CAP	CAPACITOR
CPU	MICROPROCESSOR AND CIRCUITRY
EAC-1	ELECTRONIC AIR CLEANER CONNECTION (115 VAC 1.5 AMP MAX.)
EAC-2	ELECTRONIC AIR CLEANER CONNECTION (COMMON)
FRS	FLAME ROLLOUT SW. -MANUAL RESET, SPST-(N.C.)
FSE	FLAME PROVING ELECTRODE
FU1	FUSE, 3 AMP, AUTOMOTIVE BLADE TYPE, FACTORY INSTALLED
FU2	FUSE OR CIRCUIT BREAKER CURRENT INTERRUPT DEVICE (FIELD INSTALLED & SUPPLIED)
GV	GAS VALVE-REDUNDANT OPERATORS
GVR	GAS VALVE RELAY, DPST-(N.O.)
HI/LO	BLOWER MOTOR SPEED CHANGE RELAY, SPDT
HSI	HOT SURFACE IGNITOR (115 VAC)
HSIR	HOT SURFACE IGNITOR RELAY, SPST-(N.O.)
HUM	24VAC HUMIDIFIER CONNECTION (.5 AMP. MAX.)
IDM	INDUCED DRAFT MOTOR
IDR	INDUCED DRAFT RELAY, SPST-(N.O.)
ILK	BLOWER ACCESS PANEL INTERLOCK SWITCH, SPST-(N.O.)
JB	JUNCTION BOX
LED	LIGHT-EMITTING DIODE FOR STATUS CODES
LGPS	LOW GAS PRESSURE SWITCH, SPST-(N.O.)
LS	LIMIT SWITCH, AUTO RESET, SPST(N.C.)
OL	AUTO-RESET INTERNAL MOTOR OVERLOAD TEMP. SW.
PCB	PRINTED CIRCUIT BOARD

PL1	9-CIRCUIT CONNECTOR
PL2	2-CIRCUIT PCB CONNECTOR
PL3	3-CIRCUIT IDM CONNECTOR
PL4	3-CIRCUIT IDM EXTENSION CONNECTOR
PL5	2-CIRCUIT HSI/PCB CONNECTOR
PRS	PRESSURE SWITCH, SPST-(N.O.)
SW1 & 2	BLOWER OFF DELAY
TEST/TWIN	COMPONENT TEST & TWIN TERMINAL
TRAN	TRANSFORMER-115VAC/24VAC

—●—	JUNCTION
○	UNMARKED TERMINAL
▬	PCB TERMINAL
———	FACTORY WIRING (115VAC)
———	FACTORY WIRING (24VAC)
- - -	FIELD WIRING (115VAC)
- - -	FIELD WIRING (24VAC)
	CONDUCTOR ON PCB
	FIELD WIRING SCREW TERMINAL
	FIELD GROUND
	EQUIPMENT GROUND
	FIELD SPLICE
	PLUG RECEPTACLE

Courtesy of Carrier Corporation

7. If a technician determines that an electronic module is faulty on a heat pump or gas furnace, what action should he or she take?

8. What troubleshooting procedure would a technician use to determine if a multifunction electronic module was functioning properly?

9. How has the advancement of electronic modules changed control systems on heat pumps and gas furnaces?

Exercise CON-13

Electric Heating Controls

Name	Date	Grade

OBJECTIVES: Upon completion of this exercise, you should be able to make the electrical connections on a practice wiring board, and wire an electric furnace.

INTRODUCTION: Although electric heating is expensive in most parts of the country, it is still used occasionally as a primary heating source. More often, though, it is used as the supplementary or emergency heat source on heat pump applications. No matter where the technician finds these applications, there are many similarities in the control systems. When electricity is used as the primary heat source, sequencers are used to control the resistance heaters by staging the starting and stopping of the heaters. In heat pump applications, electric resistance heaters are controlled by the indoor and outdoor thermostats, if the system is so equipped. The heaters and safety controls will all be very similar in design.

TOOLS, MATERIALS, AND EQUIPMENT:

Electric furnace with installation instructions, sequencer, transformer, relay, thermostat, other miscellaneous electrical devices, digital multimeter with amp clamp, plywood to be used for practice wiring board, miscellaneous wiring supplies, basic electrical hand tools, colored light bulbs, sockets, assorted mounting hardware, wire, solderless connectors, wire nuts, electrical tape, anemometer or velometer, digital thermocouple thermometer, and tape measure.

 SAFETY PRECAUTIONS: Make certain that the electrical source is disconnected when making electrical connections. In addition:

- Make sure all connections are tight.
- Make sure no bare current-carrying conductors are touching metal surfaces except the grounding conductor.
- Make sure the correct voltage is being supplied to the circuits.
- Make sure body parts do not come in contact with live electrical conductors.
- Keep hands and materials away from moving parts.

STEP-BY-STEP PROCEDURES

1. Study the following schematic diagram for an electric furnace that operates with two electric resistance heaters and a blower motor. Pay particular attention to the following safety components:

- Fuses to protect the heaters from current overload
- Temperature limits for each heater to ensure that overheating does not occur

FIGURE **CON-13.1**

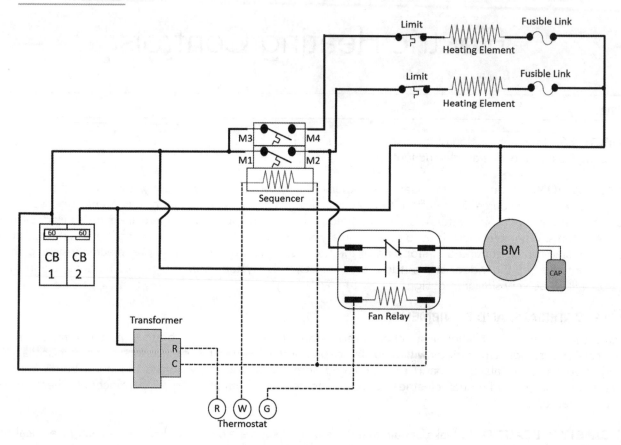

2. Create a list of the components and materials that will be needed to install the circuits on the practice board. (Note: Use light bulbs to represent the electric heaters and blower motor.)

3. Obtain the necessary components and supplies from your instructor.

4. Mount the components on the practice board, and make the necessary electrical connections to complete your diagram.

5. Have your instructor check your wiring board.

6. Connect the wiring board to a 115 V power source.

7. Operate the control system.

8. Record the operational sequence of the electric heating system.

9. Disconnect the electrical board from the power source.

10. Remove the electrical components from the board, and return them to their proper location.

11. Obtain a unit assignment (electric furnace) from your instructor. Record the following information from the appliance's nameplate:

Make: _____

Model: _____

Serial: _____

System capacity: _____

12. Read the installation instructions for the electric furnace.

13. Determine and record the correct wire size for the electric furnace.

Wire size: _____

14. Compile a list of materials that will be needed to install the assigned electric furnace.

15. Make the necessary electrical connections for the proper installation of the electric furnace.

16. Have your instructor check your installation.

17. Operate the electric furnace, and check the following electrical characteristics of the system.

Supply voltage: _____

Control voltage: _____

Operating current: _____

18. Write a brief explanation of the operation of the electric furnace.

Note: Steps 19 through 26 involve airflow and capacity calculations and can be omitted at the discretion of the instructor.

19. With the system operating, obtain the following temperature readings:

Return air temperature: _____

Supply air temperature: _____

Delta T (Supply Air Temperature − Return Air Temperature): _____

20. Obtain the measurements of the supply duct on the furnace, and record them here: _____ in. by _____ in.

21. Using the measurements from step 20, calculate the cross-sectional area of the duct, and record the value here: _____ ft^2.

22. Traverse the supply duct to obtain air velocity readings, and record them here:

Reading 1: _____ fpm   Reading 7: _____ fpm

Reading 2: _____ fpm   Reading 8: _____ fpm

Reading 3: _____ fpm   Reading 9: _____ fpm

Reading 4: _____ fpm   Reading 10: _____ fpm

Reading 5: _____ fpm   Reading 11: _____ fpm

Reading 6: _____ fpm   Reading 12: _____ fpm

23. Calculate the average air velocity, and record the value here: _____ fpm.

24. Calculate the airflow in cfm using the following formula (cfm = Velocity × Area) and the results from steps 21 and 23.

25. Using the formula QT = 1/08 × cfm × Delta T, calculate the capacity, in Btu/h, of the furnace.

26. Compare the nameplate capacity from step 11 with the calculated value in step 25. Discuss any discrepancies that exist between these two values.

_____

_____

27. Follow the instructor's directions for disassembling your installation.

MAINTENANCE OF WORK STATION AND TOOLS: Clean all tools, and return them to their proper location(s). Replace all equipment covers. Clean up the work area.

SUMMARY STATEMENT: Why are sequencers commonly used instead of contactors to control the electric resistance heaters in an electric furnace?

TAKEAWAY: Electric heat, although expensive in many geographic areas, is the most efficient source of heat when compared to fossil fuels. Even so, there are many applications where electric heat is used, so it is important to become familiar with its operation and installation.

QUESTIONS

1. How are supplementary electric resistance heaters controlled in a heat pump system?

2. How can fuses adequately protect electric resistance heaters?

3. If fuses protect the heaters, why are limit switches needed in electric resistance heater applications?

4. What is the Btu/hr output of a 5 kW resistance heater?

5. What is the advantage of using sequencers in electric furnaces?

6. What is the difference between an electric furnace and a supplementary heater used in heat pump installations?

7. Why is it important to have some type of fan interlock on electric resistance heater applications?

Exercise CON-14

Gas Heating Controls

Name	Date	Grade

OBJECTIVES: Upon completion of this exercise, you should be able to correctly install and wire gas controls on gas-fired, warm-air furnaces in the shop.

INTRODUCTION: There are several types of ignition controls used on gas-fired, warm-air furnaces: standing pilot, intermittent pilot, and direct spark ignition. The technician will find all three types of ignition control systems in the field and should understand the operation of each. The electrical circuitry of most gas furnaces is similar, with the exception of the ignition control system. Many gas furnaces are now equipped with some type of vent fan, which also must be controlled by the control system. Although newer gas-fired appliances are not equipped with standing pilots, these systems are still in operation and will require periodic service.

TOOLS, MATERIALS, AND EQUIPMENT:

Gas furnace with standing pilot ignition system, gas furnace with intermittent ignition system, gas furnace with direct ignition system, gas valves, thermocouples, pilot burners, pilot assemblies, hot surface igniter, intermittent ignition module, direct ignition module, VOM meter, basic electrical hand tools, tools to remove gas valve, and miscellaneous wiring supplies.

⚠ **SAFETY PRECAUTIONS:** Make certain that the electrical source is disconnected when making electrical connections. In addition:

- Make sure all connections are tight.
- Make sure no bare current-carrying conductors are touching metal surfaces except grounding conductors.
- Make sure the correct voltage is being supplied to the circuits or equipment.
- Make sure body parts do not come in contact with live electrical conductors.
- Keep hands and materials away from moving parts.

When installing components in gas lines, make sure there are no leaks. When igniting gas furnaces, do not stand directly in front of the combustion chamber. Do not move ignition control components from the manufacturer's location. Make sure the ignition control is correct for the application in which it is being used.

STEP-BY-STEP PROCEDURES

1. Position yourself at a gas furnace as directed by your instructor.

2. Make sure the power is off and the electrical disconnect switch is appropriately tagged.

3. Remove the front cover(s) of the furnace and the junction box cover(s).

4. Examine the components and wiring of the furnace. If you do not understand any component and its operation, refer to the appliance literature. If additional assistance is needed, consult with your instructor. Once you completely understand the wiring and operation of the furnace, study the wiring diagram on the appliance.

5. Provide the sequence of operations for the furnace.

6. Replace the blower compartment cover.

7. Operate the furnace and briefly explain in writing the function of each component.

8. Have your instructor check your work.

9. Turn off the power supplying the furnace.

10. Remove front panel of furnace and panel covering the combustion chamber.

11. Install a thermocouple adapter where the thermocouple is attached to the gas valve.

12. Light the furnace pilot. (Note: Follow the manufacturer's instructions for lighting the pilot.)

13. Record the millivoltage produced by the thermocouple when the pilot dial is depressed (unloaded) and when the pilot dial is released (loaded).
    Millivoltage (unloaded): _____
    Millivoltage (loaded): _____

14. Restore power to the furnace and set the thermostat to a call for heat.

15. Record the millivoltage produced by the thermocouple with the main burner operating.
    _____

16. Replace all covers and panels.

17. Position yourself at a standing pilot gas furnace as directed by your instructor.

18. Turn off the power supplying the furnace, and make certain that the disconnect switch is appropriately tagged.

19. Remove the panels from the furnace covering the burner compartment and combustion chamber.

20. Study the wiring and location of the components.

21. Make certain that the gas supply to the unit is off and that the main gas valve is locked and appropriately tagged.

22. Examine the gas piping and determine the disassembly procedure. Look for a union or flare connection at the inlet of the gas valve.

23. Remove the pilot line connections from the gas valve. Use the correct tools for the job. If you are unsure how to properly use pipe wrenches, ask your instructor for guidance.

24. Disassemble the piping to the gas valve. Use the correct tools for the job.

25. Examine the gas valve application to determine the type of pilot ignition. Pay careful attention to where the ignition components connect. Remove the ignition components from the gas valve.

26. Remove the gas valve from the gas manifold. Record the data from the valve's nameplate here:

    Make: _____
    Model/part: _____
    Valve type: _____
    Pipe size: _____

27. Install the gas valve obtained from your instructor back to the gas manifold. Use joint compound or Teflon tape on all external pipe threads.

28. Connect the gas supply back to the inlet of the gas valve.

29. Connect the ignition components back to the gas valve.

30. Make all electrical connections on the gas valve. Make sure connections are correct; refer to your diagram.

31. Have your instructor check your gas valve installation.

32. Turn the gas supply on.

33. Check the gas supply piping for leaks using soap bubbles.

34. Restore the power supply to the furnace.

35. Light the pilot if the furnace has a standing pilot ignition system.

36. Set the thermostat to a call for heat. When ignition of the burner is accomplished, immediately check the remaining piping for leaks, including the pilot lines.

37. When you have determined that no leaks exist, turn off the furnace.

38. Clean soap solution and residue from gas valve and piping.

39. Replace all panels on the furnace.

40. Position yourself at a gas furnace with an intermittent pilot ignition system as directed by your instructor.

41. Assume that the intermittent ignition module on the assigned furnace is faulty and must be replaced.

42. Remove any furnace panels necessary to obtain access to the ignition module.

43. Perform preinstallation safety inspection.

    a. Test gas piping for gas leaks.

    b. Visually inspect venting system for proper installation and size.

    c. Inspect burners and crossovers for blockage and corrosion.

    d. Inspect heat exchanger in warm-air furnace for damage such as cracks and excessive corrosion. If it is a boiler, check for water leaks and combustion gas leaks.

    e. Make any other safety checks required by state and local codes.

44. Obtain a new intermittent ignition module from your instructor.

45. Compare the rating of the old ignition module with the new ignition module to make sure it is suitable for your application.

46. Disconnect the power supply, and make certain that the electrical disconnect switch is appropriately tagged.

47. Disconnect, label, and tag the wires from the old module.

48. Remove the old module from its mounting location.

49. Mount the new module in the same location as the old module, if possible.

50. Wire the module, checking the wiring diagram furnished by the manufacturer and the module installation instructions.

51. Visually inspect the module installation, making sure all wiring connections are clean and tight.

52. Verify the control system ground. The igniter, flame sensor, and ignition module must share a common ground with the main burner.

53. Have your instructor check your wiring.

54. Restore power to the furnace.

55. Review the normal operating sequence and module specifications.

56. Check the safety shutoff operation. Turn off the gas supply. Set the thermostat to a call for heat. Watch for a spark at the pilot burner, and time the spark from start to shutoff; this time should be 90 seconds maximum. Open the manual gas valve, and make sure no gas is flowing to the pilot or main burner.

57. Set the thermostat to the lowest setting, and wait 1 minute.

58. Check the normal operation of the ignition module by moving the thermostat to a call for heat. Make sure the pilot lights smoothly. Make sure the main burner lights smoothly without flashback and that the main burner is correctly adjusted.

59. Return the thermostat to a setting below room temperature. Make sure the main burner goes out.

60. Replace all furnace panels.

61. Return replaced ignition module to your instructor.

62. Clean up the work area.

63. Position yourself at a gas furnace that is equipped with a direct ignition system as directed by your instructor.

64. Assume that the direct ignition module on the assigned furnace is faulty and must be replaced.

65. Remove any furnace panels necessary to obtain access to the ignition module.

66. Perform preinstallation safety inspection.

   a. Test gas piping for gas leaks.

   b. Visually inspect venting system for proper installation and size.

   c. Inspect burners and crossovers for blockage and corrosion.

   d. Inspect heat exchanger in warm-air furnace for damage such as cracks and corrosion. If it is a boiler, check for water leaks and combustion gas leaks.

   e. Make any other safety checks required by state and local codes.

67. Obtain a new direct ignition module from your instructor.

68. Compare the rating of the old ignition module with the new ignition module, and make sure it is suitable for your application.

69. Disconnect the power supply, and tag the switch "SERVICE IN PROGRESS" before doing any work on the unit.

70. Disconnect and tag the wires from the old module.

71. Remove the old module from its mounting location.

72. Mount the new module in the same location as the old module, if possible.

73. Wire the module, checking the wiring diagram furnished by the manufacturer and the module installation instructions.

74. Verify the thermostat heat anticipator setting.

75. Visually inspect the module installation, making sure all wiring connections are clean and tight.

76. Verify the control system ground. The igniter, flame sensor, and ignition module must share a common ground with the main burner.

77. Have your instructor check your wiring.

78. Restore power to the furnace.

79. Review the normal operating sequence and module specifications.

80. Set the thermostat to a temperature setting higher than room temperature for at least 1 minute.

81. Check safety shutoff operation. First, turn the gas supply off, and set the thermostat above the room temperature to a call for heat. Observe the operation of the spark igniter or hot surface igniter. (A spark igniter will spark following prepurge, and a hot surface igniter will begin to glow several seconds after prepurge.)

Determine the trial ignitions from the ignition module installation instructions to determine the number of trial ignitions that the control should allow. Open the manual knob on the gas valve, and make sure no gas flows to the burner. Set the thermostat below room temperature.

82. Check the normal operation of the ignition module by setting the thermostat above room temperature. Observe the lighting sequence, and make sure the main burner lights smoothly and without flashback. Make sure the burner operates smoothly without floating, lifting, or flame rollout.

83. Return the thermostat to a setting below room temperature. Make sure the main burner goes out.

84. Replace all furnace panels.

85. Return replaced ignition module to your instructor.

86. Clean up the work area.

MAINTENANCE OF WORKSTATION AND TOOLS: Clean all tools, and return them to their proper location(s). Replace all equipment covers. Clean up the work area.

SUMMARY STATEMENT: What is the difference between a standing pilot ignition system, an intermittent ignition system, and a direct ignition system?

TAKEAWAY: When arriving on a no-heat service call, you are not going to know what type of ignition system the appliance is equipped with. For this reason, it is important to become familiar with as many different ignition modules as possible to ensure success in the field.

QUESTIONS

1. What is the purpose of a limit switch in the control system of a gas furnace?

2. What is a thermocouple?

3. Explain the operation of an intermittent ignition module.

4. What should be the millivolt output of a good thermocouple?

5. What is a hot surface igniter?

6. What is the purpose of a prepurge cycle in a direct ignition system?

7. Why should the safety lockout of gas ignition systems always be checked when new components are installed?

8. What is the response time between the lockout of a standing pilot ignition system and an intermittent pilot ignition system?

9. Why should a technician always label the wiring when removing wires from an ignition module that is to be replaced?

10. Why is it important for a technician always to replace ignition components in the location where they were placed by the manufacturer?

Oil Heating Controls

Name	Date	Grade

OBJECTIVES: Upon completion of this exercise, you should be able to correctly install oil burner primary controls on oil-fired, warm-air furnaces in the shop.

INTRODUCTION: Oil-fired, warm-air furnaces are equipped with primary controls that turn the burner off and on in response to the thermostat action and monitor the oil burner flame. There are basically two types of primary controls used in the industry, the stack switch and the cad cell control. The stack switch uses the heat of the flue gases to monitor and prove that a flame has been established, while the cad cell primary control uses a light-sensitive cad cell to prove and monitor the flame of an oil burner. The primary control is the heart of the oil burner control system, but other controls are also important to the safe operation of an oil burner, such as fan and limit switches. Although stack relays are not, for the most part, installed any more, they are still in operation on many systems and require periodic service.

TOOLS, MATERIALS, AND EQUIPMENT:

Oil-fired furnace, stack switch primary oil burner controls, cad cell primary oil burner controls, digital multimeter, basic electrical hand tools, and miscellaneous wiring supplies.

SAFETY PRECAUTIONS: Make sure that the electrical source is disconnected when making electrical connections. In addition:

- Make sure all connections are tight.
- Make sure no bare current-carrying conductors are touching metal surfaces except the grounding conductor.
- Make sure the correct voltage is being supplied to the circuits or equipment.
- Make sure body parts do not come in contact with live electrical conductors.
- Keep hands and materials away from moving parts.

When igniting oil furnaces, do not stand directly in front of the combustion chamber access door. Make certain that there are no leaks in the oil piping. Make sure the oil burner primary control is the correct type for the application in which it is being used.

STEP-BY-STEP PROCEDURES

1. Position yourself at an oil-fired furnace equipped with a tack switch primary control as directed by your instructor.

2. Assume that the stack switch primary control on the assigned oil-fired furnace is faulty and must be replaced.

3. Remove any furnace panels necessary to obtain access to the oil burner and controls.

4. Perform preinstallation safety inspection.

   a. Make sure oil piping has no leaks.

   b. Visually inspect venting system for proper installation and size.

   c. Inspect oil burner for air blockage and corrosion.

d. Inspect heat exchanger in warm-air furnace for damage such as cracks and corrosion. If it is a boiler, check for water leaks and combustion gas leaks.

e. Make any other safety checks required by state and local codes.

5. Obtain a new stack switch primary control from your instructor.

6. Compare the rating of the old stack switch with the new stack switch, and make sure it is suitable for your application.

7. Disconnect the power supply, and make certain that the electrical disconnect switch is appropriately tagged.

8. Disconnect, label, and tag the wires from the stack switch.

9. Remove the old stack switch from its mounting location.

10. Mount the new stack switch in the same location as the old stack switch.

11. Wire the stack switch, checking the wiring diagram furnished by the appliance manufacturer and the stack switch installation instructions.

12. Verify the thermostat heat anticipator setting.

13. Visually inspect the stack switch installation, making sure all wiring connections are clean and tight.

14. Have your instructor check your wiring.

15. Restore power to the furnace.

16. Review the normal operating sequence and stack switch specifications. If the new stack switch will not close, put the stack switch in step by pulling the drive shaft lever forward 1/4 in., then slowly releasing.

17. Set the thermostat to a temperature setting lower than room temperature for at least 1 minute.

18. Close the thermostat and check the safety shutoffs of the stack switch.

a. Check the flame failure function by shutting off the oil supply to the burner. The stack switch should lock the oil burner out on safety. This safety switch must be reset manually.

b. Check the scavenger timing of the stack switch by operating the burner normally; then open and immediately close the line-voltage switch. The oil burner should stop immediately, and after recycling time (usually 1 to 3 minutes) it should restart automatically.

19. Check for normal operation of the stack switch by setting the thermostat above room temperature. Observe the lighting sequence and make sure the burner lights smoothly. Make sure the burner operates smoothly with the proper flame.

20. Return the thermostat setting to below room temperature. Make sure the oil burner goes out.

21. Replace all furnace panels.

22. Return the faulty stack switch to your instructor.

23. Clean up the work area.

24. Position yourself at an oil-fired furnace equipped with a cad cell primary control as directed by your instructor.

25. Assume that the cad cell primary control on the assigned furnace is faulty and must be replaced.

26. Remove any furnace panels necessary to obtain access to the oil burner and controls.

27. Perform preinstallation safety inspection.

⬤

   a. Make sure oil piping has no leaks.

   b. Visually inspect venting system for proper installation and size.

   c. Inspect oil burner for air blockage and corrosion.

   d. Inspect heat exchanger in warm-air furnace for damage such as cracks and corrosion. If it is a boiler, check for water leaks and combustion gas leaks.

   e. Make any other safety checks required by state and local codes.

28. Obtain a new cad cell and cad cell primary control from your instructor.

29. Plug the cad cell into the holder. Aim the cad cell toward a bright light; measure and record the resistance of the cad cell. Next, cover the cad cell; measure and record the resistance.

Lighted cad cell resistance: _____ Ω
Dark cad cell resistance: _____ Ω

30. Compare the rating of the old cad cell primary control with the new cad cell primary control, and make sure it is suitable for your application.

31. Disconnect the power supply, and make certain that the electrical disconnect switch is appropriately tagged.

32. Disconnect and tag the wires from the cad cell primary control.

33. Remove the cad cell primary control and cad cell from its mounted location.

34. Mount the new cad cell and cad cell primary control in the same location as the old cad cell and cad cell primary control. (Note: Do not move location of cad cell.)

⬤

35. Wire the cad cell primary control, checking the wiring diagram furnished by the unit manufacturer and the cad cell primary control installation instructions.

36. Verify the thermostat heat anticipator setting.

37. Visually inspect the cad cell and cad cell primary control installation, and make sure all wiring connections are clean and tight.

38. Have your instructor check your wiring.

39. Restore power to the furnace.

40. Review the normal operating sequence and cad cell primary control specifications.

41. Set the thermostat to a temperature setting lower than room temperature for at least 1 minute.

42. Close the thermostat and check the safety shutoff of the cad cell primary control by stopping the oil flow to the oil burner. The cad cell primary control should lock out in switch timing (15 to 70 seconds). The oil burner should stop. Most cad cell primary controls must be manually reset.

43. Check for normal operation of the stack switch by setting the thermostat above room temperature. Observe the lighting sequence, and make sure the oil burner lights. Make sure the burner operates smoothly with the proper flame.

44. Return the thermostat setting to below room temperature. Make sure the oil burner goes out.

45. Replace all furnace panels.

⬤

46. Return the faulty cad cell and cad cell primary control to your instructor.

47. Clean up the work area.

MAINTENANCE OF WORKSTATION AND TOOLS: Clean all tools, and return them to their proper location(s). Replace all equipment covers. Clean up the work area.

SUMMARY STATEMENT: What is the purpose of an oil burner primary control?

TAKEAWAY: When working on any fossil fuel heating appliance, it is important to verify that the safety devices are operational and that components be replaced properly. This exercise provides the opportunity to replace primary controls, both earlier and later generation, on live equipment.

QUESTIONS

1. What senses combustion when a cad cell primary control is used?

2. What senses combustion when a stack switch is used as the primary control?

3. Where is the cad cell located in an oil burner when used with a cad cell primary control?

4. What would be the results if the cad cell was not placed in line with the oil burner flame?

5. Explain the operation of a cad cell primary control.

6. Explain the operation of a stack switch primary control.

7. Where should the element of a stack switch be installed?

8. Why should a service technician check the safety functions of a primary control?

9. What would a service technician do if a new stack switch is installed and will not work?

10. When an oil burner does not ignite the oil being supplied to the combustion chamber, why is it important for the primary control to stop the oil burner?

Troubleshooting Simple Electric Circuits

Name	Date	Grade

OBJECTIVES: Upon completion of this exercise, using a multimeter and an ammeter, you should be able to properly evaluate a simple series circuit and determine whether the circuit is operational. As part of this exercise, you will determine whether system components are power-consuming or power-passing devices.

INTRODUCTION: A major part of the troubleshooting process involves locating the problem in a circuit as quickly and efficiently as possible. By understanding how electric circuits operate and how voltage and amperage readings indicate the condition of various circuit components, the troubleshooting process can be mastered with a fair amount of practice and dedication.

TOOLS, MATERIALS, AND EQUIPMENT: A power cord (length of SJ cable with a male plug attached to one end), a manual single-pole, single-throw switch, fuses (both good and defective, if possible), a line-voltage thermostat, 75 W light bulbs (good and defective, if possible), a light socket, wire, wire nuts, electrical tape, goggles, wire cutters, wire strippers, screwdrivers, a multimeter, and an ammeter.

⚠ **SAFETY PRECAUTIONS:** When working on or around electric circuits, make every attempt to avoid becoming part of the circuit. Avoid wearing metallic jewelry. Wear rubber-soled work boots. Whenever possible, work on circuits that are de-energized to avoid receiving an electric shock. When working on live circuits, be sure that you do not touch bare conductors. Always follow all of your teacher's instructions.

STEP-BY-STEP PROCEDURES

1. Wire up the circuit as shown here, Figure CON-16.1. When wiring the circuit, use a good fuse and a good light bulb.

FIGURE **CON-16.1**

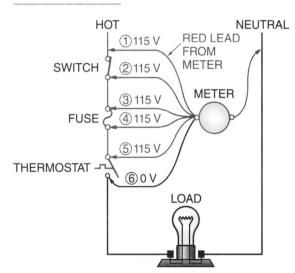

2. Identify each of the circuit components as either a power-passing or power-consuming device, and circle the appropriate term here:

Switch:	power-passing	power-consuming
Fuse:	power-passing	power-consuming
Thermostat:	power-passing	power-consuming
Light bulb:	power-passing	power-consuming

3. Have your instructor inspect the circuit to make certain that all connections are correct.

4. Place the switch in the open (off) position and the thermostat in the closed position.

5. Energize the circuit.

6. Take a voltage reading between the neutral line and point 1 in the circuit. Point 1 is the line side of the manual switch. Record the voltage here: _____ V.

7. Take a voltage reading between the neutral line and point 2 in the circuit. Point 2 is the load side of the manual switch. Record the voltage here: _____ V.

8. Take a voltage reading between point 1 and point 2, and record the voltage reading here: _____ V.

9. From the readings taken in steps 6, 7, and 8, you should be able to verify that the switch is in the open (off) position.

10. Take a voltage reading across the light bulb, and record it here: _____ V.

11. The reading obtained in step 10 should indicate why the bulb is not on.

12. Turn the switch to the closed (on) position. The bulb should light up.

13. Repeat the voltage readings in steps 6 and 7, and record them here:

   • Voltage between neutral and point 1: _____ V.

   • Voltage between neutral and point 2: _____ V.

   These readings should indicate that there is power passing through the switch because there is power between neutral and point 2.

14. Take a voltage reading between neutral and point 4, and record it here: _____ V.

15. Take a voltage reading between neutral and point 6, and record it here: _____ V.

   A positive voltage reading between neutral and point 6 indicates that all power-passing devices in the circuit are in the closed position and power is being passed to the circuit load.

16. Take a voltage reading across the bulb, and record it here: _____ V.

17. Take an amperage reading of the circuit, and record it here: _____ A.

   A positive voltage reading across the load indicates that the light bulb is being supplied power.

   A positive amperage reading indicates not only that the load is being supplied power but also that the load is good because it is consuming power.

18. Disconnect power to the circuit.

19. Remove the good fuse, and replace it with a defective one.

20. With the manual switch in the open (off) position and the thermostat in the closed position, re-energize the circuit.

21. Turn the switch to the closed (on) position.

22. Take a voltage reading between neutral and point 1, and record it here: _____ V.

23. Take a voltage reading between neutral and point 2, and record it here: _____ V.

24. Take a voltage reading between neutral and point 3, and record it here: _____ V.

25. Take a voltage reading between neutral and point 4, and record it here: _____ V.

26. Take a voltage reading between point 3 and point 4, and record it here: _____ V.

A positive voltage reading between point 3 and point 4 indicates that the fuse is no good and is not passing power on to the rest of the circuit.

27. Take an amperage reading of the circuit, and record it here: _____ A.

An amperage reading of 0 indicates that no current is flowing in the circuit and that no power is being consumed. A positive amperage reading indicates that there is a complete circuit, current is flowing, and that power is being consumed.

28. Based on the previous steps in this exercise, make a prediction about how much voltage is being supplied to the light bulb, and record it here: _____ V.

29. Take an actual voltage reading across the light bulb, and record it here: _____ V.

30. If the recorded values in steps 28 and 29 are different, re-evaluate the circuit, and repeat this portion of the exercise.

31. De-energize the circuit, and replace the defective fuse with the good one.

32. With the circuit still de-energized, replace the good light bulb with a defective one.

33. With the manual switch in the open (off) position and the thermostat in the closed position, restore power to the circuit.

34. Turn the manual switch to the closed (on) position.

35. Take a voltage reading between neutral and point 1, and record it here: _____ V.

36. Take a voltage reading between neutral and point 2, and record it here: _____ V.

37. Take a voltage reading between neutral and point 3, and record it here: _____ V.

38. Take a voltage reading between neutral and point 4, and record it here: _____ V.

39. Take a voltage reading between neutral and point 5, and record it here: _____ V.

40. Take a voltage reading between neutral and point 6, and record it here: _____ V.

41. Take a voltage reading across the bulb, and record it here: _____ V.

42. Take an amperage reading of the circuit, and record it here: _____ A.

43. What conclusions can be drawn from the information recorded in steps 35–42?

44. Now open the thermostat contacts, and take the following voltage measurements:

- Voltage between neutral and point 5: _____ V.

- Voltage between neutral and point 6: _____ V.

45. What conclusion can be drawn about the thermostat based solely on the information obtained in step 44?

MAINTENANCE OF WORKSTATION AND TOOLS: Make certain that all circuits are de-energized before disconnecting the components. Return all tools to their proper places. Leave the work area neat and clean.

SUMMARY STATEMENT: Multimeters and ammeters serve different functions when used to evaluate an electric circuit. When used together, these pieces of test equipment allow the service technician to quickly identify the circuit component that is preventing the system from operating properly. It is important for the technician to identify whether a particular circuit component is a load or a switch, as the measured voltage across it will need to be interpreted differently in each case.

TAKEAWAY: One of the major responsibilities of a service technician is to quickly and accurately troubleshoot electric circuits. This exercise provides the opportunity to evaluate simple circuits by taking and interpreting voltage and amperage readings.

QUESTIONS

1. Why is it that we often see multiple switches wired in series with a load?

2. If a load is wired in a circuit that has multiple switches in series with it, what conditions must be present in order for the load to be energized?

3. If a circuit has a current draw of 0 A, is the load energized? Explain your answer.

4. If a circuit is comprised of a load and multiple switches wired in series with the load and the load is operating, is it possible that one or more of the switches is in the open position?

5. If voltage is being supplied to a load and the load is not operating, what can be said about the switches in the circuit?

6. If voltage is being supplied to a load and the load is not operating, what can be said about the load in the circuit?

7. If voltage is being supplied to a load and the load is not operating, what can be said about the current flow in the circuit?

8. If voltage is being supplied to a load and the load is not operating, what further testing can be done to check the condition of the load in the circuit?

Exercise CON-17

Troubleshooting Electric Circuits

Name	Date	Grade

OBJECTIVES: Upon completion of this exercise, you should be able to properly evaluate a series/parallel circuit using a multimeter.

INTRODUCTION: A major part of the troubleshooting process involves locating the problem in a circuit as quickly and efficiently as possible. By understanding how electric circuits operate and how voltage and amperage readings indicate the condition of various circuit components, the troubleshooting process can be mastered with a fair amount of practice and dedication.

TOOLS, MATERIALS, AND EQUIPMENT: No tools or materials required.

⚠ **SAFETY PRECAUTIONS:** When working on or around electric circuits, make every attempt to avoid becoming part of the circuit. Avoid wearing metallic jewelry. Wear rubber-soled work boots. Whenever possible, work on circuits that are de-energized to avoid receiving an electric shock. When working on live circuits, be sure that you do not touch bare conductors. Always follow all of your teacher's instructions.

STEP-BY-STEP PROCEDURES

Using the circuit in Figure CON-17.1, what would be the voltage measurements if the meter leads were put across:

2 and 4 of the timer motor contacts: _____ V

1 and 3 of the timer motor contacts: _____ V

The timer motor: _____ V

The LPC: _____ V

The HPC: _____ V

One evaporator fan: _____ V

The defrost heaters: _____ V

The contactor coil: _____ V

The condenser fan motor: _____ V

The overload: _____ V

The fan cycle control: _____ V

The start winding: _____ V

The run winding: _____ V

Either contactor contacts: _____ V

The defrost limit switch: _____ V

SUMMARY STATEMENT: It is important for the technician to identify whether a particular circuit component is a load or a switch, as the measured voltage across it will need to be interpreted differently in each case. A schematic diagram can be used to help identify the system components.

TAKEAWAY: Being a successful troubleshooter involves being able to take voltage, amperage, and resistance readings, as well as being able to interpret these readings. This exercise provides some insight into the process of evaluating obtained readings to establish the condition of system components.

FIGURE **CON-17.1**

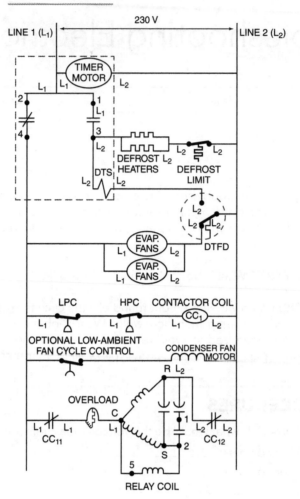

QUESTIONS

1. Explain why it is important to know the difference between reading voltage across a switch and reading voltage across a load.

2. Explain what two conditions might be present if a reading of 0 V is obtained across the two terminals of a switch.

3. Explain what two conditions might be present if a reading of 0 V is obtained across a fuse.

4. If there is a 120 V reading across a light bulb and the light bulb does not light up, what can be correctly concluded?

Exercise CON-18

Troubleshooting Thermostats

Name	Date	Grade

OBJECTIVES: Upon completion of this exercise, you should be able to correctly troubleshoot line-voltage and low-voltage thermostats.

INTRODUCTION: The main function of most refrigeration and conditioned air systems is to maintain a specific temperature of an object or space. The thermostat is the electrical device that controls the major loads in an HVACR system. The service technician must be able to troubleshoot line-voltage and low-voltage thermostats in order to effectively service HVACR systems.

TOOLS, MATERIALS, AND EQUIPMENT:

Line-voltage thermostat, single-stage cooling and heating low-voltage thermostat and subbase, two-stage heating, two-stage cooling low-voltage thermostat and subbase, thermostat kit, HVACR equipment, electrical meters, miscellaneous electrical supplies, and basic electrical hand tools.

SAFETY PRECAUTIONS: Make certain that the electrical source is disconnected when making electrical connections. In addition:

- Make sure all connections are tight.
- Make sure no bare current-carrying conductors are touching metal surfaces except the grounding conductor.
- Make sure the correct voltage is being supplied to the circuit.
- Make sure body parts do not come in contact with live electrical conductors.
- Keep hands and materials away from moving parts.

STEP-BY-STEP PROCEDURES

1. Obtain line-voltage heating and cooling thermostats from the supply room. Record the data from the devices here:

   Make and model: _____ Thermostat type: _____

   Make and model: _____ Thermostat type: _____

2. Using a multimeter, determine the action of the contacts of the line-voltage cooling thermostat as the temperature setting is increased and decreased. What is the action of the thermostat contacts on a rise and fall in temperature?

3. Using a multimeter, determine the action of the contacts of the line-voltage heating thermostat as the temperature setting is increased and decreased. What is the action of the thermostat contacts on a rise and fall in temperature?

4. Obtain a heating and cooling low-voltage thermostat and subbase from the supply room.

5. Connect thermostat wires from the R, W, Y, and G terminals of the low-voltage heating and cooling subbase.

6. Route the thermostat wires through the opening in the thermostat subbase.

7. Attach the thermostat to the subbase.

8. Record the actions of the thermostat when the following selections are made. Check the continuity of the circuit in the thermostat using a multimeter.

    a. Move the fan switch from "auto" to "on."

    _____

    b. Fan switch in "auto" position and system switch set to "off."

    _____

    c. Fan switch in "auto" position, system switch set to "cool" position, and thermostat set point decreased to a temperature that is lower than room temperature.

    _____

    d. Fan switch in "auto" position, system switch set to "heat" position, and thermostat set point increased to a temperature that is higher than room temperature.

    _____

9. Obtain a two-stage heating, two-stage cooling low-voltage thermostat and subbase from the supply room.

10. Connect thermostat wires from R, W1, W2, Y1, Y2, and G terminals of the two-stage heating, two-stage cooling subbase.

11. Route the thermostat wires through the opening in the thermostat subbase.

12. Attach the thermostat to the subbase.

13. Record the actions of the thermostat when the following selections are made. Check the continuity of the thermostat using a multimeter.

    a. Move the fan switch from "auto" to "on."

    _____

    b. Fan switch in "auto" position and system switch set to "off."

    _____

    c. Fan switch in "auto" position, system switch set to "cool" position, and thermostat set at least 5°F below room temperature.

    _____

    d. Fan switch in "auto" position, system switch set to "heat" position, and thermostat set at least 5°F above room temperature.

    _____

14. Return all thermostats to their appropriate location.

15. Obtain a thermostat kit from your instructor.

16. Troubleshoot and record the condition of the thermostats.

Line-voltage thermostat #1: _____

Line-voltage thermostat #2: _____

Line-voltage thermostat #3: _____

Low-voltage thermostat #1: _____

Low-voltage thermostat #2: _____

Low-voltage thermostat #3: _____

17. Return the thermostat kit to your instructor.

18. Troubleshoot the three thermostats on live equipment assigned by your instructor.

Thermostat	Location	Thermostat Type	Condition of Thermostat
#1	_____	_____	_____
#2	_____	_____	_____
#3	_____	_____	_____

MAINTENANCE OF WORKSTATION AND TOOLS: Clean all tools, and return them to their proper location(s). Replace all equipment covers. Clean up the work area.

SUMMARY STATEMENT: What is the difference between troubleshooting a line-voltage thermostat and troubleshooting a low-voltage thermostat?

TAKEAWAY: Thermostats are the main interface between the equipment owner/operator and the equipment itself. The troubleshooting process almost always starts at the thermostat, so familiarization with various types of thermostats is important to achieve.

QUESTIONS

1. Name some common applications in which low-voltage thermostats are used.

2. Name some common applications in which line-voltage thermostats are used.

3. Explain the action of a heating and cooling thermostat.

4. What is the proper procedure for troubleshooting a two-stage heating, two-stage cooling thermostat?

5. How is the fan operated on a gas heat, electric air-conditioning system?

6. Why is a low-voltage thermostat more difficult to troubleshoot than a line-voltage thermostat?

7. How would a service technician determine if a low-voltage thermostat is faulty?

8. How would a service technician jumper a low-voltage thermostat to determine if it is operating properly?

9. What is the best action to take if a thermostat has inaccurate cut-in and cut-out temperatures?

10. What would happen if the transmission line connecting the thermostat to the bulb was broken?

Exercise CON-19

Troubleshooting Pressure Switches

Name	Date	Grade

OBJECTIVES: Upon completion of this exercise, you should be able to correctly troubleshoot various pressure switches on equipment in the shop.

INTRODUCTION: Pressure switches are used as safety controls and operating controls in the control systems of refrigeration, heating, and air-conditioning systems. The service technician must know the function of the pressure switch in the control system before attempting to troubleshoot it. When troubleshooting pressure switches, the technician must know at what point the pressure switch should be opening or closing and at what point it actually is opening or closing. Pressure switches are used to protect the refrigeration system from operating at pressures that are unsafe.

TOOLS, MATERIALS, AND EQUIPMENT:

Low-pressure switches, high-pressure switches, dual-pressure switches, kit of pressure switches, HVACR equipment, gauge manifold set, basic hand tools, electrical meters, miscellaneous electrical supplies, basic electrical hand tools, nitrogen tank, nitrogen regulator, and gauge manifold.

SAFETY PRECAUTIONS: Make certain that the electrical source is disconnected when making electrical connections. In addition:

- Make sure all connections are tight.
- Make sure no bare current-carrying conductors are touching metal surfaces except the grounding conductor.
- Make sure the correct voltage is being supplied to the circuits.
- Make sure body parts do not come in contact with live electrical conductors.
- Keep hands and materials away from moving parts.
- Be sure to wear safety glasses

STEP-BY-STEP PROCEDURES

1. Obtain a low-pressure, a high-pressure, and a dual-pressure switch from the supply room.

2. Set the low-pressure switch (opens on a decrease in pressure) to cut in at 30 psig and cut out at 15 psig.

3. Attach a pressure source to the pressure switch and, by varying the pressure acting on the control, determine the actual cut-in and cut-out pressures. Record these values here:

   Cut-in pressure: _____ psig
   Cut-out pressure: _____ psig

4. Discuss any discrepancies that exist between the control's set points and the actual pressures at which the control opens and closes.

   _____

   _____

5. Set the high-pressure switch (opens on a rise in pressure) to cut out at 100 psig and cut in at 75 psig.

6. Attach a pressure source to the pressure switch and, by varying the pressure acting on the control, determine the actual cut-in and cut-out pressures.

Cut-in pressure: _____ psig

Cut-out pressure: _____ psig

7. Discuss any discrepancies that exist between the control's set points and the actual pressures at which the control opens and closes.

_____

_____

8. Set the low-pressure switch of the dual-pressure switch to cut out at 15 psig and cut in at 30 psig. Set the high-pressure switch of the dual-pressure switch to cut out at 150 psig with a fixed differential.

9. Attach a pressure source to both sides of the dual-pressure switch and, by varying the pressure acting on the control, determine the actual cut-in and cut-out pressures.

Low Pressure

Cut-in pressure: _____ psig

Cut-out pressure: _____ psig

High Pressure

Cut-in pressure: _____ psig

Cut-out pressure: _____ psig

10. Discuss any discrepancies that exist between the control's set points and the actual pressures at which the control opens and closes.

_____

_____

11. Return pressure switches to the tool room.

12. Obtain a pressure switch kit from your instructor.

13. Troubleshoot the five pressure switches in the kit, and record their condition.

Pressure Switch	**Condition**
#1	_____
#2	_____
#3	_____
#4	_____
#5	_____

14. Return the pressure switches to your instructor.

15. Troubleshoot the three pressure switches on live equipment assigned by your instructor.

Pressure Switch	**Location**	**Pressure Switch Type**	**Condition of Pressure Switch**
#1	_____	_____	_____
#2	_____	_____	_____
#3	_____	_____	_____

MAINTENANCE OF WORKSTATION AND TOOLS: Clean all tools, and return them to their proper location(s). Replace all equipment covers. Clean up the work area.

SUMMARY STATEMENT: What is the most important factor that a technician must consider when troubleshooting a pressure switch?

TAKEAWAY: In order for air-conditioning and refrigeration systems to operate properly, the pressure switches must be set properly. Quite often, improperly adjusted or set pressure controls are the cause for system malfunction. Knowing how to properly set pressure controls is a valuable skill to possess.

QUESTIONS

1. How can a pressure switch be used as an operating control to control temperature?

2. What is a dual-pressure switch?

3. Why must a technician know what the pressure is in a refrigeration system before troubleshooting a pressure switch?

4. What would be the switching action of a low-pressure switch used as a safety control?

5. Why is differential important when setting pressure switches?

6. Why are some pressure switches nonadjustable?

7. What would be the switching action of a low-pressure switch used as an operating control?

8. What would be the switching action of a high-pressure switch used to maintain a constant head pressure by stopping and starting a condenser fan motor?

9. What would be the approximate setting of a low-pressure switch used to maintain 45°F in a walk-in cooler?

10. What procedure would a technician use to troubleshoot pressure switches?

Exercise CON-20

Troubleshooting Heating Controls

Name	Date	Grade

OBJECTIVES: Upon completion of this exercise, you should be able to correctly troubleshoot electric, gas, and oil heating controls.

INTRODUCTION: The air conditioning of structures during the winter months requires a source of heat to maintain the desired temperature level. This heating source must be safely controlled while maintaining the desired temperature in the structure. Heating controls are designed to take care of both of these functions in the system. The service technician must be able to locate and correct problems in heating appliances while making certain that the safety controls are operating properly.

TOOLS, MATERIALS, AND EQUIPMENT:

Electric furnace, gas furnace with standing pilot, gas furnace with intermittent ignition, gas furnace with direct ignition, oil furnace with stack switch, oil furnace with cad cell primary control, electrical meters, miscellaneous electrical supplies, basic electrical hand tools, parts to repair furnaces, and troubleshooting charts.

⚠ **SAFETY PRECAUTIONS:** Make certain that the electrical source is disconnected when making electrical connections. Make certain that no gas leaks exist. Locate and make sure the gas supply cutoff valve is operating correctly. Shut off the gas supply to the device when installing, modifying, or repairing it. Allow at least 5 minutes for any unburned gas to leave the area before beginning work. Remember that LP gas is heavier than air and does not vent upward. When working on an oil-fired furnace, make sure there are no leaks and no excess fuel oil exists around the furnace or combustion chamber. In addition:

- Make sure all electrical connections are tight.
- Make sure no bare current-carrying conductors are touching metal surfaces except the grounding conductor.
- Make sure the correct voltage is being supplied to the circuit or appliance.
- Make sure body parts do not come in contact with live electrical conductors.
- Keep hands and materials away from moving parts.
- Make sure no leaks are present in gas or oil lines.

STEP-BY-STEP PROCEDURES

1. Obtain from your instructor an assignment of an electric furnace that is not operating properly.

2. Turn the power supply off, and check the system wiring for any loose or broken electrical connections.

3. Turn the power supply on, and check for line voltage at the circuit breaker or fuse block. If line voltage is not available, locate and repair the problem. Describe your findings and any corrective actions you took here.

4. With line voltage available to the electric furnace, set the thermostat to call for heat. There should be 24 V available to the sequencer or relay controlling the resistance heaters. If 24 V is not detected, check the thermostat and transformer, and as a last resort, check the thermostat wire connecting the furnace and thermostat. Repair or replace any faulty component(s). Describe your findings and any corrective actions you took here. _____

_____

5. If 24 V are supplied to the electric furnace controls, the first-stage heating elements and the blower motor should come on together. The remaining heating elements should come on as the sequencer closes the contacts. Check the voltage being supplied to each element and the current draw of each element. If voltage is available to the heating element but there is no current draw, the heating element is probably broken or there is an open fuse link or limit switch. If no voltage is available to the heating element(s), check across the contacts of the sequencer or relay with a multimeter; if voltage is read, the contacts are bad. Repeat the procedure for each heating element in the appliance. Replace any electrical device found to be faulty. Describe your findings and any corrective actions you took here. _____

_____

6. Increase the setting of the thermostat to determine if the electric heating appliance is operating properly. Describe your findings and any corrective actions you took here. _____

7. Check the voltage to each element and the current draw of each element to make sure all elements and the blower are operating properly. Describe your findings and any corrective actions you took here. _____

_____

8. Decrease the setting of the thermostat. Make sure that all elements and the blower are turned off.

9. Replace all covers on equipment, and clean up the work area.

10. Obtain from your instructor an assignment of a gas furnace with a standing pilot that is not operating properly.

11. Visually inspect the gas furnace to determine if there are any loose or broken electrical connections.

12. Use an appropriate troubleshooting chart to proceed with your conclusions and repair.

13. Replace any electrical devices found to be faulty. Describe your findings and any corrective actions you took here. _____

14. Increase the setting of the thermostat to a call for heat. Observe the operation of the gas furnace to determine if it is operating properly. Check all safety controls for proper operation. Describe your findings and any corrective actions you took here. _____

15. Decrease the setting of the thermostat, and make sure the main burner cuts off. Make sure the fan cools the combustion chamber before stopping. Describe your findings and any corrective actions you took here. _____

16. Replace all equipment covers, and clean up the work area.

17. Obtain from your instructor an assignment of a gas furnace with an intermittent ignition system.

18. Visually inspect the gas furnace to determine if any loose or broken electrical connections are present. Describe your findings and any corrective actions you took here. _____

_____

19. Use an appropriate troubleshooting chart to isolate the problem.

20. Replace any electrical devices found to be faulty. Describe your findings and any corrective actions you took here. _____

21. Increase the setting of the thermostat to a call for heat. Observe the operation of the gas furnace to determine if it is operating properly. Check all safety controls for proper operation. Describe your findings and any corrective actions you took here. _____

_____

22. Decrease the setting of the thermostat, and make sure the burners cut off. Make sure the fan cools the combustion chamber before stopping. Describe your findings and any corrective actions you took here.

_____

23. Replace all equipment covers, and clean up the work area.

24. Obtain from your instructor an assignment of a gas furnace with a direct ignition system.

25. Visually inspect the gas furnace to determine if any loose or broken electrical connections are present. Describe your findings and any corrective actions you took here. _____

_____

26. Use an appropriate troubleshooting chart to isolate the problem. Describe your findings and any corrective actions you took here. _____

_____

27. Replace any electrical devices found to be faulty. Describe your findings and any corrective actions you took here. _____

28. Increase the setting of the thermostat to a call for heat. Observe the operation of the gas furnace to determine if it is operating properly. Check all safety controls for proper operation. Describe your findings and any corrective actions you took here. _____

_____

29. Decrease the setting of the thermostat and make sure the burners cut off. Make sure the fan cools the combustion chamber before stopping.

30. Replace all equipment covers, and clean up the work area.

31. Obtain from your instructor an assignment of an oil-fired furnace with a stack switch primary control.

32. Visually inspect the oil furnace and the stack switch for loose and broken electrical connections. Describe your findings and any corrective actions you took here. _____

_____

33. To completely troubleshoot an oil burner installation, both the burner and ignition systems, as well as the primary control, must be checked for proper operation and condition.

In this troubleshooting section, only the electrical components are possible faults.

34. If the trouble does not seem to be in the burner or ignition systems, check all limit switches to make sure they are closed. Describe your findings and any corrective actions you took here. _____

_____

35. Reset the safety switch on the primary control.

36. Make sure that line voltage is available to the primary control.

37. Check the thermostat to make sure it is closed.

38. Set the thermostat to call for heat.

39. Put the contacts of the bimetal in step by pulling the drive shaft lever out 1/4 in. and releasing.

40. Troubleshoot the primary control. Describe your findings and any corrective actions you took here.

_____

41. Replace any electrical devices found to be faulty. Describe your findings and any corrective actions you took here. _____

42. Increase the setting of the thermostat to a call for heat. Observe the operation of the oil furnace to determine if it is operating properly. Check all safety controls for proper operation. Describe your findings and any corrective actions you took here. _____

_____

43. Decrease the setting of the thermostat, and make sure the oil burner cuts off. Make sure the fan cools the combustion chamber before stopping.

44. Replace all equipment covers, and clean up the work area.

45. Obtain from your instructor an assignment of an oil-fired furnace with a cad cell primary control.

46. Visually inspect the oil furnace, cad cell, and primary control for loose and broken electrical connections. Describe your findings and any corrective actions you took here. _____

_____

47. To completely troubleshoot an oil burner installation, both the burner and ignition system, as well as the primary control, must be checked for proper operation and condition.

In this troubleshooting section, only the electrical components are possible faults.

48. If the trouble does not seem to be in the burner or ignition system, check all limit switches to make sure they are closed. Describe your findings and any corrective actions you took here. _____

_____

49. Inspect the position and cleanliness of the cad cell. Describe your findings and any corrective actions you took here. _____

50. Reset the safety switch on the primary control.

51. Make sure that line voltage is available to the primary control. Describe your findings and any corrective actions you took here. _____

52. Set the thermostat to a call for heat.

53. Troubleshoot the cad cell primary control. Describe your findings and any corrective actions you took here.

_____

54. Replace any electrical device found to be faulty. Describe your findings and any corrective actions you took here. _____

55. Increase the setting of the thermostat to a call for heat. Observe the operation of the oil furnace to determine if it is operating properly. Check all safety controls for proper operation. Describe your findings and any corrective actions you took here. _____

_____

56. Decrease the setting of the thermostat, and make sure the oil burner cuts off. Make sure the fan cools the combustion chamber before stopping.

57. Replace all equipment covers, and clean up the work area.

MAINTENANCE OF WORKSTATION AND TOOLS: Clean all tools, and return them to their proper location(s). Replace all equipment covers. Clean up the work area.

SUMMARY STATEMENT: Why is it important for the technician to check the operation of the safety controls on a fossil fuel installation?

TAKEAWAY: A systematic troubleshooting process helps ensure that all aspects of the system are checked and that nothing is overlooked. There is more than one correct way to troubleshoot HVACR equipment, so each and every technician will develop a unique troubleshooting strategy.

QUESTIONS

1. Why are electric resistance heaters wired so they are not all energized at the same time?

2. Explain the operation of a sequencer used to control an electric furnace with four resistance heaters and a blower motor.

3. Why is the location of the cad cell important in an oil furnace using a cad cell primary control?

4. How many devices would have to be checked on a standing pilot ignition system, and what is the procedure for checking each?

5. What is a redundant gas valve, and how is it checked?

6. What two methods are used to prove ignition in an oil-fired furnace?

7. What could be the problem in an electric furnace if voltage is being supplied to the resistance heater but the heater is not producing heat? How would a technician determine this action?

8. What two methods are used to ignite a gas burner on a direct ignition system, and how is each checked?

9. Why are limit switches important to the safe operation of warm-air furnaces?

10. What procedure would a technician use to troubleshoot direct ignition and intermittent ignition systems on gas furnaces?

Troubleshooting Refrigeration, Heating, or Air-Conditioning Systems

Name	Date	Grade

OBJECTIVES: Upon completion of this exercise, you should be able to correctly troubleshoot basic air-conditioning problems assigned by your instructor.

INTRODUCTION: One of the most important jobs of air-conditioning technicians is to be able to locate and repair a system problem and return the system to normal operation. Approximately 85% of the problems in air-conditioning and heating systems will be electrical. Service technicians must be able to use electrical meters and read schematic diagrams in order to troubleshoot air-conditioning and heating systems.

TOOLS, MATERIALS, AND EQUIPMENT:

Operating heating and air-conditioning systems, electrical meters, and basic electrical hand tools.

⚠ SAFETY PRECAUTIONS: Make certain that the electrical source is disconnected when making electrical connections. In addition:

- Make sure all connections are tight.
- Make sure no bare conductors are touching metal surfaces except the grounding conductor.
- Make sure the correct voltage is being supplied to the unit.
- Make sure body parts do not come in contact with live electrical conductors.
- Keep hands and materials away from moving parts.
- Make sure all covers on the equipment are replaced.

STEP-BY-STEP PROCEDURES

1. You answer a complaint of "no cooling." The condenser fan motor and evaporator fan are operating. You check the schematic diagram (see Figure CON-21.1) and correctly determine that the problem is _____. Explain how you reached this conclusion.

   a. an open internal overload in the compressor

   b. blown fuses in the disconnect

   c. a bad transformer

   d. a bad thermostat

FIGURE **CON-21.1**

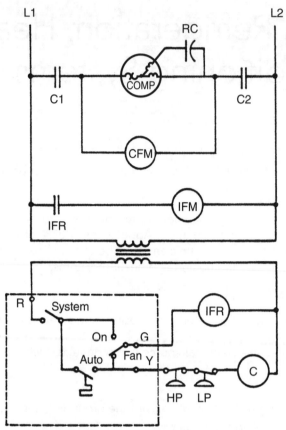

Legend

C:	Contactor
COMP:	Compressor
CFM:	Condenser fan motor
IFM:	Indoor fan motor
IFR:	Indoor fan relay
HP:	High-pressure switch
LP:	Low-pressure switch
RC:	Run capacitor

Thermostat (R, G, and Y are markings on thermostat terminals)

2. You answer a complaint of "not enough cooling." All unit components are operating, but after about 5 minutes, the compressor cuts out. You check the operating pressures of the system and find that the discharge pressure is 260 psig and the suction pressure is 42 psig. The refrigerant in the system is R-410A. You check the schematic diagram (see Figure CON-21.1) and correctly determine that the problem is _____. Explain how you reached this conclusion.

   a. a bad transformer

   b. a bad thermostat

   c. an open low-pressure switch

   d. an open high-pressure switch

3. You answer a complaint of "no cooling." The compressor and condenser fan motor are operating, but the indoor fan motor is not (refer to Figure CON-21.1). The probable cause is _____. Explain how you reached this conclusion.

a. a bad indoor fan motor

b. a bad transformer

c. a bad indoor fan relay

d. both a and c

4. You answer a complaint of "no heating." The compressor and outdoor fan motor are not operating but are good. The indoor fan is operating properly. After checking the schematic diagram (see Figure CON-21.2), you correctly determine that the problem is a bad _____. Explain how you reached this conclusion.

a. transformer

b. thermostat

c. contactor

d. supplementary heater

5. You answer a complaint of "insufficient heating." The compressor, the condenser fan motor, and indoor fan motor are operating, and there is a heavy coating of ice on the outdoor unit coil. The defrost thermostat is closed. (See Figure CON-21.2 for the schematic diagram.) The probable cause is _____. Explain how you reached this conclusion.

a. a bad defrost board

b. a bad transformer

c. a bad defrost thermostat

d. both a and c

6. You answer a complaint of "no heating." The combustion chamber and blower section is extremely hot and the LC is open. The blower motor is good, and there is no restriction in the supply air distribution system. The blower relay contacts and coil are good. The schematic diagram of the unit is shown in Figure CON-21.3. The probable cause is a bad _____. Explain how you reached this conclusion.

a. gas valve

b. pilot relight control

c. flame rollout switch

d. fan cycle control

FIGURE **CON-21.2**

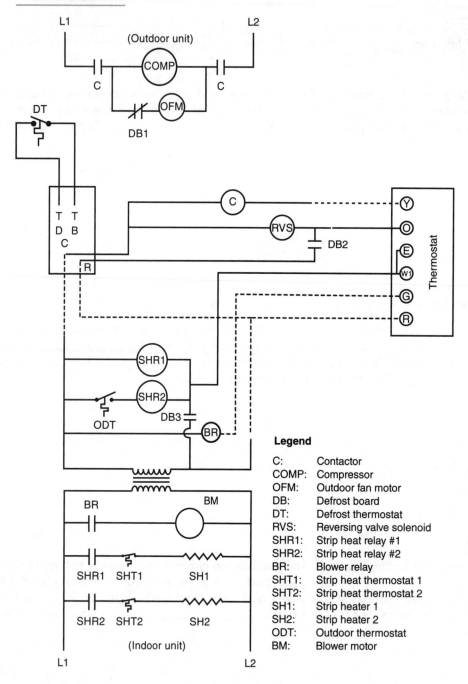

Legend

C:	Contactor
COMP:	Compressor
OFM:	Outdoor fan motor
DB:	Defrost board
DT:	Defrost thermostat
RVS:	Reversing valve solenoid
SHR1:	Strip heat relay #1
SHR2:	Strip heat relay #2
BR:	Blower relay
SHT1:	Strip heat thermostat 1
SHT2:	Strip heat thermostat 2
SH1:	Strip heater 1
SH2:	Strip heater 2
ODT:	Outdoor thermostat
BM:	Blower motor

7. You answer a complaint of "no heating." The pilot ignites, but the main burner does not. You check, and 24 V are available to the PRC, but no voltage is available to the GV. The schematic diagram is shown in Figure CON-21.3. What is the probable cause? Explain how you reached this conclusion.

FIGURE **CON-21.3**

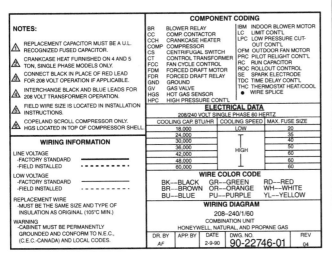

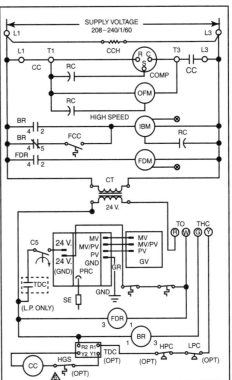

a. Gas valve

b. CS

c. ROC

d. PRC

8. You answer a complaint of rising temperature in a walk-in freezer. The evaporator is completely covered with frost. The schematic of the unit is shown in Figure CON-21.4. The unit is being supplied the correct voltage. The DTC contacts are good, and the DTM and the DH are good. The probable cause of the problem is _____. Explain how you reached this conclusion.

a. LLS

b. open DT

c. closed DT

d. CC

9. You answer a complaint of a walk-in freezer not cycling off properly, and the temperature of the freezer is −55°F. The schematic of the unit is shown in Figure CON-21.4. The thermostat is set at 0°F. The probable cause of the problem is _____. Explain how you reached this conclusion.

FIGURE **CON-21.4**

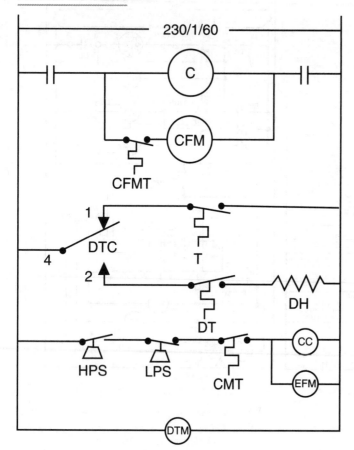

230/1/60

Legend

C:	Compressor
CC:	Compressor contactor
CFMT:	Condenser fan motor thermostat
CFM:	Condenser fan motor
DT:	Defrost timer motor
DTC:	Defrost timer contacts
T:	Thermostat
LLS:	Liquid line solenoid
DH:	Defrost heater
HPS:	High-pressure switch
LPS:	Low-pressure switch
CMT:	Compressor motor thermostat
EFM:	Evaporator fan motor

a. DTC

b. LPS

c. CMT

d. None of the above

10. You answer a complaint of a walk-in-freezer temperature of 40°F. The schematic of the unit is shown in Figure CON-21.4. You discover that the discharge pressure reaches 500 psig when the HPS opens, stopping the compressor. The condenser fan motor is good. The probable cause is _____.
    Explain how you reached this conclusion.

a. CFMT

b. LLS

c. DH

d. CMT

11. Your instructor will assign five systems for you to troubleshoot.

12. Troubleshoot and record in the following table the problems found with the five systems.

Unit	System Problems	
#1	_____	Explain how you reached this conclusion.
#2	_____	Explain how you reached this conclusion.
#3	_____	Explain how you reached this conclusion.
#4	_____	Explain how you reached this conclusion.
#5	_____	Explain how you reached this conclusion.

MAINTENANCE OF WORKSTATION AND TOOLS: Clean all tools, and return them to their proper location(s). Replace all equipment covers. Clean up the work area.

SUMMARY STATEMENT: Briefly explain the procedure that you used for troubleshooting the five systems.

TAKEAWAY: A systematic troubleshooting process helps ensure that all aspects of the system are checked and that nothing is overlooked. There is more than one correct way to troubleshoot HVACR equipment, so each and every technician will develop a unique troubleshooting strategy.

QUESTIONS

1. If no part of a unit is operating, what is the first check that a service technician should make?

2. What is hopscotching?

3. When a service technician arrives on the job, what are the steps that should be taken first?

4. A hermetic compressor in a small residential condensing unit is not operating, but voltage is available to the compressor terminals. What is the probable cause?

5. What would be some possible causes of a compressor motor humming when the contactor is closed?

6. What safety control would be likely to open if the unit had an extremely dirty condenser coil?

7. What are some common problems that occur with low-voltage thermostats?

8. What are some common causes of contactor failures?

9. Why are electric meters important to the service technician?

10. Why are schematic wiring diagrams important to the service technician?

Exercise CON-22

Smart Thermostats

Name	Date	Grade

OBJECTIVES: Upon completion of this exercise, you should be able to properly setup and configure a smart thermostat by navigating its menus and connecting it to a wireless network.

INTRODUCTION: You will be recording the visual data available and then installing a smart thermostat. You will then walk through the setup menus to configure the thermostat for the equipment that it is connected to. Finally, you will use a smart device to access the thermostat and adjust its temperature set points.

TOOLS, MATERIALS AND EQUIPMENT: A smart thermostat, an operating system to connect it to, a smart device with applicable app, and an available Wi-Fi network.

⚠ **SAFETY PRECAUTIONS:** Care should be taken not to damage any of the thermostat connectors or clips. Take care not to scratch the touchscreen display. Follow all manufacturers guidelines regarding the safe handling and usage of the thermostat.

STEP-BY-STEP PROCEDURES

1. With the thermostat disassembled, record the following information from the device:

   - Manufacturer _____
   - Model number: _____
   - Serial number: _____

2. Place a check next to each thermostat terminal that should be wired on the subbase:
   - R_C _____
   - R_H _____
   - G _____
   - Y1 _____
   - Y2 _____
   - O/B _____
   - C _____
   - W1 _____
   - W2 _____
   - AUX1 _____
   - AUX2 _____
   - R _____

3. Inspect the thermostat and determine if there are any dual in-line package (DIP) switches on the devices. Are there any DIP switches?   YES NO

4. If the answer to the question in step 3 is yes, refer to the thermostat's manual to determine the proper position for each of the DIP switches based on the type of equipment being worked on. If there are no DIP switches on the thermostat, proceed to step 5. Record the DIP switch information here:

DIP Switch Number/Letter/Identifier	DIP Switch Function	Position Options	Position Selected

5. Mount the thermostat's subbase to the wall.

6. Make certain that the system power is off.

7. Connect the thermostat wires to the subbase according to the installation manual.

8. Connect the thermostat to the subbase, making certain that all connection pins, screws, and connectors between the thermostat and subbase line up properly. DO NOT FORCE THE THERMOSTAT ONTO THE SUBBASE. If you are having difficulty mounting the thermostat onto the subbase, ask your instructor for assistance.

9. Restore power to the system, and allow the thermostat to power up. It may take several minutes for the thermostat to power up and go through its initial internal setup.

10. Set the following: date, time, time zone, zip code, climate zone (different thermostat models will have different options). List the specific setup parameters for the thermostat being used here:

_____   _____   _____

_____   _____   _____

_____   _____   _____

If required by your instructor, take a video of the thermostat setup process, and be prepared to submit it as part of this exercise.

11. Follow the onscreen instructions to set up the thermostat so that it will properly control the operation of the system being worked on. If required by your instructor, take a video of the thermostat setup process, and be prepared to submit it as part of this exercise.

12. If applicable, configure the thermostat to control the continuous fan.

13. In the menus, locate the Wi-Fi settings, and connect the thermostat to the network.

14. Once the installation is completed and the thermostat is configured, use the smart device to access the thermostat.

15. Use the smart device to change the temperature settings on the thermostat so that the system will energize.

16. De-energize the system, power the unit down, and store the wireless device.

MAINTENANCE OF WORKSTATION AND TOOLS: Leave the thermostat and related materials as directed by instructor.

SUMMARY STATEMENT: What are some of the differences between a smart thermostat and a conventional, non-communicating thermostat. What additional options are available on a smart thermostat? Use a separate sheet of paper for your answer if necessary.

TAKEAWAY: As more advanced and efficient HVACR equipment is engineered, the devices used to control these systems must advance to take full advantage of their capabilities. Smart thermostats are continuously adding more functionality and communication with the system to which it is connected.

QUESTIONS

1. How would a thermostat that requires a "C" wire be installed if no "C" wire was present?

2. What information is displayed on the home screen of the thermostat used in this exercise?

3. On the thermostat used in this exercise, what thermostat settings were adjustable from the wireless device app?

4. Does the thermostat used in this exercise incorporate remote room sensors?

5. What types of equipment can be controlled by a smart thermostat's auxiliary terminals?

6. Explain how using smart thermostats and other smart control devices can help increase a system's operating efficiency and reduce power consumption.

SECTION 12

Domestic Appliances (DOM)

SUBJECT AREA OVERVIEW

The term domestic appliance, within the scope of this manual, can refer to a refrigerator/freezer, stand-alone freezer, and both window and wall unit air conditioners. The refrigeration cycle components are the same four major components that have been discussed in other units. Domestic appliances are typically package units that are plug-in-type systems. Domestic refrigerators and freezers are typically 120 V appliances, while window and wall air-conditioning units are available in both 120 V and 240 V varieties.

A typical household refrigerator has two compartments, a freezer section and a higher-temperature refrigerator section. Most refrigerators/freezers maintain both sections with one compressor that operates at the lowest box temperature conditions. Some higher-end appliances have two compressors, one for the medium-temperature box and one for the low-temperature box. For single-compressor appliances, the evaporator must operate at the low-temperature condition as well as maintain the refrigerator compartment. This may be accomplished by allowing some of the air from the freezer compartment to flow into the refrigerator compartment or by using two evaporators piped in series, one for the freezer and one for refrigerator. The air movement over the evaporators may be either natural or mechanical draft. Natural draft relies on natural convection currents to move air across the evaporator coil, while mechanical draft utilizes blowers to move the air across the coil's surface. Mechanical draft can be either forced-draft, where air is pushed through the coil, or induced-draft, where air is pulled through the coil. Most evaporators on refrigerators/freezers are of the forced-draft variety.

Natural-draft evaporators are normally the flat plate type with the refrigerant passages stamped into the plates. They may have automatic or manual defrost. Mechanical-draft refrigerators often have two fans, one for the evaporator and one for the condenser. The condenser fan is usually of the propeller type, and the evaporator fan may be a small squirrel cage blower or propeller-type fan.

The domestic stand-alone freezer is different from the refrigerator in that it is a single low-temperature system. Food to be frozen must be packaged properly or else dehydration, commonly referred to as freezer burn, will occur. In the refrigerator/freezer, a small fan circulates air inside the appliance across a cold, refrigerated coil. The air gives up sensible heat to the coil, and the air temperature is lowered. The air also gives up latent heat (from moisture in the air) to the coil, causing dehumidification and frost to form on the coil. Once the air has given up heat to the coil, it is distributed back to the box so that it can absorb more heat and humidity. This process continues until the desired temperature in the box is reached.

Domestic refrigerators/freezers are configured as vertical appliances with the refrigerator and freezer compartments either side-by-side or one on top of the other. The domestic freezer may be configured as a horizontal chest with a lid that raises up or an upright, vertical appliance with a door that opens outward. The upright does not take up as much floor space but may not be as efficient because cold air falls out of the box every time the door is opened. In the chest freezer, the cold air stays in the box.

The refrigerant piping circuits on most domestic refrigerated appliances are hermetically sealed and do not have gauge ports. It is poor practice

(continued)

to routinely install gauges on these appliances, as they only contain a small amount of refrigerant, some of which is lost when gauges are installed and removed. Gauges should be installed only as a last resort, and when they are installed, the charge must be adjusted to ensure proper system operation.

The evaporators on domestic refrigerators and freezers must operate at temperatures that will lower the air temperature to about 0°F. Compressors on domestic refrigeration appliances are either air- or refrigerant-cooled. Air-cooled compressors have fans; refrigerant-cooled compressors are cooled by suction gas.

Single-room air conditioners may be window or through-the-wall units, front discharge or top-air discharge. Most units range from 4000 to 24,000 Btu/h.

Some of these units have electric resistance heating, and some are heat pumps. Room units are designed for either window installation or wall installation. Window units can be installed and removed without damaging the window. A hole must be cut through the wall for the installation of wall units, so the installation of this type of unit is considered to be permanent. Controls for room units are mounted within the cabinet. A typical room unit will have a selector switch to control the fan speed and provide power to the compressor circuit. These units have only one fan motor, so a reduction in fan speed slows both the indoor and outdoor fans. Slowing the fan motor reduces the noise level of the unit but also reduces the capacity of the unit.

KEY TERMS

Automatic defrost
Capillary tube
Condensate
Critical charge
Current relay
Defrost timer

Domestic freezer
Domestic refrigerator
Door gasket
Forced-draft evaporator
Front discharge
Hot pulldown

Latent heat
Manual defrost
Natural-draft evaporator
Package unit
Stand-alone freezer
Static condenser

Through-the-wall unit
Top-air discharge
Window unit

REVIEW TEST

Name	Date	Grade

Circle the letter that indicates the correct answer.

1. A household refrigerator may have more than one evaporator.
A. True
B. False

2. One of the best ways to remove ice from the evaporator in a manual-defrost system is to scrape it with a sharp knife.
A. True
B. False

3. An accumulator at the outlet of the evaporator is there to:
A. provide a place for liquid refrigerant to vaporize before returning to the compressor.
B. provide for the optimum amount of liquid refrigerant in the evaporator.
C. provide space for refrigerant vapor to collect.
D. accumulate foreign particles and other contaminants.

4. The condensate from the evaporator coil is typically evaporated using heat from the compressor discharge line or by:
A. heated air from the condenser.
B. air from the evaporator fan.
C. heat from the evaporator defrost cycle.
D. heat from the crankcase fins.

5. When a substance changes state from a solid to a vapor, the process is known as:
A. freezer burn.
B. specific heat.
C. sublimation.
D. superheat.

6. A plate- or tube-type evaporator is generally used with:
A. natural draft.
B. forced-air draft.

7. The two types of compressors used most frequently in domestic refrigerators are the reciprocating and:
A. scroll.
B. centrifugal.
C. screw.
D. rotary.

8. Natural-draft condensers are normally located at the _____ of the refrigerator.
A. top
B. bottom
C. back

9. An air-cooled chimney-type condenser is generally a _____ draft type.
A. natural
B. forced-air

10. Induced-draft condensers are typically used with units that have:
A. manual defrost.
B. automatic defrost.

11. The type of metering device most commonly used in household freezers is the:
A. capillary tube.
B. thermostatic expansion valve.
C. electronic expansion valve.
D. automatic expansion valve.

12. The capillary tube metering device on a domestic refrigerator is generally fastened to the _____ to facilitate a heat exchange.
A. liquid line
B. compressor discharge line
C. condenser coil
D. suction line

13. The heaters located near the refrigerator door that keep the temperature above the dew point are called _____ heaters.
    A. defrost cycle
    B. mullion or panel
    C. condensate
    D. defrost limiter

14. Compressor running time may be used as a factor to _____ automatic defrost.
    A. initiate
    B. terminate

15. If the compressor is sweating around the suction line, there is:
    A. too much refrigerant.
    B. not enough refrigerant.
    C. probably a refrigerant leak.

16. If a suction-line filter drier is added to a system, it _____ require an extra charge of refrigerant.
    A. will
    B. will not

17. With an induced-draft condenser, it is important that:
    A. all cardboard partitions, baffles, and backs are in place.
    B. the refrigerator is not located under a cabinet.
    C. the refrigerant pressures are checked when servicing the refrigerator.
    D. the system is swept with nitrogen on a regular basis.

18. Quickly freezing food:
    A. makes it mushy.
    B. is not recommended.
    C. is not cost effective.
    D. preserves the flavor and texture.

19. The blood and water you find when you thaw out a steak:
    A. is caused by quick freezing.
    B. is caused by freezing too slowly.
    C. is normal.
    D. is good.

20. When moving a heavy freezer in a house, you should:
    A. use a forklift.
    B. use refrigerator hand trucks.
    C. use four people, one on each corner.
    D. do all of the above.

21. Ice buildup on the inside of a chest-type freezer is due to:
    A. the thermostat being set too low.
    B. a defective gasket.
    C. food stacked too closely together.
    D. the unit running all the time.

22. Single-room air-conditioning units generally range in capacity from _____ Btu/h.
    A. 1500 to 15,000
    B. 3000 to 20,000
    C. 4000 to 24,000
    D. 5000 to 30,000

23. Most room air conditioners utilize _____ metering device.
    A. an electronic expansion valve
    B. a thermostatic expansion valve
    C. an automatic expansion valve
    D. a capillary tube

24. The evaporator typically operates _____ the dew point temperature of the room air.
    A. below
    B. above

25. The compressor in a room air conditioner is _____ type.
    A. an open
    B. a serviceable hermetic
    C. a fully sealed hermetic

26. The condensate in a window unit is generally directed to:
    A. a place where it will drain into the soil.
    B. a storm drain.
    C. the condenser area where it evaporates.
    D. a holding tank.

27. A heat pump room air conditioner has the four major components in a cooling-only unit plus:
    A. a three-way valve.
    B. a four-way valve.
    C. an automatic expansion valve.
    D. an indoor coil.

28. _____are used in heat pump room air conditioners to ensure the refrigerant flows through the correct metering device at the correct time.
    A. Filter-driers
    B. Check valves
    C. Globe valves
    D. Electronic relays

29. During the heating cycle of a heat pump system, the hot gas from the compressor is first directed to the:
    A. indoor coil.
    B. outdoor coil.
    C. metering device.
    D. quid line filter drier.

30. In the heating cycle of a heat pump system, the refrigerant condenses in the:
    A. indoor coil.
    B. outdoor coil.
    C. reversing valve.
    D. metering device.

31. All 230 V power cords have the same plug.
    A. True
    B. False

32. The fin patterns on a coil:
    A. increase the heat exchange rate.
    B. are for decoration.
    C. make the air sweep the room.
    D. do none of the above.

33. Two types of compressors commonly used for window and room units are:
    A. centrifugal and rotary.
    B. rotary and reciprocating.
    C. reciprocating and screw.
    D. screw and centrifugal.

34. A regular window air conditioner can be used in:
    A. a double-hung window.
    B. a casement or fixed-sash window.

35. The thermostat bulb located close to the fins in the return airstream:
    A. keeps the unit running for long cycles.
    B. stops the unit when it is overheating.
    C. keeps the fins clean.
    D. helps prevent frost or ice buildup.

Exercise DOM-1

Domestic Refrigerator Familiarization

Name	Date	Grade

OBJECTIVES: Upon completion of this exercise, you should be able to identify and describe various parts of a domestic refrigerator.

INTRODUCTION: You will move a refrigerator from the wall and remove enough panels to have access to the compressor compartment and identify all components.

TOOLS, MATERIALS, AND EQUIPMENT: Straight-slot and Phillips-head screwdrivers, ¼ in. and 5⁄16 in. nut drivers, a flashlight, and a domestic refrigerator.

⚠ **SAFETY PRECAUTIONS:** Make sure the unit is unplugged before opening the compressor compartment.

STEP-BY-STEP PROCEDURES

1. Unplug the refrigerator, and pull it far enough from the wall to remove the covers from the compressor compartment and the evaporator. Remove the cover, and fill in the following information:

Refrigerator Nameplate:

- Manufacturer: _____
- Model number: _____
- Serial number: _____
- Operating voltage: _____ V
- Full-load current: _____ A
- Power (if available): _____ W

Compressor Compartment:

- Discharge-line size: _____ in.
- Suction-line size: _____ in.
- Type of start assist for the compressor (relay, capacitor, or PTC device): _____

Refrigerator Components and Features:

- Side-by-side or over and under: _____
- Evaporator location: _____
- Forced-draft or natural-draft: _____
- Evaporator coil material: _____
- Condenser coil material: _____
- Type of condenser (induced-air or natural-draft): _____
- If induced-draft, where does the air enter the condenser? _____
- If natural-draft, where does the air enter the condenser? _____

Defrost Method:

- Type of defrost (manual or automatic): _____
- How is the condensate evaporated? _____
- Where is the defrost timer located? _____

MAINTENANCE OF WORKSTATION AND TOOLS: Replace all panels with the correct fasteners, and move the refrigerator back to its correct location. Put all tools in their proper places. Restart the refrigerator if your instructor advises.

SUMMARY STATEMENT: Describe the flow of refrigerant through the entire cycle of the refrigerator that you worked on. Use an additional sheet of paper if more space is necessary for your answer.

TAKEAWAY: Domestic refrigerators are manufactured in a variety of configurations; however, they all operate in a similar manner. They all utilize the same refrigeration cycle and the same general electrical components.

QUESTIONS

1. What type of metering device was used on the refrigerator you inspected?

2. Can a domestic refrigerator ever have two separate evaporators?

3. What is the difference between a natural- and induced-draft condenser?

4. How is heat from the compressor used to evaporate condensate?

5. What is the purpose of wrapping the liquid line or capillary tubing around the suction line in a refrigeration circuit?

6. What is full-load amperage (FLA)?

Exercise DOM-2

Domestic Freezer Familiarization

Name	Date	Grade

OBJECTIVES: Upon completion of this exercise, you should be able to recognize the different features of a domestic freezer.

INTRODUCTION: You will remove enough panels to be able to identify the compressor, evaporator, and condenser of a freezer.

TOOLS, MATERIALS, AND EQUIPMENT: Straight-slot and Phillips-head screwdrivers, ¼ in. and ⁵⁄₁₆ in. nut drivers, a flashlight, and a stand-alone domestic freezer.

⚠ **SAFETY PRECAUTIONS:** Be sure the unit is unplugged before removing panels.

STEP-BY-STEP PROCEDURES

1. Unplug a freezer, and pull it far enough from the wall to be able to remove the compressor compartment cover.

2. Remove the compressor compartment cover, and fill in the following information:

   - Compressor suction-line size: _____ in.
   - Discharge-line size: _____ in.
   - Suction-line material: _____
   - Discharge-line material: _____
   - Type of compressor relay (potential, current, or PTC): _____
   - Pressure ports? _____
   - Type of condenser (natural- or induced-draft): _____
   - Where is the condenser located (in the bottom of the box, at the back, or in the sides)? _____
   - Condenser material: _____
   - Does the compressor have an oil cooler circuit? _____
   - How is the compressor mounted, on springs or rubber feet? _____
   - Type of freezer (upright or chest): _____
   - Type of evaporator (forced, shelf, or wall): _____

   - Manual or automatic defrost: _____
   - Where is the thermostat located? _____
   - Does the power cord have a ground leg? _____
   - Does the box have a lock on the door? _____
   - What holds the door closed, a latch or magnetic gasket? _____
   - Does the box have a wiring diagram on the back? _____
   - Are there any heaters in the diagram? _____
   - Is there a light bulb; if so, how is it controlled? _____
   - If the unit has electric defrost, where is the timer located? _____
   - What is the full-load amperage draw of the freezer? _____ A
   - The voltage rating: _____ V

3. Replace all panels with the correct fasteners, and move the freezer back to the correct location.

MAINTENANCE OF WORKSTATION AND TOOLS: Return all tools to their correct places, and make sure the workstation is clean and neat.

SUMMARY STATEMENT: Describe why a defrost cycle is used and the different types of defrost methods that may be encountered. Use a separate sheet of paper if more space is needed for your answer.

TAKEAWAY: Domestic freezers, like refrigerators, are designed with various component configurations and locations. However, they all utilize the same refrigerant cycle and the same basic electrical components.

QUESTIONS

1. What is the primary problem with induced-draft condensers?

2. At what temperature should a freezer operate to keep ice cream hard?

3. What refrigerant is commonly used in new domestic freezers?

4. How can the heat from the compressor be used to evaporate condensate in the condensate pan?

5. A high-efficiency condenser (induced-draft) will condense refrigerant at how many degrees above the ambient temperature?

6. How does a compressor that is operating normally feel to the touch?

7. What conditions may produce a heavy load on a freezer?

8. What are two controls located on a freezer that does not utilize an automatic defrost?

9. Describe two ways to manually defrost a domestic freezer.

10. What substance may be used to keep food frozen temporarily while a freezer is being serviced?

Charging a Refrigerator Using the Frost-Line Method

Name	Date	Grade

OBJECTIVES: Upon completion of this exercise, you should be able to establish the correct operating charge in a refrigerator by using the frost-line method.

INTRODUCTION: Domestic refrigerators are critical charge systems, typically charged by weight. The frost-line method is a way of estimating the charge without having to recover the refrigerant and charge it from scratch. You will fasten gauges to a refrigerator and charge the unit to the correct level if refrigerant is needed.

TOOLS, MATERIALS, AND EQUIPMENT: A refrigerator with service ports, a thermometer, a gauge manifold, appropriate refrigerant for the appliance being worked on, and straight-slot and Phillips-head screwdrivers.

⚠ **SAFETY PRECAUTIONS:** Gloves and goggles should be worn when transferring refrigerant.

STEP-BY-STEP PROCEDURES

1. Unplug the refrigerator, and pull it from the wall.

2. Place a thermometer sensor in the refrigerator section. With goggles and gloves on, fasten a high-pressure and a low-pressure gauge to the proper gauge ports. If the refrigerator does not have gauge ports, the instructor will provide taps and/or instructions for installing access ports or line-tap valves on the unit.

3. Briefly loosen the gauge hose connections at the unit and the center hose connection on the gauge manifold's blank port to purge any air from the hoses.

4. Replace any panels that may cause air to bypass the condenser if the condenser is forced-air.

5. Locate the point on the suction line just as it leaves the evaporator and before the capillary tube, Figure DOM-3.1.

6. Turn the thermostat to the off position so the unit will not start, and plug the refrigerator into the power supply.

7. Grip the refrigerant line between your thumb and first finger at the point indicated in step 5 so that you can feel the suction-line temperature.

8. Start the compressor while holding the line.

   Note: If the line flashes cold for a few seconds just after the compressor starts, the system charge is close to correct.

FIGURE **DOM-3.1**

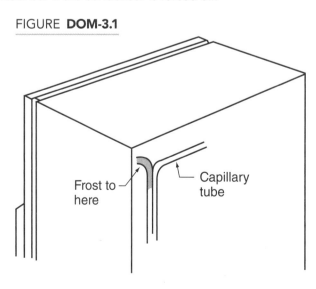

Frost to here — Capillary tube

9. Let the system run long enough so that the box begins to cool inside, to about 50°F, and observe the frost line leaving the evaporator. As the box continues to cool toward 40°F, this line should sweat and begin to frost.

10. The charge is correct when this part of the line has frost at the end of the cycle, at the time the thermostat shuts the unit off. Note the gauge pressures as the box temperature nears 40°F. Record them here:

    Suction: _____ psig.
    Discharge: _____ psig.

    Note: The frost should not extend down the line where the capillary tube is fastened. If it does, the refrigerator has an overcharge of refrigerant, and some should be recovered.

11. Shut the compressor off and let the system stand for 10 minutes so that the pressures equalize in the system. Repeat step 8 to get the feel of a suction line on a fully charged system at start-up.

12. When the charge is established as correct, unplug the box and remove the gauges. Remove the thermometer sensor, and replace all panels. Move the box to the proper location.

MAINTENANCE OF WORKSTATION AND TOOLS: Make sure the workstation is clean and in good order. Put all tools in their proper places.

SUMMARY STATEMENT: Describe how the refrigerant line felt when the refrigerator was started with a complete charge. Use an additional sheet of paper if more space is necessary for your answer.

TAKEAWAY: The charge for a domestic refrigerator can be estimated by observing the changes in the line temperatures and taking note of where and when frost develops on them.

QUESTIONS

1. Why is it necessary for all panels to be in place on a refrigerator with a forced-draft condenser?

2. Will the refrigerant head pressure be higher or lower than normal if the ambient temperature is below 65°F?

3. If part of the condenser airflow is blocked, will the head pressure be higher or lower?

4. What is the refrigerant most commonly used in new household refrigerators?

5. Why should gauges be kept under pressure with clean refrigerant when they are not being used?

6. Why is it good practice to remove Schrader valve depressors from the gauge hoses before starting the evacuation procedure?

7. How many fans would typically be found in a refrigerator with an induced-draft condenser?

8. Where would these fans be located?

Exercise DOM-4

Charging a Freezer by the Frost-Line Method

Name	Date	Grade

OBJECTIVES: Upon completion of this exercise, you should be able to charge a domestic freezer using the frost-line method.

INTRODUCTION: Domestic freezers are critical charge systems, typically charged by weight. The frost-line method is a way of estimating the charge without having to recover the refrigerant and charge it from scratch. You will be provided a unit from which some refrigerant has been removed so that the charge is low. You will then add refrigerant until the charge is correct according to the frost-line method.

TOOLS, MATERIALS, AND EQUIPMENT: Straight-slot and Phillips-head screwdrivers, a cylinder of refrigerant (ensure that it is the same type as in the freezer you will be working with), a gauge manifold, a thermometer, safety goggles, a charging scale, and a freezer with gauge ports.

 SAFETY PRECAUTIONS: Gloves and goggles should be worn while working with refrigerant.

STEP-BY-STEP PROCEDURES

1. Unplug the freezer, and pull it far enough from the wall to access the compressor compartment. Remove the compressor compartment cover.

2. Put on your goggles and gloves. Fasten gauges to the suction and discharge service ports. Replace any panels that may affect condenser airflow if the unit is a forced-draft unit. Suspend the thermometer in the freezer compartment so that it is not touching anything in the box.

3. Plug in the unit, and start the compressor. If the box was plugged in and cold, you may need to open the door for a few minutes or turn the thermostat down to start the compressor.

4. Purge any air from the suction and discharge hoses through the center gauge hose, and fasten it to the refrigerant cylinder. If the unit has a correct charge, you may recover some refrigerant from the unit to reduce the charge in the unit.

5. Locate a point on the suction line where it leaves the evaporator before the capillary tube heat exchange. If the unit has the evaporator in the wall of the box, you will probably not be able to find this point. In this case, locate the evaporator in the wall so that you can follow the frost pattern as it develops in the evaporator. You will need to monitor the point on the suction line or the frost pattern of the evaporator in the wall. The suction line leaving the evaporator may be inside the box also. In these cases, the door must remain closed except for brief periods when you are checking the frost line.

   Note: You will have to monitor this spot. It may be inside the box. The door must remain closed unless you are checking the frost line.

6. Record the suction and discharge pressures.

   Note: The suction pressure may be in a vacuum if the charge is low.

   Suction pressure: _____ psig; Discharge pressure: _____ psig

7. Slowly add vapor in short bursts with about 15 minutes between bursts until there is frost on the suction line leaving the evaporator or, if the evaporator is in the wall, until there is a pattern of frost toward the end of the evaporator. If the box temperature is near 0°F, the suction pressure should rise when refrigerant is added. As the space temperature drops, the suction pressure will drop. The compressor should be shut off when the box temperature is at about 0 to −5°F, and the suction pressure should be about 0.7 Psig with r-134a.

8. When the frost line is correct on the suction line or when the evaporator is frosted on the wall unit and the suction pressure is near correct, you have the correct charge. Shut the unit off, allow the pressures to equalize, and then remove the gauges. Replace all panels with the correct fasteners, and return the box to its assigned location.

MAINTENANCE OF WORKSTATION AND TOOLS: Return all tools to their respective places, and clean the workstation.

SUMMARY STATEMENT: Describe why the suction pressure rises when adding refrigerant and falls as the box temperature falls.

TAKEAWAY: Similar to a refrigerator, the charge for a domestic freezer can be estimated by observing the changes in the line temperatures and taking note of where and when frost develops on them. When the lines are located inside of the unit, the door should remain closed during operation, only opening it for brief periods to observe the frost lines.

QUESTIONS

1. How is a forced-draft evaporator fan in a freezer wired with respect to the compressor?

2. Is there ever a fan switch? If so, where is it located?

3. What is typically wrong with a fan motor when it does not run?

4. What may happen if an owner becomes impatient with the manual defrost and attempts to chip away the frost and ice?

5. What will the ohmmeter read when checking an open motor winding?

6. What would happen if the compressor were sweating down the side with an overcharge of refrigerant?

7. How is the condensate dealt with in the unit used for this exercise?

Exercise DOM-5

Charging a Freezer Using a Measured Charge

Name	Date	Grade

OBJECTIVES: Upon completion of this exercise, you should be able to charge a freezer using a measured charge.

INTRODUCTION: You will evacuate a freezer to a deep vacuum and properly charge the unit by weight using a digital scale.

TOOLS, MATERIALS, AND EQUIPMENT: Straight-slot and Phillips-head screwdrivers, ¼ in. and ⁵⁄₁₆ in. nut drivers, a refrigerant recovery system, a cylinder of refrigerant containing the type of refrigerant used in the freezer you are working with, a gauge manifold, a vacuum pump, a digital scale, gloves, goggles, and a freezer with at least a suction gauge port.

 SAFETY PRECAUTIONS: Gloves and goggles should be worn while transferring refrigerant.

STEP-BY-STEP PROCEDURES

1. Move the freezer to where you can access the compressor compartment.

2. Put your gloves and goggles on. Fasten the gauge manifold to the service port or ports. If there is any refrigerant in the system, recover it the way your instructor advises.

3. When the system pressure is reduced to the level indicated by your instructor, fasten the center gauge line to the vacuum pump, and start it.

4. Open the low-side manifold gauge valve to the low-side gauge line. Allow the vacuum pump to run while you get the scale and refrigerant tank set up. Write the refrigerant charge from the freezer nameplate in ounces here: _____.

5. When the vacuum pump has lowered the system pressure to the required vacuum level, add refrigerant to the system.

6. Shut the gauge manifold valve, and transfer the center gauge line to the refrigerant cylinder. Open the valve on the refrigerant cylinder, and allow refrigerant to enter the center gauge line.

   Note: Purge the center line with refrigerant from the cylinder to remove any air that entered the gauge line when disconnected from the vacuum pump.

7. With the refrigerant tank in place, zero out the scale so that it will display only the amount being charged.

8. Allow the vapor refrigerant to flow into the refrigeration system by opening the low-side valve on the gauge manifold. When refrigerant has stopped flowing, you may start the compressor. Refrigerant will start to flow again. You will have to watch your scale and shut off the gauge manifold valve when the correct weight has been reached. You may have to add a few ounces over what the data plate calls for to account for the gauge manifold and hoses. Your instructor will give you the proper information for doing this.

Note: If flow has stopped before the complete charge has entered the system, it may be because the vapor boiling from your storage cylinder has lowered and is as low as the system pressure. Ask your instructor for advice.

9. When the correct charge has been added to the system, remove the hoses from the service ports. This should be done while the system is running in order to keep the correct charge in the system.

10. Cover all gauge ports with the correct caps, and replace all panels. Move the freezer to its correct location.

MAINTENANCE OF WORKSTATION AND TOOLS: Return all tools to their correct storage places.

SUMMARY STATEMENT: Explain why it is necessary to measure the charge placed in a freezer.

TAKEAWAY: Measuring the charge into a system by weight, known as a critical charge, is the most accurate way to charge any refrigeration system. The charge weight is determined by the manufacturer and usually printed on the unit nameplate. When utilizing this method, it is important to account for any refrigerant that may fill up the hoses and manifold. Failure to add for the hoses may result in an undercharged system.

QUESTIONS

1. What is meant by the term "critical charge"?

2. What is the advantage of using an electronic scale?

3. Why may it become necessary to heat the refrigerant cylinder when charging a system?

4. Why are refrigerant cylinders color coded?

5. How is automatic defrost normally started in a domestic freezer?

6. What type of component is often used to accomplish automatic defrost?

Exercise DOM-6

Refrigerator Operating Conditions During a Hot Pulldown

Name	Date	Grade

OBJECTIVES: Upon completion of this exercise, you should be able to state the temperatures of the refrigerated box fresh-food and frozen-food compartments, suction and discharge pressures, and compressor amperage at various intervals during a pulldown from room temperature.

INTRODUCTION: You will install a gauge manifold on the suction and discharge lines, an ammeter on the compressor common line, and thermometer sensors or individual thermometers in the refrigerator and freezer compartments of a refrigerator at room temperature. You will record data from these instruments at various intervals as the refrigerator is cooling down.

TOOLS, MATERIALS, AND EQUIPMENT: An electronic thermometer, hand tools, a clamp-on ammeter, a gauge manifold, and a correctly charged refrigerator with service ports.

⚠ **SAFETY PRECAUTIONS:** Goggles and gloves should be worn while making gauge connections. Unplug the refrigerator while attaching the clamp-on ammeter. Be sure to have the correct airflow over the condenser.

STEP-BY-STEP PROCEDURES

1. Unplug and move a refrigerated box with the inside at room temperature far enough from the wall to fasten gauges and a clamp-on ammeter. Purge the air from hoses and manifold.

   Note: Be sure to have the correct airflow over the condenser.

2. Place a thermometer in the air space in the freezer and another in the refrigerator compartment.

3. Before starting the refrigerator, record the following:

   • Suction pressure: _____ psig

   • Discharge pressure: _____ psig

   • Temperature of the refrigerator compartment: _____°F

   • Temperature of the freezer compartment: _____°F

4. Start the compressor, and record the following every 10 minutes:

   • Suction pressure: _____ psig _____ psig _____ psig _____ psig _____ psig _____ psig

   • Discharge pressure: _____ psig _____ psig _____ psig _____ psig _____ psig _____ psig

   • Temperature of the refrigerator compartment: _____°F _____°F _____°F _____°F _____°F _____°F

   • Temperature of the freezer compartment: _____°F _____°F _____°F _____°F _____°F _____°F

   • Compressor amperage: _____ A _____ A _____ A _____ A _____ A _____ A

5. If the box temperature has not reached 35°F, do not be concerned. You may continue to monitor the box or stop the exercise. It may take several hours for the box to reach 35°F.

6. Turn the system off, and allow the pressures to equalize. Remove the gauges and meter, and return the system to its proper location.

MAINTENANCE OF WORKSTATION AND TOOLS: Return all tools to their correct places, and make sure the workstation is clean.

SUMMARY STATEMENT: Describe the relationship of the evaporator temperature converted from suction pressure to box temperature as the box cooled down.

TAKEAWAY: During normal operation, the temperatures of both the refrigerator and the freezer sections should drop together at a somewhat steady pace until the set point is reached, around 0°F in the freezer and 35°F to 40°F in the refrigerator. The heat from the refrigerated space(s) is transferred/discharged into the area where the appliance is located.

QUESTIONS

1. What would the saturation temperature of the refrigerant in the evaporator be if the refrigerant low-side pressure was 23.7 psig for an R-134a system?

2. What would the saturation temperature of the refrigerant in the condenser be if the high-side pressure was 135 psig on an R-134a system?

3. What are four ways of determining a refrigerant leak?

4. What might happen if a low-side line tap valve were installed when the system was operating in a vacuum?

5. Why should gauges be installed only when absolutely necessary on a domestic refrigerator?

6. What type of defrost was utilized on the system used for this exercise?

Exercise DOM-7

Testing a Compressor Electrically

Name	Date	Grade

OBJECTIVES: Upon completion of this exercise, you should be able to identify common, run, and start terminals on a compressor and determine whether the compressor is grounded.

INTRODUCTION: You will use an ohmmeter to determine the resistances across the run and start windings and from the compressor to ground. You can then use the mathematical ohm relationship of (C+S) + (C+R) = (S+R) to determine the terminal configurations.

TOOLS, MATERIALS, AND EQUIPMENT: A refrigerator compressor (it may be installed in a refrigerator), a digital ohmmeter, needle-nose pliers, straight-slot and Phillips-head screwdrivers, and ¼ in. and ⁵⁄₁₆ in. nut drivers.

⚠ **SAFETY PRECAUTIONS:** Make sure the unit is unplugged before starting this exercise. Properly discharge the capacitor if there is one.

STEP-BY-STEP PROCEDURES

1. Unplug the refrigerator if the compressor is installed in it and move it from the wall far enough to reach comfortably into the compressor compartment.

2. Remove the cover over the compressor compartment. Properly discharge the capacitor if there is one.

3. Remove the compressor terminal cover.

4. Using needle-nose pliers, remove the terminal connectors from the compressor terminals, one at a time, and LABEL THE WIRES. DO NOT PULL ON THEM. In a current relay type, two of the terminals may be covered by a plug. Ensure the wires are labeled so that you will not make a mistake when replacing them.

5. Using the terminal layout in Figure DOM-7.1, record the resistance, in ohms, from terminal to terminal, as indicated.

   - 1 to 2: _____ Ω
   - 1 to 3: _____ Ω
   - 2 to 3: _____ Ω

FIGURE **DOM-7.1**

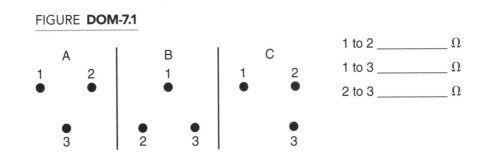

6. Based on your resistance readings, identify the following: Start windings: terminal: _____ to _____; Run windings: terminal _____ to _____.

7. Using the ohmmeter, touch one of the meter's test leads to a copper line and the other to each compressor terminal, one by one, and record the readings: _____ Ω.

   Note: A typical reading should be OL or infinity. If a resistance is measured, ask your instructor for further directions.

8. Replace the motor terminal wires or relay to the compressor, and fasten the terminal cover back in place.

9. Replace the compressor compartment cover with the proper fasteners, and return the refrigerator to its proper location.

MAINTENANCE OF WORKSTATION AND TOOLS: Turn the ohmmeter to the off position to prevent battery drain, and return all tools to their places. Make sure the workstation is clean and orderly.

SUMMARY STATEMENT: Describe what the symptoms would be if the compressor windings to ground reading were 500 Ω. Use an additional sheet of paper if more space is necessary for your answer.

TAKEAWAY: When wiring a compressor, it is imperative that each wire is connected to the proper terminal. Mixing up the wires could result in compressor damage or failure. An ohmmeter can be used to determine which terminal is which.

QUESTIONS

1. Which has the most resistance, the start or run winding?

2. Does the start winding have a smaller- or larger-diameter wire than the run winding?

3. What is the purpose of the start winding?

4. At what approximate percentage of the normal run speed is the start winding disconnected from the circuit?

5. What is the purpose of the current relay in the start circuit?

6. What two types of compressors are used in household refrigerators?

7. What would be the evaporator saturation pressure in a system using R-134a if the saturation temperature of the refrigerant were −3°F?

Exercise DOM-8

Freezer Gasket Examination

Name	Date	Grade

OBJECTIVES: Upon completion of this exercise, you should be able to determine whether the gaskets on a freezer are sealing correctly.

INTRODUCTION: Depending on the type of gasket on the freezer you are working with, you will use one of two different procedures to determine whether the gasket is sealing correctly to prevent air and moisture from entering the freezer. Most late-model upright freezers will have magnetic gaskets and will be checked with a light. Chest-type freezers may have compressible gaskets and will be checked with a crisp piece of paper.

TOOLS, MATERIALS, AND EQUIPMENT: Straight-slot and Phillips-head screwdrivers, tape, a trouble light, and a crisp piece of paper such as a dollar bill.

⚠ **SAFETY PRECAUTIONS:** Do not place the trouble light in the freezer next to any plastic parts because it may cause them to melt. Use caution while shutting the door on the trouble light cord.

STEP-BY-STEP PROCEDURES

Choose a freezer for the gasket test. Most new upright freezers have magnetic gaskets while chest-type freezers may have compressible-type gaskets. One of each type should be used.

Freezer with Magnetic Gaskets

1. Place the trouble light inside the box, and turn it on.

   Note: Do not place the light bulb next to plastic components.

2. Gently shut the door on the cord. If too much pressure must be applied to the door, move the cord to another location.

3. Turn the room lights off, and examine the door gasket all around its perimeter for light leaks. Light will leak around the cord, so it should be moved to a different location, and the gasket should be checked where the cord was first placed.

4. If leaks are found, make a note and mark the places with tape.

5. Open the door and examine any leak spots closely to see whether the gasket is dirty or worn.

Freezer with Compressible Gaskets

1. Close the lid on the piece of paper (a dollar bill) repeatedly around the perimeter of the door, and gently pull it out. The paper should have some drag as it is pulled out. If it has no drag, air will leak in. Mark with tape any places that may leak.

2. After you have proceeded around the perimeter of the door and marked any questionable places, open the lid or door and examine each place for dirt or deterioration.

3. Remove all the tape and return each freezer to its permanent location. Your instructor may want you to make a repair on any gaskets found defective.

MAINTENANCE OF WORKSTATION AND TOOLS: Return all tools to their designated places.

SUMMARY STATEMENT: Describe the repair or replacement procedures for a magnetic gasket that is defective. Use an additional sheet of paper if more space is necessary for your answer.

TAKEAWAY: The door gasket is an integral component in the design of any refrigerating box. It prevents the infiltration of the air outside into the refrigerated space. Defects in the gasket can lead to poor operation or even component failure. It should be visually inspected regularly.

QUESTIONS

1. What type of material is used as insulation in the walls of a modern freezer?

2. What type of conditions will a worn gasket cause?

3. How should a refrigerator or freezer be prepared for storage, and how should it be placed in the storage area?

4. What type of gasket is used in most modern upright refrigerators and freezers?

5. What will be the condition of the outside of the compressor if liquid refrigerant is returning to it through the suction line?

Troubleshooting Exercise: Refrigerator

Name	Date	Grade

OBJECTIVES: Upon completion of this exercise, you should be able to troubleshoot the defrost circuit of a typical household refrigerator.

INTRODUCTION: This unit should have a specified problem introduced by your instructor. You will locate the problem, using a logical approach and instruments. THE INTENT IS NOT FOR YOU TO MAKE THE REPAIR, BUT TO LOCATE THE PROBLEM AND ASK YOUR INSTRUCTOR FOR DIRECTION BEFORE COMPLETION.

TOLLS, MATERIALS, AND EQUIPMENT: Straight-slot and Phillips-head screwdrivers, safety goggles, ¼ in. and ⁵⁄₁₆ in. nut drivers, a digital multimeter, and a drop light.

⚠ **SAFETY PRECAUTIONS:** Use care while working with energized electrical circuits.

STEP-BY-STEP PROCEDURES

1. The refrigerator should be running, and the evaporator should be frozen solid. Name at least two things that can cause a frozen evaporator.

2. Put on your safety goggles. Let's assume we have a defrost timer problem. Locate the timer. You may have to follow the wiring diagram to find it.

   Note: If you have to remove the compressor panel in the back, do not allow air to bypass the condenser or else high head pressure will occur and overload the compressor.

   Describe the location of the defrost timer.

3. Examine the timer. Does it have a viewing window to show the timer turning? If not, it should have some sort of dial that turns over once every hour. If it has a dial that turns over once per hour, mark the dial with a pencil, and wait 15 minutes to see whether it is turning. You may measure the voltage to the timer coil during the wait. Is the timer turning?

4. If it is not turning, does it have voltage to the coil? _____

5. Describe your opinion of the problem.

6. What materials do you think it would take to make the repair?

7. Ask your instructor for directions at this point.

MAINTENANCE OF WORKSTATION AND TOOLS: Repair the unit if your instructor directs you to do so; otherwise, leave the unit as directed. Return all tools to their correct places.

SUMMARY STATEMENT: Describe the complete defrost cycle, including the electrical circuit.

TAKEAWAY: The evaporator of a domestic refrigerator typically operates below freezing and, as such, requires periodic defrosting due to the accumulation of ice. Refrigerator defrost timers generally initiate a defrost cycle based on the compressor run time. Without this cycle, the evaporator will eventually freeze solid.

QUESTIONS

1. List the components of the electrical circuit for the defrost cycle.

2. What causes the timer motor to turn?

3. Did this refrigerator have an ice maker?

4. If yes, was it still making ice with a frozen evaporator?

5. Were the wires color coded or numbered on this refrigerator?

6. Was the wiring diagram easy to read? Why or why not?

7. What type of condenser did this refrigerator have, static or induced-draft?

8. How would high head pressure affect the operation of this refrigerator?

Exercise DOM-10

Removing the Moisture from a Refrigerator After an Evaporator Leak

Name	Date	Grade

OBJECTIVES: Upon completion of this exercise, you should be able to remove moisture from a refrigerator (or freezer) after an evaporator leak.

INTRODUCTION: You will use a vacuum pump and the correct triple evacuation procedures to remove moisture from the refrigeration circuit and oil of a refrigerator. You will then add driers, pressure check, evacuate again, and charge the system.

TOOLS, MATERIALS, AND EQUIPMENT: Straight-slot and Phillips-head screwdrivers, a Schrader valve tool, a nitrogen tank and regulators, ¼ in. and ⁵⁄₁₆ in. nut drivers, two 100-watt drop lights and a 150-watt spotlight lamp, a torch kit and solder, two ¼ in. tees, two ¼ in. Schrader valve fittings, a liquid-line drier, a suction-line drier, a gauge manifold, safety goggles, a valve wrench, a vacuum pump, refrigerant, a digital refrigerant scale, and a refrigerator that has had its charged recovered.

⚠ **SAFETY PRECAUTIONS:** Use care with any heat lamps applied to the refrigerator to avoid melting the plastic. DO NOT START THE REFRIGERATOR COMPRESSOR WHILE THE SYSTEM IS IN A DEEP VACUUM OR ELSE DAMAGE TO THE REFRIGERATOR COMPRESSOR MOTOR MAY OCCUR.

STEP-BY-STEP PROCEDURES

1. Verify that the system charge has been recovered.

2. Open the refrigerant circuit to the atmosphere by cutting the suction line to the compressor.

3. Cut the liquid line.

4. Prepare the copper lines, and braze or solder the tees and Schrader valves in place, whichever your instructor has determined to be proper for this exercise. Remove the valve cores from the Schrader valves prior to applying heat.

5. Connect the core removal tools to the service valves.

6. Connect the gauge manifold to the Schrader ports. Remove the core depressors from the hoses, or use copper tubing for gauge lines to the gauge manifold for the best vacuum. It will reduce the evacuation time.

7. Start the vacuum pump.

8. Place a light bulb in each compartment of the refrigerator. LEAVE THE DOOR CRACKED ABOUT 2 IN. OR ELSE THE PLASTIC WILL MELT, Figure DOM-10.1.

9. Place the 150 W spotlight against the compressor shell. This will boil any moisture out of the oil.

FIGURE **DOM-10.1**

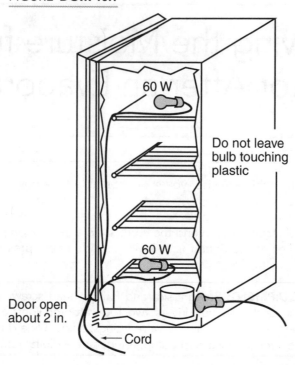

60 W

Do not leave
bulb touching
plastic

60 W

Door open
about 2 in.

← Cord

10. The vacuum pump should run for several hours like this. If there is a great deal of water in the refrigerant circuit, it will begin to condense in the vacuum pump crankcase. You will see the vacuum pump oil level rise, and the oil color will turn milky. Change the vacuum pump oil when the color changes. You may have to do this several times for a large amount of water. If the vacuum pump is left on overnight, make sure that the power supply will not be interrupted. Be sure to change the vacuum pump oil as needed.

11. The vacuum should be verified with an electronic vacuum gauge. The correct vacuum level must be reached and held. The vacuum may then be broken with nitrogen to atmospheric pressure. IT IS RECOMMENDED TO START THE COMPRESSOR FOR ABOUT 20 SECONDS AT THIS TIME. IF ANY MOISTURE IS TRAPPED IN THE COMPRESSOR CYLINDERS, IT WILL BE PUMPED OUT. Evacuate the system again.

12. After the second vacuum, break the vacuum to atmospheric pressure with nitrogen and install the driers, one in the liquid line and one in the suction line. DO NOT ALLOW THE DRIERS TO BE OPEN TO THE ATMOSPHERE FOR MORE THAN 5 MINUTES.

    Note: The liquid-line drier should not be oversized. The suction-line drier will handle any excess moisture.

13. Leak check the system with nitrogen and evacuate for the last time to a deep vacuum.

14. When a deep vacuum is reached, break the vacuum with the nitrogen to about 5 psig.

15. Remove the gauge lines, one at a time, and replace the valve cores and depressors.

16. Prepare to measure the refrigerant into the system, start the vacuum pump, and pull one more vacuum. Remember, all refrigerant should be measured into the system from a vacuum. If the volume of the liquid-line drier has not been changed, the charge on the nameplate should be correct.

17. Charge the refrigerant into the system, and start the unit.

18. It is good practice to allow a box that has had moisture in the refrigerant circuit to run in the shop for several days before returning it to the customer. This may prevent a callback.

MAINTENANCE OF WORKSTATION AND TOOLS: Return all tools to their proper places. Be sure to clean the entire work area, making certain that any oil that might have spilled is cleaned up thoroughly.

SUMMARY STATEMENT: Describe triple evacuation.

TAKEAWAY: Evacuation of a refrigeration system is required any time the system is exposed to the atmosphere. Refrigerant oil is hygroscopic and will absorb moisture that can damage the system and lead to compressor failure. A thorough evacuation, along with a clean filter/drier, can effectively rid the system of air and moisture.

QUESTIONS

1. Why does it help to start the compressor after the first vacuum?

2. Why does the evacuation process take so long when there is moisture in the system?

3. Why is it good field practice to frequently change the vacuum pump oil?

4. Why is a digital vacuum gauge recommended when pulling a deep vacuum?

Changing a Compressor on a Sealed Refrigeration System

Name	Date	Grade

OBJECTIVES: Upon completion of this exercise, you should be able to change a compressor on a household refrigerator.

INTRODUCTION: You will recover the refrigerant from a refrigerator, change the compressor, leak-check, evacuate, and charge the refrigerator.

TOOLS, MATERIALS, AND EQUIPMENT: Hand tools, ¼ in. and ⁵⁄₁₆ in. nut drivers, a liquid-line drier, two ¼ in. tees, two ¼ in. Schrader fittings, a torch kit, two 6 in. adjustable wrenches, a set of sockets, two piercing valves, a digital charging scale, slip-joint pliers, a clamp-on ammeter, a gauge manifold, a refrigerant recovery system, a vacuum pump, refrigerant, a refrigerator with a sealed refrigeration circuit, goggles, and gloves.

⚠ **SAFETY PRECAUTIONS:** Use care while using the torch. Wear goggles and gloves when attaching gauges to a system and transferring refrigerant.

STEP-BY-STEP PROCEDURES

1. Put on your goggles and gloves. Fasten the piercing valves to both the suction and discharge lines, and leak-check the valve before proceeding.

2. Attach the gauge manifold hoses to the piercing valves. Open the piercing valves, and purge the hoses.

3. Connect the center port of the gauge manifold to the recovery system, and recover the refrigerant.

4. Once the refrigerant has been removed from the system, open the system to the atmosphere.

5. Mark the compressor terminal wiring, and remove the wiring.

6. Heat the compressor suction, and discharge lines with the torch. Use pliers to gently pull them from their connectors when they are hot enough. USE A HEAT SHIELD TO PREVENT OVERHEATING ADJACENT PARTS.

7. Remove the compressor mounting bolts, and remove the compressor from the compartment. This compressor may be reinstalled as if it were a new one.

8. Clean the suction and discharge stubs on the "new" compressor. They may be steel and must be perfectly clean. There can be no rust, paint, or pits in the lines or else reconnection will be impossible. Clean the pipe stubs on the refrigerator the same way. A small file is sometimes necessary to clean out the pits and the small solder balls from the pipe. Make certain that, if a file is used, no filings get into the compressor or the refrigerant lines.

9. When all is cleaned correctly, replace the compressor in the compartment. There will likely be solder balls inside the fittings that cannot be removed. They will have to be heated and the pipe connection made hot. USE A HEAT SHIELD TO PREVENT OVERHEATING ADJACENT PARTS.

10. With the compressor soldered back in place, remove the piercing valves, cut the copper lines where they were pierced, and install the tees and Schrader valves.

11. Cut the liquid line, and sweat the liquid-line drier in place. If the same size liquid-line drier is used, the charge will not need to be adjusted.

12. Pressurize and leak-check the system.

13. Triple-evacuate the system, and on the third evacuation, prepare to charge the system using a digital scale.

    Note: Use a micron gauge and be sure to evacuate to the required level. Be sure to use nitrogen to break the vacuum when performing a triple evacuation.

    While the system is being evacuated, reconnect the wiring and fasten the compressor mounting bolts.

14. With the system evacuated, add the correct charge to the system.

15. When the correct charge is in the system, start the system and observe its operation. Be sure to check the amperage.

MAINTENANCE OF WORKSTATION AND TOOLS: Leave the refrigerator as your instructor directs you. Return all tools to their places.

SUMMARY STATEMENT: Describe why it takes longer to evacuate from the low-pressure side of the system only. Use an additional sheet of paper if more space is necessary for your answer.

TAKEAWAY: Compressor replacement is a typical task performed in the field. Technicians need to follow specific procedures in order to ensure a successful repair. Proper evacuation and charging procedures will ensure that the appliance will operate as intended after the repair has been made.

QUESTIONS

1. Describe the gas ballast on a vacuum pump.

2. Describe a two-stage rotary vacuum pump, and state how it works.

3. Why is using a rotary vacuum pump better than utilizing a reciprocating compressor?

4. What are the advantages of using a digital vacuum gauge during the evacuation process?

5. Why should a filter/drier be cut out rather than be removed with a torch?

Exercise DOM-12

Window Air-Conditioner Familiarization

Name	Date	Grade

OBJECTIVES: Upon completion of this exercise, you should be able to identify the various parts of a window air conditioner.

INTRODUCTION: Your instructor will assign you a window unit. You will remove it from its case and then identify and describe the major components.

TOOLS, MATERIALS, AND EQUIPMENT: Straight-slot and Phillips-head screwdrivers, ¼ in. and 5⁄16 in. nut drivers, a flashlight, and a window or room air conditioner.

⚠ **SAFETY PRECAUTIONS:** Make sure the unit is unplugged before starting the exercise. Be careful while working around the coil fins because they are sharp. Gloves and safety glasses should be worn during this exercise.

STEP-BY-STEP PROCEDURES

1. If the unit is installed in a window in a slide-out case, remove it from its case, and set it on a level bench. If the unit is in a case fastened with screws, set the unit on a bench, and remove the case.

2. When all parts of the unit are visible, fill in the following information:

Indoor Section
- Type of fan blade: _____
- What is the evaporator tubing made of? _____
- What are the fins made of? _____
- Where does the condensate drain to from the evaporator? _____

Outdoor Section
- Type of fan blade: _____
- What is the condenser tubing made of? _____
- What are the fins made of? _____
- Does the condenser fan blade have a slinger ring for evaporating condensate? _____
- Does the condenser coil appear to be clean for a good heat exchange? _____

Fan Motor Information
- Size of fan shaft: _____
- Looking at the lead end of the motor, what is the rotation, clockwise or counterclockwise? _____
- Fan motor voltage: _____ V
- Fan motor current: _____ A
- Type of fan motor, PSC or shaded pole: _____
- If PSC, what is the microfarad rating of the capacitor? _____ μF
- If available on the nameplate of the motor, what are its speeds? _____

Unit Nameplate
- Manufacturer: _____
- Serial number: _____
- Model number: _____
- Unit voltage: _____
- Unit full-load current: _____ A
- Test pressure: _____ psig

General Unit Information
- What ampere rating does the power cord have? _____ A
- What type of air filter is furnished with the unit, fiberglass or foam rubber? _____

3. Assemble the unit.

MAINTENANCE OF WORKSTATION AND TOOLS: Return all tools to their proper locations. Place the window or room unit in a storage area if instructed to do so. Leave your work area clean.

SUMMARY STATEMENT: Describe how condensate is evaporated in the condenser portion of the window unit you worked with. Use an additional sheet of paper if more space is necessary for your answer.

TAKEAWAY: Window units operate much the same as a split-system air-conditioning system, only smaller. All of the system components are packaged in a single housing, which can be accessed by removing the cover.

QUESTIONS

1. How many fan motors are normally included in a window cooling unit?

2. What are the two types of designs of window units relative to covers and access for service?

3. What is the most common refrigerant used in window units?

4. What is the most common metering device used in window units?

5. Why must window cooling units operate above freezing?

6. At what temperature do window-cooling-unit evaporators typically boil the refrigerant?

7. What is the evaporator tubing in these units usually made of?

8. What are the evaporator fins made of?

9. How does the unit dehumidify the conditioned space?

Checking the Charge Without Gauges

Name	Date	Grade

OBJECTIVES: Upon completion of this exercise, you should be able to evaluate the refrigerant charge for a window air conditioner without disturbing the refrigerant circuit with gauge connections.

INTRODUCTION: You will operate a window unit and use temperature drop across the evaporator and temperature rise across the condenser to evaluate the refrigerant charge without gauges.

TOOLS, MATERIALS, AND EQUIPMENT: A digital, multiprobe thermocouple thermometer, straight-slot and Phillips-head screwdrivers, ¼ in. and ⁵⁄₁₆ in. nut drivers, and a window air-conditioning unit.

⚠️ **SAFETY PRECAUTIONS:** Make sure that when the unit panels are removed for the purpose of observing the suction line, the condenser and evaporator have full airflow over them. Ensure that all leads, clothing, and your hands are kept away from fan blades.

STEP-BY-STEP PROCEDURES

1. Remove the unit from its case, and place it on a bench with a correct power supply nearby.

2. Make sure that the correct airflow passes across each coil before starting the compressor. Cardboard may be placed over any place where a panel has been removed to direct the airflow across the coil.

3. Place a thermometer at the inlet and outlet of the evaporator, and start the compressor. DO NOT LET THE THERMOMETER LEAD GET IN THE FAN.

4. Cover about ⅔ of the condenser surface, Figure DOM-13.1. This will cause the head pressure to rise. DO NOT COVER THE ENTIRE CONDENSER. When the head pressure rises, the condenser then operates with the correct charge, pushing the correct refrigerant charge to the evaporator. The suction line leading to the compressor should then sweat.

FIGURE **DOM-13.1**

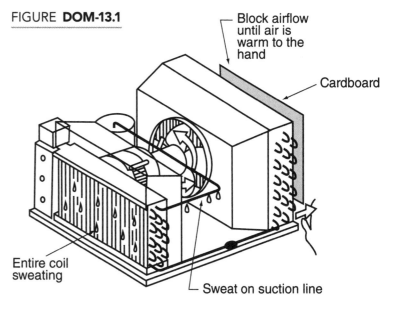

Block airflow until air is warm to the hand

Cardboard

Entire coil sweating

Sweat on suction line

5. Let the unit run for about 15 minutes; then go to step 6.

6. How much of the evaporator is sweating—¼, ½, ¾, or all of it? _____ How much of the suction line to the compressor is sweating? _____

7. Record the air temperature entering the evaporator: _____°F

Record the air temperature leaving the evaporator: _____°F

Record the temperature difference: _____°F

8. Now record the temperature difference across the condenser: _____°F. It will not be significant, but it is worth noting.

9. Shut the unit off and unplug it. Replace all panels and place the unit in its permanent location.

MAINTENANCE OF WORKSTATION AND TOOLS: Clean up the workstation, and return all tools to their proper places.

SUMMARY STATEMENT: Evaluate the unit you worked on, and tell how the preceding test indicates whether the charge is correct. Use an additional sheet of paper if more space is necessary for your answer.

TAKEAWAY: Window units are typically sealed systems that lack valves for gauge connection. The charge can be verified by observing the system's performance and operating conditions. By using air temperatures as well as the line temperatures, technicians can reasonably assume the level of charge contained in the unit.

QUESTIONS

1. What is the primary maintenance needed for room cooling units?

2. How do some manufacturers provide evaporator freeze protection?

3. Why should gauges be installed only when absolutely necessary?

4. What two types of compressors may be used in room units?

5. What are the four major components in the room unit refrigeration cycle?

6. What are two purposes of the condenser?

7. How is the compressor heat used to evaporate condensate?

8. Describe how the condenser fan slinger provides for condensate evaporation.

9. How can the exhaust or fresh air options be controlled?

10. What options does the selector switch control?

Charging a Window Unit by the Sweat-Line Method

Name	Date	Grade

OBJECTIVES: Upon completion of this exercise, you should be able to add a partial charge to a window unit using the sweat-line method.

INTRODUCTION: You will add refrigerant to a window unit through the suction-line service port until a correct pattern of sweat has formed on the suction line to the compressor.

TOOLS, MATERIALS, AND EQUIPMENT: Straight-slot and Phillips-head screwdrivers, ¼ in. and $\frac{5}{16}$ in. nut drivers, a gauge manifold, a cylinder of refrigerant, gloves, goggles, and a window unit with low- and high-side pressure service ports.

⚠ **SAFETY PRECAUTIONS:** Gloves and goggles should be worn while making gauge connections and transferring refrigerant. A high-side gauge should be used while adding refrigerant.

STEP-BY-STEP PROCEDURES

1. Unplug the unit and remove it from its case. Set the unit on a bench with a power supply for the unit.

2. Put on your goggles and gloves. Fasten the gauge hoses to the high- and low-side connections. Purge the gauge hoses of any air through the center hose, and connect it to the refrigerant cylinder.

3. Make sure the airflow across the evaporator and condenser is correct. Start the unit and observe the pressures. Record the following after 15 minutes: Suction pressure: _____ psig. Discharge pressure: _____ psig. Where is the last point that the suction line is sweating? _____

4. Restrict the airflow across the condenser, as in Exercise DOM-13, until about ⅔ of the condenser is blocked. Wait 15 minutes, and record the following: Suction pressure: _____ psig. Discharge pressure: _____ psig. Where is the last point that the suction line is sweating? _____

5. If the unit has the correct charge, recover some refrigerant until the suction pressure is about 30 psig for R-22 or 58 psig for R-410A.

6. Record the following after 15 minutes of running time at the pressure reached in step 5: Suction pressure: _____ psig. Discharge pressure: _____ psig. Where is the last point that the suction line is sweating? _____

7. Start adding refrigerant in short bursts through the suction line. Add a little and wait for about 5 minutes. Keep adding in short bursts until the sweat line moves to the compressor, waiting about 5 minutes after each addition of refrigerant. When the correct charge is reached, record the following: Suction pressure: _____ psig. Discharge pressure _____ psig. See Figure DOM-14.1.

8. Unplug the unit and remove the gauges. Return the unit to its permanent location. Be sure to seal the unit gauge ports.

MAINTENANCE OF WORKSTATION AND TOOLS: Return all tools to their proper locations, and make sure the workstation is clean.

SUMMARY STATEMENT: Explain why the condenser airflow must be blocked in order to correctly add a partial charge to a unit. Use an additional sheet of paper if more space is necessary for your answer.

TAKEAWAY: By observing the sweat line, as well as other system operating conditions, a technician can estimate the charge for small, package-type systems such as domestic refrigerators, freezers, and window units.

FIGURE **DOM-14.1**

113 psig   402 psig R-410A

Coil sweating all over

Suction line sweating to and on the compressor shelf

Condenser airflow blocked to maintain 402 psig discharge pressure

QUESTIONS

1. Does the fan motor normally need periodic lubrication?

2. What will happen to the suction pressure if the indoor coil becomes dirty?

3. If the low-side pressure becomes too low, what will happen to the condensate on the evaporator coil?

4. Why should gauges be installed only when absolutely needed?

5. What are line tap valves?

6. In this exercise, why is the airflow through the condenser blocked?

7. If only part of the evaporator gets cold in this test, what could be the problem?

8. If the charge is correct but the evaporator coil is only cool, what could be the problem?

Charging a Unit Using Scales

Name	Date	Grade

OBJECTIVES: Upon completion of this exercise, you should be able to add an accurate charge to a window unit using a digital scale.

INTRODUCTION: You will evacuate a unit to a low vacuum and add a measured charge to the unit using a digital scale.

TOOLS, MATERIALS, AND EQUIPMENT: Straight-slot and Phillips-head screwdrivers, ¼ in. and ⁵⁄₁₆ in. nut drivers, a gauge manifold, a cylinder of refrigerant correct for the system, a vacuum pump, goggles, gloves, a digital refrigerant scale, and a room cooling unit.

⚠ **SAFETY PRECAUTIONS:** Goggles and gloves should be worn while making gauge connections and transferring refrigerant.

STEP-BY-STEP PROCEDURES

1. Remove the unit to a bench with power. Put on your goggles and gloves. Fasten the gauge hoses to the gauge ports, and purge the hoses of any air that might be present in them.

   Note: If the unit only has a suction port, you can still do this exercise.

2. If the unit has a charge of refrigerant, your instructor may advise you to recover the refrigerant from the system using an approved recovery system.

3. When all of the refrigerant is removed from the system, unplug the unit, and connect the vacuum pump. Check the oil in the vacuum pump. If it is not at the correct level, ask your instructor for directions. Start the pump to evacuate the system.

4. While the vacuum pump is running, prepare the tools and materials needed to charge the system. Determine the correct charge, make necessary calculations, and be ready to add the refrigerant as soon as the correct vacuum has been obtained. Be sure to account for any refrigerant that will be contained in the hoses during the charging procedure.

5. When refrigerant is allowed to enter the system, the system pressure will soon equalize with the cylinder pressure, and no more refrigerant will flow. You may start the unit and charge the rest of the refrigerant into the system.

6. Let the unit run long enough to establish that it is running correctly. You may want to cover part of the condenser. The suction line should sweat, and the suction pressure should be about 65 psig for R-22 and 114 psig for R-410A.

7. When the correct charge is in the system, shut the unit off, and disconnect the low-side gauge hose from the unit. Place caps on all gauge connections and replace the panels.

8. Return the unit to its permanent location.

MAINTENANCE OF WORKSTATION AND TOOLS: Return all tools to their proper locations.

SUMMARY STATEMENT: Describe what you found to be the most difficult step in the preceding charging sequence.

TAKEAWAY: Measuring the charge into a system by weight, known as a critical charge, is the most accurate way to charge any refrigeration system. The charge weight is determined by the manufacturer and is usually printed on the unit nameplate. When utilizing this method, it is important to account for any refrigerant that may fill up the hoses and manifold. Failure take into account the hoses may result in an undercharged system.

QUESTIONS

1. Why are larger ports better when evacuating a system?

2. What is the problem if, when charging a system, the head pressure rises and the suction pressure does not?

3. If there is a problem with the capillary tube, what other types of metering devices may be installed in a room cooling unit?

4. What must the compressor motor have for starting devices if one of the above metering devices is installed?

5. Why must the motor have one of the previously listed starting devices?

6. Why is it important to check for a leak if the refrigerant is found to be low?

7. Why should an accurate dial or electronic scale be used when weighing in the refrigerant?

Situational Service Ticket

Name	Date	Grade

Note: Your instructor must have placed the correct service problem in the system for you to successfully complete this exercise.

Technician's Name: _____ Date: _____

CUSTOMER COMPLAINT: Refrigerator compartment is not cold.

CUSTOMER COMMENTS: Unit runs all the time and the refrigerator compartment is not cold. The milk is not cold.

TYPE OF SYSTEM:

  MANUFACTURER _____   MODEL NUMBER _____   SERIAL NUMBER _____

COMPRESSOR (where applicable):

  MANUFACTURER _____   MODEL NUMBER _____   SERIAL NUMBER _____

TECHNICIAN'S REPORT

1. SYMPTOMS: _____

2. DIAGNOSIS: _____

3. ESTIMATED MATERIALS FOR REPAIR: _____

4. ESTIMATED TIME TO COMPLETE THE REPAIR: _____

SERVICE TIP FOR THIS CALL

1. Let the system run, and record the temperatures of both the refrigerator and freezer sections. Take note of the cycle times, and check for excessive heat loads.

Situational Service Ticket

Name	Date	Grade

Note: Your instructor must have placed the correct service problem in the system for you to successfully complete this exercise.

Technician's Name: _____ Date: _____

CUSTOMER COMPLAINT: No cooling.

CUSTOMER COMMENTS: Unit stopped in the middle of the night last night for no apparent reason.

TYPE OF SYSTEM:

    MANUFACTURER _____   MODEL NUMBER _____   SERIAL NUMBER _____

COMPRESSOR (where applicable):

    MANUFACTURER _____   MODEL NUMBER _____   SERIAL NUMBER _____

TECHNICIAN'S REPORT

1. SYMPTOMS: _____

2. DIAGNOSIS: _____

3. ESTIMATED MATERIALS FOR REPAIR: _____

4. ESTIMATED TIME TO COMPLETE THE REPAIR: _____

SERVICE TIP FOR THIS CALL

1. Use your digital multimeter to follow the electrical circuits.

SECTION 13

Installation and Start-Up (ISU)

SUBJECT AREA OVERVIEW

The expected lifespan of a unit, as well as the efficiency at which it operates, can be traced back to its installation. A quality installation, coupled with regular maintenance, can ensure that an HVACR system will operate as intended for its expected life cycle. A poor installation and/or lack of regular maintenance can lead to higher energy bills, poor system performance, and premature failure of system components. Using the proper materials and procedures during system installation, start-up, and service will benefit the technician, the customer, and the equipment in the long run.

The use of correct piping materials and installation procedures are necessary for an HVACR system to operate properly. Copper tubing is the most commonly used piping material. Tubing used for air-conditioning and refrigeration applications is dehydrated for cleanliness and is classified as air conditioning and refrigeration (ACR) piping. Copper piping material can be purchased in rolls or straight lengths. Copper tubing that is supplied in rolls is referred to as soft-drawn tubing, while straight-length material is referred to as hard-drawn copper. Soft-drawn tubing can be easily bent in the field, but hard-drawn piping materials cannot. Copper tubing can be soldered or brazed, with soldering temperatures being below 800°F and brazing temperatures over 800°F. Flare joints with fittings may also be used to join copper tubing.

Sheet metal duct is fabricated in sections, which can be wrapped with insulation and connected on the job site. Vibration eliminators or canvas connectors should be used between the air handler and the duct to reduce noise transmission through the system. Fiberglass ductboard is popular in some areas, where allowed by code, because it can be formed into many configurations on the jobsite. Ductboard absorbs sound and does not transmit blower noise. Flexible duct is round and may be used for both supply and return ducts. Sharp bends should be avoided, as this may cause the flexible duct to close and greatly reduce the airflow.

Electrical service to a system should be run by a licensed electrician. HVACR technicians are typically responsible for installing the low-voltage control circuits, including system thermostats. Low-voltage cables can have anywhere from two to eight color-coded conductors, which are typically, 18 or 20 gauge, in the same sheath.

Package air-conditioning systems house all system components in a single cabinet and do not required field-installed refrigerant lines. These units should be set on a firm foundation, with rooftop units set on curbs, which help maintain the integrity of the roof, provide a means to reduce noise transmission, and also keep the units off the roof. This reduces the effect that snow accumulation might have on system operation.

In a split system, the evaporator is located near the blower section in a furnace or an air handler. The condensing unit should be placed where it has power and air circulation and is convenient for future servicing. To keep system efficiency as high as possible, the field-installed refrigerant lines should be as short as possible, so the condensing unit should be positioned as close as possible to the indoor unit.

Although the refrigerant charge for the system is typically furnished by the manufacturer and stored in the condensing unit, in most cases, the charge will need to be adjusted depending

(continued)

on the length of the refrigerant lines. Ensure that all manufacturer's recommendations are followed when installing equipment, including piping, and equipment start-up.

The Air-Conditioning, Heating, and Refrigeration Institute (AHRI) has developed ratings for equipment. These ratings include EER, or energy efficiency ratio, and SEER, or seasonal energy efficiency ratio. These ratings compare a system's output in Btu/h to the power requirements, in watts. The EER rating is a "snapshot in time" and does not take into account the energy lost during system start-up and shutdown. The SEER rating, on the other hand, does take into consideration the start-up and shutdown for each cycle over the course of an entire cooling season.

There are typically three grades of equipment: economy, standard efficiency, and high efficiency. Economy and standard grades have about the same efficiency. The high-efficiency grade has different operating characteristics. A standard condenser will condense the refrigerant at a temperature of approximately 30°F above the outside ambient temperature. In the high-efficiency condenser, the condensing temperature may be as low as 10°F–20°F above the outside ambient air. Air-conditioning equipment is designed to operate at its rated capacity and efficiency at a specified design condition. This condition is generally accepted to be at an outside temperature of 95°F and an inside temperature of 80°F with a relative humidity of 50%. This rating has been established by the AHRI.

Charging a system refers to the process of adding refrigerant to a system. Refrigerant may be added to a system in the vapor state or, under certain conditions, in the liquid state. Although vapor charging is the safest way to charge, liquid charging is necessary when working with blends. The liquid refrigerant should be throttled into the low side of the system, taking care not to slug the compressor.

KEY TERMS

Air-acetylene
Brazing
Bubble point
Burr
Butane
Capillary attraction
Charging scale
CPVC (chlorinated polyvinyl chloride)
Degassing
Dehydration
Dew point
Electronic leak detector
Energy efficiency ratio (EER)
Evacuation
Flaring
Graduated charging cylinder
Halide leak detector
Line sets
MAPP gas
Micron gauge
Oxyacetylene
PE (polyethylene)
Propane
PVC (polyvinyl chloride) reamer
Reamer
Season energy efficiency ratio (SEER)
Soldering
Swaging
System performance
System start-up
Triple evacuation
Vacuum

REVIEW TEST

Name	Date	Grade

Circle the letter that indicates the correct answer.

1. Types K and L copper tubing have different:
   A. inside diameters.
   B. outside diameters.
   C. wall thicknesses.
   D. atomic structures.

2. ACR tubing is measured by its:
   A. inside diameter.
   B. outside diameter.
   C. wall thickness.
   D. none of the above.

3. Tubing used for air-conditioning and refrigeration installations is usually insulated when used on the:
   A. suction line.
   B. discharge line.
   C. liquid line.
   D. all of the above.

4. When tubing is cut with a tubing cutter, a _____ is often produced from excess material pushed into the end of the pipe from the pressure of the cutter.
   A. burr   C. filing
   B. chip   D. flux

5. The maximum temperature at which soft soldering is done is about:
   A. 400°F.   C. 600°F.
   B. 212°F.   D. 800°F.

6. Brazing is done at _____ soldering.
   A. the same temperature as
   B. lower temperatures than
   C. higher temperatures than
   D. temperatures that can be higher or lower than

7. A common filler material used to solder lines containing potable water is composed of:
   A. 50/50 tin–lead.
   B. 95/5 tin–antimony.
   C. 15%–60% silver.
   D. both A and B.

8. A common filler material used for brazing is composed of:
   A. 50/50 tin–lead.
   B. 95/5 tin–antimony.
   C. 5%–60% silver.
   D. cast steel.

9. Which of the following heat sources is often used for soldering and brazing?
   A. Natural gas
   B. Liquified oxygen
   C. Air-acetylene
   D. Nitrogen

10. A flux is used when soldering to:
    A. minimize oxidation while the joint is being heated.
    B. allow the tubing to fit easily into the fitting.
    C. keep the filler metal from dripping on the floor.
    D. help the tubing and fitting heat faster.

11. A flare joint:
    A. is made by expanding tubing to fit over other tubing.
    B. uses a flare on the end of a piece of tubing against an angle on a fitting, secured with a flare nut.
    C. uses a flare on the end of a piece of tubing soldered to a flare on a mating piece of tubing.
    D. is constructed using a tubing bending spring.

12. A swaged joint:
    A. uses a flare on one piece of tubing against the end of another piece of tubing.
    B. is made by expanding one piece of tubing to fit over another piece of tubing of the same diameter.
    C. is constructed using a tubing bending spring.
    D. is a fitting brazed to a piece of tubing.

13. Steel pipe:
    A. is joined with threaded fittings.
    B. may be joined by welding.
    C. is often joined with a flare fitting.
    D. both A and B.

14. When used on pipe that will contain liquid or vapor under pressure, pipe threads:
    A. are generally tapered.
    B. should have seven perfect threads.
    C. generally have a V-shape.
    D. have all of the above attributes.

15. In the thread specification ½–14 NPT, the 14 stands for:
    A. thread diameter.
    B. pipe size.
    C. number of threads per inch.
    D. length of the taper.

16. The end of a piece of steel pipe is reamed after it is cut:
    A. to remove the oil from the cut.
    B. to remove the burr from the inside of the pipe.
    C. to begin the threading process.
    D. to make sure that the thread will be the correct length.

17. Plastic pipe is often used for:
    A. natural gas.
    B. oil burners.
    C. installations where it will be exposed to extreme heat and pressure.
    D. plumbing, venting, and condensate applications.

18. A digital micron gauge:
    A. is used to measure a vacuum in a refrigeration system.
    B. is used to measure atmospheric pressure.
    C. is used to measure the pressure in a refrigeration system under normal operating conditions at the compressor discharge line.
    D. is used to measure atmospheric pressure at sea level.

19. There are _____ microns in 1 in. Hg.
    A. 100
    B. 25,400
    C. 10,000
    D. 24,500

20. A deep vacuum involves reducing the pressure in a refrigeration system to approximately:
    A. 200 to 500 microns.
    B. 400 to 800 microns.
    C. 28 in. Hg.
    D. 20 psig.

21. A multiple evacuation means that:
    A. you connect three vacuum pumps simultaneously to a refrigeration system and pull a vacuum.
    B. you pull a vacuum on three different systems at the same time.
    C. you pull a vacuum on a refrigeration system more than once, allowing some dry nitrogen back into the system between each procedure.
    D. there must be more than one system operating before a vacuum can be pulled.

22. If a vacuum has been pulled on a system and there is a leak, the vacuum gauge will:
    A. begin to rise.
    B. stay where it was when the vacuum pump was valved off.
    C. begin to fall.
    D. vibrate back and forth.

23. The best leak detection test is when the refrigeration system is:
    A. in a vacuum.
    B. open to the atmosphere.
    C. operating with all fans activated.
    D. not operating but in a standing pressure condition.

24. Moisture in a liquid state must be _____ before it can be removed with a vacuum pump.
    A. vaporized
    B. condensed
    C. cooled
    D. solidified

25. Which of the following may be used to protect the vacuum pump when excess moisture is in a system to be evacuated?
    A. Cold trap
    B. Changing the vacuum pump oil often
    C. Using the gas ballast feature of the vacuum pump
    D. All of the above

26. To pull a vacuum means to:
    A. raise the pressure in a refrigeration system.
    B. lower the pressure in a refrigeration system.
    C. allow refrigerants to escape into the atmosphere.
    D. condense the gases in a system.

27. A compound gauge:
   A. indicates pressures above 14.696 psia.
   B. indicates pressures below atmospheric pressure.
   C. is one of two gauges on a typical gauge manifold.
   D. has all of the above attributes.

28. Noncondensable gases:
   A. are desirable in a refrigeration system.
   B. are undesirable in a refrigeration system.
   C. condense readily under normal atmospheric conditions.
   D. help in pulling a vacuum.

29. The most common package air-conditioning system used in residential applications is the _____ system.
   A. air-to-air
   B. air-to-water
   C. water-to-water
   D. water-to-air

30. The trap in the condensate drain line prevents:
   A. air and foreign materials from being drawn in through the drain line.
   B. the air handler from forcing air out the drain line.
   C. a heat exchange from taking place through the drain line.
   D. excessive noise from traveling through the drain line.

31. The crankcase heater is used to:
   A. keep the temperature of the compressor at a predetermined level for better heat transfer.
   B. keep the oil from deteriorating.
   C. keep the motor relay from sticking.
   D. boil the liquid refrigerant out of the crankcase.

32. What two fasteners are often used with sheet metal square and rectangular duct?
   A. S-fasteners and drive cleats
   B. Staples and tape
   C. Self-tapping sheet metal screws and rivets
   D. Tabs and glue

33. A boot in a duct system is used:
   A. at the plenum.
   B. between the main and branch duct lines.
   C. between the branch duct and register.
   D. to eliminate vibration between the fan and duct.

34. Vapor refrigerant is added to a system through the:
   A. receiver king valve.
   B. liquid-line service valve.
   C. compressor discharge line.
   D. low-pressure side of the system.

35. As vapor refrigerant is removed from a cylinder, the pressure inside the cylinder will:
   A. rise.
   B. fall.
   C. stay the same.
   D. will do none of the above.

36. When charging a system with liquid refrigerant using a refrigerant cylinder that is not equipped with a dip tube, the cylinder should be:
   A. turned upside down.
   B. turned right side up.
   C. rested on its side.
   D. cooled to speed the charging process.

37. A programmable electronic charging scale:
   A. will calculate the amount of charge for you.
   B. has a microprocessor that will provide a readout of the condition of the refrigerant.
   C. has a solid-state microprocessor that stops the charging process when the programmed weight of refrigerant has been dispensed.
   D. has all of the above attributes.

38. A graduated charging cylinder is a device that:
   A. measures refrigerant by volume for precise refrigerant charging.
   B. has a provision for dialing in the temperature of the refrigerant.
   C. often has a heater to keep the refrigerant temperature from dropping when vapor is released into the system.
   D. does all of the above.

39. Charging liquid refrigerant into a system must be done very carefully because:
A. refrigerant is expensive.
B. damage to the compressor may occur.
C. it takes too long.
D. all of the above.

40. When charging vapor refrigerant into a system and the cylinder pressure drops to the point that refrigerant will not move into the system, you should:
A. warm the cylinder in warm water, not exceeding 90°F.
B. wait until the next day when the cylinder pressure rises.
C. cool the refrigerant cylinder to increase its pressure.
D. do none of the above.

41. The line voltage should be within _____ of the rated voltage on the air-conditioning unit.
A. ±10%
B. ±15%
C. ±20%
D. ±25%

42. The control voltage in an air-conditioning system comes directly from:
A. a tap on the line voltage.
B. a terminal on the compressor.
C. the secondary winding of a step-down transformer.
D. the circuit breaker panel.

43. The standard wire size for the control circuit is _____ gauge.
A. 16
B. 18
C. 24
D. 28

44. What wiring is installed by the installation technician?
A. Code wiring
B. Strip wiring
C. Field wiring
D. None of the above

45. Which of the following components is not a safety control in a condensing unit used for a residential air-conditioning system?
A. High-pressure switch
B. Low-pressure switch
C. Contactor
D. Compressor internal overload

46. What is the line and control voltage required by most residential condensing units?
A. 115 V/24 V
B. 115 V/12 V
C. 230 V/24 V
D. 460 V/24 V

47. How many electrical power connections must be made when installing a condensing unit and a furnace?
A. One
B. Two
C. Three
D. None of the above

48. How many electrical power connections must be made when installing a gas pack?
A. One
B. Two
C. Three
D. None of the above

49. What guide(s) should the installation technician use when sizing the power wire for an air-conditioning installation?
A. National Electrical Code
B. Manufacturer's installation instructions
C. State and local codes
D. All of the above

50. Which of the following letter designation supplies the reversing valve with 24 V that energizes in the cooling mode of operation?
A. R
B. O
C. W
D. B

Exercise ISU-1

Flaring Tubing

Name	Date	Grade

OBJECTIVES: Upon completion of this exercise, you should will be able to cut copper tubing, make a flare connection, and perform a simple leak check to ensure that the tubing connections are leak-free and that they may be used for refrigerant.

INTRODUCTION: You will cut tubing of different sizes, make a flare on the end, assemble it with fittings into a single piece, pressurize it, and check it for leaks.

TOOLS, MATERIALS, AND EQUIPMENT: A tube cutter with reamer, flaring tool set, refrigerant oil, goggles, adjustable wrenches, a gauge manifold, a cylinder of nitrogen with regulator, soap bubbles, and the following materials:

    2 each flare nuts, ¼ in. , ⅜ in., and ½ in.

    1 each reducing flare unions, ¼ in. × ⅜ in. and ⅜ in. × ½ in.

    1 ½ in. flare plug

    1 ¼ in. × ¼ in. flare union

    Soft copper tubing ¼ in., ⅜ in., and ½ in. outside diameter (OD)

See Figure ISU-1.1 for a photo showing these materials.

FIGURE **ISU-1.1**

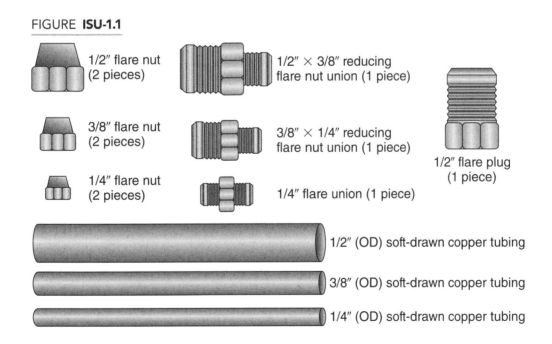

1/2" flare nut (2 pieces)

1/2" × 3/8" reducing flare nut union (1 piece)

3/8" flare nut (2 pieces)

3/8" × 1/4" reducing flare nut union (1 piece)

1/2" flare plug (1 piece)

1/4" flare nut (2 pieces)

1/4" flare union (1 piece)

1/2" (OD) soft-drawn copper tubing

3/8" (OD) soft-drawn copper tubing

1/4" (OD) soft-drawn copper tubing

⚠ **SAFETY PRECAUTIONS:** Care must be used while cutting and flaring tubing. Your instructor should teach you to use the tube cutter, reamer, flaring tools, leak detector, and gauge manifold before you begin this exercise. Always wear goggles while working in a lab environment.

STEP-BY-STEP PROCEDURES

1. Cut a 10 in. length of each of the tubing sizes: ¼ in., ⅜ in., and ½ in.

2. Carefully ream the end of each piece of tubing to prepare it for flaring, Figure ISU-1.2.

FIGURE **ISU-1.2**

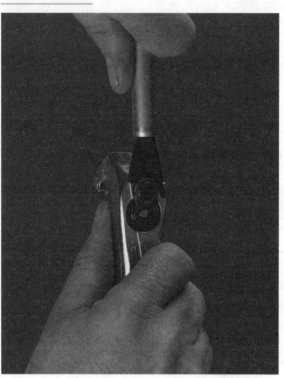

3. Slide the respective flare nuts over each piece of tubing before the flares are made, Figure ISU-1.3. Be sure to apply some refrigerant oil to the flare cone before the flare is made. Make the flares.

FIGURE **ISU-1.3**

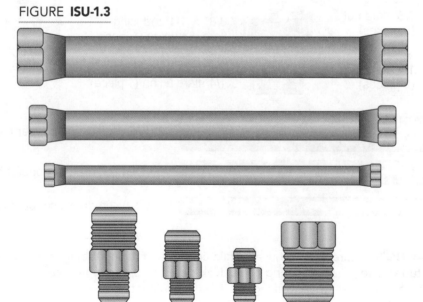

4. Assemble the connections as shown in Figure ISU-1.4, and tighten all connections using two adjustable wrenches.

FIGURE **ISU-1.4**

5. Pressurize the assembly using the nitrogen tank, nitrogen regulator, and the gauge manifold, Figure ISU-1.5.

Note: Make certain your instructor is supervising this process to avoid potential damage or injury.

FIGURE **ISU-1.5**

Photo by Eugene Silberstein

6. Using the soap bubble leak detection solution, check for leaks around each fitting, Figure ISU-1.6.

FIGURE **ISU-1.6**

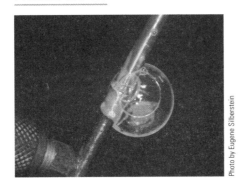

Photo by Eugene Silberstein

7. Examine any connection that leaks, and repair it by either replacing the brass fitting or flaring the tubing again. A scratch on the brass fitting may be a possible leak. A ridge around the flare is a sign that the tubing was not reamed correctly and will probably leak, Figure ISU-1.7.

FIGURE **ISU-1.7**

← Ridge

Photo by Bill Johnson

8. After all fittings have been proven leak-free, take the assembly apart except for the end plug and the ¼ in. × ¼ in. flare union at the ends. Cut all flare nuts off except the two on the end. They will be used for pressure-testing in the next exercise. Save the pieces of tubing for soldering in the next exercise.

MAINTENANCE OF WORKSTATION AND TOOLS: Wipe any oil off the tools, and return all tools and supplies to their places. Store the leftover tubing lengths for the next exercise.

SUMMARY STATEMENT: Describe, step-by-step, the flaring process. Use an additional sheet of paper if more space is necessary for your answer.

TAKEAWAY: Flaring is a method commonly used in the HVACR field to join copper tubing. Refrigerant oil should be used during the flaring process, and all joints should be leak-checked after assembly.

QUESTIONS

1. Why should you apply refrigerant oil to the flare cone before making a flare?

2. What happens if the tubing cutter is tightened down too fast while cutting?

3. What is the purpose of reaming the tube before flaring?

4. Why is thread-sealing compound not used on flare connections?

5. What actually seals the connection in a flare connection?

6. Why is a flare connection sometimes preferred to a solder connection?

7. What is the angle on a flare connection?

8. Give two reasons why a leak may occur in a flare connection.

9. Name two methods of leak-testing a flare connection.

10. When a flare connection leaks due to a bad flare, what is the recommended repair?

Soldering and Brazing

Name	Date	Grade

OBJECTIVES: Upon completion of this exercise, you should be able to make connections with copper tubing using both low-temperature solder and high-temperature brazing material.

INTRODUCTION: You will first join together the tubing sections used in Exercise ISU-1 using low-temperature solder, and then cut the connections apart and join them again, this time using high-temperature brazing filler material.

TOOLS, MATERIALS, AND EQUIPMENT: An air-acetylene kit with a medium tip, a striker, clear safety goggles, adjustable wrenches, light gloves, low-temperature solder (50/50, 95/5, or low-temperature silver), appropriate flux, high-temperature brazing filler material and the proper flux, sand cloth, wire brushes (¼ in., ⅜ in., and ½ in.), a vise to hold the tubing, two of each reducing couplings (sweat type, ¼ in. × ⅜ in., ⅜ in. × ½ in.), bubble-type leak detector solution, a tubing cutter, a gauge manifold, and a nitrogen cylinder with regulator. (An alternate method of conducting this exercise will not use the reducing couplings but will require swaging tools, a flaring block, and a hammer.)

⚠ **SAFETY PRECAUTIONS:** You will be working with acetylene gas and heat. Your instructor will provide you with instructions and safety considerations regarding the torch assembly as well as proper procedures for soldering and brazing before you begin this exercise. Wear light gloves while soldering or brazing. Always wear goggles when working in the lab.

STEP-BY-STEP PROCEDURES

1. Disassemble the piping arrangement from Exercise ISU-1, leaving the ½ in. flare plug in place, Figure ISU-2.1.

FIGURE **ISU-2.1**

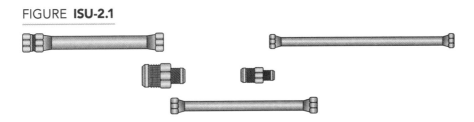

2. Slide the flare nuts away from the ends of the tubing and cut the flared portion from the tubing, Figure ISU-2.2.

FIGURE **ISU-2.2**

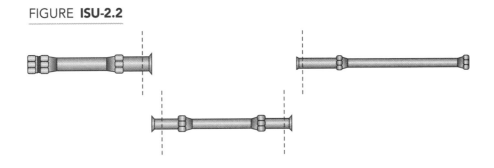

3. Remove the flare nuts from the tubing; they will not be used in this exercise.

4. Clean the ends of the tubing with the sand cloth.

5. Clean the inside of the fittings using an appropriate fitting brush.

6. Fit the assembly together using the correct flux, Figure ISU-2.3. Apply flux to the male portion of the joint only.

FIGURE **ISU-2.3**

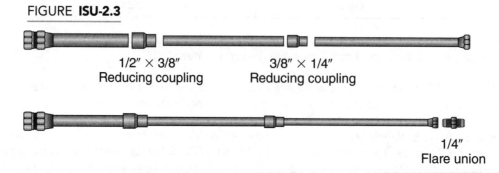

1/2" × 3/8"
Reducing coupling

3/8" × 1/4"
Reducing coupling

1/4"
Flare union

7. Place the assembly in a vertical position using the vise. This will give you practice in soldering both up and down in the vertical position, Figure ISU-2.4.

FIGURE **ISU-2.4**

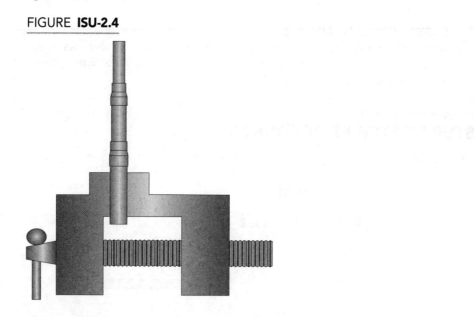

8. Start with the top joint at the top connection; then solder the joint underneath, Figure ISU-2.5. After completing this, solder the top joint of the second connection; then solder underneath. Be sure not to overheat the connections. IF THE TUBING BECOMES RED OR DISCOLORS BADLY, THE CONNECTION IS TOO HOT.

FIGURE **ISU-2.5**

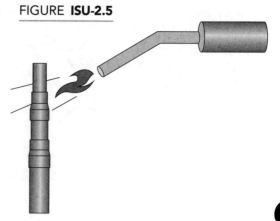

9. Let the assembly cool before you disturb it, or else the connections may come loose.

10. After the assembly has cooled, perform a leak test as in Exercise ISU-1. You may also want to submerge the pressurized assembly in water for a test. If bubbles appear when the assembly is submerged, a leak is present.

11. Cut the reducing connectors out of the assembly, and reclean the tubing ends, Figure ISU-2.6.

FIGURE **ISU-2.6**

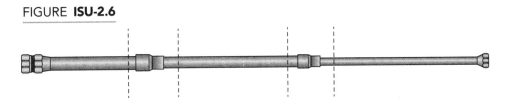

12. Clean the second set of reducing couplings.

13. Assemble the pipes and connectors as before, and hold them upright in the vise.

14. Heat the connections as in the previous exercise, but this time you are to use a high-temperature brazing material.

Caution: THE PIPE AND FITTINGS WILL HAVE TO BE CHERRY RED TO MELT THE FILLER METAL, SO BE CAREFUL.

15. After the fittings have cooled, leak-test them as before.

16. Use a hacksaw to cut into the fittings and examine the penetration of the filler material into the joint.

17. Upon completion of the leak test, cut off the flare nuts, and scrap the tubing.

MAINTENANCE OF WORKSTATION AND TOOLS: CLEAN ANY FLUX OFF THE TOOLS OR FITTINGS AND THE WORK BENCH. Turn off the torch, turn off the tank, and bleed the acetylene from the torch hose. Place all extra fittings and tools in their proper places.

SUMMARY STATEMENT: Discuss the advantages and disadvantages of low-temperature and high-temperature solder.

TAKEAWAY: Brazing and soldering are two methods commonly employed in the HVACR field to join copper tubing. Brazing is done at high temperatures, usually just under the melting point of the copper, while soldering is done at much lower temperatures, typically around 400°F. Both processes require that the copper be mechanically cleaned with sand cloth or a wire brush to ensure proper bonding of the filler metal.

QUESTIONS

1. What is the approximate melting temperature of 50/50 solder?

2. What is the approximate melting temperature of a 45% silver brazing rod?

3. What is the approximate melting temperature of a 15% silver brazing rod?

4. During the soldering or brazing process, what causes the filler material to melt?

5. What is the composition (makeup) of 50/50 solder?

6. What is the composition (makeup) of 95/5 solder?

7. Which filler material is stronger, 95/5 soft solder or a 15% silver (brazing) filler material? Why?

8. Which filler material would be the best choice for the discharge line, 95/5 or 15% silver solder? Why?

9. What pulls the filler material into the connection when it is heated? Explain this process.

10. Why is a special sand cloth used on hermetic compressor systems?

11. What are the typical gases used for soldering copper tubing?

12. What are the typical gases used for brazing copper tubing?

Exercise ISU-3

Joining PVC Pipe Sections

Name	Date	Grade

OBJECTIVES: Upon completion of this exercise, you should be able to properly join sections of PVC pipe.

INTRODUCTION: In this exercise, you will follow a diagram to create a piping project that meets the required measurement specifications while gaining experience joining sections of PVC using fittings.

TOOLS, MATERIALS, AND EQUIPMENT: PVC cutter, tape measure, PVC primer, PVC cement, safety goggles, rags, pencil, and the following materials:

- 24 in. of ¾ in. schedule-40 PVC piping material
- 2 ¾ in. PVC 90° elbows
- 1 ¾ in. PVC coupling

⚠ **SAFETY PRECAUTIONS:** Make certain that the area is well ventilated; inhaling the PVC primer and cement fumes is dangerous. Exercise caution when using the PVC cutter to avoid personal injury.

STEP-BY-STEP PROCEDURES

1. Dry fit the PVC piping material into one side of the PVC coupling and mark the pipe at the edge, Figure ISU-3.1.

FIGURE **ISU-3.1**

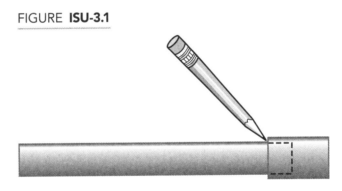

2. Remove the section of PVC pipe from the fitting and measure the distance from the mark to the end of the pipe, Figure ISU-3.2.

FIGURE **ISU-3.2**

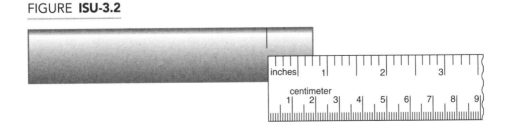

3. Record this measurement here: _____

4. Repeat steps 1 and 2 with the 90° elbow.

5. Record this measurement here: _____

6. Using the measurements documented in steps 3 and 5 as well as the measurements in Figure ISU-3.3, calculate the required lengths for the four pipe sections needed to complete the piping project.

FIGURE **ISU-3.3**

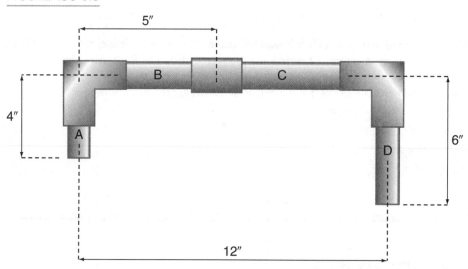

7. Record the calculated pipe lengths here:

   A. Section A: _____

   B. Section B: _____

   C. Section C: _____

   D. Section D: _____

8. Using the PVC cutter, cut four pieces of PVC pipe to the proper lengths, as recorded in step 7.

9. Dry-fit the pipes and fittings, as shown in Figure ISU-3.3, to ensure that the overall measurements of the project are correct.

10. Disconnect all connections.

11. Ream the ends of the four pieces of PVC.

12. Using PVC primer, clean both ends of pipe sections B and C.

13. Using PVC primer, clean one end of pipe sections A and D.

14. Using PVC primer, clean the female ends of the 90° elbows and the coupling.

15. Securely close the PVC primer container.

16. Using PVC cement, connect the pipe sections and fittings according to Figure ISU-3.3, making sure that the PVC cement is applied to only the male portions of the joints. Press the pieces together, one at a time, and hold them together firmly until the glue dries, usually about 10–15 seconds.

17. Securely close the PVC cement container.

MAINTENANCE OF WORKSTATION AND TOOLS: Make certain that the PVC primer and cement cans are closed tightly. Return all tools to their proper location. Clean the work area.

SUMMARY STATEMENT 1: Explain why it is important to measure the piping material to determine how far into the fitting the pipe extends.

SUMMARY STATEMENT 2: Explain why it is important to use primer on PVC piping connections. Be sure to include examples of what can happen if the pipes are not primed prior to connecting.

TAKEAWAY: Joining PVC is a task frequently performed by HVACR technicians in the field. Accurate measurements and proper technique are imperative to leak-free connection. There are a variety of different primers and glues available. Always check manufacturer's specifications to ensure that the proper ones are used.

QUESTIONS

1. Why is it recommended that PVC cement be applied to only the male portion of the fitting?

2. Why is it a good idea to dry-fit the pipe sections and fitting prior to permanently joining them?

3. In what applications is PVC found in the HVACR industry?

Bending Copper Tubing

Name	Date	Grade

OBJECTIVES: Upon completion of this exercise, you should be able to properly bend copper tubing to create three different piping projects.

INTRODUCTION: In this exercise, you will follow a diagram to create three separate piping projects that meet the required measurement specifications.

TOOLS, MATERIALS, AND EQUIPMENT: Tubing cutter, tape measure, lever-type tubing bender, pencil, and the following materials:

  5 feet of ⅜ in. OD soft-drawn copper tubing

⚠ **SAFETY PRECAUTIONS:** Exercise caution when using the tubing cutter to avoid personal injury. Be careful when handling copper tubing. Burrs on the ends of the tubing are sharp and can cause injury.

STEP-BY-STEP PROCEDURES

1. Using Figure ISU-4.1 as a reference, determine the lengths of tubing that will be required for the three projects.

FIGURE **ISU-4.1**

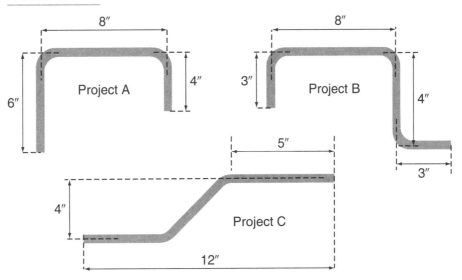

2. Record the lengths of the three pieces of tubing here:

   A. Project A: _____

   B. Project B: _____

   C. Project C: _____

3. Cut three pieces of tubing that will be used to create the three projects.

4. Using Figure ISU-4.1 as a reference, create Project A.

5. Bend the tubing in such a manner that the measurements of the completed project correspond to those in the drawing.

   Note: The drawings of the piping projects are not drawn to scale.

6. For any leftover pieces of copper tubing from the creation of project A, record the measurement(s) here:

   A. Leftover piece 1 (If there is any): _____

   B. Leftover piece 2 (If there is any): _____

7. Using Figure ISU-4.1 as a reference, create Project B.

8. Bend the tubing in such a manner that the measurements of the completed project correspond to those in the drawing. Note: The drawings of the piping projects are not drawn to scale.

9. For any leftover pieces of copper tubing from the creation of project B, record the measurement(s) here:

   A. Leftover piece 1 (If there is any): _____

   B. Leftover piece 2 (If there is any): _____

10. Using Figure ISU-4.1 as a reference, create Project C.

11. Bend the tubing in such a manner that the measurements of the completed project correspond to those in the drawing. Note: The drawings of the piping projects are not drawn to scale.

12. For any leftover pieces of copper tubing from the creation of project C, record the measurement(s) here:

    A. Leftover piece 1 (If there is any): _____

    B. Leftover piece 2 (If there is any): _____

13. Compare the measurements of the three completed projects to the measurements indicated in Figure ISU-4.1.

MAINTENANCE OF WORKSTATION AND TOOLS: Return all tools to their proper location. Clean the work area and recycle the copper.

SUMMARY STATEMENT: Explain any discrepancies that existed between the lengths of tubing that were cut in step 3 and the actual amount of piping material required to create the three projects. Include information from step 13 in your response.

TAKEAWAY: When bending copper tubing, measurements must be precise, or else the workpiece may have to be redone. Knowledge of both the tape measure and the particular bender being used are required for successful completion of the task.

QUESTIONS

1. Explain why bending copper tubing is often a better choice than using fittings when creating piping circuits for air-conditioning and refrigeration systems.

2. Explain why lever-type tubing benders are more accurate than spring-type tubing benders.

3. What types of copper can be bent with a tubing bender?

Standing Pressure Test

Name	Date	Grade

OBJECTIVES: Upon completion of this exercise, you should be able to perform a standing pressure test on a vessel using dry nitrogen.

INTRODUCTION: Before a gauge manifold can be used with confidence in a high-vacuum situation, it must be leak-free. You will use dry nitrogen to test your gauge manifold to ensure that it is leak-free. The standing pressure test is the same for any vessel, including a refrigeration system.

TOOLS, MATERIALS, AND EQUIPMENT: A gauge manifold, plugs for the gauge manifold lines, a cylinder of dry nitrogen with regulator, and goggles.

⚠ SAFETY PRECAUTIONS: Goggles should be worn at all times while working with pressurized gases. NEVER USE DRY NITROGEN WITHOUT A PRESSURE REGULATOR. THE PRESSURE IN THE TANK CAN BE IN EXCESS OF 2000 PSI. Your instructor must give you instruction in the use of the gauge manifold and a dry nitrogen setup before you start this exercise.

STEP-BY-STEP PROCEDURES

1. Check the gauges for calibration. With the gauge opened to the atmosphere, it should read 0 psig. If it does not, use the calibration screw under the gauge glass and calibrate it to 0 psig.

2. Check the gaskets in both ends of the gauge hoses. These are often tightened too much, which ruins the gaskets. If the gasket is worn or defective, replace it. Figure ISU-5.1 shows good and defective gaskets.

   Note: Some hoses will have a core depressor in the center of the gasket. It must be carefully removed before the gasket can be replaced.

FIGURE **ISU-5.1**

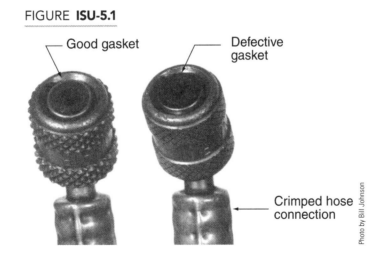

Good gasket   Defective gasket

Crimped hose connection

Photo by Bill Johnson

3. Fasten the center gauge line to the dry nitrogen regulator and open both gauge manifold handles. This opens the high- and low-side gauges to the center port, Figure ISU-5.2.

FIGURE **ISU-5.2**

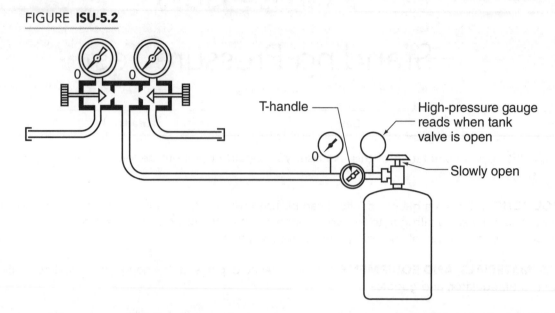

T-handle

High-pressure gauge reads when tank valve is open

Slowly open

4. Place plugs in the high- and low-side gauge lines.

Caution: TURN THE NITROGEN REGULATOR T-HANDLE ALL OF THE WAY OUT, Figure ISU-5.2. Now slowly open the tank valve, and allow tank pressure into the regulator. DO NOT STAND IN FRONT OF THE REGULATOR T-HANDLE.

5. Slowly turn the T-handle inward, and allow pressure to fill the gauge manifold. Allow the pressure in the manifold to build up to 150 psig; then shut off the tank valve, and turn the T-handle outward. The regulator is now acting as a plug for the center gauge line. See Figure ISU-5.3.

FIGURE **ISU-5.3**

150 psig   150 psig

150 psig

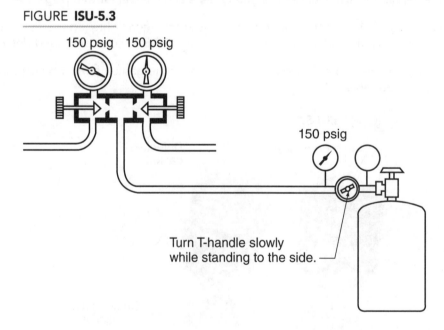

Turn T-handle slowly while standing to the side.

6. Allow the pressurized gauge manifold to set, undisturbed, for 15 minutes. THE PRESSURE SHOULD NOT DROP.

7. If the pressure drops, pressurize the manifold again and valve off the manifold by closing the gauge manifold valves (turn toward the center). A drop in pressure in either gauge will help locate the leak.

Note: If you cannot make the manifold leak-free by gasket replacement, you may want to submerge the gauge lines in water to determine where the pressure is escaping. Gauge lines often leak around the crimped ends or become porous along the hoses and need to be replaced. A bubble solution can be used around the area where the gauges are connected to the manifold to check for possible leaks there.

8. If you cannot stop the leak in a gauge manifold, replace the gauge lines with ¼ copper lines used in evacuation of a system.

MAINTENANCE OF WORKSTATION AND TOOLS: Put all tools in their proper places. Your instructor may suggest that you leave the gauge manifold under pressure for a period of time by placing plugs in the lines. If you do this, you will know whether the gauge manifold is leak-free the next time you use it.

SUMMARY STATEMENTS: Discuss the reason that dry nitrogen is used in a pressure test instead of a refrigerant. Discuss why a leak test while the system is in a positive pressure is more effective than using a vacuum for leak checking.

TAKEAWAY: Gauge manifolds that are used in the service and installation of HVACR equipment must be leak-free. Loss of refrigerant, inaccurate readings, or system contamination can result if a set of leaky gauges is used on the equipment. A standing pressure test, done with nitrogen, can help identify if there are any leaks and where they are.

QUESTIONS

1. What can cause a gauge line to leak?

2. What is the gauge manifold made of?

3. What fastens the gauge line to the end fittings of a gauge line?

4. What seals the gauge line fitting to the gauge manifold?

5. What can be done for repair if a gauge line leaks at the fitting?

6. What would happen if a leaking gauge manifold were used while trying to evacuate a system to a deep vacuum?

7. Why is refrigerant not used for a standing pressure test?

8. How much pressure does the atmosphere exert on a gauge line under a deep vacuum?

9. How is a gauge line tightened on a fitting?

10. What may be done if the flexible gauge lines cannot be made leak-free?

Exercise ISU-6

Performing a Deep Vacuum

Name	Date	Grade

OBJECTIVES: Upon completion of this exercise, you should be able to perform a deep vacuum test using a high-quality vacuum pump and an electronic vacuum gauge.

INTRODUCTION: You will perform a deep-vacuum (500 microns) evacuation on a typical refrigeration system that has service valves. A deep-vacuum evacuation is the same whether the system is small or large, the only difference being the time it takes to evacuate the system. Larger vacuum pumps are typically used on large systems to save time. It is always best to give a system a standing pressure test using nitrogen before a deep vacuum is pulled to make sure that the system is leak-free. An overnight pressure test with no pressure drop will ensure that the vacuum test will go as planned. Performing a vacuum test on just the gauge manifold and gauge lines will also help ensure that the test will be performed satisfactorily.

TOOLS, MATERIALS, AND EQUIPMENT: A leak-free gauge manifold, a cylinder of refrigerant, a nitrogen cylinder with regulator, a high-quality vacuum pump (two-stage is preferred), an electronic micron vacuum gauge with some arrangement for valving it off from the system, a refrigeration system that has had its charge recovered, goggles, and gloves.

SAFETY PRECAUTIONS: Goggles and gloves are safety requirements when transferring refrigerant. A manifold arrangement such as that in Figure ISU-6.1 may be used when performing a system evacuation with a micron gauge. The electronic vacuum gauge sensing element must be protected from any refrigerant oil that may enter it. Your instructor must give you instruction in using a gauge manifold, refrigerant recovery system, vacuum pump, and an electronic vacuum gauge before you start this exercise.

FIGURE **ISU-6.1**

Courtesy of Ferris State University. Photo by John Tomczyk.

STEP-BY-STEP PROCEDURES

1. Fasten the gauge manifold lines to the refrigeration system to be evacuated. Fasten the lines to the service valves at the high and the low sides of the system, if possible. DO NOT overtighten the gauge line connections or else you will damage the gaskets.

2. Position the service valve stems in the mid-position, and tighten the packing glands on the valve stems. Be sure to replace the protective caps on the stems before starting the evacuation. Many service technicians have been deceived because of leaks at the service valves. If the system has Schrader valves, you should remove the valve stems prior to evacuation and replace them after completing the test. This will speed up the evacuation procedure.

3. The system pressure should have been reduced to 0 psi prior to this exercise. If there is pressure in the system, it must be recovered before proceeding.

4. Check the oil level in the vacuum pump, and add vacuum pump oil if needed.

5. After all of the refrigerant has been transferred from the system using the recovery machine, connect the gauge manifold to the vacuum pump, Figure ISU-6.2.

FIGURE **ISU-6.2**

Photo by Jason Obrzut

6. Put on your goggles, and start the vacuum pump.

7. When the vacuum pump has operated long enough that the low-side compound gauge has pulled down to 25 in. Hg (inches of mercury), open the valve to the electronic vacuum gauge sensing element. Observe the compound gauge as the system pressure reduces. It depends on the type of vacuum gauge you have as to when it will start to indicate. Some of them do not start to indicate until around 1000 microns and will not indicate until very close to the end of the vacuum. Some gauges start indicating at about 12,000 microns.

8. Operate the vacuum pump until the indicator reads between 200 and 500 microns.

   Note: The deeper the vacuum, the slower it pulls down. It takes much more time to obtain the last few microns than the first part of the vacuum.

9. When the vacuum reaches between 200 and 500 microns, you have obtained a deep vacuum, one that will satisfy almost any manufacturer's requirements.

   Note: There is no industry standard for a deep vacuum. However, most manufacturers require a deep vacuum to be between 200 and 500 microns.

10. If the system will not reach 500 microns, valve the vacuum pump off and let the system stand. If the vacuum rises, either there is a leak or liquid water or refrigerant is still boiling out of the system. You must find the problem. A starting point is to put pressure back into the system to above atmosphere and remove the gauge manifold. Evacuate the gauge manifold as in a system evacuation, and perform a leak test on it. It must be leak-free, or else you will never obtain a vacuum.

11. Upon successful completion of an evacuation, shut off the valve to the micron gauge and charge enough refrigerant or nitrogen into the system to get the pressure above atmospheric. IT IS NEVER GOOD PRACTICE TO LEAVE A SYSTEM IN A DEEP VACUUM FOR LONG PERIODS. A LEAK COULD CONTAMINATE THE SYSTEM AND DRIERS.

MAINTENANCE OF WORKSTATION AND TOOLS: Check the oil in the vacuum pump and change it if necessary. Return the refrigeration system to the condition in which you found it. Wipe any oil off the tools, and return them to their proper places.

SUMMARY STATEMENT: In your own words, explain how your evacuation procedure went. Do not leave out any mistakes you made. Use an additional sheet of paper if more space is necessary for your answer.

TAKEAWAY: During the installation, maintenance, and service of a system, air and moisture can enter the refrigerant circuit. An evacuation can rid the system of both the air and moisture when done properly. Deep vacuums are most accurately measured in microns. The deeper the vacuum, the more contaminants are removed.

QUESTIONS

1. Why should the electronic micron vacuum gauge sensing element always be kept upright?

2. At what point did the electronic micron vacuum gauge start indicating?

3. Which is the most accurate for checking a deep vacuum, an electronic micron vacuum gauge or a compound gauge?

4. What is the purpose of system evacuation?

5. What happens to a system that leaks while in a vacuum?

6. Is a vacuum the best method of removing large amounts of moisture from a system?

7. What will the results be if a fitting is leaking and the system is evacuated to a deep vacuum?

8. What is the approximate boiling pressure of refrigerant oil in a system?

9. Why is vacuum pump oil used in a vacuum pump instead of refrigerant oil?

10. How may water be released from below the oil level in a compressor crankcase?

Triple Evacuation

Name	Date	Grade

OBJECTIVES: Upon completion of this exercise, you should be able to perform a triple evacuation on a typical refrigeration system using an electronic micron gauge.

INTRODUCTION: You will be performing a triple evacuation on a refrigeration system equipped with services valves. Each vacuum will be "broken" with nitrogen before the next one begins. You will use a digital vacuum gauge to monitor your progress.

TOOLS, MATERIALS, AND EQUIPMENT: A gauge manifold, gloves, goggles, a high-quality two-stage vacuum pump, an electronic micron gauge, a refrigerant recovery system, and a cylinder of dry nitrogen.

⚠ **SAFETY PRECAUTIONS:** Wear gloves and goggles. Check the oil in the vacuum pump. Do not begin this exercise without proper instruction in the use of the gauge manifold, electronic micron gauge, refrigerant recovery system, and a two-stage vacuum pump.

STEP-BY-STEP PROCEDURES

1. Put on your goggles and gloves. Fasten the gauge lines to the service valves on the low and high sides of the refrigeration system.

2. Recover all refrigerant from the system down to the current level required by EPA. Your instructor will tell you how to recover the refrigerant.

3. Check the oil in the vacuum pump; change it if necessary.

4. Fasten the electronic micron gauge to the system in such a manner that it may be valved off, Figure ISU-7.1.

5. Fasten the center gauge line of the gauge manifold to the vacuum pump. Open the gauge manifold valves, and start the vacuum pump.

6. When the needle on the compound gauge starts into a vacuum, you may open the valve to the electronic micron gauge. REMEMBER: It will not start to indicate until the vacuum is down to about 25 in. Hg on the compound gauge.

7. Allow the vacuum pump to run until the electronic micron gauge reads about 1500 microns.

8. IMPORTANT: Valve off the electronic micron vacuum gauge.

9. Allow a small amount of dry nitrogen to enter the system, making certain that the system remains in a vacuum.

FIGURE **ISU-7.1**

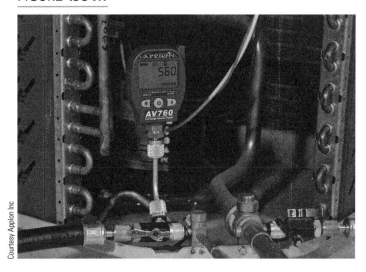

Courtesy Applon Inc

10. Now restart the vacuum procedure with the two-stage vacuum pump, and open the valve to the electronic micron gauge. Allow the vacuum to reach 1500 microns again, and repeat steps 8 and 9.

11. Allow the electronic micron gauge to reach 200–500 microns for a deep vacuum.

12. Valve off the vacuum pump, and allow the vacuum to stand for 10 to 15 minutes without increasing the pressure. This will assure the technician that there are no leaks in the system or connecting gauges. Valve off the electronic vacuum gauge, and allow refrigerant to flow through the system by adding liquid to the high side of the system. After a short time, start the compressor and finish charging.

MAINTENANCE OF WORKSTATION AND TOOLS: Check the oil in the vacuum pump, and change it if necessary. Return the refrigeration system to the condition in which you found it. Wipe any oil off the tools, and return them to their places.

SUMMARY STATEMENT: Describe, step-by-step, the triple evacuation process. Use a separate sheet of paper if more space is necessary for your answer.

TAKEAWAY: An evacuation is performed to rid a system of noncondensable gases and moisture. A triple evacuation is a method of evacuating a system more completely. A series of deep vacuums (usually to around 1500 microns) is broken with nitrogen until the final evacuation, where the system is pulled down to around 500 microns and a "time to rise test" is observed.

QUESTIONS

1. What is a deep vacuum?

2. Why must no foreign vapors be left in a refrigeration system?

3. What vapors does a deep vacuum pull out of the system?

4. What type of oil is used in a vacuum pump?

5. What comes out of the exhaust of a vacuum pump during the early part of an evacuation?

6. Why must vacuum pump oil only be used for a deep vacuum pump?

7. Why is a large vacuum pump required to remove moisture from a system that has been flooded with water?

8. How often should vacuum pump oil be changed?

Exercise ISU-8

Charging a System With a Charging Cylinder

Name	Date	Grade

OBJECTIVES: Upon completion of this exercise, you should be able to add refrigerant to a charging cylinder and then charge that refrigerant into a typical refrigeration system.

INTRODUCTION: You will install gauges on a typical capillary tube refrigeration system, recover the existing charge, and evacuate the system to a deep vacuum. You will then add enough charge to a charging cylinder, connect it to the system, and measure in the correct charge. For this exercise, a small self-contained unit with service ports should be used.

TOOLS, MATERIALS, AND EQUIPMENT: A vacuum pump, gauge manifold, refrigerant recovery equipment, safety goggles, gloves, refrigerant that is the same as the refrigerant in the system, graduated charging cylinder, straight-slot screwdriver, ¼ in. and ⁵⁄₁₆ in. nut drivers, and a capillary tube refrigeration system, as described before.

⚠ **SAFETY PRECAUTIONS:** Wear gloves and goggles while transferring refrigerant to and from a system. DO NOT overcharge the charging cylinder. Follow all instructions carefully. Your instructor must provide prior instruction to ensure that the gauge manifold, refrigerant recovery equipment, charging cylinder, and vacuum pump are all used correctly.

STEP-BY-STEP PROCEDURES

1. Connect the gauge manifold to the system using the appropriate service valves or Schrader valve ports on the high and low sides of the system.

2. Using the refrigerant recovery equipment, recover all of the refrigerant from the unit to be charged. Your instructor will instruct you where and how to recover the refrigerant.

3. Check the oil level in the vacuum pump, add oil if needed, connect it to the center line of the manifold, and start the pump. Open the gauge manifold valves.

4. From the unit's nameplate, determine the correct charge for the system, and write it here for reference: _____ oz.

5. While the refrigerant is being recovered, connect the charging cylinder to the cylinder of refrigerant, Figure ISU-8.1.

6. Open the tank of refrigerant to allow refrigerant to flow from the tank to the charging cylinder.

7. Before opening the valve on the charging cylinder, briefly loosen the hose connection on the charging cylinder to vent any air that might be present in the hose.

8. Open the valve on the charging cylinder to allow liquid refrigerant to flow from the tank into the charging cylinder.

FIGURE **ISU-8.1**

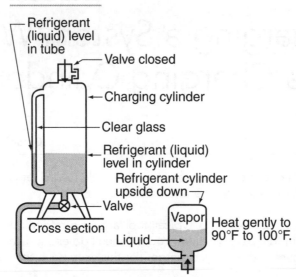

9. It may be necessary to recover vapor refrigerant from the top of the charging cylinder in order to get enough liquid refrigerant into the charging cylinder. Note: Recovering vapor from the charging cylinder will lower the pressure in the charging cylinder, allowing more liquid refrigerant to flow into the cylinder. When transferring any refrigerant from a refrigerant cylinder to a charging cylinder, it is illegal under Section 608 of the Clean Air Act to vent vapor from the top of the charging cylinder when it is being filled. The service technician can gently heat the refrigerant cylinder to a temperature of 90°F to increase its pressure before refrigerant can be successfully transferred to the charging cylinder. The best way to heat a cylinder is to put it in a container of warm water.

10. When the charging cylinder has enough refrigerant to charge the system and the vacuum has reached the correct level (a deep vacuum is best determined by a digital micron gauge), the refrigerant is ready to be charged into the system.

11. Close the isolation valve on the vacuum pump and then turn the vacuum pump off.

12. Disconnect the vacuum pump, and connect the charging cylinder, Figure ISU-8.2.

FIGURE **ISU-8.2**

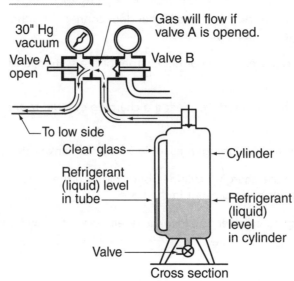

13. Open the vapor valve at the top of the charging cylinder.

14. Briefly loosen the center hose connection on the gauge manifold to purge any air that might be present in the center hose.

15. Turn the outside dial on the charging cylinder to correspond to the pressure in the cylinder. For example, if the pressure gauge at the top of the cylinder reads 125 psig (this would probably be for R-22), turn the dial to the 125 psig scale and read the exact number of ounces on the sight glass of the charging cylinder, Figure ISU-8.3. Record it here: _____ oz.

FIGURE **ISU-8.3**

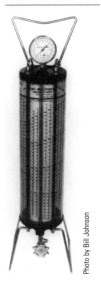

16. Subtract the amount of the unit charge from the charging cylinder level and record here: _____ oz. (This is the stopping point of the charge.)

17. Open the low-side valve on the gauge manifold. The charge will begin to fill the system. Remember, when charging with a refrigerant blend that has a potential to fractionate, the service technician must also throttle liquid refrigerant from the charging cylinder into the refrigeration system to avoid fractionation.

18. Keep adding refrigerant vapor until the correct low level appears on the sight glass on the charging cylinder, Figure ISU-8.4, and then close the low-side valve on the gauge manifold as well as the vapor valve on the charging cylinder.

FIGURE **ISU-8.4**

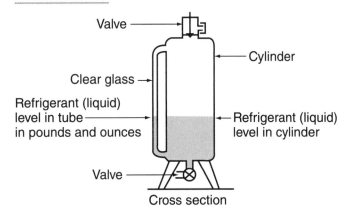

19. After the system has been properly charged, recover any remaining refrigerant from the charging cylinder.

MAINTENANCE OF WORKSTATION AND TOOLS: Wipe up any oil in the work area. Return all tools to their proper places. Replace panels on the unit.

SUMMARY STATEMENT: Describe the calculations you had to make to determine and make the correct charge for the system. Use an additional sheet of paper if more space is necessary for your answer.

TAKEAWAY: Charging cylinders were commonly used to accurately charge "critical charge" systems before digital scales were invented. When using a charging cylinder, it is important to observe the regulations set forth by EPA Section 608.

QUESTIONS

1. What are the graduations on the charging cylinder?

2. Why is it necessary to turn the dial to the correct pressure reading and use this figure for the charge?

3. In what scale or graduation was the charge expressed in the system charged, pounds or pounds and ounces?

4. How many ounces are in one-fourth (0.25) of a pound?

5. How may refrigerant be charged into the system if the cylinder pressure drops?

6. How do some charging cylinders keep the pressure from dropping as the charge is added?

7. What type of metering device normally requires an exact charge?

8. Why do blended refrigerants have to be charged as liquid?

Charging a System with Refrigerant Using a Digital Scale

Name	Date	Grade

OBJECTIVES: Upon completion of this exercise, you should be able to charge a typical refrigeration system using a scale to weigh the refrigerant.

INTRODUCTION: You will recover refrigerant, if any, from a refrigeration system, evacuate the system to prepare it for charging using the refrigerant weight listed on the unit data plate, then charge the system to the correct operating charge.

TOOLS, MATERIALS, AND EQUIPMENT: A gauge manifold, gloves, safety goggles, refrigerant recovery equipment, a vacuum pump, an electronic charging scale, and a cylinder of the proper refrigerant for the system to be charged. You will need a refrigeration system that has service valves or ports and that has a known required charge.

⚠ **SAFETY PRECAUTIONS:** Wear gloves and goggles any time you are transferring refrigerant. Your instructor must give you instruction in recovering and transferring refrigerant before starting this exercise.

STEP-BY-STEP PROCEDURES

1. Recover any refrigerant that may be in the system according to EPA guidelines. Your instructor will provide instruction on how to recover the refrigerant properly.

2. While wearing goggles, connect the gauges to the high- and the low-side service valves on the system.

3. After inspecting the vacuum pump oil, connect the pump to the center port hose and turn it on.

4. Determine the correct charge for the system from the nameplate or the manufacturer's specifications and record it here: _____ oz.

5. Set the refrigerant cylinder on the charging scale with the tank positioned to allow vapor refrigerant to leave the tank. This assumes that the refrigerant being used is not a blend. If the refrigerant is a blend, be sure to look at the side of the refrigerant cylinder to determine the proper position for liquid removal.

FIGURE **ISU-9.1**

6. Make provisions to secure the gauge hose when it is disconnected from the vacuum pump and connected to the cylinder on the scale so it will not move and affect the reading. A long gauge hose is preferred for this, Figure ISU-9.1.

7. When the vacuum pump has evacuated the system down to the recommended level, close the isolation valve on the vacuum pump and the valves on the gauge manifold, turn off the vacuum pump, disconnect the center port hose from the vacuum pump, and connect it to the refrigerant cylinder.

8. Open the valve on the refrigerant cylinder, and briefly loosen the center hose connection on the gauge manifold to purge any air that might have been pulled into the hose when it was disconnected from the vacuum pump.

Photo by Jason Obrzut

9. Record the cylinder weight in ounces here: _____ lb _____ oz.

   Note: Weights are often displayed in pounds and tenths of a pound. Convert the decimal to ounces to make calculations easier. For instance, 0.4 lb equals 6.4 oz (16 oz × 0.4 = 6.4).

10. Once the tank has been placed on the scale and its weight recorded, zero the scale. When charging begins, the scale will display how much refrigerant left the tank.

    Note: It may be necessary to add an extra ounce or so due to the refrigerant contained in the gauge manifold that did not get charged into the system.

11. Open the low-side gauge manifold valve, and allow refrigerant to start entering the system. Remember, when charging with a refrigerant blend that has a potential to fractionate, the service technician must also throttle liquid refrigerant from the charging cylinder into the refrigeration system to avoid fractionation.

12. If all of the charge will not enter the system, the system may be started to lower the system pressure so that the rest of the charge will be pulled in. Do not let the low-side pressure exceed normal operating low-side pressure.

13. When the complete charge is in the system according to the scale, disconnect the low-side line, and allow the system to stabilize. Ensure the system is operating normally.

MAINTENANCE OF WORKSTATION AND TOOLS: Wipe any oil off the tools and work area. Check the oil in the vacuum pump, and add or change the oil if needed. Replace any panels on the unit. Place all tools in their respective places.

SUMMARY STATEMENTS: Describe the procedure for determining the amount of refrigerant to be charged into the system. Describe the calculations made regarding the use of the scales. Include any conversions from pounds to ounces.

TAKEAWAY: Digital scales are precise instruments used to charge systems by weight, known as critical charge systems. The precision is similar to using a graduated charging cylinder, but with less refrigerant loss. Before charging, a technician should become familiar with the scale being used, its operation, and the units of measurement that it is capable of.

QUESTIONS

1. Why is the line to the refrigerant cylinder purged after evacuation and before charging?

2. Why is the gauge line secured when the refrigerant cylinder is on the scale while transferring refrigerant?

3. Why would the cylinder pressure drop when charging vapor into a system?

4. What can be done to keep refrigerant moving into the system if the cylinder pressure drops and the cylinder must remain on the scale?

5. Why is only vapor recommended for charging a system?

6. What would happen if liquid refrigerant were to enter the compressor cylinder while it was running?

7. How many ounces are in three-tenths (0.3) of a pound?

8. When charging with a scale, how should the refrigerant in the manifold and hoses be accounted for?

Exercise ISU-10

Charging a Fixed-Bore Split-Type Air-Conditioning System

Name	Date	Grade

OBJECTIVES: Upon completion of this exercise, you should be able to properly charge a split-type air-conditioning system that is equipped with a capillary tube or other fixed-bore metering device.

INTRODUCTION: As part of this lab exercise, you will install gauges on a split-type air-conditioning system that is equipped with a fixed-bore metering device. Because the equipment is field-installed, the exact refrigerant charge for the system is not found on the nameplate of the system. To properly charge this system, you will calculate the evaporator superheat and compare it to a superheat chart or use a charging app on a smart device.

TOOLS, MATERIALS, AND EQUIPMENT: Gauge manifold, refrigerant tank, high-quality thermocouple thermometer, temperature/pressure chart, digital psychrometer, goggles, gloves, split-system air conditioner as described previously, a smart device with charging app, and a refrigerant scale.

SAFETY PRECAUTIONS: SAFETY PRECAUTIONS: When handling and working with refrigerants, be sure to protect yourself from frostbite, as escaping refrigerant is extremely cold. Wear safety goggles as well as gloves and other pieces of personal protection equipment. Make certain that the charging process is performed slowly and carefully to ensure that excessive system pressures do not result. Always follow all of your teacher's instructions.

STEP-BY-STEP PROCEDURES

1. Make certain that the system you are working on has been properly evacuated to the required vacuum level determined by your instructor. At this point, the gauges should be on the system, with the low-side hose connected to the suction-line service port on the condensing unit and the high-side hose connected to the liquid-line service port on the condensing unit. Both valves on the gauge manifold should be in the closed position.

2. If possible, connect a low-side compound gauge to the suction line at the outlet of the evaporator. By taking the pressure reading as close as possible to the outlet of the evaporator, superheat calculations will be more accurate.

3. Inspect the system and make certain that the suction line is properly insulated. If the insulation is missing or loosely installed, add and resecure insulation as needed.

4. Determine the correct refrigerant for the system. This information is provided on the nameplate of the unit.

5. Obtain the proper tank of refrigerant.

6. Place the refrigerant tank on the charging scale, and turn the scale on.

7. Connect the center hose from the gauge manifold to the refrigerant tank.

8. Open the valve on the refrigerant tank.

9. Very briefly, loosen the center hose connection on the gauge manifold to purge any air that might be in the center hose.

10. Retighten the center hose connection on the gauge manifold.

11. On refrigerant scales equipped with a zero-adjust feature, reset the scale so that the display reads "0 lb, 0 oz." If the scale does not have this feature, write down the weight of the cylinder in pounds and ounces.

12. Make certain that the air-conditioning system is off.

13. Open the high-side valve on the gauge manifold. Refrigerant will flow from the tank through the gauge manifold and into the high side of the air-conditioning system. Note that although refrigerant is being added to the high-side system, the pressure reading on the low side will rise. This is because refrigerant is flowing through the capillary tube into the low side of the system.

14. Continue to allow refrigerant to flow into the system until the high- and low-side system pressures are equal and no more refrigerant is flowing from the refrigerant tank.

15. Close the valves on the gauge manifold.

16. Secure the sensing bulb of the thermocouple thermometer to the top of the suction line at the outlet of the evaporator. Make certain that the sensing bulb is well insulated by wrapping the sensing element and the suction line with foam insulation tape. Poor thermal contact between the sensing element and the suction line will result in a system undercharge, as the calculated evaporator superheat will be higher than the actual evaporator superheat.

17. Inspect the system and make certain that there are no restrictions or obstructions in the air paths through the evaporator and condenser coils. This means checking for dirty air filters, dirty coils, closed supply registers, blocked returns, and other possible restrictions to airflow that might be present.

18. Energize the air-conditioning system, and allow it to operate for a minimum of 5 minutes.

19. Obtain and record the following information:

   - Ambient temperature surrounding the condensing unit: _____ °F
   - Indoor ambient temperature (return air to the evaporator coil): _____ °F
   - Indoor air wet bulb temperature: _____ °F
   - Supply air temperature coming off the evaporator coil: _____ °F
   - Evaporator saturation pressure: _____ psig
   - Evaporator saturation temperature: _____ °F
   - Suction-line temperature: _____ °F
   - Condenser saturation pressure: _____ psig
   - Condenser saturation temperature: _____ °F
   - Compressor amperage: _____ A
   - Temperature differential between return- and supply-air temperatures: _____ °F
   - Temperature differential between outside ambient and condenser saturation temperatures: _____ °F

20. Calculate the evaporator superheat:

   Suction-line temperature: _____ °F

   − Evaporator saturation temperature: _____ °F

   Evaporator superheat: _____ °F

21. Using either a superheat chart or a charging app on a smart device, compare the actual superheat to the recommended superheat. If the charge needs to be adjusted, do so and allow it operate for at least 5 minutes; then return to step 20. If the evaporator superheat is within the recommended range, the system is properly charged. Proceed to step 22.

22. Now that the system is charged, obtain and record the following information:

- Ambient temperature surrounding the condensing unit: _____ °F
- Indoor ambient temperature (return air to the evaporator coil): _____ °F
- Indoor air wet bulb temperature: _____ °F
- Supply-air temperature coming off the evaporator coil: _____ °F
- Evaporator saturation pressure: _____ psig
- Evaporator saturation temperature: _____ °F
- Suction-line temperature: _____ °F
- Condenser saturation pressure: _____ psig
- Condenser saturation temperature: _____ °F
- Evaporator superheat: _____ °F
- Compressor amperage: _____ A
- Temperature differential between return- and supply-air temperatures: _____ °F
- Temperature differential between outside ambient and condenser saturation temperatures: _____ °F
- Total amount of refrigerant added to the system: _____ lb _____ oz

23. Compare the recorded values from step 19 to those recorded in step 22. For each of the listed parameters, determine whether it increased, decreased, or remained the same.

24. Explain the changes noted in step 23.

MAINTENANCE OF WORKSTATION AND TOOLS: Make certain that the gauge manifold is properly removed from the system and that the hoses are connected to the blank ports on the manifold to prevent hose contamination. Make certain that the service port caps are replaced on the system service valves. Return all tools to their proper places. Make certain that refrigerant tanks are closed and properly stored.

SUMMARY STATEMENT: Explain how evaporator superheat changes with respect to the refrigerant charge in an operating air-conditioning system. Explain the effects of an overcharge/undercharge on system operation and compressor amperage. Use an additional sheet of paper if more space is necessary for your answer.

TAKEAWAY: Fixed-bore split systems that are field-installed are typically charged by calculating the system superheat at the evaporator. Charging these systems requires a technician to make precise temperature and pressure measurements. As with any system, airflow at both the evaporator and the condenser should be verified.

QUESTIONS

1. Why is it necessary to purge the center hose of the gauge manifold before adding refrigerant to the system?

2. In plain terms, explain what evaporator superheat is.

3. In plain terms, explain how evaporator superheat is calculated.

4. Why is it important to secure the loose ends of the service hoses to the blank ports on the gauge manifold?

5. Why does the sensor on the thermocouple thermometer need to be insulated?

6. Why do you have to wait at least 5 minutes before adding additional refrigerant to the system?

7. Explain why excessive evaporator superheat is an indication of an underfed evaporator coil.

8. Explain why low evaporator superheat is an indication of an overfed evaporator coil.

9. Why is it more accurate to take the suction-line temperature at the evaporator rather than at the condensing unit?

Charging a TVX Air-Conditioning System With Refrigerant

Name	Date	Grade

OBJECTIVES: Upon completion of this exercise, you should be able to properly charge a split-type air-conditioning system that is equipped with a thermostatic expansion valve.

INTRODUCTION: As part of this lab exercise, you will install gauges on a split-type air-conditioning system that is equipped with a thermostatic expansion valve. Because the equipment is field-installed, the exact refrigerant charge for the system is not found on the nameplate of the system. To properly charge this system, we will rely on the condenser subcooling to ensure that the refrigerant charge is correct.

TOOLS, MATERIALS, AND EQUIPMENT: Gauge manifold, refrigerant tank, high-quality thermocouple thermometer, temperature/pressure chart, goggles, gloves, refrigerant scale, an air-conditioning system equipped with a TXV, and foam insulation tape.

⚠ **SAFETY PRECAUTIONS:** When handling and working with refrigerants, be sure to protect yourself from frostbite, as escaping refrigerant is extremely cold. Wear goggles as well as gloves and other pieces of personal protection equipment. Make certain that the charging process is performed slowly and carefully to ensure that excessive system pressures do not result. Always follow all of your teacher's instructions.

STEP-BY-STEP PROCEDURES

1. Make certain that the system being worked on has been properly evacuated to the required vacuum level, determined by the instructor. At this point, the gauges should be on the system, with the low-side hose connected to the suction-line service port on the condensing unit and the high-side hose connected to the liquid-line service port on the condensing unit. Both valves on the gauge manifold should be in the closed position.

2. Inspect the system and make certain that the thermal bulb on the thermostatic expansion valve is properly secured to the suction line and that the bulb is properly insulated. If the insulation is missing or if the bulb is loosely installed, add insulation and resecure the bulb as needed.

3. Determine the correct refrigerant type and subcooling for the system. This information is provided on the nameplate of the unit. If the recommended subcooling is not listed, a subcooling chart or charging app on a smart device may be used to determine the correct subcooling. See your instructor for direction and recommendations.

4. Obtain the proper tank of refrigerant.

5. Place the refrigerant tank on the charging scale, and turn the scale on.

6. Connect the center hose from the gauge manifold to the refrigerant tank.

7. Open the valve on the refrigerant tank.

8. Very briefly, loosen the center hose connection on the gauge manifold to purge any air that might be in the center hose.

9. Retighten the center hose connection on the gauge manifold.

10. On refrigerant scales equipped with a zero-adjust feature, reset the scale so that the display reads "0 lb, 0 oz." If the scale does not have this feature, write down the weight of the cylinder in pounds and ounces.

11. Make certain that the system is off.

12. With the refrigerant tank properly connected, open both the high-side and the low-side valves on the gauge manifold. Refrigerant will flow from the tank through the gauge manifold and into the system.

    Note: On systems that operate with blended refrigerants, the refrigerant must leave the tank as a liquid. In this case, position the tank so that liquid refrigerant will leave the tank and introduce refrigerant only to the high-pressure side of the system.

13. Continue to allow refrigerant to flow into the system until the system pressure equalizes with the pressure in the refrigerant tank.

14. Close the valves on the gauge manifold.

15. Secure the sensor of your thermocouple thermometer to the liquid line at the outlet of the condenser coil. Make certain that the sensor is well insulated by wrapping it, along with the liquid line, with foam insulation tape. Poor thermal contact between the sensor and the liquid line will result in improper system charging, as the liquid-line temperature reading will be different from the actual liquid-line temperature.

16. Inspect the system and make certain that there are no restrictions or obstructions in the air paths through the evaporator and condenser coils. This means checking for dirty air filters, dirty coils, closed supply registers, blocked returns, and other possible restrictions to airflow that might be present.

17. Energize the system and allow it to operate for a minimum of 5 minutes.

18. Obtain and record the following information:

    - Ambient temperature surrounding the condensing unit: _____ °F
    - Indoor ambient temperature (return air to the evaporator coil): _____ °F
    - Supply-air temperature coming off of the evaporator coil: _____ °F
    - Evaporator saturation pressure: _____ psig
    - Evaporator saturation temperature: _____ °F
    - Condenser saturation pressure: _____ psig
    - Condenser saturation temperature: _____ °F
    - Liquid-line temperature: _____ °F
    - Compressor amperage: _____ A
    - Temperature differential between return- and supply-air temperatures: _____ °F
    - Temperature differential between outside ambient and condenser saturation temperatures: _____ °F

19. Calculate the condenser subcooling:

    Condenser saturation temperature: _____ °F

    − Liquid-line temperature: _____ °F

    Condenser subcooling: _____ °F

20. If the condenser subcooling is lower than the manufacturer's recommended range, add refrigerant to the system, allow it to operate for at least 5 minutes, and return to step 19. If the condenser subcooling is greater than the recommended range, some refrigerant will need to be recovered from the system; allow it operate for at least 5 minutes, and return to step 19. If the condenser subcooling is within the recommended range, the system is properly charged. Proceed to step 21.

Note: See your instructor for which method is to be used to determine the proper subcooling range.

21. Now that the system is charged, obtain and record the following information:

- Ambient temperature surrounding the condensing unit: _____ °F
- Indoor ambient temperature (return air to the evaporator coil): _____ °F
- Supply-air temperature coming off the evaporator coil: _____ °F
- Evaporator saturation pressure: _____ psig
- Evaporator saturation temperature: _____ °F
- Condenser saturation pressure: _____ psig
- Condenser saturation temperature: _____ °F
- Liquid-line temperature: _____ °F
- Condenser subcooling: _____ °F
- Compressor amperage: _____ A
- Temperature differential between return- and supply-air temperatures: _____ °F
- Temperature differential between outside ambient and condenser saturation temperatures: _____ °F
- Total amount of refrigerant added to the system: _____ lb _____ oz

22. Compare the recorded values from step 18 to those recorded in step 21. For each of the listed parameters, determine whether it increased, decreased, or remained the same.

23. Explain the changes noted in step 22.

MAINTENANCE OF WORKSTATION AND TOOLS: Make certain that the gauge manifold is properly removed from the system and that the hoses are connected to the blank ports on the manifold to prevent hose contamination. Make certain that the service port caps are replaced on the system service valves. Return all tools to their proper places. Make certain that refrigerant tanks are closed and properly stored.

SUMMARY STATEMENT: Explain how condenser subcooling changes with respect to the refrigerant charge in an operating air-conditioning system. Explain the effects of an overcharge/undercharge on system operation, subcooling, and compressor amperage. Use an additional sheet of paper if more space is necessary for your answer.

TAKEAWAY: On systems that are equipped with a thermostatic expansion valve, the superheat will remain constant whether refrigerant is added or recovered. This is because the TXV is designed to maintain a constant evaporator superheat. Using evaporator superheat as a method to charge these systems can be difficult and problematic. For this reason, subcooling and condenser split are used.

QUESTIONS

1. Explain the function of the thermostatic expansion valve.

2. Explain, in plain terms, what condenser subcooling is.

3. Explain how condenser subcooling is calculated.

4. Explain why we use condenser subcooling to charge air-conditioning and refrigeration systems that are equipped with thermostatic expansion valves.

5. Why is it important to secure the loose ends of the service hoses to the blank ports on the gauge manifold?

6. Why does the sensor on the thermocouple thermometer need to be insulated?

7. Why do you have to wait at least 5 minutes before adding additional refrigerant to the system?

8. Explain why low condenser subcooling is an indication of an undercharged system.

9. Explain why high condenser subcooling is an indication of an overcharged system.

10. Where and how should a TXV's sensing bulb be installed to ensure proper operation?

Power Circuit Wiring

Name	Date	Grade

OBJECTIVES: Upon completion of this exercise, you·should be able to interpret the wiring diagram for the power, line-voltage circuit for a package air-conditioning system and a split-type air-conditioning system.

INTRODUCTION: The wiring diagram for an air-conditioning system is often divided into two parts. One part of the circuit is the line-voltage or power circuit, while the other is the low-voltage control circuit. In this exercise, you will choose the diagram that illustrates the power circuit components correctly. Compressor overloads, for example, are part of the power circuit, while pressure switches are typically part of the low-voltage, control circuit.

TOOLS, MATERIALS, AND EQUIPMENT: There are no tools or materials required for this exercise.

SAFETY PRECAUTIONS: There are no safety precautions required for this exercise.

STEP-BY-STEP PROCEDURES

1. Which of the following package AC unit diagrams illustrates the correct line-voltage diagram? _____

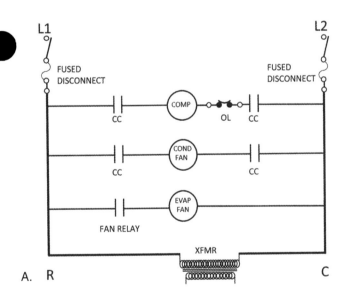

A.

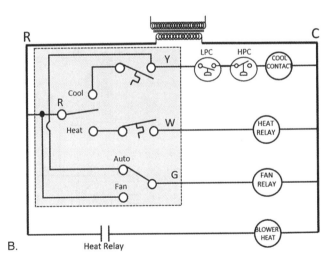

B.

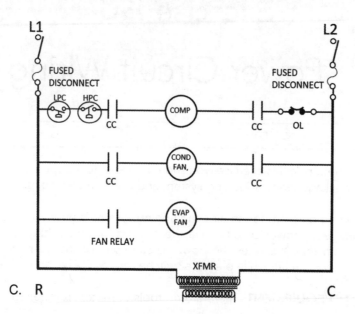

C. R

2. Which of the following split-system AC unit diagrams illustrates the correct line-voltage diagram?_____

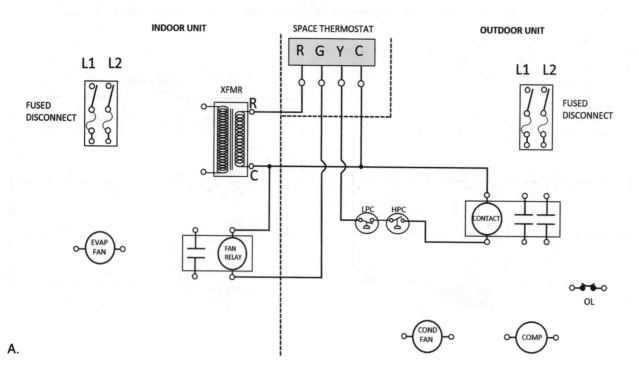

A.

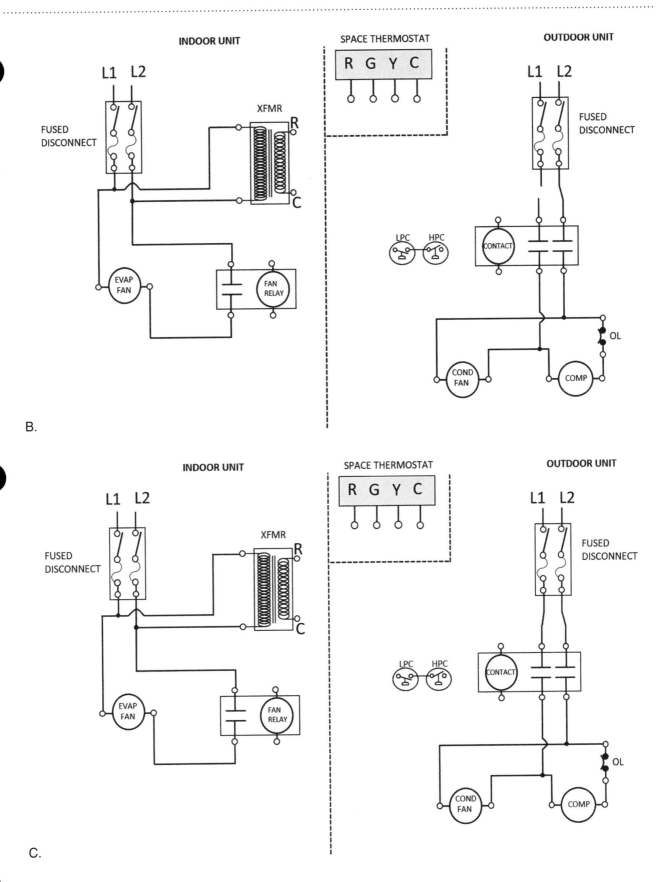

B.

C.

MAINTENANCE OF WORKSTATION AND TOOLS: There are no tools required for this exercise.

SUMMARY STATEMENT: Explain how the power circuit for a package air-conditioning system differs from the power circuit for a split-type air-conditioning system.

TAKEAWAY: Wiring diagrams are an important tool that can be used to both understand a system's sequence of operations and to troubleshoot it. Being able to differentiate between low-voltage components and line-voltage components can help gain an overall understanding of the diagram that, at first glance, can look confusing or intimidating.

QUESTIONS

1. With regard to the line-voltage power circuit, the compressor overload:
   A. is wired in series with both the compressor and the condenser fan motor.
   B. is wired in series with the compressor but in parallel with the condenser fan motor.
   C. is wired in parallel with both the compressor and the condenser fan motor.
   D. is wired in series with the condenser fan motor but in parallel with the compressor.

2. With regard to the line-voltage power circuit, the high- and low-pressure switches:
   A. are wired in parallel with each other.
   B. are both wired in series with the compressor overload.
   C. are both wired in series with the condenser fan motor.
   D. are not in the circuit at all.

3. The contactors in this exercise have:
   A. one normally open set of contacts, one normally closed set of contacts, and a line-voltage coil.
   B. one normally open set of contacts, one normally closed set of contacts, and a low-voltage coil.
   C. two sets of normally open contacts and a line-voltage coil.
   D. two sets of normally open contacts and a low-voltage coil.

4. What types of switches, when used, are usually located in series on the low-voltage circuit with the contactor coil?
   A. Rollout switch, vent switch, and an auxiliary limit switch.
   B. Sail switch, negative pressure switch, and a compressor overload switch.
   C. Low-pressure switch, high-pressure switch, and a time-delay relay.
   D. Float switch, flow switch, and a fan-cycling switch.

Exercise ISU-13

Low-Voltage Control Circuit Wiring

Name	Date	Grade

OBJECTIVES: Upon completion of this exercise, you should be able to interpret the wiring diagram for the low-voltage control circuit for a package air-conditioning system and a split-type air-conditioning system.

INTRODUCTION: The wiring diagram for an air-conditioning system is often divided into two parts. One part of the circuit is the line-voltage or power circuit, while the other is the low-voltage control circuit. In this exercise, you will choose the diagram that illustrates the low-voltage control circuit components correctly. Compressor overloads, for example, are part of the power circuit, while pressure switches are typically part of the low-voltage, control circuit.

TOOLS, MATERIALS, AND EQUIPMENT: There are no tools or materials required for this exercise.

⚠ **SAFETY PRECAUTIONS:** There are no safety precautions required for this exercise.

STEP-BY-STEP PROCEDURES

1. Which of the following package AC unit diagrams illustrates the correct low-voltage diagram? _____

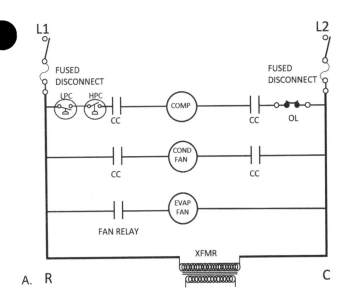

A.

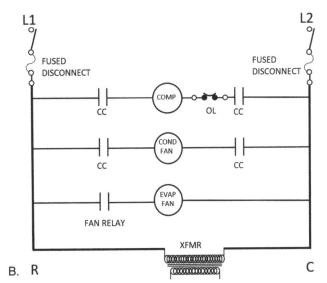

B.

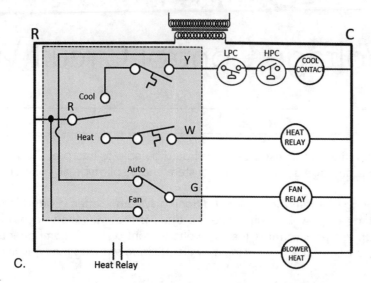

C.

2. Which of the following split-system AC unit diagrams illustrates the correct low-voltage diagram?

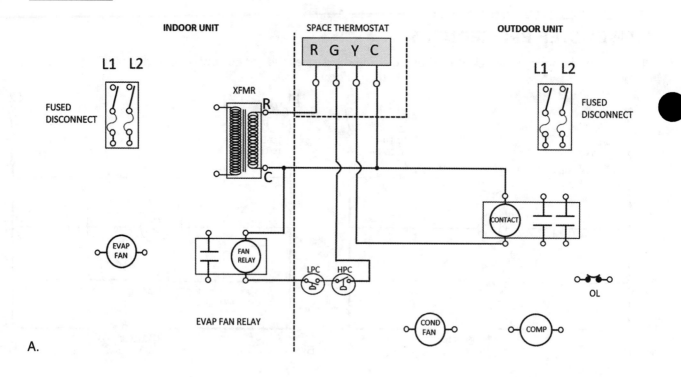

A.

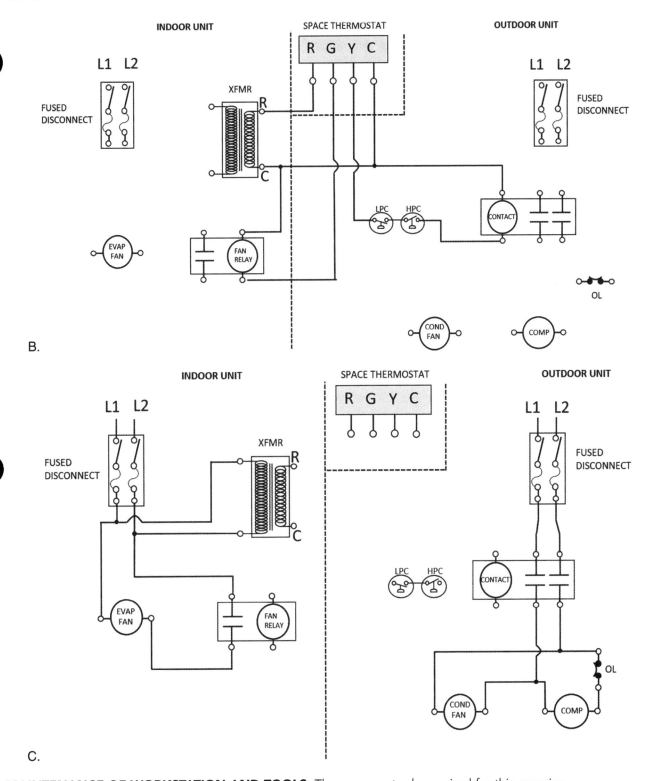

B.

C.

MAINTENANCE OF WORKSTATION AND TOOLS: There are no tools required for this exercise.

SUMMARY STATEMENT: Explain how the low-voltage control circuit for a package air-conditioning system differs from the low-voltage control circuit for a split-type air-conditioning system.

TAKEAWAY: Wiring diagrams are an important tool that can be used to both understand a system's sequence of operations and to troubleshoot it. Being able to differentiate between low-voltage components and line-voltage components can help gain an overall understanding of the diagram. Package units, though similar to split systems, differ in that all of the components are located in one outdoor unit.

QUESTIONS

1. The C terminal on the thermostat is used to:
   A. provide power to the fan relay coil.
   B. provide power to the contactor coil.
   C. allow the thermostat to be powered all the time.
   D. allow the compressor to be powered all the time.

2. If the system is functioning properly, and the compressor, condenser fan motor, and evaporator fan motor are all operating, the voltage measured between terminals R and C on the thermostat should be:
   A. 120 VAC.
   B. 220 VAC.
   C. 24 VAC.
   D. 0 VAC.

3. If the system is functioning properly, and the compressor, condenser fan motor, and evaporator fan motor are all operating, the voltage measured between terminals R and G on the thermostat should be:
   A. 120 VAC.
   B. 220 VAC.
   C. 24 VAC.
   D. 0 VAC.

4. If the system is functioning properly, and the compressor, condenser fan motor, and evaporator fan motor are all operating, the voltage measured between terminals R and Y on the thermostat should be:
   A. 120 VAC.
   B. 220 VAC.
   C. 24 VAC.
   D. 0 VAC.

5. If the system is functioning properly, and the compressor, condenser fan motor, and evaporator fan motor are all operating, the voltage measured between terminals G and C on the thermostat should be:
   A. 120 VAC.
   B. 220 VAC.
   C. 24 VAC.
   D. 0 VAC.

6. If the system is functioning properly, and the compressor, condenser fan motor, and evaporator fan motor are all operating, the voltage measured between terminals Y and C on the thermostat should be:
   A. 120 VAC.
   B. 220 VAC.
   C. 24 VAC.
   D. 0 VAC.

7. In the low-voltage control circuit, the high-pressure switch and the low-pressure switch:
   A. are both wired in series with the contactor coil.
   B. are both wired in series with the indoor fan relay coil.
   C. are wired in parallel with each other.
   D. are wired in series with the G terminal on the thermostat.

Exercise ISU-14

Wiring an Air-Conditioning Packaged Unit

Name	Date	Grade

OBJECTIVES: Upon completion of this exercise, you should be able to correctly size line-voltage conductors for a packaged unit installation; formulate a list of supplies that will be needed to make a packaged unit installation; make the electrical connection for a packaged unit installation; operate the packaged unit after installation is complete; and complete a check, test, and start checklist on installation.

INTRODUCTION: Technicians in the HVACR industry will be called on to install air-conditioning packaged units. The installation instructions usually furnished with new equipment along with local and state codes will be a guide for the installation of conditioned air equipment. The technician should be able to correctly size the line-voltage conductors using the National Electrical Code or manufacturer's installation instructions. A list of materials that will be needed to complete the installation should be compiled by the installation technician. The technician will be required to install and operate conditioned air equipment. Once equipment is installed and operating, the technician should complete a check, test, and start form.

TOOLS, MATERIALS, AND EQUIPMENT: Package air-conditioning system, installation manual for the unit being used, low-voltage thermostat, wire strippers, a digital multimeter with an amp clamp, digital thermometer, straight-slot and Phillips-head screwdrivers, safety goggles and gloves. Your instructor will provide the electrical supplies necessary to do the install.

⚠️ **SAFETY PRECAUTIONS:** Make certain that the electrical source is disconnected when making electrical connections. In addition:

- Make sure all electrical connections are tight.
- Make sure no current-carrying conductors are touching metal surfaces except the grounding conductor.
- Make sure the correct voltage is being supplied to the equipment.
- Make sure all equipment covers are replaced.
- Make sure body parts do not come in contact with live electrical conductors.
- Keep hands and materials away from moving parts.

STEP-BY-STEP PROCEDURES

Install Package Air-Conditioning Unit

1. Obtain the installation manual for the packaged air-conditioning unit being used from your instructor.

2. Read the section in the installation instructions on wiring the unit.

3. Determine the correct wire size for the assigned unit. Take note of the distance between the packaged unit and the electrical supply panel. (Note: The type of insulation on the conductor plays an important part in its current-carrying ability.)

4. Complete the Data Sheet.

DATA SHEET

Package unit model number: _____

Packaged unit serial number: _____

Distance from power supply to unit: _____

Size of wire to be installed: _____

5. Complete the Materials List, including electrical supplies needed to complete the installation.

MATERIALS LIST

_____

_____

_____

_____

_____

_____

_____

6. Have your instructor check your wire sizing and materials list for the assigned packaged unit installation. Obtain the necessary materials from your instructor.

7. Make the necessary electrical connections to the packaged unit assigned by the instructor.

8. Operate the unit and complete the Check, Test, and Start Checklist.

CHECK, TEST, AND START CHECKLIST

Voltage being supplied to unit: _____ V

Amp draw of compressor: _____ A

Amp draw of outdoor fan motor: _____ A

Amp draw of indoor fan motor: _____ A

Supply-air temperature: _____ °F

Return-air temperature: _____ °F

CHECKLIST

All electrical connections tight	☐ YES	☐ NO
Electrical circuits properly labeled	☐ YES	☐ NO
Equipment properly grounded	☐ YES	☐ NO
All equipment level	☐ YES	☐ NO
All equipment covers in place and properly attached	☐ YES	☐ NO
Installation area clean	☐ YES	☐ NO

Note: The technician should also check the mechanical refrigeration cycle characteristics when performing check, test, and start procedures.

9. Have instructor check your installation.

10. Clean up the work area, and return all tools and supplies to their correct locations.

MAINTENANCE OF WORKSTATION AND TOOLS: Clean and return all tools to their proper locations.

● **SUMMARY STATEMENT:** What procedure is used to size the conductor supplying an air-conditioning unit with more than one motor? Why should the technician perform a check, test, and start procedure on a new installation?

TAKEAWAY: When performing an installation of any HVACR system, it is important to read through the installation manual to become familiar with the manufacturer's recommendations and the specific installation codes that apply to the unit. Making checklists is a good way to ensure that you have all the materials needed prior to beginning the installation as well as checking system performance afterward.

QUESTIONS

1. Explain the construction of a packaged unit.

2. Why are disconnects used in the installation of equipment?

3. What are two installation applications for packaged units?

4. How many electrical connections must a technician make when installing a packaged unit?

5. What line-voltage connections are required to complete the installation of a cooling-only packaged unit?

6. What low-voltage connections are required to complete the installation of a cooling-only packaged unit?

7. Why is it important for the technician to make sure that the unit is level?

8. Why should equipment be grounded?

9. Why should the installing technician check the amp draw of all electrical motors in a new installation?

10. What is the technician's responsibility to the customer at the completion of the installation?

Exercise ISU-15

Wiring a Split System With a Fan Coil Unit

Name	Date	Grade

OBJECTIVES: Upon completion of this exercise, you should be able correctly size line-voltage conductors for a condensing unit and fan coil unit installation; formulate a list of supplies that will be needed to complete the installation of a condensing unit with a fan coil unit; make the electrical connection for a condensing unit with a fan coil unit; operate the installed conditioned air system; and complete a check, test, and start checklist on the installation.

INTRODUCTION: Technicians will oftentimes have to install split air-conditioning systems with a fan coil unit that houses the fan used as the air source and the evaporator coil. The installation instructions usually furnished with new equipment along with local and state codes will be the guide for the installation of conditioned air equipment. The technician should be able to correctly size the line-voltage conductors using the National Electrical Code or manufacturer's installation instructions. A list of materials that will be needed to complete the installation should be compiled by the installation technician. The technician will be required to install and operate conditioned air equipment. Once equipment is installed and operating, the technician should complete a check, test, and start form.

TOOLS, MATERIALS, AND EQUIPMENT: AC condensing unit, fan coil unit, installation manual for the units being used, low-voltage thermostat, wire strippers, a digital multimeter with an amp clamp, digital thermometer, straight-slot and Phillips-head screwdrivers, safety goggles, and gloves. Your instructor will provide the electrical supplies necessary to do the install.

⚠ SAFETY PRECAUTIONS: Make certain that the electrical source is disconnected when making electrical connections. In addition:

- Make sure all electrical connections are tight.
- Make sure no current-carrying conductors are touching metal surfaces except the grounding conductor.
- Make sure the correct voltage is being supplied to the equipment.
- Make sure all equipment covers are replaced.
- Make sure body parts do not come in contact with live electrical conductors.
- Keep hands and materials away from moving parts.

STEP-BY-STEP PROCEDURES

Install Conditioned Air Split System (Condensing Unit and Fan Coil Unit)

1. Obtain the installation manual for the split conditioned air system including condensing unit and fan coil unit from your instructor.

2. Read the section in the installation instructions on wiring the units.

3. Determine the correct wire size for the assigned unit components (condensing unit and fan coil unit). Take note of the distance between the equipment and the electrical supply panel. (Note: The type of insulation on the conductor plays an important part in its current-carrying ability.)

4. Complete the Data Sheet.

DATA SHEET

Condensing unit model number: _____

Condensing unit serial number: _____

Distance from power supply to condensing unit: _____

Size of wire to be installed for condensing unit: _____

Fan coil unit model number: _____

Fan coil unit serial number: _____

Distance from power supply to unit: _____

Size of wire to be installed for fan coil unit: _____

5. Complete the Materials List including electrical supplies needed to complete installation.

MATERIALS LIST

_____

_____

_____

_____

_____

_____

6. Have your instructor check your wire sizing and material list for the assigned conditioned air system installation. Obtain the necessary materials from your instructor.

7. Make the necessary electrical connections to the system assigned by your instructor.

8. Operate the system and complete the Check, Test, and Start CheckList.

CHECK, TEST, AND START CHECKLIST

Voltage being supplied to condensing unit: _____

Amp draw of compressor: _____

Amp draw of condenser fan motor: _____

Voltage being supplied to the fan coil unit: _____

Amp draw of blower motor: _____

Supply air temperature: _____ °F

Return air temperature: _____ °F

CHECKLIST

All electrical connections tight	☐ YES	☐ NO
Electrical circuits properly labeled	☐ YES	☐ NO
Equipment properly grounded	☐ YES	☐ NO
All equipment level	☐ YES	☐ NO
All equipment covers in place and properly attached	☐ YES	☐ NO
Installation area clean	☐ YES	☐ NO

Note: The technician should also check the mechanical refrigeration cycle characteristics when performing check, test, and start procedures.

9. Have instructor check your installation.

10. Clean up the work area, and return all tools and supplies to their correct locations.

MAINTENANCE OF WORKSTATION AND TOOLS: Clean and return all tools to their proper locations.

SUMMARY STATEMENT: What is the purpose of a disconnect switch on a condensing unit installation? Why should a technician check the amp draw of loads when a conditioned air system is installed?

TAKEAWAY: When performing an installation of any HVACR system, it is important to read through the installation manual to become familiar with the manufacturer's recommendations and the specific installation codes that apply to the unit. Making checklists is a good way to ensure that you have all the materials needed prior to beginning the installation as well as checking system performance afterward.

QUESTIONS

1. In what application would a fan coil system be used?

2. Where would a split-system conditioned air system be used?

3. How many power supply connections must a technician make when installing a conditioned air system using a fan coil unit and a condensing unit?

4. What control wires are connected between the fan coil unit control and the thermostat?

5. What control wires are connected between the fan coil unit control and the condensing unit?

6. What type of thermostat would be used on a cooling-only installation of a condensing unit and a fan coil unit?

7. Why should the installation technician make certain that the unit is level?

8. Why should the installation technician check the supply and return air temperature upon completion of an installation?

9. What is the purpose of labeling the power supply circuits of a new installation?

10. Why is it important for an installation technician to communicate with the customer on a new installation in an existing home?

Exercise ISU-16

Wiring a Split-System Air-Conditioning Unit With a Gas Furnace

Name	Date	Grade

OBJECTIVES: Upon completion of this exercise, you should be able to correctly size line-voltage conductors for a condensing unit and gas furnace installation; formulate a list of supplies that will be needed for the installation of a condensing unit with a gas furnace; make the electrical connection for a condensing unit with a gas furnace; operate the conditioned air system after the installation is completed; and complete a check, test, and start checklist on the installation.

INTRODUCTION: Technicians will oftentimes have to install split air-conditioning systems with a gas furnace that houses the fan used as the air source. The installation instructions usually furnished with new equipment, along with local and state codes, will be the guide for the installation of conditioned air equipment. The technician should be able to correctly size the line-voltage conductors using the National Electrical Code or manufacturer's installation instructions. A list of materials that will be needed to complete the installation should be compiled by the installation technician. The technician will be required to install and operate conditioned air equipment. Once equipment is installed and operating, the technician should complete a check, test, and start form.

TOOLS, MATERIALS, AND EQUIPMENT: AC condensing unit, a gas furnace, installation manual for the units being used, low-voltage thermostat, wire strippers, a digital multimeter with an amp clamp, digital thermometer, straight-slot and Phillips-head screwdrivers, safety goggles, and gloves. Your instructor will provide the electrical supplies necessary to do the installation.

⚠️ **SAFETY PRECAUTIONS:** Make certain that the electrical source is disconnected when making electrical connections. In addition:

- Make sure all electrical connections are tight.
- Make sure no current-carrying conductors are touching metal surfaces except the grounding conductor.
- Make sure the correct voltage is being supplied to the equipment.
- Make sure all equipment covers are replaced.
- Make sure body parts do not come in contact with live electrical conductors.
- Keep hands and materials away from moving parts.

STEP-BY-STEP PROCEDURES

Install Conditioned Air Split System (Condensing Unit and Gas Furnace)

1. Obtain the installation manual for the split-system air-conditioning unit and the gas furnace from your instructor.

2. Read the section in the installation instructions on wiring the units.

3. Determine the correct wire size for the assigned unit components (condensing unit and gas furnace). Take note of the distance between the equipment and the electrical supply panel. (Note: The type of insulation on the conductor plays an important part in its current-carrying ability.)

4. Complete the Data Sheet.

DATA SHEET

Condensing unit model number: _____

Condensing unit serial number: _____

Distance from power supply to condensing unit: _____

Size of wire to be installed for condensing unit: _____

Gas furnace model number: _____

Gas furnace serial number: _____

Distance from power supply to gas furnace: _____

Size of wire to be installed for gas furnace: _____

5. Complete the Materials List including electrical supplies needed to complete installation.

MATERIALS LIST

_____

_____

_____

_____

_____

_____

6. Have your instructor check your wire sizing and materials list for the assigned conditioned air system installation. Obtain the necessary materials from your instructor.

7. Make the necessary electrical connections to the system assigned by your instructor.

8. Operate the system and complete the Check, Test, and Start Checklist.

CHECK, TEST, AND START CHECKLIST

Voltage being supplied to condensing unit: _____

Amp draw of compressor: _____

Amp draw of condenser fan motor: _____

Voltage being supplied to the fan coil unit: _____

Amp draw of blower motor: _____

Supply air temperature: _____ °F

Return air temperature: _____ °F

CHECKLIST

All electrical connections tight	☐ YES	☐ NO
Electrical circuits properly labeled	☐ YES	☐ NO
Equipment properly grounded	☐ YES	☐ NO
All equipment level	☐ YES	☐ NO
All equipment covers in place and properly attached	☐ YES	☐ NO
Installation area clean	☐ YES	☐ NO

Note: The technician should also check the mechanical refrigeration cycle characteristics and gas heating operation when performing check, test, and start procedures.

9. Have the instructor check your installation.

10. Clean up the work area, and return all tools and supplies to their correct locations

MAINTENANCE OF WORKSTATION AND TOOLS: Clean and return all tools to their proper locations.

SUMMARY STATEMENT: What is the heating source of the conditioned air system installed in this lab exercise? Where is the evaporator located in this application?

TAKEAWAY: When performing an installation of any HVACR system, it is important to read through the installation manual to become familiar with the manufacturer's recommendations and the specific installation codes that apply to the unit. Making checklists is a good way to ensure that you have all the materials needed prior to beginning the installation as well as checking system performance afterward.

QUESTIONS

1. Where is the connection point for the control wiring from the thermostat and condensing unit?

2. What controls the operation of the gas furnace in modern units?

3. What controls the blower motor in modern gas furnaces?

4. What are the electrical components of a basic condensing unit?

5. What do the following letter designations represent in a split system utilizing a gas furnace as an air source and a condensing unit: "R," "G," "W," and "Y"?

6. What connections on a gas furnace terminal board would be supplied to the condensing unit?

7. What controls the condenser fan motor on a basic conditioned air-condensing unit?

8. What controls the gas valve on a modern gas furnace?

9. How many power connections are required in the installation of a gas furnace and a condensing unit?

10. What is the control voltage of a conditioned air split system utilizing a gas furnace as the air source?

Exercise ISU-17

Wiring a Split-System Heat Pump

Name	Date	Grade

OBJECTIVES: Upon completion of this exercise, you should be able to correctly size line-voltage conductors for the outdoor unit, indoor unit, and supplementary electrical resistance heat of a split-system heat pump; formulate a list of supplies that will be needed to install the outdoor unit, indoor unit, and supplementary electrical resistance heat of a split-system heat pump; make the electrical connection for the outdoor unit, indoor unit, and supplementary electrical resistance heat of a split-system heat pump; operate the conditioned air system after the installation is completed; and complete a check, test, and start checklist on the installation.

INTRODUCTION: Technicians will oftentimes have to install split-system heat pumps along with some type of supplementary heat. The installation instructions usually furnished with new equipment, along with local and state codes, will be the guide for the installation of conditioned air equipment. The technician should be able to correctly size the line-voltage conductors using the National Electrical Code or manufacturer's installation instructions. A list of materials that will be needed to complete the installation should be compiled by the installation technician. The technician will be required to install and operate conditioned air equipment. Once equipment is installed and operating, the technician should complete a check, test, and start form.

TOOLS, MATERIALS, AND EQUIPMENT: Indoor and outdoor components of a split-system heat pump, installation manual for the units being used, low-voltage thermostat, wire strippers, a digital multimeter with an amp clamp, digital thermometer, straight-slot and Phillips-head screwdrivers, safety goggles, and gloves. Your instructor will provide the electrical supplies necessary to do the installation.

⚠️ **SAFETY PRECAUTIONS:** Make certain that the electrical source is disconnected when making electrical connections. In addition:

- Make sure all electrical connections are tight.
- Make sure no current-carrying conductors are touching metal surfaces except the grounding conductor.
- Make sure the correct voltage is being supplied to the equipment.
- Make sure all equipment covers are replaced.
- Make sure body parts do not come in contact with live electrical conductors.
- Keep hands and materials away from moving parts.

STEP-BY-STEP PROCEDURES

Install Split-System Heat Pump with Supplementary Heat

1. Obtain the installation manuals for the indoor and outdoor units of the split-system heat pump from your instructor.

2. Read the section in the installation instructions on wiring the units.

3. Determine the correct wire size for the assigned unit components (outdoor unit and indoor unit). Take note of the distance between the equipment and the electrical supply panel. (Note: The type of insulation on the conductor plays an important part in its current-carrying ability.)

4. Complete the Data Sheet.

DATA SHEET

Outdoor unit model number: _____

Outdoor unit serial number: _____

Distance from power supply to condensing unit: _____

Size of wire to be installed for condensing unit: _____

Indoor unit model number: _____

Resistance heat model number: _____

Distance from power supply to indoor unit: _____

Size of wire to be installed for indoor unit with supplementary heat: _____

5. Complete the Materials List including electrical supplies needed to complete installation.

MATERIALS LIST

_____

_____

_____

_____

_____

_____

6. Have your instructor check your wire sizing and materials list for the assigned conditioned air system installation. Obtain the necessary materials from your instructor.

7. Make the necessary electrical connections to the heat pump system assigned by your instructor.

8. Operate the heat pump system and complete the Check, Test, and Start Checklist.

CHECK, TEST, AND START CHECKLIST

Voltage being supplied to outdoor unit: _____

Amp draw of compressor: _____

Amp draw of outdoor fan motor: _____

Voltage being supplied to the indoor unit: _____

Amp draw of blower motor: _____

Amp draw of resistance heaters: _____

Supply air temperature (without supplementary heat): _____ °F

Supply air temperature (with supplementary heat): _____ °F

Return air temperature: _____ °F

CHECKLIST

All electrical connections tight	☐ YES	☐ NO
Electrical circuits properly labeled	☐ YES	☐ NO
Equipment properly grounded	☐ YES	☐ NO
All equipment level	☐ YES	☐ NO
All equipment covers in place and properly attached	☐ YES	☐ NO
Installation area clean	☐ YES	☐ NO

Note: The technician should also check the mechanical refrigeration cycle characteristics and gas heating operation when performing check, test, and start procedures.

9. Have instructor check your installation.

10. Clean up the work area, and return all tools and supplies to their correct locations

MAINTENANCE OF WORKSTATION AND TOOLS: Clean and return all tools to their proper locations.

SUMMARY STATEMENT: Why is supplementary heat necessary on an air-to-air heat pump? What type of low-voltage thermostat is used on the heat pump in this lab exercise?

TAKEAWAY: When performing an installation of any HVACR system, it is important to read through the installation manual to become familiar with the manufacturer's recommendations and the specific installation codes that apply to the unit. Making checklists is a good way to ensure that you have all the materials needed prior to beginning the installation as well as checking system performance afterward.

QUESTIONS

1. Why is supplementary heat required on most air-to-air heat pumps?

2. What is the difference between the mechanical refrigeration cycle of a heat pump and a regular air conditioner?

3. What are the common letter designations of most heat pumps and what modes of operation are related to these letter designations?

4. What two sources of supplementary heat could be used by modern heat pumps?

5. What is the purpose of the reversing valve in the mechanical refrigeration cycle of a heat pump?

6. What controls the operation of modern heat pumps?

7. What line-voltage electrical connections must be made when installing an air-to-air split-system heat pump using an electrical resistance heater as the supplementary heat?

8. What control voltage connections must be made when installing an air-to-air split-system heat pump using electrical resistance heaters as the supplementary heat?

9. How are electrical resistance heaters mounted in most indoor blower units in a split-system heat pump installation?

10. Where is the indoor coil positioned in a split-system heat pump installation when using a gas furnace as the supplementary heat?

Exercise ISU-18

Using the Halide Leak Detector

Name	Date	Grade

OBJECTIVES: Upon completion of this exercise, you should be able to use a halide torch to leak check a system for refrigerant leaks.

INTRODUCTION: Using gauges, you will check to determine that there is refrigerant pressure in an air-conditioning system. You will then check all field and factory connections with a halide leak detector to determine whether there are leaks. If the leak is from a flare connection, tighten it. If it is from a solder connection, ask your instructor what should be done.

TOOLS, MATERIALS, AND EQUIPMENT: A system that is charged with an HCFC refrigerant, a halide leak detector, a gauge manifold, a digital multimeter, straight-slot and Phillips-head screwdrivers, ¼ in. and ⁵⁄₁₆ in. nut drivers, goggles, and light gloves. If a leak is found, you will need additional tools to repair it.

⚠️ **SAFETY PRECAUTIONS:** Turn the electrical power off. Lock and tag the disconnect boxes to both the indoor and outdoor units before beginning the exercise. Check with the multimeter to ensure that the power to the system is off. Wear goggles and light gloves when attaching or removing gauges. Beware of toxic fumes from the halide torch. Ventilate the area if necessary. DO NOT USE A HALIDE LEAK DETECTOR ON A SYSTEM THAT IS CHARGED WITH A FLAMMABLE REFRIGERANT.

STEP-BY-STEP PROCEDURES

You will need instruction in the use of the halide leak detector before beginning this exercise. Ensure that there is adequate ventilation, and beware of toxic fumes.

1. With the power off and locked out and while wearing goggles and gloves, fasten the gauge hoses to the high- and low-side gauge ports. Check the gauge readings to ensure that there is pressure.

2. With the multimeter, check the electrical disconnect to both units to ensure that the electrical power is off.

3. With the halide leak detector, check all joints (solder and flare) in the indoor and outdoor piping. Check the joints under the insulation by making a small hole in the insulation with a small screwdriver and holding the sensing tube to the hole.

4. Remove any additional necessary panels, and check all factory connections.

5. If a leak is found in a flare joint, tighten it. If one is found in a solder joint, ask for instructions.

6. When you have established that there is no leak, or when you have finished, remove the gauges and replace the panels and service valve caps.

MAINTENANCE OF WORKSTATION AND TOOLS: Return all tools, equipment, and materials to their proper storage places. Clean your workstation.

SUMMARY STATEMENT: Describe the halide leak-check procedure.

TAKEAWAY: Halide leak detectors are capable of picking up very small refrigerant leaks, as small as about 1 oz a year. If a leak is detected, the flame color will change. However, halid leake detectors cannot pick HFC refrigerants, as they do not contain chlorine. They also cannot be used around flammable vapors because of the open flame.

QUESTIONS

1. What pulls the refrigerant to the sensor in a halide leak detector?

2. What is the normal color of the gas flame in a halide leak detector?

3. What is the color of the gas flame in a halide leak detector when refrigerant is present?

4. Can a halide leak detector be used to detect natural gas? Why?

5. What does the term *halide* refer to?

6. Name two gases that are commonly used to produce the flame in a halide leak detector.

7. Describe how a halide leak detector discovers leaks.

8. Does the halide leak detector work best in the light or in the dark?

9. Why will a halide leak detector not be able to locate a leak on an R-410A air-conditioning system?

10. How small of a leak can a halide leak detector typically pick up?

Exercise ISU-19

Using Electronic Leak Detectors

Name	Date	Grade

OBJECTIVES: Upon completion of this exercise, you should be able to use an electronic leak detector to find refrigerant leaks.

INTRODUCTION: You will check an air-conditioning system with gauges to determine that there is refrigerant pressure. You will then check all field and factory connections with an electronic leak detector to determine if there are refrigerant leaks.

TOOLS, MATERIALS, AND EQUIPMENT: A system charged with refrigerant, an electronic leak detector, a gauge manifold, a digital multimeter, straight-slot and Phillips-head screwdrivers, ¼ in. and ⁵⁄₁₆ in. nut drivers, goggles, and light gloves. If a leak is found, you will need additional tools to repair it.

⚠ **SAFETY PRECAUTIONS:** Turn the electrical power to both the indoor and outdoor units off before beginning this exercise. Lock and tag the disconnect box and keep the key in your possession. Check with the digital multimeter to ensure that the system is off. Wear goggles and light gloves when attaching or removing gauges. Do not allow liquid refrigerant to contact your skin. Ventilate the area if necessary.

STEP-BY-STEP PROCEDURES

You will need instruction in the use of the electronic leak detector before beginning this exercise.

1. With the power off and locked out and while wearing goggles and gloves, fasten the gauge hoses to the high- and low-side gauge ports. Check the gauge readings to ensure that there is pressure. If there is no pressure, check with your instructor for the next step.

2. With the multimeter, check the electrical disconnect to both units to ensure that the power is off.

3. Using the electronic leak detector, check all connections (flare and solder) at the indoor unit. Check under fittings.

4. Check all connections at the outdoor unit.

5. Remove necessary panels and check the factory connections.

6. Use a small screwdriver to make a small hole in the insulation on the tubing at each joint, and insert the probe in the holes to check for leaks.

7. Tighten any flare connection where you have found a leak. Ask for instructions before repairing solder connections.

8. When you have finished, carefully remove the gauge lines, and replace all panels with proper fasteners.

MAINTENANCE OF WORKSTATION AND TOOLS: Place all tools, equipment, and materials in their proper storage places. Ensure that your workstation is left clean.

SUMMARY STATEMENT: Describe how the electronic leak detector operates. What are some of its good and bad points? Use an additional sheet of paper if more space is necessary for your answer.

TAKEAWAY: Electronic leak detectors can be highly sensitive, detecting leaks as small as about ¼ oz a year. They are capable of picking CFC, HCFC, and HFC refrigerants, depending on the model. When using a leak detector, check all field and factory connections on the system.

QUESTIONS

1. Can natural gas leaks be located with an electronic leak detector?

2. How can the refrigerant be moved from the point of the leak to the detector sensor?

3. What can be done to keep wind from affecting leak detection?

4. What is the power source for an electronic leak detector?

5. How can soap bubbles be used to further pinpoint the leak when it is detected?

6. What is the purpose of the filter under the tip of the sensor on the detector?

7. Does the electronic leak detector work best in a light or dark place?

8. How can an electronic leak detector be used to pinpoint the source of a large leak?

Refrigerant Recovery

Name	Date	Grade

OBJECTIVES: Upon completion of this exercise, you should be able to use one of two methods for recovering refrigerant from a system.

INTRODUCTION: You will use two methods of removing refrigerant from a system. The first is to pump all available liquid into an approved refrigerant cylinder using the refrigeration system containing the refrigerant. The second is to use refrigerant recovery equipment commercially manufactured for refrigerant recovery.

TOOLS, MATERIALS, AND EQUIPMENT: Gauge manifold, gloves, goggles, a service valve wrench, two 8 in. adjustable wrenches, an empty DOT-approved refrigerant cylinder in a vacuum, a digital scale, a refrigerant recovery system, and a refrigeration system from which refrigerant is to be recovered.

⚠ **SAFETY PRECAUTIONS:** DO NOT OVERFILL ANY REFRIGERANT CYLINDER; USE ONLY APPROVED CYLINDERS. Do not exceed the net weight of the cylinder. Wear gloves and goggles while transferring refrigerant.

STEP-BY-STEP PROCEDURES

Pumping the Liquid from the System Using Passive Recovery

1. Fasten the low-side gauge line to the suction connection on the system used.

2. Fasten the high-side gauge line to the king valve or a liquid-line connection.

   Note: Not all systems have a liquid-line connection. If this system does not have one, you will need to use the second method of recovering refrigerant. Wear goggles and gloves.

3. Loosely connect the center line of the gauge manifold to the refrigerant cylinder. Note: More refrigerant can be transferred from the system to the tank if the tank is empty and in a vacuum.

4. Bleed a small amount of refrigerant from each line through the center line, and allow it to seep out at the cylinder connection. This will purge any contaminants from the gauge manifold.

5. Set the cylinder on the scale and determine the maximum weight printed on the cylinder. DO NOT ALLOW THE CYLINDER WEIGHT TO EXCEED THIS MAXIMUM. (The maximum weight is typically determined by multiplying the tank's water capacity [WC] by 80% [0.80]. Then the tare weight [TW] of the tank is added to determine the cylinder's maximum allowable weight.)

6. Start the system and open the valve arrangement between the liquid line and the cylinder. Liquid refrigerant will start to move toward the cylinder.

7. Watch the high-pressure gauge, and do not allow the system pressure to rise above the working pressure of the refrigerant cylinder.

8. Watch the suction pressure, and do not allow it to fall below 0 psig. When it reaches 0 psig, you have removed all the refrigerant from the system that you can using this method.

9. Remove the refrigerant cylinder. You may need to recover the remaining refrigerant using recovery equipment. Recovery can be sped up by setting the recovery cylinder in a bucket of ice, thus reducing the pressure in the cylinder.

Refrigerant Transfer With a Self-Contained Recovery Machine

1. Read the manufacturer's directions closely. Some units recover refrigerants only, and some recover and filter the refrigerant.

2. Connect the recovery unit to the system following the manufacturer's directions for the particular type of system you have.

3. Operate the recovery system as per the manufacturer's directions.

4. Upon completion of the operation, the refrigerant will be recovered in an approved container.

MAINTENANCE OF WORKSTATION AND TOOLS: Return the system to the condition your instructor requests you to. Return all tools and materials to their proper places.

SUMMARY STATEMENT: Describe the difference between recovery only and recycling. Use an additional sheet of paper if more space is necessary for your answer.

TAKEAWAY: Recovery of refrigerant is required by EPA Section 608. Passive recovery is accomplished by using the compressor in the system to pump the liquid refrigerant into an empty cylinder. Active recovery is done using a separate recovery machine that is capable of removing all of the refrigerant from the system and pumping it into a cylinder.

QUESTIONS

1. After a hermetic motor burn, what contaminant might be found in the recovered refrigerant?

2. What would the result be if a recovery cylinder were filled with liquid and allowed to warm up?

3. Why must liquid and vapor refrigerant be present for the pressure/temperature relationship to apply?

4. If an air-conditioning unit using R-22 were located totally outside and the temperature were 85°F, what would the pressure in the system be with the unit off? (*Hint:* It would follow the P and T relationship.) _____

5. What effect can noncondensables in the recovery cylinder have on the recovery process?

6. What is the generally recommended safe fill level of a recovery cylinder?

7. When the suction pressure of an R-22 system is 60 psig, at what temperature is the refrigerant boiling? _____°F

8. When the suction pressure of an R-404A system is 20 psig, at what temperature is the refrigerant boiling? _____°F

9. What is the condensing temperature for a head pressure of 175 psig for an R-134a system? _____°F

10. What is the condensing temperature for a head pressure of 296 psig for an R-22 system? _____°F

Evaluating an Air-Conditioning Installation

Name	Date	Grade

OBJECTIVES: Upon completion of this exercise, you should be able to look at an air-conditioning system and evaluate the installation for good workmanship and completeness.

INTRODUCTION: You will use this lab exercise as a checklist to ensure a split-system air-conditioning installation is complete and ready for start-up.

TOOLS, MATERIALS, AND EQUIPMENT: Straight-slot and Phillips-head screwdrivers, a digital multimeter equipped with an amp clamp, ¼ in. and ⁵⁄₁₆ in. nut drivers, a tape measure, and a flashlight.

⚠ **SAFETY PRECAUTIONS:** Turn the power off and verify with the digital multimeter that it is off. Lock and tag the panel or disconnect the box where the power is turned off. You should have the single key. Keep it in your possession while the power is off.

STEP-BY-STEP PROCEDURES

Use the following checklist to ensure the installation is in good order. Indicate yes or no. Turn off and lock out the power.

1. Evaporator section:
   - air handler level _____,
   - vibration eliminator installed _____,
   - duct fastened tightly _____,
   - connections taped _____,
   - filter in place _____,
   - auxiliary drain pan if evaporator in attic _____,

2. Condenser section:
   - condensing unit level _____,
   - a firm foundation under the condensing unit _____,
   - service panel in correct location _____,

   - secondary condensate drain line piped correctly _____,
   - primary drain pan piped correctly and trap with clean-out plug _____,
   - condensate pump, if needed, to pump the condensate to the drain _____,

   - airflow correct and not recirculating _____,
   - refrigerant piping in good order and insulated as required _____,
   - electrical connections tight _____,

   - electrical connections in the air handler tight, high-voltage correct _____,
   - low-voltage correct _____,
   - blower motor and wheel can be removed easily _____.

   - high-voltage correct _____,
   - low-voltage correct _____,
   - service valves open, if any _____.

3. Condensate drain termination: in a dry well _____, on the ground _____, in a drain _____.

4. Voltage check:
   - line voltage to indoor fan section _____V,

   - to condensing unit _____V,

   - low-voltage power supply voltage _____ V.

5. Start the indoor blower motor and verify the voltage and current:
   - rated voltage _____ V,

   - actual voltage _____ V,
   - rated current _____ A,

   - actual current _____ A.

6. Start the condensing unit and verify the voltage and amperage:
   - rated voltage _____ V,
   - actual voltage _____ V,
   - compressor rated current _____ A,
   - actual current _____ A,
   - fan motor rated current _____ A,
   - actual current _____ A.

7. Replace all panels with the correct fasteners.

MAINTENANCE OF WORKSTATION AND TOOLS: Return all tools to their proper places.

SUMMARY STATEMENTS: Describe the installation you inspected as it compares to the checklist. Describe any improvements you would have made.

TAKEAWAY: How long a unit will last and how efficient it will operate begin with the quality of its installation. A poor installation can condemn a unit to poor efficiency, high energy bills, and a short life span. After installation, a technician should verify that a system operates as designed to ensure that it will last for many years.

QUESTIONS

1. What are vibration isolators for?

2. Describe a dry well for condensate.

3. What is the purpose of insulation on the suction line?

4. What is the purpose of an auxiliary drain pan when the evaporator is installed in the attic?

5. What is the purpose of the trap in the condensate drain line?

6. Why should water from the roof not drain into the top of a top-discharge condenser?

7. How can an evaporator section be suspended from floor joists in a crawl space?

8. Why should air not be allowed to recirculate back into a condenser?

9. What would the symptoms be if air was permitted to recirculate through the condenser coil?

10. What would the symptoms of reduced airflow be at the evaporator?

11. What should be done if the line voltage to the unit is too low?

Exercise ISU-22

Start-Up and Checkout of an Air-Conditioning System

Name	Date	Grade

OBJECTIVES: Upon completion of this exercise, you should be able to check out components of an air-conditioning system for an orderly system start-up.

INTRODUCTION: You will start up an air-conditioning system, one component at a time, and check each one to ensure that it is operating correctly. You will record the operating characteristics for evaluation.

TOOLS, MATERIALS, AND EQUIPMENT: A system charged with refrigerant, a gauge manifold, a digital multimeter with an amp clamp, a thermometer, straight-slot and Phillips-head screwdrivers, ¼ in. and 5⁄16 in. nut drivers, a flashlight, electrical tape, light gloves, and goggles.

⚠ **SAFETY PRECAUTIONS:** Exercise caution while taking voltage and amperage readings on live circuits. Wear gloves and goggles while attaching and removing gauges and taking pressure readings. When the power is turned off, lock and tag the panel.

STEP-BY-STEP PROCEDURES

With the unit off:

1. Check the line voltage to the indoor and outdoor units: indoor unit: _____ V, outdoor unit: _____ V.

2. Record the rated current of the indoor blower: _____ A, outdoor fan: _____ A, compressor: _____ A.

3. With goggles and gloves on, fasten the gauge hoses to the condensing unit gauge ports. Put one temperature lead in the return air and one in the outdoor air to record the ambient temperature.

4. Disconnect the common wire going to the compressor, and tape the end to insulate it.

5. Set the thermostat to the off position, and turn the power to the fan section and the condenser section on.

6. Turn the fan switch on to start the indoor blower. Check and record the voltage and amperage of the motor: _____ V, _____ A. Compare with manufacturer's specifications. (Ensure that are panels are in place when recording measurements.)

7. Verify that air is coming out of all registers and that they are open.

8. Turn the thermostat to call for cooling, and check the voltage and amperage of the fan motor at the condensing unit: _____ V, _____ A. Compare with the manufacturer's specifications.

9. Turn the power off, and reconnect the compressor wire.

10. Turn the power on, and the compressor will start. Record the voltage and current: _____ V, _____ A. Compare with manufacturer's specifications.

11. Let the unit run and record the following information:

- Indoor air temperature: _____ °F, outdoor: _____ °F
- Suction pressure: _____ psig, discharge pressure: _____ psig
- Suction-line temperature: _____ °F, liquid-line temperature: _____ °F
- Superheat: _____°F, subcooling: _____°F

12. Carefully remove the gauges. Fasten the service valve caps.

13. Replace all panels with the correct fasteners.

MAINTENANCE OF WORKSTATION AND TOOLS: Return all tools to their correct places. Leave your workstation clean.

SUMMARY STATEMENT: Compare the temperature for the suction pressure on the preceding system in relation to the inlet air temperature. Indicate whether the head pressure was correct for the conditions.

TAKEAWAY: A proper refrigeration system evaluation should include checking each component, voltage and current readings of all electrical loads, and pressure and temperature readings of the refrigeration cycle. All of the readings should be compared to the manufacturer's recommended operating characteristics.

QUESTIONS

1. Why is it wise to disconnect the compressor while going through a start-up?

2. What should be done if the line voltage is too high before start-up?

3. What should be done if the line voltage is too low before start-up?

4. How does the outdoor temperature affect the head pressure?

5. How does the indoor temperature affect the suction pressure?

6. What should be done if there is no refrigerant in the system before start-up?

7. What type of service valve ports did the unit you worked on have: service valves or Schrader valves?

8. What is the typical refrigerant used in new residential central air-conditioning systems?

SECTION 14

Building Sciences (BSC)

SUBJECT AREA OVERVIEW

Green awareness, global warming, energy efficiency, energy savings, sustainability, and high-performance buildings are at the environmental and energy forefront. Whether in residential, commercial, or industrial applications, energy audits have become an integral part of evaluating and assessing an existing building's energy performance. Higher efficiency standards for new buildings have been established. Higher levels of training and certification have been developed for auditors, contractors, and HVACR technicians in an effort to meet the requirements of more sophisticated, energy-efficient buildings and equipment. Residential energy audits not only provide homeowners with an opportunity to lower their monthly utility bills, they also can help our nation become less dependent on oil by conserving energy. The ability to conserve natural resources is an added attraction for having an energy audit performed on homes. Homeowners can cut utility bills by 15% or more by following energy audit recommendations. According to the United States Department of Energy (DOE), 43% of the energy used by the average home goes toward space heating and space cooling.

The DOE estimates that a home's heating and cooling costs can be reduced by as much as 30% through proper insulation and air-sealing techniques. The DOE also estimates that 12% of a home's total energy bill is attributed to heating water for household use. Grants for improving energy efficiency may be available from state and local utilities or from low-income housing programs. Homeowners can check with their local utility, city, or the DOE-sponsored Database of State Incentives for Renewables and Efficiency (DSIRE). Much research has been conducted by the DOE's Building America and Home Performance with ENERGY STAR, sponsored by the U.S. Environmental Protection Agency (EPA) and the DOE. Home Performance with ENERGY STAR offers a comprehensive whole-house approach to improving the energy efficiency and comfort of existing homes and the testing of combustion products. The DOE's Building America program has worked with some of the nation's leading building scientists and more than 300 production builders on over 40,000 new homes at the time of this writing. Building America research applies building science to the goal of achieving efficient, comfortable, healthy, and durable homes. There are a number of organizations that offer certification for home auditors and contractors, including the following:

- Association for Energy Engineers (AEE)
- Building Performance Institute (BPI)
- Green Mechanical Council
- Residential Energy Services Network (RESNET)

Structural heat gain and heat loss calculations are used to properly size heating and air-conditioning equipment. Heat gain is the heat that enters a structure from the outside during the warmer months. Heat loss is the heat that leaks from a structure in the cooler months. Air-conditioning and heating systems should be sized to match the rates at which heat leaks into and out of the structure. Oversized equipment costs more to purchase, install, and maintain. Undersized equipment will not be able to meet the temperature requirements of the structure. There

(continued)

are many factors that determine how much heat enters or leaves a structure. These factors include, in part, the construction materials that are used, the shading on the structure, and the direction the building is facing. Structural materials have heat values associated with them, such as U-values and R-values. R-values are used for insulation materials and represent the resistance to heat transfer, or thermal resistance. U-values are often used to describe the thermal conductivity of construction panels such as windows, ceilings, and walls. Factors that affect the rate of heat transfer through a construction panel include the U-value, the area of the panel, and the temperature differential across the panel. Other factors that go into a heat gain and heat loss calculation include losses and gains associated with ductwork, pipes, people, plants, fish tanks, lights, appliances, and structural leaks.

KEY TERMS

Air changes per hour (ACH$_{50}$)
Air tightness
Annual fuel utilization Efficiency (AFUE)
Atmospheric-draft
Backdrafting
Barometric damper
Base loads
Building Performance Institute (BPI)
CFM$_{50}$
Chimney
Combustion air
Combustion blower motor
Conditioned space
Cubic feet per minute (cfm)

Dilution
Dilution air
Direct insulation contact
Draft
Draft diverter
Draft hood
Duct blower
Emittance
Energy conservation measure (ECM)
Energy index
Energy recovery ventilator (ERV)
Excess air
Exfiltration
Flue draft
Forced draft
Heat gain
Heat loss

HERS index
Included draft
Infiltration
Intermediate zones
Lighting, appliance, and miscellaneous electrical loads (LAMELS)
Net stack temperature
Open combustion devices
Over-the-fire-drafts
Pascal
Power draft
Reflectivity
Residential Energy Analysis and Rating Software (REM/Rate)
Residential Energy Services Network (RESNET)

Shell
Stack draft
Stack temperature
Steady-static efficiency (SSE)
Thermal boundary
Thermal break
Thermal bridging
Thermal bypass
Unconditioned space
Unintentionally conditioned space
Updraft
Weatherization
Zip Code Insulation Calculator tool

REVIEW TEST

Name	Date	Grade

Circle the letter that indicates the correct answer.

1. Oversized air-conditioning systems:
   A. cost more to operate, install, and maintain.
   B. are more efficient because they cool faster.
   C. use smaller blower and compressor motors.
   D. provide excessive dehumidification.

2. Undersized air-conditioning or heating systems:
   A. will cycle on and off constantly.
   B. will run continuously and not be able to satisfy the thermostat.
   C. cost more to install than larger systems.
   D. utilize larger blowers and duct systems.

3. As the temperature differential across a construction panel increases:
   A. the rate of heat transfer through the panel will increase.
   B. the rate of heat transfer through the panel will decrease.
   C. the rate of heat transfer through the panel will remain unchanged.
   D. The rate of heat transfer through a construction panel is not affected by the temperature differential across it.

4. The thermal resistance of a construction material is identified by its:
   A. U-value.   C. R-value.
   B. P-value.   D. B-value.

5. The thermal conductivity of a construction panel is identified by its:
   A. U-value.   C. R-value.
   B. P-value.   D. B-value.

6. If the R-value of a certain construction material is 5, what is the corresponding U-value?
   A. 5   C. 0.75
   B. 2.5   D. 0.2

7. A wall that separates two conditioned spaces is called a(n):
   A. interior wall.   C. partition wall.
   B. exterior wall.   D. below-grade wall.

8. A wall that separates a conditioned space from the outside is called a(n):
   A. interior wall.
   B. exterior wall.
   C. partition wall.
   D. below-grade wall.

9. A wall that separates an interior conditioned space from an interior unconditioned space is called a(n):
   A. interior wall.
   B. exterior wall.
   C. partition wall.
   D. below-grade wall.

10. What is the heat transfer rate through a 10' × 20' wall that has a U-value of 0.5 and a temperature differential across it of 50°F?
   A. 2500 Btu/h
   B. 2500 Btu/lb
   C. 5000 Btu/h
   D. 5000 Btu/lb

11. What is the heat transfer rate through a 10' × 20' wall that has a U-value of 0.4 and a temperature differential across it of 0°F?
   A. 0 Btu/h
   B. 0 Btu/lb
   C. 500 Btu/h
   D. 500 Btu/lb

12. What can be correctly concluded about the heat gain and heat loss calculations for two identical houses located in different cities?
   A. The heat gain and heat loss calculations will be the same because the houses are identical.
   B. The heat gain and heat loss calculations will be different because the orientations of the houses might be different.
   C. The heat gain and heat loss calculations will be different because the design temperatures of the cities might be different.
   D. Both B and C are correct.

13. What is the gross wall area of a 10' × 30' wall with a 3' × 7' door and a 4' × 6' window?
    A. 21 ft²   C. 255 ft²
    B. 24 ft²   D. 300 ft²

14. What is the net wall area of a 10' × 30' wall with a 3' × 7' door and a 4' × 6' window?
    A. 21 ft²   C. 255 ft²
    B. 24 ft²   D. 300 ft²

15. What is the approximate U-value of a single-pane window with a storm?
    A. 0.1   C. 2
    B. 0.5   D. 5

16. Name the three categories of a residential energy audit.

17. Name two nationally recognized energy certifications for home auditors and contractors.

18. One Pascal is equal to how many inches of water column?

19. One inch of water column is equal to how many Pascals?

20. What does an infrared scanning camera do?

21. Briefly describe a blower door test.

22. Briefly describe a duct blower's function in energy audits.

23. Define draft as it applies to combustion appliances.

24. What is the function of a flame safeguard control?

25. What is the difference between a base load and seasonal energy usage as they apply to a residence?

Gross Wall Area and Net Wall Area Calculations

Name	Date	Grade

OBJECTIVE: Upon completion of this exercise, you should be able to calculate the gross wall area and net wall area for various walls. This information can then be used later to help calculate heat gain and/or heat loss in a future lab exercise.

INTRODUCTION: Using tape measures and/or job prints or plans, you will become familiar with the procedures for determining the net and gross areas for the walls and partitions that make up part of a structure's envelope.

TOOLS, MATERIALS, AND EQUIPMENT: Tape measures, pencils, rulers, calculators, and possibly ladders.

⚠ **SAFETY PRECAUTIONS:** If ladders are used to obtain measurements, make certain that all ladder-related safety issues are reviewed and understood.

STEP-BY-STEP PROCEDURES

1. Identify a wall in your classroom that has windows, doors, or a combination of both.

2. Measure the height of the wall and enter the measurement here: _____

3. Measure the length of the wall and enter the measurement here: _____

4. Measure the width of Door 1, if there is one, and enter the measurement here: _____

5. Measure the height of Door 1, if there is one, and enter the measurement here: _____

6. Measure the width of Door 2, if there is one, and enter the measurement here: _____

7. Measure the height of Door 2, if there is one, and enter the measurement here: _____

8. Measure the width of Window 1, if there is one, and enter the measurement here: _____

9. Measure the height of Window 1, if there is one, and enter the measurement here: _____

10. Measure the width of Window 2, if there is one, and enter the measurement here: _____

11. Measure the height of Window 2, if there is one, and enter the measurement here: _____

12. Measure the width of Window 3, if there is one, and enter the measurement here: _____

13. Measure the height of Window 3, if there is one, and enter the measurement here: _____

14. Determine the gross wall area.

A. If the measurements in steps 2 and 3 are only in feet, multiply the length and the height of the wall to obtain the gross wall area.

Wall Height (in feet) × Wall Length (in feet) = Gross Wall Area (in square feet)

(Height) _____ ft × (Length) _____ ft = _____ ft²

B. If the measurements in steps 2 and 3 are in feet and inches, convert the measurements to either feet (using decimals) or inches.

Example: 10'6" = 10.5 ft

Example: 15'9" = 15.75 ft

Then, multiply the length and the height of the wall to obtain the gross wall area.

(Height) _____ ft × (Length) _____ ft = _____ ft²

C. If you are converting the measurements to inches:

Example: 10'5" = (10' × 12"/ft) + 5" = 120" + 5" = 125"

Then, multiply the length and the height of the wall to obtain the gross wall area.

(Height) _____ in. × (Length) _____ in. = _____ in².

Once you have obtained the gross wall area in square inches, you must divide the area in square inches by 144 to get the gross wall area in square feet.

Gross Wall Area (ft²) = Gross Wall Area (in².) ÷ 144

Gross Wall Area (ft²) = _____ ÷ 144

Gross Wall Area (ft²) = _____ ft²

15. Using the same methods used in step 14, determine the area of all windows and doors that are on the wall you are working with and record them here:

Area of Door 1: _____ ft²
Area of Door 2: _____ ft²
Area of Window 1: _____ ft²
Area of Window 2: _____ ft²
Area of Window 3: _____ ft²

16. Determine the net wall area.

A. Enter the gross wall area (from step 14) here: _____ ft²

B. Add up the areas of all windows and doors and enter the value here: _____ ft²

C. Subtract "B" from "A" to get the net wall area: _____ ft²

MAINTENANCE OF WORKSTATION AND TOOLS: Return all tools to their respective places.

SUMMARY STATEMENT: Describe any difficulties that were encountered during the measurement process. Describe any factors that you feel might have affected the accuracy of the measurements you obtained.

TAKEAWAY: Knowing the gross wall area and net wall area is extremely important! These values are required to obtain accurate heat gain and heat loss results, which help ensure that air-conditioning and heating systems are properly sized.

QUESTIONS

1. What is the difference between the gross wall area and the net wall area?

2. State a general formula for gross wall area.

3. State a general formula for net wall area.

4. Why is it important for all of the measurements to be in the same units?

5. Why did you divide by 144 in step 14C?

6. For *final* heat gain and heat loss calculations, which is likely to be used more often, gross wall area or net wall area? Why?

7. What do you think would happen to the total heat gain or heat loss calculation if you used gross wall area instead of net wall area?

Determining the Heat Transfer Rate Through a Wall

Name	Date	Grade

OBJECTIVE: Upon completion of this exercise, you should be able to determine the approximate rate of heat transfer through a wall.

INTRODUCTION: In this exercise, you can use the wall in Exercise BSC-1 to estimate the rate of heat transfer through a construction panel (wall), by determining the U-value of the panel and the temperature differential across it.

TOOLS, MATERIALS, AND EQUIPMENT: Tape measures, pencils, rulers, calculators, high-quality thermocouple thermometers, and possibly ladders.

SAFETY PRECAUTIONS: If ladders are used to obtain measurements, make certain that all ladder-related safety issues are reviewed and understood.

STEP-BY-STEP PROCEDURES

1. Record the net wall area from Exercise BSC-1 here: _____ ft^2

2. Determine the U-value for the wall using the following table (Figure BSC-2.1), which contains values for commonly used exterior walls:

FIGURE **BSC-2.1**

WALL CONSTRUCTION	STUD SIZE (WOOD)	BOARD INSULATION	INSULATION R-VALUE	U-VALUE
Frame	2 x 4	R-2	R-11	0.086
Frame	2 x 4	R-2	R-13	0.081
Frame	2 x 4	R-4	R-11	0.073
Frame	2 x 4	R-4	R-13	0.069
Frame	2 x 4	R-6	R-11	0.064
Frame	2 x 4	R-6	R-13	0.060
Frame	2 x 6	R-2	R-15	0.077
Frame	2 x 6	R-2	R-19	0.063
Frame	2 x 6	R-4	R-15	0.066
Frame	2 x 6	R-4	R-19	0.055
Frame	2 x 6	R-6	R-15	0.058
Frame	2 x 6	R-6	R-19	0.049

Table derived from data contained in the ASHRAE manuals and ACCA Manual J charts

Record the U-value of the wall here: _____

3. If the information for the wall you are working on is not contained in that table, you can use the following chart (Figure BSC-2.2) to determine the R-values of the construction elements that make up the wall:

FIGURE **BSC-2.2**

MATERIAL	R-VALUE	MATERIAL	R-VALUE
4" Common Brick	0.80	Fiberglass Blown Insulation (attic)	2.2-4.3 per inch
4" Brick Face	0.44	Fiberglass Blown Insulation (wall)	3.7-4.3 per inch
4" Concrete Block	0.80	Cellulose Blown Insulation (attic)	3.13 per inch
8" Concrete Block	1.11	Cellulose Blown Insulation (wall)	3.7 per inch
12" Concrete Block	1.28	3/4" Plywood Flooring	0.93
Lumber (2 x 4)	4.38	5/8" Particleboard (Underlayment)	0.82
Lumber (2 x 6)	6.88	3/4" Hardwood flooring	0.68
½" Plywood Sheathing	0.63	Linoleum Tiles	0.05
5/8" Plywood Sheathing	0.77	Carpet with fibrous pad	2.08
3/4" Plywood Sheathing	0.94	Carpet with rubber pad	1.23
1/2" Hardboard Siding	0.34	Wood Door (hollow core)	2.17
5/8" Plywood Siding	0.77	Wood Door (1.75" solid)	3.03
3/4" Plywood Siding	0.93	Wood Door (2.25" solid)	3.70
Hollow-backed aluminum siding	0.61	Single pane windows	0.91
Insulating board-backed aluminum siding	1.82	Single pane windows w/storm	2.00
Insulating board-backed aluminum siding (foil-backed)	2.96	Double pane windows with 0.25" air space	1.69
1/2" Drywall	0.45	Double pane windows with 0.50" air space	2.04
5/8" Drywall	0.56	Triple pane windows with 0.25" air space	2.56
Fiberglass Batt Insulation	3.14 – 4.3 per inch	Asphalt Shingles	0.44
Rock Wool Batt Insulation	3.14 – 4.0 per inch	Wood Shingles	0.97

Table derived from data contained in ASHRAE manuals and ACCA Manual J charts

Wall construction element 1: _____, R-value: _____

Wall construction element 2: _____, R-value: _____

Wall construction element 3: _____, R-value: _____

Wall construction element 4: _____, R-value: _____

Add up all the individual R-values to get the total R-value for the wall.

R-value of the wall = _____

To determine the U-value of the wall, take the inverse of the R-value.

Example: If the R-value is 10, the U-value = 1 ÷ 10 = 0.1.

U-value of the wall = _____

4. Determine the temperature differential across the panel.

A. Record the temperature on one side of the panel and record it here: _____ °F

B. Record the temperature on the other side of the panel and record it here: _____ °F

C. Take the difference between the two temperatures and record it here: _____ °F

5. Determine the rate of heat transfer by using the following formula:

$$Q = \text{U-value} \times \text{Temperature Differential} \times \text{Net Wall Area}$$

$$Q = \underline{\hspace{1.5cm}} \times \underline{\hspace{1.5cm}} °F \times \underline{\hspace{1.5cm}} ft^2$$

$$Q = \underline{\hspace{1.5cm}} \text{Btu/h}$$

MAINTENANCE OF WORKSTATION AND TOOLS: Return all tools to their respective places.

SUMMARY STATEMENT: Describe any difficulties that were encountered using the tables or charts. Explain the process of determining the elements that made up the wall you were working with. Describe any issues that you feel might have affected the accuracy of your U-value calculation.

TAKEAWAY: To determine the rate of heat flow through the envelope of a structure, each construction element must be evaluated separately. This exercise provides practice evaluating such construction elements.

QUESTIONS

1. In step 3, why do we add all of the individual R-values to get the total R-value for the wall?

2. If a wall element has an R-value of 20, what is its U-value?

3. What are the units of the U-value?

4. What will happen to the rate of heat transfer through a wall if the temperature differential is zero?

5. In a certain area of the country, heating equipment is designed for an outdoor temperature of 15°F and cooling equipment is designed for an outdoor temperature of 85°F. If we use indoor design temperatures of 75°F for cooling and 70°F for heating, what can be said about the rates of heat transfer through a wall during the winter months when compared to the summer months?

Heat Loss Exercise

Name	Date	Grade

OBJECTIVES: Upon completion of this exercise, you should be able to determine the total heat loss through the side of a structure for various scenarios.

INTRODUCTION: Using a calculator and the appropriate mathematical relationships, you will complete the charts provided.

TOOLS, MATERIALS, AND EQUIPMENT: Calculator.

⚠ **SAFETY PRECAUTIONS:** There are no safety precautions for this exercise.

STEP-BY-STEP PROCEDURES

Using your calculator and information from Figure BSC-3.1, complete the following chart by determining the correct values for all of the missing readings. Keep the following in mind:

- GWA—Gross Wall Area
- NWA—Net Wall Area
- TD—Temperature Difference Across the Wall
- Total Heat Loss through the wall includes the heat loss through the doors and windows.
- The GWA in the following chart represents the entire side of a structure.
- The side of the structure has only wall, window, and door elements.

FIGURE **BSC-3.1**

GWA	NWA	TOTAL WINDOW AREA	TOTAL DOOR AREA	TD	U-VALUE OF WALL MATERIAL	U-VALUE OF WINDOW	U-VALUE OF DOOR	TOTAL HEAT LOSS THROUGH WALL (Q)
400 ft²	355 ft²	24 ft²	21 ft²	50°F	0.05	0.50	0.40	
500 ft²	418 ft²	40 ft²	42 ft²	20°F	0.06	1.25	0.42	
300 ft²	244 ft²	35 ft²	21 ft²	60°F	0.04	0.40	0.39	
500 ft²	428 ft²	30 ft²	42 ft²	35°F	0.05	0.85	0.27	
500 ft²	439 ft²	20 ft²	21 ft²	40°F	0.055	0.6	0.45	
600 ft²	533 ft²	25 ft²	42 ft²	70°F	0.06	0.7	0.50	

MAINTENANCE OF WORKSTATION AND TOOLS: With the exception of the calculator, there are no tools required for this exercise.

SUMMARY STATEMENT: Explain how the total heat loss through the structure is affected by the geographic location of the structure.

TAKEAWAY: This exercise provides the opportunity to practice using the formulas for determining heat transfer through construction elements.

QUESTIONS

1. Based on the information in the completed chart, which line represents the structure that is located in the coldest climate? Explain your answer.

2. Based on the information in the completed chart, which line represents the structure that is located in the warmest climate? Explain your answer.

3. Based on the information in the completed chart, which line represents the structure that has the most efficient windows? Explain your answer.

4. Based on the information in the completed chart, which line represents the structure that has the most energy-efficient wall construction? Explain your answer.

Commercial and Industrial Systems (COM)

SUBJECT AREA OVERVIEW

The term "commercial and industrial HVACR systems" covers many types of equipment, ranging from commercial ice-making systems to rooftop packaged units to 1000-plus ton systems that provide cooling and/or heating for comfort, industrial processes, load management, and other functions. Commercial air-conditioning systems can be configured as package-type equipment, where all of the main system components are housed in one cabinet, or systems that are built-up on the job. Some systems are designed to cool water, which is then circulated throughout the building to provide comfort cooling. These systems are referred to as chilled-water systems.

Chilled-water systems are very popular in commercial and industrial buildings. They utilize equipment called chillers to cool water to a temperature of about 45°F, which is then circulated, via pumps to various locations throughout the building. This chilled water is then used to cool air that circulates within the occupied space. There are two basic categories of chillers, the vapor-compression cycle and the absorption chiller. The vapor-compression cycle chiller uses a compressor to provide the pressure differences in the system, while the absorption chiller uses a combination of heat, salt, and water to accomplish the same results. Vapor-compression chillers can be classified as either high-pressure or low-pressure chillers, depending on the refrigerant that is used in the system. Low-pressure chillers use centrifugal compressors and operate with the low-pressure side of the system in a vacuum.

Absorption chillers utilize heat as the driving force instead of a compressor. This type of equipment utilizes water for the refrigerant and certain salt solutions that have enough attraction for water to create the needed pressure difference. Chillers can be air-cooled or water-cooled. Water-cooled systems utilize cooling towers to provide the water needed to condense the refrigerant.

The cooling tower is the part of a water-cooled system that ultimately rejects heat from the system to the atmosphere. Cooling towers reduce the temperature of the water by means of evaporation. Water from the cooling tower is pumped to the air-conditioning equipment, where it flows through a water-cooled condenser, absorbing system heat. The water then flows back to the cooling tower where it is cooled down and then circulated back to the unit. Two types of cooling towers are natural-draft and mechanical-draft. Mechanical-draft cooling towers can be either forced- or induced-draft and rely on fans or blowers to move air through them. Natural-draft towers rely on the prevailing winds to increase the evaporation of the water. All cooling towers need service on a regular basis to keep them operating properly and free from disease-causing bacteria.

Many commercial air-conditioning systems are equipped with economizers. Economizers are system components that provide mechanical ventilation and, when conditions are desirable, can provide free-cooling to the conditioned space. Return-air conditions and outside-air conditions

(continued)

are compared to determine the best position for the economizer dampers. ASHRAE Standard 62 addresses the quality and quantity of ventilation air in commercial structures. A strategy that is growing in popularity to control the rate of ventilation air is called demand-controlled ventilation, or DCV. DCV controls the amount of ventilation for a particular space based on the level of carbon dioxide in the space. Carbon dioxide levels rise as occupancy rises and drop as occupancy levels drop.

Larger commercial and industrial systems generally require more sophisticated and complex control systems that are designed specifically to help meet the needs of the structure, its occupants and the functions they are intended to perform. Complex control systems often include a series of interlocking circuits that ensure that one system component or function is operational before energizing another component. Newer buildings typically have some type of direct digital (electronic) control, while many older buildings utilize pneumatic (air) control systems. Many buildings have some type of hybrid control system, using a combination of both DCC and pneumatic control elements.

These sophisticated control systems must, in part, help ensure that the electric loads are operating in a safe and efficient manner because of their high operating costs. Most commercial and industrial systems use some method of capacity control to allow the equipment to match the cooling and heating requirements for various processes and in various areas. There are many strategies being used to vary the capacity of equipment.

These strategies include varying the volume of air being moved through the system, referred to as variable air volume (VAV) systems, and varying the amount of refrigerant being pumped through the system, referred to as variable refrigerant flow (VRF) systems. VAV systems modulate the flow of air to the occupied space depending on the heating, ventilating, or cooling requirements of the area. VRF systems sense the saturation temperatures and pressures of the system to determine if the system load is increasing or decreasing and adjust the refrigerant flow accordingly.

Commercial ice-making equipment operates with evaporator temperatures at about 10°F. Ice making is accomplished by allowing ice to form on some type of evaporator surface. This ice must then be removed (harvested) and stored. Machines that make flake ice, for example, use a rotating auger inside a cylinder on which the ice has formed. Ice can be harvested by other methods as well, the most popular being hot gas. Just as with other commercial pieces of equipment, the operation of icemaker must be precisely controlled. Microprocessors are often the main controller on modern ice machines. Knowing the system's sequence of operations and how to troubleshoot the microprocessor are important things for the service technician to know. When water is frozen, minerals from the water are left behind and often become attached to the walls of the evaporator. For this reason, periodic preventive maintenance and cleaning are required. Preventive maintenance involves the inspecting, cleaning, sanitizing, and servicing of the ice machine and external water filtration/treatment system.

KEY TERMS

Absorber
Auger
Blocked suction
Butterfly valve
Centrifugal pump
Chiller barrel
Counterflow
Crossflow
Crystallization

Cylinder unloaders
Demand-controlled ventilation (DCV)
Direct-expansion evaporator
Economizer
Free-cooling
Glycol
Harvest cycle

High-side float
Lithium bromide (LiBr)
Low-side float
Purge unit
Rigging
Rupture disc
Strainer
Suction valve lifting
Sump

Unloading
Variable air volume (VAV)
Variable frequency drive (VFD)
Variable refrigerant flow (VRF)
Water-cooled

REVIEW TEST

Name	Date	Grade

Circle the letter that indicates the correct answer.

1. Most commercial and industrial HVACR systems have some type of capacity control because:
   A. the load on the structure is always constant.
   B. the load on the structure is always changing.
   C. ASHRAE Standard 62 requires capacity control.
   D. ASHRAE Standard 26 requires capacity control.

2. A pneumatic control system uses which of the following fluids as its driving force?
   A. Water
   B. Oil
   C. Air
   D. Refrigerant

3. If the compressor on a chilled-water system is operating and the chilled-water pump is not operating:
   A. the system's operating pressures will increase.
   B. the water in the chiller barrel can freeze.
   C. the water in the condenser can boil.
   D. the system's high-pressure switch will open.

4. The control device that ensures that one portion of a system is operational before another portion is energized is referred to as an:
   A. overload.
   B. interlock.
   C. interface.
   D. overflow.

5. A control device that uses air pressure to control the operation of an electric/electronic/digital control device would most likely be found on what type of control system?
   A. Pneumatic
   B. Direct digital
   C. Hybrid
   D. Electromagnetic

6. If the evaporator coil on a commercial icemaker has scale (mineral) deposits accumulated on its surface, which of the following conditions will be observed?
   A. High head pressure
   B. Low suction pressure
   C. Increased ice production
   D. Improved ice quality

7. A flush cycle on a commercial icemaker is intended to:
   A. filter the refrigerant in the piping circuit.
   B. replace the grease in the icemaker's gear box.
   C. add more water into the icemaker's sump.
   D. decrease the concentration of minerals in the water.

8. In a commercial ice-making system, an infrared eye is most often used to determine:
   A. whether or not the refrigerant charge is correct.
   B. the level of water in the sump.
   C. the level of ice in the storage bin.
   D. whether or not the storage bin door is open.

9. Which of the following materials is desirable for use on commercial icemaker evaporators because of its durability and ability to resist corrosion?
   A. Plastic
   B. Copper
   C. Brass
   D. Stainless steel

10. The thermal cooling capacity of ice is referred to as:
   A. coolness.
   B. hardness.
   C. compactness.
   D. density.

11. A chiller is a piece of equipment that is designed to cool:
   A. water.
   B. refrigerant.
   C. oil.
   D. air.

12. What type of compressor is commonly used on an absorption chiller?
   A. Centrifugal
   B. Screw
   C. Reciprocating
   D. None of the above.

13. Which of the following terms describes a method that can be used for system capacity control?
   A. Loading
   B. Feeding
   C. Packing
   D. Bracing

14. Lithium bromide (LiBr) is most likely to be found on what type of system?
   A. High-pressure chiller
   B. Low-pressure chiller
   C. Absorption chiller
   D. Centrifugal chiller

15. Centrifugal compressors rotate at speeds as high as:
   A. 15,000 rpm.
   B. 20,000 rpm.
   C. 30,000 rpm.
   D. 50,000 rpm.

16. A purge unit is used on low-pressure chillers to:
   A. increase the subcooling in the condenser.
   B. increase the superheat in the chiller barrel.
   C. remove noncondensables from the system.
   D. remove condensables from the system.

17. A low-pressure refrigerant is so designated because it:
   A. has a high boiling temperature at atmospheric pressure.
   B. has a low boiling temperature at atmospheric pressure.
   C. does not follow a pressure/temperature relationship at low pressures.
   D. only follows a pressure/temperature relationship at low pressures.

18. The heat that is rejected by the cooling tower is the heat that is:
   A. absorbed from the structure and generated in the compressor.
   B. rejected to the structure and generated in the compressor.
   C. absorbed from the structure and absorbed by the condenser.
   D. rejected to the structure and absorbed by the condenser.

19. Cooling towers are generally designed to reduce the temperature of the water entering the tower to a point that is within how many degrees of the wet-bulb temperature of the ambient air?
   A. 7°F   C. 21°F
   B. 14°F   D. 28°F

20. The "blowdown" cycle on a cooling tower is intended to:
   A. balance water flow to all condenser circuits.
   B. provide extra draft for greater evaporation.
   C. reduce mineral concentration in the tower.
   D. prevent excessive vortexing in the tower.

21. The cooling tower bypass valve is intended to:
   A. direct debris away from the tower sump.
   B. prevent pump cavitation from occurring.
   C. provide additional cold water to the sump.
   D. maintain the temperature of tower water.

22. Which type of water pump is commonly used in conjunction with cooling towers?
   A. Rotary
   B. Centrifugal
   C. Propeller
   D. Screw

23. Algae levels in cooling towers are controlled most effectively by:
   A. multistage filtration.
   B. periodic water changes.
   C. hydraulic distillation.
   D. chemical treatment.

24. One psi of pressure will support a column of water that is how high?
   A. 1.11 ft
   B. 2.31 ft
   C. 10.6 ft
   D. 14.7 ft

25. Before starting a chiller, it is important to verify that:
   A. the compressor is set to operate fully loaded upon start-up.
   B. there is water flow through the chiller barrel and the condenser.
   C. all safety devices and backup controls have been jumped out.
   D. the marine boxes have been removed from the heat exchangers.

26. In terms of chilled-water systems, a flow switch is a moving paddle in a(n):
   A. water stream that opens and closes a switch.
   B. air stream that opens and closes a switch.
   C. refrigerant stream that opens and closes a switch.
   D. oil stream that opens and closes a switch.

27. All of the following are positive displacement compressor except the:
   A. scroll compressor.
   B. rotary screw compressor.
   C. scroll compressor.
   D. centrifugal compressor.

28. It is often possible to identify failing tubes in a shell-and-tube heat exchanger before major problems arise by using a(n):
   A. megohmmeter.
   B. acid test kit.
   C. eddy current test.
   D. moisture test kit.

29. The instrument used to check for small electrical leaks to ground on large motors is the:
   A. megohmmeter.
   B. clamp-on ammeter.
   C. analog volt-ohm-milliammeter.
   D. digital wattmeter.

30. Water-cooled chilled-water systems must have a water treatment plan in place to:
   A. increase the pH of the water in the system.
   B. decrease the pH of the water in the system.
   C. increase the rate of algae growth in the system.
   D. decrease the rate of algae growth in the system.

31. Which of the following is true regarding the refrigerant piping on a rooftop package unit?
   A. Both the discharge line and the liquid line must be field installed.
   B. Both the suction line and the liquid line must be field installed.
   C. Neither the suction line nor the liquid line must be field installed.
   D. The suction, liquid, and discharge lines must all be field installed.

32. Which of the following best describes the process of mechanically removing air from a building?
   A. Exfiltration
   B. Exhausting
   C. Infiltration
   D. Ventilating

33. The mixed air stream entering the coiling coil of a commercial air-conditioning system is made up of:
   A. return air and supply air.
   B. exhaust air and return air.
   C. outside air and return air.
   D. outside air and supply air.

34. Forward-curved centrifugal blowers are desirable for use in conjunction with duct systems because they:
   A. are able to overcome the pressures in the duct system.
   B. compensate for the low pressures in the duct system.
   C. cause the pressures in the duct system to drop.
   D. ensure that all supply registers get the correct cfm.

35. The term "free-cooling" is made possible by a(n):
   A. economizer.
   B. solar cell.
   C. DC compressor.
   D. return fan.

36. A bonded roof is one that has a:
   A. steel-reinforced support.
   B. written insurance policy.
   C. leak-free, rubber membrane.
   D. slope of less than 30°.

Exercise COM-1

Icemaker Data and Familiarization

Name	Date	Grade

OBJECTIVES: Upon completion of this exercise, you should become familiar with ice machine controls, settings, and ice production.

INTRODUCTION: By filling out the icemaker information sheet, you will record important data.

TOOLS, MATERIALS, AND EQUIPMENT: A pencil and eraser, safety glasses, a compound gauge set, two thermometers, a flashlight, straight-slot and Phillips-head screwdrivers, an adjustable wrench, ¼ in. and ⁵⁄₁₆ in. nut drivers, a valve wrench, scale, temperature/pressure chart, and an icemaker of any type with an instruction and specification manual.

⚠ **SAFETY PRECAUTIONS:** Wear safety glasses and be alert to any electrical shock hazard, especially with wet surfaces near the ice machine. Make sure the ice machine is properly grounded.

STEP-BY-STEP PROCEDURES

Complete the information sheet for icemakers below.

Information Sheet for Icemakers

- Unit manufacturer: _____
- Date of manufacture: _____
- Model number: _____
- Serial number: _____
- Refrigerant type: _____
- Refrigerant charge: _____
- Type of system: _____
- Make of compressor: _____
- Compressor model number: _____
- Metering device type: _____
- Low-pressure control settings: Range _____ Differential _____ Cut-in/out ___/___
- Condenser-type: Water-cooled ___ Air-cooled ___
- Proper level of water in sump: _____ in.
- Inlet/City water temperature: _____ °F
- Pounds of ice per day (measured): _____
- Type of harvest: _____

- Harvest time: _____:_____
- Freeze time: _____:_____
- Ice condition: Clear _____   Cloudy _____
- Recommended cleaning interval: _____
- Evaporator superheat: _____
- Ice bin capacity: _____ lb
- Rated ice production per day: _____

MAINTENANCE OF WORKSTATION AND TOOLS: Make sure all panels are placed back on the ice machine properly. Clean up any water spills on the ice machine and floor area. Return all tools to their respective places.

SUMMARY STATEMENT: Describe why it is important to know the data you recorded in the Information Sheet for Icemakers.

_____

TAKEAWAY: Although they all make ice, all commercial icemakers are not the same. It is, therefore, important to be aware of the type of machine you are working on, the refrigerant being used and the type/configuration of the harvest cycle. This exercise is intended to provide the student with practice in identifying the key features and characteristics of a particular type of commercial icemaker.

QUESTIONS

1. Explain what factors could cause the freeze time of an icemaker to increase.

2. What would happen if the icemaker with a hot gas harvest got stuck in the harvest mode?

3. Why is most nameplate data on ice machines important to the service technician?

4. What may cause cloudy ice?

Icemaking Methods

Name	Date	Grade

OBJECTIVES: Upon completion of this exercise, you should be able to follow the ice-making cycle of a particular commercial ice machine.

INTRODUCTION: You will use gauges and a thermometer or digital thermocouple thermometer to monitor the ice-making process. You will record the temperature at which ice is being made.

TOOLS, MATERIALS, AND EQUIPMENT: A gauge manifold, goggles, gloves, straight-slot and Phillips-head screwdrivers, an adjustable wrench, a valve wrench, thermometer or thermistor, and package icemaker of any type.

⚠ **SAFETY PRECAUTIONS:** Turn off and lock out the power while installing gauges. Wear goggles and gloves.

STEP-BY-STEP PROCEDURES

1. Turn off the power to the icemaker. Lock and tag the panel and keep the single key in your possession. Put on your goggles and gloves. Fasten the high- and low-side gauge lines to the service valves or Schrader ports. Make sure that the pressure is indicated on the gauges.

2. Remove enough panels from the unit so that you may fasten a thermometer or thermistor lead to the suction line leaving the evaporator but before a heat exchanger, if there is one.

3. Place a probe from the thermocouple thermometer in the ice storage bin.

4. Replace any panels that may be required for normal operation, making certain not to damage the thermometer's probe.

5. Start the icemaker and record the following information:

   • Type of ice made (flake or cube): _____

   • Type of refrigerant used: _____

   • Refrigerant charge required: _____

6. When ice is being made at a steady rate, record the following information. (Cube makers make ice in batches about 5 to 10 minutes apart. Wait for at least three batches to be made before establishing a steady rate.)

   • Suction pressure at harvest (for cube makers only): _____ psig

   • Suction pressure while making ice: _____ psig

   • Suction-line temperature, making ice: _____ °F

   • Superheat, making ice: _____ °F

   • Discharge pressure: _____ psig

   • Temperature in ice storage bin: _____°F

7. Remove gauges and temperature probe and replace panels with the correct fasteners.

759

MAINTENANCE OF WORKSTATION AND TOOLS: Return all tools to their respective places and make sure that your work area is left orderly and clean.

SUMMARY STATEMENT: Describe the process by which ice is made in the equipment you worked with. Use an additional sheet of paper if more space is necessary for your answer.

TAKEAWAY: Monitoring a properly operating ice-making process will help service technicians determine when there is a problem more quickly. Knowing the sequence of operations of any system is an extremely important aspect of the troubleshooting process.

QUESTIONS

1. How is flake ice made?

2. How is cube ice made?

3. What is the difference between an ice-holding bin and an icemaker?

4. What is the purpose of a flush cycle in an ice machine?

5. What is the approximate evaporator temperature when ice is harvested in a cube icemaker?

6. How is ice cut into cubes when it is made in sheets?

7. Explain why mineral deposits on an evaporator of an ice machine will cause low suction pressure.

8. How is ice kept refrigerated in the holding bin of a typical icemaker?

9. Explain the difference between cleaning and sanitizing an ice machine.

10. Why does an icemaker have a drain line?

Cleaning and Sanitizing an Ice Machine

Name	Date	Grade

OBJECTIVES: Upon completion of this exercise, you should have become familiar with cleaning and sanitizing an ice machine.

INTRODUCTION: Under instructor supervision, and using the cleaning solution instructions and the icemaker's service manual, you will clean and sanitize an icemaker.

TOOLS, MATERIALS, AND EQUIPMENT: A pencil and eraser, goggles, a flashlight, straight-slot and Phillips-head screwdrivers, an adjustable wrench, ¼ in. and ⁵⁄₁₆ in. nut drivers, rubber gloves, soft bottle scrubbing brushes, clean dry rags, an icemaker of any type, ice machine cleaner and sanitizer, and an instruction and specification manual.

⚠ **SAFETY PRECAUTIONS:** Wear goggles and rubber gloves and be alert to any electrical shock hazard, especially with wet surfaces near the ice machine. Make sure the ice machine is properly grounded. Make sure the area is properly ventilated when using cleaning and sanitizing solutions. Always read the instructions on the cleaning and sanitizing solution bottles. Always refer to the service and maintenance manual for any specific cleaning and sanitizing instructions.

STEP-BY-STEP PROCEDURES

Note: Many ice machine manufacturers recommend cleaning and sanitizing every 6 months to prevent harvest problems and consumption of unsafe ice. However, conditions may exist where the ice machine will require more or less frequent cleaning. Ice machines located in bar environments may be exposed to a number of airborne contaminants and may have to be cleaned more frequently. However, never clean and sanitize an ice machine until it needs to be cleaned and/or sanitized.

Cleaning

1. Remove the ice machine's front panel. Make sure all ice is off the evaporator. You may have to initiate a harvest cycle or wait for the ice-making cycle to end. Turn the machine to the OFF position.

2. Remove any ice that may be in the ice machine's bin.

3. Following the manufacturer's instructions, add the proper amount of ice machine cleaner to the water trough.

4. Start a WASH cycle at the three-position ICE/OFF/WASH switch by moving the switch to the WASH position. This should start the water pump but leave the compressor off. The cleaner must circulate for at least 15 to 20 minutes in order for it to be effective on any mineral deposits.

5. After 15 to 20 minutes, the diluted cleaning solution must be washed down the icemaker's drain and replaced by fresh incoming water. If the icemaker has a PURGE switch, press and hold this switch until the cleaner has been flushed down the drain and has been replaced by fresh incoming water.

6. End the WASH cycle by switching the three-position ICE/OFF/WASH switch to the OFF position. Remove the splash curtain and inspect the evaporator and associated spillways for any mineral deposits. Many times, mineral deposits will not show up until the evaporator and associated part have been air-dried. A blow drier and/or a soft absorbent cloth will help the drying procedure.

7. Sometimes, the water distribution tubes, spillway, and header will have to be removed, disassembled, and cleaned with a bottle brush. They can be soaked in a solution of ice machine cleaner and water mixed in a 3- to 5-gallon bucket. Always follow the manufacturer's directions or instructions on the cleaner container label for the proper concentrations of cleaner.

8. The water trough and float assembly area may need special cleaning, especially if scale or slime buildup is visible. Disassembling, soaking, and scrubbing with a soft bottle brush is recommended.

9. Reassemble the icemaker.

Sanitizing

1. Mix the proper amount and concentration of approved sanitizing solution according to the manufacturer's recommendations or instructions on the sanitizing solution's container label.

2. Add enough sanitizing solution to the water trough for it to overflow the trough.

3. Place the three-position ON/OFF/WASH switch to the WASH position and circulate the sanitizing solution for 10 to 15 minutes or whatever time interval the manufacturer's instructions recommend.

4. During the sanitizing time period in step 3, use the remaining sanitizing solution to wipe down the ice machine's bin interior surface, ice deflector, door, and splash zones. Once finished, rinse all these areas with fresh water.

5. Inspect to make sure that all parts and assemblies are in their proper place and are leak-free.

6. When the 10- to 15-minute sanitizing period is over, flush the sanitizing solution down the drain. If the icemaker has a PURGE switch, press and hold this switch until the sanitizer solution has been flushed down the drain and has been replaced by fresh incoming water. Flush the ice machine for another 3 to 4 minutes to ensure all sanitizing solution has been removed from the icemaker.

7. Place the ICE/OFF/WASH switch to the ICE position and replace the front cover.

8. Discard the first four ice harvests.

MAINTENANCE OF WORKSTATION AND TOOLS: Return all tools to their proper places and make sure that your work area is left orderly and clean.

SUMMARY STATEMENT: Describe the difference between cleaning and sanitizing an ice machine. Use an additional sheet of paper if more space is necessary for your answer.

TAKEAWAY: Ice produced in equipment that is not properly cleaned can lead to illness. The number of people that can be affected by unsafe ice from a single icemaker can easily extend into the hundreds. Proper icemaker cleaning and sanitizing is not an option and should never be ignored.

QUESTIONS

1. Define "scale" as it applies to icemakers.

2. Define "preventive maintenance" as it applies to icemakers.

3. What is meant by the term "ice hardness"?

4. Why is it important to clean an icemaker?

5. What does sanitizing do to an icemaker that cleaning does not?

6. How can cleaning an icemaker help keep the harvest cycles operating smoothly and trouble-free?

Commercial Ice Cube Machines

Name	Date	Grade

OBJECTIVES: Upon completion of this exercise, you should be able to become familiar with a commercial type of ice cube machine, understand its sequence of operation and perform a service on the machine.

TOOLS, MATERIALS, AND EQUIPMENT: Manifold gauge set, ammeter, digital multimeter, Phillips-head and straight-slot screw driver, digital thermocouple thermometer, an operating ice cube machine, and the machine's service and troubleshooting manual.

⚠ **SAFETY PRECAUTIONS:** Make sure the ice cube machine is properly grounded. Certain lines on the ice cube machine may be hot. Be careful when working around rotating fans blades. Use care in fastening the refrigerant gauges. Wear goggles and light gloves. Do not proceed with this exercise until your instructor has given you instructions in the use of the gauge manifold.

STEP-BY-STEP PROCEDURES

1. Give a brief description of the electrical sequence of operation for the ice cube machine being used in this exercise. The sequence can be in paragraph form, bullet listed, or numbered. Refer to Figure COM-4.1.

FIGURE **COM-4.1**

ICE MAKING PROGRAM

When the selector switch is turned to the ICE position, in addition to displaying the evaporator and condenser temperatures, the computer will also display the part of the program that the computers is in.

ICE 1	ICE 1 is the untimed portion of the freeze cycle. In this part of the program, the computer is waiting for the evaporator temperature to reach 14°F (–10°C) before it starts the timer.
ICE 2	ICE 2 is the timed portion of the freeze cycle. The machine is still in freeze, but the timer in the computer is now running.
ICE 3	ICE 3 occurs during the last 20 seconds (12 seconds on single evaporator units) of the timed portion of the freeze cycle. During this time the purge valve is energized.
ICE 4	Once the amount of time set on the timer has passed, the program enters ICE 4 for 20 seconds. The WATER PUMP is now switched off and the HOT GAS VALVE is energized (open).
ICE 5	During ICE 5, the harvest assist motor(s) are now switched on and remain energized until the cam switch(es) return to the N.O. position, at which time the program returns to ICE 1 or ICE 0.
ICE 0	If the splash curtain(s) are held open when the cam switch(es) opens at the End of ICE 5, the machine will shut off and ICE 0 will display. When the curtain(s) Closes the program returns to ICE 1.

Courtesy Ice-O-Matic

2. **Record the following information:**
   Refrigerant type _____
   Refrigerant charge _____
   Ampere reading when in the middle of an ice making cycle _____
   RLA _____ LRA _____
   Type of compressor _____ Brand of compressor _____
   Starting relay type _____
   Start capacitor   yes   no   Run capacitor   yes   no
   Volts and phase of the incoming power supply _____
   Freeze cycle operating pressures: Low-pressure _____ psig   High-pressure _____ psig
   Harvest cycle operating pressures: Low-pressure _____ psig   High-pressure _____ psig
   Type of harvest _____
   Type of bin control _____
   Inlet water temperature _____°F
   Type of ice thickness control _____
   Sump water temperature _____°F
   What type of cube does the ice machine make? _____

Is the ice machine's condenser water- or air-cooled?   Water   Air

Does the ice machine have a microprocessor?   Yes   No

3. If the ice machine has a microprocessor, list the "inputs" to the microprocessor and tell if they are analog or digital, refer to Figure COM-4.2.

1. _____ analog or digital

2. _____ analog or digital

3. _____ analog or digital

4. _____ analog or digital

5. _____ analog or digital

6. _____ analog or digital

7. _____ analog or digital

List the "outputs" from the microprocessor and tell if they are analog or digital.

1. _____ analog or digital

2. _____ analog or digital

3. _____ analog or digital

4. _____ analog or digital

5. _____ analog or digital

6. _____ analog or digital

7. _____ analog or digital

FIGURE **COM-4.2**

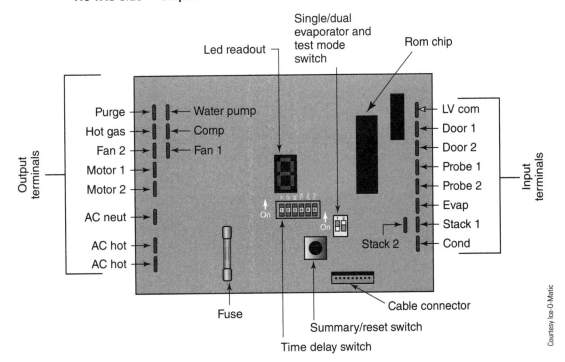

Courtesy Ice-O-Matic

Your instructor will put a fault in the electrical system of the icemaker. Using the appropriate pieces of test instrumentation, you are to troubleshoot the icemaker. Once the fault is found, have your instructor put the ice machine back in proper running order.

MAINTENANCE OF WORKSTATION AND TOOLS: Remove the gauges and wipe up any oil that may have accumulated on the unit. Disconnect power from the ice machine. Replace all tools. Replace all panels and leave the unit as you found it.

TAKEAWAY: Working with solid-state control boards can be very intimidating. Becoming familiar with the board's inputs, outputs, and sensory data will help you more fully understand the purpose of the boards and how to effectively evaluate them.

QUESTIONS

1. How would you go about testing each input to the microprocessor to see if they are good or bad?

2. How would you go about testing each output to the microprocessor to see if they are good or bad?

3. What type of signal do the inputs to the microprocessor produce?

4. What should the signals be for the outputs of the microprocessor?

5. Explain the self-diagnostic function of the ice machine's microprocessor.

Exercise COM-5

Chilled-Water System Familiarization

Name	Date	Grade

OBJECTIVES: Upon completion of this exercise, you should be able to recognize the different features of a vapor compression chilled-water system.

INTRODUCTION: You will perform a visual inspection of a chiller and be able to identify the compressor, chiller barrel, the condenser, the purge unit (if there is one), the suction line, and the discharge line. In addition, you will be able to determine whether the system is air-cooled or water-cooled and particular information about the equipment.

TOOLS, MATERIALS, AND EQUIPMENT: A vapor-compression chiller, notebook, pencils/pens, and flashlights.

⚠ **SAFETY PRECAUTIONS:** Do not turn or adjust any valves on the system. Do not push any buttons, remove any covers, disconnect any wires, or climb on the equipment.

STEP-BY-STEP PROCEDURES

1. Identify the compressor location on the chiller.

2. With what type of compressor is this chiller operating?

3. Identify the compressor's suction and discharge ports.

4. Visually, how were you able to differentiate one from the other?

5. Is this chiller equipped with an air-cooled condenser or a water-cooled condenser? _____. Explain how you are able to determine this. If air-cooled, continue with step 6. If water-cooled, skip to step 9.

6. What two refrigerant lines are connected between the condenser and the rest of the chiller? _____ and the _____

7. What is the state of the refrigerant in each of these two interconnecting lines? _____ and _____

8. Where is the condenser located with respect to the rest of the chilled-water system? _____

9. Locate the condenser on the chilled-water system.

10. Identify the type/configuration of the condenser. _____

11. How many passes does this condenser operate with? _____

12. Identify the water lines that are carrying water into and out of the condenser.

13. What is the temperature of the water entering the condenser? _____

14. What is the temperature of the water leaving the condenser? _____

15. What is the temperature difference of the condenser water circuit? _____

16. Locate the chiller barrel on the chilled-water system.

17. How many passes is the chiller barrel operating with? _____

18. What is the temperature of the water entering the chiller barrel? _____

19. What is the temperature of the water leaving the chiller barrel? _____

20. What is the temperature difference of the chilled-water circuit? _____

21. With what type of metering device is this chilled-water system operating? _____

22. With what refrigerant is this chilled-water system operating? _____

23. Is this chiller a high-pressure or a low-pressure system? _____

24. If this system is a low-pressure chiller, describe the purge unit and where it is connected to the chilled-water system. _____

25. Explain the purpose and operation of the purge unit.

26. What is the capacity, in tons, of this chiller? _____

27. Identify the pressure-relief valves on the system and list their locations here: _____

28. Locate any refrigerant sensors located around/on the chiller.
    How many are there? _____
    Where are they located with respect to the chiller and the floor? _____
    Why are they positioned as they are? _____
    Where are these sensors ultimately connected? _____

29. Determine if there is a mechanical exhaust system in the mechanical equipment room. Describe its configuration _____.
    What is the purpose of this exhaust system? _____
    Where does the exhaust system get its intake from? _____
    Explain why this is so.

MAINTENANCE OF WORKSTATION AND TOOLS: Make certain that the area around the chilled-water system is clean and free from debris and other obstructions.

SUMMARY STATEMENT: What are the potential benefits and drawbacks of using such a large system for cooling a building? Use an additional sheet of paper if more space is necessary for your answer.

TAKEAWAY: Chilled-water systems can be very intimidating, especially for a new technician. Becoming familiar with the components of a chiller and learning about the normal operating conditions for these systems will help clear up much of the confusion that often exists.

QUESTIONS

1. How were you able to determine what type of compressor the system is operating with?

2. Why are the suction lines and discharge lines different in sizes?

3. How were you able to determine if the system was air-cooled or water-cooled?

4. Which system would likely require a larger refrigerant charge, air-cooled or water-cooled? Why?

5. If the system you looked at was air-cooled, what other piece of equipment needs to be maintained to ensure proper system operation?

6. If the system you looked at was water-cooled, what other piece of equipment needs to be maintained to ensure proper system operation?

7. If the system you looked at was water-cooled, was the calculated water temperature difference across the condenser what you expected? Why or why not?

8. Was the calculated water temperature difference across the chiller barrel what you expected? Why or why not?

9. How did you determine what refrigerant the system contained?

10. How did you determine the capacity of the chilled-water system?

Exercise COM-6

Chiller Barrel Approach Temperature

Name	Date	Grade

OBJECTIVES: Upon completion of this exercise, you should be able to calculate the missing values in a chart based on a chiller barrel's approach temperature.

INTRODUCTION: Using a calculator and the appropriate mathematical relationships, you will complete the charts provided.

TOOLS, MATERIALS, AND EQUIPMENT: Calculator.

⚠ **SAFETY PRECAUTIONS:** There are no safety precautions for this exercise.

STEP-BY-STEP PROCEDURES

Using your calculator and the information contained in your HVACR textbook, complete the following chart by determining the correct values for all of the missing readings.

	Chiller Leaving Water Temp	Chiller Entering Water Temp	Delta T Across Chiller Barrel	Refrigerant's Boiling Temp	Approach Temp	Number of Passes	Evaporator Saturation Pressure	Refrigerant
1	45°F		10°F	42°F				R-123
2	46°F	55°F			7°F			R-11
3		54°F	11°F	35°F				R-22
4	45°F	53°F				3	18.1 in. Hg	
5	42°F		8°F		5°F		64 psig	
6	47°F		9°F			4		R-11
7		56°F	10°F	100°F	5°F			R-123
8			10°F		8°F		60 psig	R-22

MAINTENANCE OF WORKSTATION AND TOOLS: With the exception of the calculator, there are no tools required for this exercise.

SUMMARY STATEMENT: Based on the results of the exercise just completed, explain how the number of passes in a chiller, the approach temperature, the refrigerant's saturation temperature, and the temperature of the water leaving the chiller barrel are all related to each other.

TAKEAWAY: To properly evaluate and troubleshoot a chilled-water system, the technician must be able to obtain, interpret, and manipulate data. This exercise provides practice in converting pressures and temperatures for various refrigerants while using this information to evaluate the performance of a chiller barrel.

QUESTIONS

1. Explain how the number of passes in a chiller barrel affects the rate of heat transfer and the approach temperature of the heat exchanger.

2. Explain the formula that was used to calculate the approach temperature in a chiller barrel.

3. Explain why approach temperature can be used to gauge the efficiency of a chilled-water system.

Mechanical-Draft Cooling Tower Familiarization

Name	Date	Grade

OBJECTIVES: Upon completion of this exercise, you should be able to identify and describe the various parts of a cooling tower.

INTRODUCTION: You will perform a visual inspection of a cooling tower and be able to identify the sump, the make-up water mechanism, the fill, the decking material, the heater, and the method used to move air through the cooling tower. In addition, you will determine if the cooling tower is a natural-draft tower a mechanical-draft tower and, if it is a mechanical-draft tower, if it is forced-draft or induced-draft.

TOOLS, MATERIALS, AND EQUIPMENT: A cooling tower, safety glasses, hand tools, work boots, notebook, pens/pencils, tape measure, flashlight. Optional tools: Tachometer, velometer/anemometer.

⚠️ **SAFETY PRECAUTIONS:** Do not open, close, turn or adjust any water valves on the system. Do not push any buttons on the equipment. If ladders are used, make certain that all ladder-related safety issues are reviewed and understood. If permitted to climb the ladder to the top of the cooling tower, do so only if you are wearing rubber-soled work boots and are under the constant supervision of your instructor. Stay away from the perimeter of the tower and, of course, the edge of the roof if the tower is so located. Make certain that the cooling tower power is turned off and that the disconnect switch is locked-out and tagged when working on or around the electrical and mechanical/rotating portions of the equipment. Make certain that the air-conditioning equipment that relies on the cooling tower is not operating. When the cooling tower is powered up to take measurements, make certain that all of your classmates are clear of the equipment and your instructor has told you that it is safe to energize the equipment.

STEP-BY-STEP PROCEDURES

1. Identify the make of the cooling tower:_____

2. Identify the model of the cooling tower: _____

3. Identify the nameplate capacity of the cooling tower: _____

4. How many cells are used to make up this cooling tower? _____

5. Number of fan/blower motors in the tower: _____
   A. Motor shaft diameter: _____
   B. Horsepower rating of motor(s): _____
   C. Voltage rating of the motor(s): _____
   D. Amperage rating of the motor(s): _____

6. How many fans or blowers are used on this tower: _____

   A. Diameter of fan blades: _____

   B. Number of blades on each fan: _____

   C. Diameter of blower wheels: _____

   D. Length of blower wheels: _____

7. Describe the method that is used to move air through the tower.

   A. Propeller-type fans or centrifugal blowers? _____

   B. Induced-draft or forced-draft? _____

8. Describe the drive assembly type.

   A. Direct-drive or belt-drive? _____

      i. Belt length and type: _____

      ii. Number of belts: _____

      iii. Diameter of the drive pulley: _____

      iv. Diameter of the driven pulley: _____

      v. Speed of the motor: _____

      vi. Calculated speed of the fan or blower: _____

      vii. Measured speed of the fan or blower: _____

   B. Gearbox? _____

      i. Inline gears? _____

      ii. 90° gears? _____

      iii. Bevel gears or spiral bevel gears? _____

9. Is this tower equipped with a variable frequency drive (VFD) to control the speed of the motor(s)? _____

   If yes, provide the following information:

      i. Make of the VFD _____

      ii. Model of the VFD _____

      iii. Maximum horsepower _____

      iv. Maximum output amperage _____

10. Traverse the airstream leaving the tower and record the air velocity values here:

A. _____		L. _____
B. _____		M. _____
C. _____		N. _____
D. _____		O. _____
E. _____		P. _____
F. _____		Q. _____
G. _____		R. _____
H. _____		S. _____
I. _____		T. _____
J. _____		U. _____
K. _____		

11. Calculate the average velocity of the air moving through the tower: _____

12. Calculate the airflow, in cfm, that is moving through the tower: _____

13. Measure the diameter of the pipes that carry water into and out of the tower: _____

14. How many water distribution pans are located on this tower? _____

15. What are the measurements of each distribution pan? _____

16. How many calibrated holes (for even water distribution) are in each distribution pan? _____

17. What is the diameter of the calibrated holes? _____

18. Inspect the holes for signs of rust or corrosion damage. Describe the condition of the holes here:

19. Locate the fill material on the tower and describe the material from which the fill is made:

20. Inspect the fill and describe its condition here:

21. Locate and inspect the strainer on the cooling tower. Describe the condition of the strainer here:

22. Locate the sump on the tower and determine if there is a heater in the sump. If so, provide the manufacturer's information for the heater here:
   A. Make: _____
   B. Model: _____

23. Locate the point in the cooling tower where the make-up water is added to the tower.

24. Describe how the addition of make-up water is controlled in this cooling tower:

MAINTENANCE OF WORKSTATION AND TOOLS: Make certain that any panels that might have been removed as part of this exercise are replaced. Return all tools to their proper location.

SUMMARY STATEMENT: Explain the importance of proper cooling tower maintenance. In your discussion, include the possible symptoms that a water-cooled air-conditioning system might experience of the cooling tower is not functioning properly.

TAKEAWAY: The cooling tower is a major portion of a water-cooled air-conditioning system and, quite often, the cooling tower is tied in to multiple systems. A malfunctioning cooling tower can cause an entire building, or at least a large portion of it, to lose cooling. Cooling tower maintenance is often overlooked, but doing so can have severe negative results. Becoming familiar with the components and operation of a cooling tower will help you better understand its importance in the system.

QUESTIONS

1. Explain the purpose of the calibrated holes in the cooling tower's distribution pans.

2. Explain why it is important to keep the strainer on a cooling tower clean. As part of your response, be sure to include the system symptoms of a clogged strainer.

3. Explain the process by which a cooling tower is able to cool water so that it can be repeatedly circulated back to the condenser.

Exercise COM-8

Mechanical-Draft Cooling Tower Evaluation

Name	Date	Grade

OBJECTIVES: Upon completion of this exercise, you should be able to evaluate the operation of a mechanical-draft cooling tower.

INTRODUCTION: As part of this exercise, you will obtain various temperature and flow readings to evaluate the operation of a mechanical-draft cooling tower.

TOOLS, MATERIALS, AND EQUIPMENT: An operational cooling tower, safety glasses, hand tools, work boots, notebook, pens/pencils, tape measure, thermocouple thermometer, and psychrometer.

⚠ **SAFETY PRECAUTIONS:** Do not turn or adjust any water valves on the system. Do not push any buttons on the equipment. If ladders are used, make certain that all ladder-related safety issues are reviewed and understood. If climbing a ladder, do so only if you are wearing rubber-soled work boots and are under the constant supervision of your instructor. Stay away from the perimeter of the tower. When the cooling tower is powered up to take measurements, make certain that all of your classmates are clear of the equipment and your instructor has told you that it is safe to energize the equipment.

STEP-BY-STEP PROCEDURES

1. Maker certain the the cooling tower is operating and that the air-conditioning equipment that relies on the cooling tower is operating.

2. If possible, provide the nameplate capacities of the individual air-conditioning systems that utilize the cooling tower:

   A. Unit 1 Location: _____

      i. Unit Model: _____

      ii. Unit Capacity: _____ Btu/h, _____ tons

   B. Unit 2 Location: _____

      i. Unit Model: _____

      ii. Unit Capacity: _____ Btu/h, _____ tons

   C. Unit 3 Location: _____

      i. Unit Model: _____

      ii. Unit Capacity: _____ Btu/h, _____ tons

   D. Unit 4 Location: _____

      i. Unit Model: _____

      ii. Unit Capacity: _____ Btu/h, _____ tons

    E. Unit 5 Location: _____

      i. Unit Model: _____

      ii. Unit Capacity: _____ Btu/h, _____ tons

    F. Unit 6 Location: _____

      i. Unit Model: _____

      ii. Unit Capacity: _____ Btu/h, _____ tons

3. Enter the total of all of the unit capacities here: _____ Btu/h, _____ tons

4. Turn all of the units in step 2 on and set the thermostat to a temperature that is far enough below the room temperature so the units remain operating.

5. Obtain the pressure at the inlet of the condenser water pump from the gauge located at the inlet of the pump. Record this pressure here: _____ psig

6. Obtain the pressure at the outlet of the condenser water pump from the gauge located at the inlet of the pump. Record this pressure here: _____ psig

7. Using the information from steps 5 and 6, determine the pressure rise through the centrifugal pump here: _____ psig

8. Referring to the pump's performance curve, use the pressure rise across the pump (step 7) to determine the flow rate, in gpm, through the pump. Record this value here: _____ gpm

9. At the cooling tower, measure the temperature of the water returning to the cooling tower. Record this temperature here: _____°F

10. At the cooling tower, measure the temperature of the water leaving the cooling tower. Record this temperature here: _____°F

11. Using the results from steps 9 and 10, calculate the Tower Range. Record this value here: _____°F

12. Using the information obtained in steps 8 and 11, use the Cooling Tower Load formula (Tower Load = 500 × gpm × Tower Range) to determine the load on the cooling tower. Record the results here: _____ Btu/h, _____ tons

13. Using a psychrometer, measure the wet-bulb temperature of the outside air, near the cooling tower. Record this temperature here: _____°F

MAINTENANCE OF WORKSTATION AND TOOLS: Make certain that any panels that might have been removed as part of this exercise are replaced. Return all tools to their proper location. Reset all unit thermostats to their previous settings.

SUMMARY STATEMENT: Explain why water-cooled air-conditioning systems are more efficient than air-cooled systems. As part of your response, include specific examples to support your reasoning.

TAKEAWAY: Becoming familiar with the components and operation of a cooling tower will help you better understand its importance in the system. This exercise provides practice with taking measurements from the system, using these measurements to perform system calculations, and evaluating the results of the calculations.

QUESTIONS

1. Compare the values in step 3 to those obtained in step 12. In your discussion, be sure to include an explanation of why these values might differ from each other.

2. Compare the reading taken in step 10 to the reading obtained in step 13. In your discussion, be sure to include an explanation of the significance of these values and what it is referred to as in the HVACR industry.

3. Explain the outside air conditions that would best be suited for efficient cooling tower operation. Provide specific examples to support your response.

4. Describe a pump's performance curve and how the pressure across the pump relates to the flow rate through the pump.

Using a Heat Exchanger Pressure-Drop Chart

Name	Date	Grade

OBJECTIVES: Upon completion of this exercise, you should be able to complete the following chart as it relates to pressure drop, flow rate, and system tonnage.

INTRODUCTION: Using a calculator and the appropriate mathematical relationships, you will complete the chart provided.

TOOLS, MATERIALS, AND EQUIPMENT: Calculator.

SAFETY PRECAUTIONS: There are no safety precautions for this exercise.

STEP-BY-STEP PROCEDURES

Using your calculator and information from Figure COM-9.1, complete the following chart by determining the correct values for all of the missing readings.

FIGURE **COM-9.1**

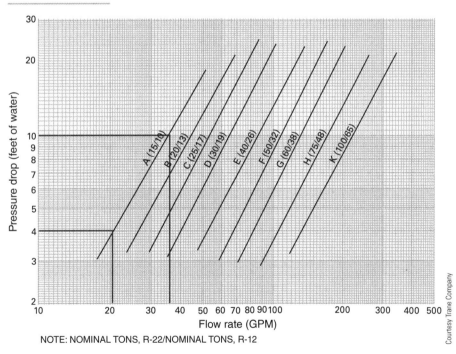

NOTE: NOMINAL TONS, R-22/NOMINAL TONS, R-12

Line	Pressure Drop (Feet of Water)	Flow Rate (gpm)	Heat Exchanger Curve	Nominal R-22 Tonnage	Pressure Drop (psig)
1		100	G		
2	10	36			
3		50	D		
4				50	3
5	8		K		
6		90		30	
7			B		5
8		70			5

MAINTENANCE OF WORKSTATION AND TOOLS: With the exception of the calculator, there are no tools required for this exercise.

SUMMARY STATEMENT: Explain the relationship that exists between the pressure drop (in psig) and the flow rate (in gpm) through a heating exchanger.

TAKEAWAY: Effective and efficient system troubleshooting should be as noninvasive as possible. By understanding the effects of pressure drop on flow rate, the service technician can, with reasonable accuracy, evaluate the flow performance of a heat exchanger without having to physically access the piping circuit.

QUESTIONS

1. If it is determined that the flow rate through a particular heat exchanger is correct, what other things can be done to check a heat exchanger on a system that is not functioning correctly? In your response, provide a possible set of system (refrigerant) pressure readings that would support your response.

2. If a particular pump is responsible for circulating water through six separate air-conditioning units, is it possible for the pump to be defective if all of the units, with the exception of one, are operating properly? Explain why this is or is not possible. Provide some other possible causes for the failure of the single unit.

Exercise COM-10

Using a Megger (Megohmmeter)

Name	Date	Grade

OBJECTIVES: Upon completion of this exercise, you should be able to check a compressor for internal grounds using a megohmmeter.

INTRODUCTION: Using a megohmmeter, you will check a compressor for an internal ground.

TOOLS, MATERIALS, AND EQUIPMENT: Chiller compressor, megohmmeter, hand tools.

⚠ **SAFETY PRECAUTIONS:** Make certain that the chiller system is powered down and that the disconnect switch is locked-out and tagged. You should have been instructed in proper use of the megohmmeter for this exercise.

STEP-BY-STEP PROCEDURES

1. Check the compressor's paperwork to determine the allowable leak rate to ground. If this information is not available, ask your instructor for guidance.

2. Disconnect the power to the compressor.

3. Lock out and tag the disconnect switch.

4. If the compressor has been running, allow it to cool.

5. Take a temperature reading of the motor. If allowed to remain off long enough, the temperature of the motor will be the same as the temperature of the ambient air surrounding the motor.

6. Check the manufacturer's literature to determine the acceptable readings for the compressor at the given temperature. The temperature of the motor will have an effect on the obtained readings.

7. Remove the electrical cover from the compressor, allowing you to gain access to the compressor terminals.

8. Carefully label and remove the power terminals from the compressor.

9. Using Figure COM-10.1 and your textbook as a guide, obtain resistance readings from winding-to-winding and from winding-to-ground. The terminals will vary depending on the motor, so you will need to identify the points you are checking between. Make certain that you are taking readings between the motor

FIGURE **COM-10.1**

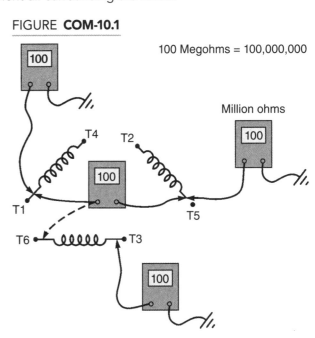

terminals and the bare metal case of the motor. Scrape any paint away to ensure accurate readings. Enter the readings here:

A. Terminal _____ to ground: _____ Ω

B. Terminal _____ to ground: _____ Ω

C. Terminal _____ to ground: _____ Ω

D. Terminal _____ to ground: _____ Ω

E. Terminal _____ to ground: _____ Ω

F. Terminal _____ to ground: _____ Ω

G. Terminal _____ to terminal _____: _____ Ω

H. Terminal _____ to terminal _____: _____ Ω

I. Terminal _____ to terminal _____: _____ Ω

J. Terminal _____ to terminal _____: _____ Ω

K. Terminal _____ to terminal _____: _____ Ω

L. Terminal _____ to terminal _____: _____ Ω

10. Carefully replace the power terminals on the compressor.

11. Replace the electrical cover on the compressor.

12. Restore power to the compressor as instructed by your teacher.

MAINTENANCE OF WORKSTATION AND TOOLS: Make certain that any panels that might have been removed as part of this exercise are replaced. Return all tools to their proper location.

SUMMARY STATEMENT: Explain what can cause the resistance readings obtained from the megohmmeter to be lower than desired. Provide as many possible reasons as possible.

TAKEAWAY: The megohmmeter is a specialized piece of test instrumentation that is able to measure very small (high resistance) paths to ground on large motors. These meters can accurately measure resistance values in the hundreds of millions of ohms.

QUESTIONS

1. Why are megohmmeters capable of measuring such high resistance values when traditional multimeters are not?

2. Explain two likely causes for a megohmmeter's readings on a particular compressor motor to drop over time.

3. How often should the compressor motor on a chiller be checked with a megohmmeter?

Exercise COM-11

● ## Packaged Air Conditioner Familiarization

Name	Date	Grade

OBJECTIVES: Upon completion of this exercise, you should be able to identify the various parts of a packaged air-conditioning system.

INTRODUCTION: Your instructor will take you to a packaged unit. You will remove service and access panels as instructed and describe the major components.

TOOLS, MATERIALS, AND EQUIPMENT: A packaged cooling unit, straight-slot and Phillips-head screwdrivers, ¼ in. and ⁵⁄₁₆ in. nut drivers, ladders, and a flashlight. Optional: Belt-tensioning tool.

⟨!⟩ **SAFETY PRECAUTIONS:**

- Make sure the power to the unit is off before starting the exercise.
- Be careful while working around the coil fins because they are sharp.
- Be careful if roof access is needed. Make sure all ladder safety issues and guidelines are followed.
- Make certain all blowers and fans have completely stopped before working.

● ## STEP-BY-STEP PROCEDURES

1. Record information from the unit's nameplate:
   A. Make: _____
   B. Model: _____
   C. Serial Number: _____
   D. Refrigerant: _____
   E. Refrigerant Charge: _____ lb, _____ oz
   F. Compressor Amperages: LRA _____, FLA _____, RLA _____
   G. Fan Motor 1 Amperages: LRA _____, FLA _____, RLA _____
   H. Fan Motor 2 Amperages: LRA _____, FLA _____, RLA _____
   I. Fan Motor 3 Amperages: LRA _____, FLA _____, RLA _____

2. Is this unit equipped to heat as well as cool? _____ If so, what is the heat source? _____

3. Is this unit equipped with an economizer? _____

4. What types of controls are used to determine the economizer damper positions? _____

5. How many fan motors are on the unit? _____

6. Describe the type and function of each of the fan motors in the unit, including the motor-type and any starting/running components that each is equipped with.

7. What type of fan blade or blower wheel is mounted on these fan motors? _____

8. Are there belt-driven drives on this unit? _____

9. What is the dimension of the drive pulley? _____

10. What is the dimension of the driven pulley? _____

11. What is the belt size? _____

12. How many belts are there? _____

13. Is the belt tension correct? _____

14. What size air filters are found on this unit? _____

15. How many filters are there in this unit? _____

16. How many independent refrigerant circuits does this system have? _____

17. Are all of the circuits designed for the same capacity? _____

18. What type of metering device is being used on this system? _____

19. Are there distributors at the outlet of the metering device? _____

20. If yes, how many circuits are connected to each evaporator coil? _____

21. Is there some type of low-ambient control on this unit? _____

22. Describe the type of low-ambient control.

MAINTENANCE OF WORKSTATION AND TOOLS:
- Return all tools to their proper locations.
- Make certain all panels are replaced and all screws are in place.
- Make certain that power has been restored to the unit before leaving the roof.

SUMMARY STATEMENT: How were you able to identify what was behind the service panels before actually removing them?

TAKEAWAY: Rooftop package units are among the most popular configuration for commercial air-conditioning applications. Becoming familiar with these units and being able to quickly determine where the system components are will help the technician evaluate and troubleshoot these systems in a more efficient manner.

QUESTIONS

1. What is meant by LRA when referring to motor amperage?

2. How was the total refrigerant charge determined?

3. Why might a packaged unit have multiple condenser fan motors?

4. Why might a packaged unit have multiple refrigerant circuits?

5. How is belt tension properly checked?

6. What are some potential problems if belt tension is too loose?

7. What are some potential problems if belt tension is too tight?

8. How do we ensure that air filters are properly installed in the unit?

9. Why is it important for all service panels to be secured in place?

Exercise COM-12

Hand Signals for Communicating with the Crane Operator

Name	Date	Grade

OBJECTIVES: Upon completion of this exercise, you should be able to identify the various hand signals used to communicate with the crane operator.

INTRODUCTION: Because the air-conditioning installation crew is often on the roof while the crane operator is on the ground, communication between the two crews can be difficult. Job site noise may make verbal communication difficult. The acceptable communication method is using hand signals.

TOOLS, MATERIALS, AND EQUIPMENT: There are no tools or materials required for this exercise.

⚠ **SAFETY PRECAUTIONS:** There are no safety precautions associated with this exercise.

STEP-BY-STEP PROCEDURES

Under each figure, you are to enter the letter that corresponds to the correct hand signal.

A. Stop
B. Stop immediately
C. Use main moist
D. Use auxiliary hoist
E. Extend the boom
F. Retract the boom

G. Raise the boom
H. Lower the boom
I. Rotate the boom
J. Raise the cable
K. Lower the cable

FIGURE **COM-12.2**

FIGURE **COM-12.1**

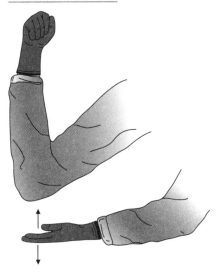

FIGURE **COM-12.3**

_____

FIGURE **COM-12.4**

_____

FIGURE **COM-12.5**

_____

FIGURE **COM-12.6**

_____

FIGURE **COM-12.7**

_____

FIGURE **COM-12.8**

_____

FIGURE **COM-12.9**

FIGURE **COM-12.11**

FIGURE **COM-12.10**

MAINTENANCE OF WORKSTATION AND TOOLS: There are no requirements for workstation maintenance associated with this exercise.

SUMMARY STATEMENT: Explain the importance of being able to communicate effectively with the crane operator.

TAKEAWAY: The symbols used by the installation team to communicate with the rigger are a universal language that must be understood in order to facilitate the proper landing of HVACR equipment.

QUESTIONS

1. Explain why it is important to discuss the desired location of the unit on the roof with the rigger before starting the process of bringing the unit to the roof.

2. Explain why it is important to finalize all the details of the rigging process before the rigger arrives at the jobsite.

3. Explain the purpose of outriggers on a crane.

Variable Air Volume (VAV) Box Familiarization

Name	Date	Grade

OBJECTIVES: Upon completion of this exercise, you should become familiar with the configuration of a variable air volume (VAV) box.

INTRODUCTION: Variable air volume systems are intended to vary the amount of air delivered to the occupied space based on the heating, cooling, and ventilation needs of the space. VAV boxes can differ greatly in their complexity depending on the functions they are intended to perform.

TOOLS, MATERIALS, AND EQUIPMENT: Safety glasses, hand tools, work boots, notebook, tape measure, pens/pencils, and various VAV boxes.

⚠ **SAFETY PRECAUTIONS:** Do not turn or adjust any valves that might be connected to the VAV box. Do not push any buttons on the equipment. Because the VAV boxes are likely positioned overhead, be sure to exercise caution when setting up and climbing ladders. If ladders are used, make certain that all ladder-related safety issues are reviewed and understood. When removing access panels from the VAV boxes, be sure that they are de-energized and that the disconnect switches are off and locked-out.

STEP-BY-STEP PROCEDURES

1. Locate the VAV box in the air distribution system. Record the following data from the VAV box.
   A. Make: _____
   B. Model: _____
   C. Duct inlet to VAV:
      i Round or rectangular: _____
      ii Diameter or measurements of duct: _____
   D. Duct outlet from VAV:
      i Round or rectangular: _____
      ii Diameter or measurements of duct: _____

2. Is the VAV box blower-assisted? _____. If yes,
   A. Parallel blower-assisted? _____
   B. Series blower-assisted? _____
   C. Describe the method by which air from the ceiling cavity is introduced to the VAV blower:

3. What type of VAV box are you working on?
   A. Cooling only? _____
   B. Cooling and heating? _____

4. Is there a reheat coil? _____. If yes,

    A. Hot water reheat? _____. If yes,

        i. Diameter of hot water pipe: _____

        ii. Number of tubes/passes in the coil: _____

        iii. Two-way or three-way valve on the reheat coil?: _____

    B. Electric reheat? _____. If yes,

        i. Voltage rating of heater: _____ V

        ii. Power rating of heater: _____ kW

MAINTENANCE OF WORKSTATION AND TOOLS: Make certain that any panels that might have been removed as part of this exercise are replaced. Return all tools to their proper location. Properly store ladders.

SUMMARY STATEMENT: Explain how using variable air volume boxes can help reduce the energy consumption of air-conditioning and heating systems.

TAKEAWAY: VAV boxes are popular pieces of equipment found on many commercial air-conditioning systems. Although they basically perform the same function, there are different types of VAV boxes. Knowing the different types of VAV boxes and how they operate will help ease the system troubleshooting process.

QUESTIONS

1. Explain the purposes of reheat coils on VAV boxes.

2. Explain how a three-way water modulating valve can be used in conjunction with a VAV box and a single-speed pump.

3. Explain why a pump that can vary its flow rate should be used in conjunction with hot water reheat coils installed with two-way modulating valves.

PART

2

Exercise Correlations

MAIN TEXT REFERENCE GUIDE

Exercise HTG-10

RACT 9E	Sections 31.10–31.12, 31.14
RACT 8E	Sections 31.17–31.20, 31.22
EforHVAC 11E	Section 13.3
EforHVAC 10E	Section 13.3
EforHVAC 9E	Section 14.3
Heat Pumps 2E	Chapter None
RCA: HVAC 2E	Chapter 24 pgs. 625–631

Exercise HTG-11

RACT 9E	Sections 31.5–31.8
RACT 8E	Sections 31.5–31.10, 31.13–31.14
EforHVAC 11E	Sections 13.2–13.3
EforHVAC 10E	Sections 13.2–13.3
EforHVAC 9E	Sections 14.2–14.3
Heat Pumps 2E	Chapter None
RCA: HVAC 2E	Chapter 24 pgs. 631–636

Exercise HTG-12

RACT 9E	Section 31.28
RACT 8E	Section 31.27
EforHVAC 11E	Sections 12.9, 13.3, 18.2
EforHVAC 10E	Sections 12.9, 13.3, 18.2
EforHVAC 9E	Sections 13.9, 14.3, 18.2
Heat Pumps 2E	Chapter None
RCA: HVAC 2E	Chapter 17

Exercise HTG-13

RACT 9E	Sections 31.4, 31.7, 31.21
RACT 8E	Sections 31.4, 31.13, 31.30
EforHVAC 11E	Sections None
EforHVAC 10E	Sections None
EforHVAC 9E	Sections None
Heat Pumps 2E	Chapter None
RCA: HVAC 2E	Chapter 24 pgs. 623–624

Exercise HTG-14

RACT 9E	Section 31.16
RACT 8E	Section 31.25
EforHVAC 11E	Sections None
EforHVAC 10E	Sections None
EforHVAC 9E	Sections None
Heat Pumps 2E	Chapter None
RCA: HVAC 2E	Chapter 24 pgs. 631–633

Exercise HTG-15

RACT 9E	Sections 31.4, 31.6
RACT 8E	Sections 31.4, 31.10
EforHVAC 11E	Sections None
EforHVAC 10E	Sections None
EforHVAC 9E	Sections None
Heat Pumps 2E	Chapter None
RCA: HVAC 2E	Chapter 24 pgs. 615–617

Exercise HTG-16

RACT 9E	Section 31.8
RACT 8E	Section 31.14
EforHVAC 11E	Sections None
EforHVAC 10E	Sections None
EforHVAC 9E	Sections None
Heat Pumps 2E	Chapter None
RCA: HVAC 2E	Chapter None

Exercise HTG-17

RACT 9E	Sections 31.3–31.4, 31.7–31.11
RACT 8E	Sections 31.3–31.4, 31.13–31.18
EforHVAC 11E	Sections 13.2–13.3
EforHVAC 10E	Sections 13.2–13.3
EforHVAC 9E	Sections 14.2–14.3
Heat Pumps 2E	Chapter None
RCA: HVAC 2E	Chapter 24 pgs. 618–637

Exercise HTG-18

RACT 9E	Unit 31
RACT 8E	Unit 31
EforHVAC 11E	Chapter 18
EforHVAC 10E	Chapter 18
EforHVAC 9E	Chapter 18
Heat Pumps 2E	Chapter None
RCA: HVAC 2E	Chapter 24 pgs. 639–640

Exercise HTG-19

RACT 9E	Section 31.6
RACT 8E	Sections 31.6–31.10
EforHVAC 11E	Section 13.3
EforHVAC 10E	Section 13.3
EforHVAC 9E	Section 14.3
Heat Pumps 2E	Chapter None
RCA: HVAC 2E	Chapter 24 pgs. 621–622, 637

Exercise HTG-20

RACT 9E	Sections 17.8, 17.9, 17.14, 17.15, 17.58, 17.29
RACT 8E	Sections 17.8–17.9, 17.14–17.15, 17.28, 17.29
EforHVAC 11E	Sections None
EforHVAC 10E	Sections None
EforHVAC 9E	Sections None
Heat Pumps 2E	Chapter None
RCA: HVAC 2E	Chapter 24 pg. 635

Exercise HTG-21

RACT 9E	Sections 31.11, 31.12, 31.14
RACT 8E	Sections 31.19–31.20, 31.22
EforHVAC 11E	Sections None
EforHVAC 10E	Sections None
EforHVAC 9E	Sections None
Heat Pumps 2E	Chapter None
RCA: HVAC 2E	Chapter 24 pgs. 627–628

Exercise HTG-22

RACT 9E	Section 31.9
RACT 8E	Sections 31.15–31.16
EforHVAC 11E	Section 13.2
EforHVAC 10E	Section 13.2
EforHVAC 9E	Section 14.2
Heat Pumps 2E	Chapter None
RCA: HVAC 2E	Chapter 24 pgs. 635–636

Exercise HTG-23

RACT 9E	Sections 32.2–32.3, 32.7–32.9
RACT 8E	Sections 32.2–32.3, 32.7–32.9

Exercise CON-17
RACT 9E	Sections 15.1–15.3
RACT 8E	Sections 15.1–15.3
EforHVAC 11E	Sections 4.1–4.4, 18.1–18.3
EforHVAC 10E	Sections 4.1–4.4, 18.1–18.3
EforHVAC 9E	Sections 4.1–4.4, 18.1–18.3
Heat Pumps 2E	Chapter 8 pgs. 326–328
RCA: HVAC 2E	Chapter 22 pgs. 558–570

Exercise CON-18
RACT 9E	Section 15.4
RACT 8E	Section 15.4
EforHVAC 11E	Sections 11.2–11.4, 14.4
EforHVAC 10E	Sections 11.2–11.4, 14.4
EforHVAC 9E	Sections 12.2–12.4, 15.4
Heat Pumps 2E	Chapter 8 pgs. 326–331
RCA: HVAC 2E	Chapter 22 pgs. 559–561

Exercise CON-19
RACT 9E	Sections 14.6–14.12
RACT 8E	Sections 14.6–14.12
EforHVAC 11E	Sections 11.5, 14.5
EforHVAC 10E	Sections 11.5, 14.5
EforHVAC 9E	Sections 12.5, 15.5
Heat Pumps 2E	Chapters 12.5–12.6
RCA: HVAC 2E	Chapter 16 pgs. 430–433

Exercise CON-20
RACT 9E	Sections 31.6–31.14
RACT 8E	Sections 31.6–31.22
EforHVAC 11E	Sections 13.2–13.5, 18.1–18.3
EforHVAC 10E	Sections 13.2–13.5, 18.1–18.3
EforHVAC 9E	Sections 14.2–14.5, 18.1–18.3
Heat Pumps 2E	Chapter None
RCA: HVAC 2E	Chapter 24 pgs. 625–637

Exercise CON-21
RACT 9E	Unit 41
RACT 8E	Unit 41
EforHVAC 11E	Chapter 18
EforHVAC 10E	Chapter 18
EforHVAC 9E	Chapter 18
Heat Pumps 2E	Chapter 10
RCA: HVAC 2E	Chapter 17

Exercise CON-22
RACT 9E	Section 16.7
RACT 8E	Section 16.7
EforHVAC 11E	Sections 12.2, 12.4–12.5
EforHVAC 10E	Sections 12.2, 12.4–12.5
EforHVAC 9E	Sections 12.2, 12.4–12.5
Heat Pumps 2E	Chapter 8 pgs. 329–334
RCA: HVAC 2E	Chapter 16 pgs. 412–415

Exercise DOM-1
RACT 9E	Sections 45.1, 45.3–45.6, 45.8–45.9
RACT 8E	Sections 45.1, 45.3–45.6, 45.8–45.9
EforHVAC 11E	Sections None
EforHVAC 10E	Sections None

EforHVAC 9E	Sections None
Heat Pumps 2E	Chapter None
RCA: HVAC 2E	Chapter None

Exercise DOM-2
RACT 9E	Sections 45.1–45.7
RACT 8E	Sections 45.1–45.7
EforHVAC 11E	Sections None
EforHVAC 10E	Sections None
EforHVAC 9E	Sections None
Heat Pumps 2E	Chapter None
RCA: HVAC 2E	Chapter None

Exercise DOM-3
RACT 9E	Sections 45.7, 45.10
RACT 8E	Sections 45.7, 45.10
EforHVAC 11E	Sections None
EforHVAC 10E	Sections None
EforHVAC 9E	Sections None
Heat Pumps 2E	Chapter None
RCA: HVAC 2E	Chapter None

Exercise DOM-4
RACT 9E	Sections 45.7, 45.10
RACT 8E	Sections 45.7, 45.10
EforHVAC 11E	Sections None
EforHVAC 10E	Sections None
EforHVAC 9E	Sections None
Heat Pumps 2E	Chapter None
RCA: HVAC 2E	Chapter None

Exercise DOM-5
RACT 9E	Sections 8.3–8.6, 8.11–8.12, 10.1–10.2, 10.4–10.5
RACT 8E	Sections 8.3–8.6, 8.11–8.12, 10.1–10.2, 10.4–10.5
EforHVAC 11E	Sections None
EforHVAC 10E	Sections None
EforHVAC 9E	Sections None
Heat Pumps 2E	Chapter None
RCA: HVAC 2E	Chapter None

Exercise DOM-6
RACT 9E	Sections 45.7, 45.10
RACT 8E	Sections 45.7, 45.10
EforHVAC 11E	Sections None
EforHVAC 10E	Sections None
EforHVAC 9E	Sections None
Heat Pumps 2E	Chapter None
RCA: HVAC 2E	Chapter None

Exercise DOM-7
RACT 9E	Sections 17.7, 20.6–20.9
RACT 8E	Sections 17.7, 20.6–20.9
EforHVAC 11E	Section 8.12
EforHVAC 10E	Section 8.12
EforHVAC 9E	Section 9.11
Heat Pumps 2E	Chapter 8 pgs. 346–347
RCA: HVAC 2E	Chapter 15 pgs. 399–400, Chapter 22 pgs. 566–569

Exercise DOM-8
RACT 9E	Section 45.1
RACT 8E	Section 45.1
EforHVAC 11E	Sections None

EforHVAC 10E .... Sections None
EforHVAC 9E .... Sections None
Heat Pumps 2E .... Chapter None
RCA: HVAC 2E .... Chapter None

Exercise DOM-9
RACT 9E .... Unit 45
RACT 8E .... Unit 45
EforHVAC 11E .... Sections None
EforHVAC 10E .... Sections None
EforHVAC 9E .... Sections None
Heat Pumps 2E .... Chapter None
RCA: HVAC 2E .... Chapter None

Exercise DOM-10
RACT 9E .... Sections 8.6–8.8,
8.10–8.15, 10.1–10.2,
10.4–10.5, 18.13–18.15, 45.10
RACT 8E .... Sections 8.6–8.8,
8.10–8.15, 10.1–10.2,
10.4–10.5, 18.13–18.15, 45.10
EforHVAC 11E .... Sections None
EforHVAC 10E .... Sections None
EforHVAC 9E .... Sections None
Heat Pumps 2E .... Chapter 6 pgs. 271–272,
Chapter 7 pgs. 299–300
RCA: HVAC 2E .... Chapter 12 pgs. 288–298,
302–305

Exercise DOM-11
RACT 9E .... Section 45.10,
Units 8, 9, 10
RACT 8E .... Section 45.10,
Units 8, 9, 10
EforHVAC 11E .... Sections None
EforHVAC 10E .... Sections None
EforHVAC 9E .... Sections None
Heat Pumps 2E .... Chapter 6 pgs. 270–273
RCA: HVAC 2E .... Chapter 10, 11, 12

Exercise DOM-12
RACT 9E .... Sections 46.1–46.5
RACT 8E .... Sections 46.1–46.5
EforHVAC 11E .... Sections None
EforHVAC 10E .... Sections None
EforHVAC 9E .... Sections None
Heat Pumps 2E .... Chapter None
RCA: HVAC 2E .... Chapter None

Exercise DOM-13
RACT 9E .... Section 46.8
RACT 8E .... Section 46.8
EforHVAC 11E .... Sections None
EforHVAC 10E .... Sections None
EforHVAC 9E .... Sections None
Heat Pumps 2E .... Chapter None
RCA: HVAC 2E .... Chapter None

Exercise DOM-14
RACT 9E .... Section 46.8
RACT 8E .... Section 46.8
EforHVAC 11E .... Sections None
EforHVAC 10E .... Sections None
EforHVAC 9E .... Sections None
Heat Pumps 2E .... Chapter None
RCA: HVAC 2E .... Chapter None

Exercise DOM-15
RACT 9E .... Sections 8.3–8.6,
9.15–9.16,
10.4–10.5, 46.8
RACT 8E .... Sections 8.3–8.6,
9.15–9.16,
10.4–10.5, 46.8
EforHVAC 11E .... Sections None
EforHVAC 10E .... Sections None
EforHVAC 9E .... Sections None
Heat Pumps 2E .... Chapter None
RCA: HVAC 2E .... Chapter 7 pg. 168

Exercise DOM-16 (Service Ticket)
RACT 9E .... To be Determined
by Instructor
RACT 8E .... To be Determined
by Instructor
EforHVAC 11E .... To be Determined
by Instructor
EforHVAC 10E .... To be Determined
by Instructor
EforHVAC 9E .... To be Determined
by Instructor
Heat Pumps 2E .... To be Determined
by Instructor
RCA: HVAC 2E .... To be Determined
by Instructor

Exercise DOM-17 (Service Ticket)
RACT 9E .... To be Determined
by Instructor
RACT 8E .... To be Determined
by Instructor
EforHVAC 11E .... To be Determined
by Instructor
EforHVAC 10E .... To be Determined
by Instructor
EforHVAC 9E .... To be Determined
by Instructor
Heat Pumps 2E .... To be Determined
by Instructor
RCA: HVAC 2E .... To be Determined
by Instructor

Exercise ISU-1
RACT 9E .... Sections 7.1–7.2, 7.5, 7.13
RACT 8E .... Sections 7.1–7.2, 7.5, 7.13
EforHVAC 11E .... Sections None
EforHVAC 10E .... Sections None
EforHVAC 9E .... Sections None
Heat Pumps 2E .... Chapter 6 pgs. 260–262,
Chapter 9 pgs. 362–363
RCA: HVAC 2E .... Chapter 6, 10

Exercise ISU-2
RACT 9E .... Sections 7.1–7.2,
7.5, 7.7–7.12
RACT 8E .... Sections 7.1–7.2,
7.5, 7.7–7.12
EforHVAC 11E .... Sections None
EforHVAC 10E .... Sections None
EforHVAC 9E .... Sections None
Heat Pumps 2E .... Chapter 6 pgs. 260–262
RCA: HVAC 2E .... Chapters 6, 10